Education Division 2016

Core Programming Area at the 2016 AIChE Annual Meeting

San Francisco, California, USA
13 - 18 November 2016

ISBN: 978-1-5108-3433-0

Printed from e-media with permission by:

Curran Associates, Inc.
57 Morehouse Lane
Red Hook, NY 12571

Printed by Curran Associates, Inc. (2017)

For permission requests, please contact AIChE
at the address below.

AIChE
120 Wall Street, FL 23
New York, NY 10005-4020

Phone: (800) 242-4363
Fax: (203) 775-5177

www.aiche.org

Additional copies of this publication are available from:

Curran Associates, Inc.
57 Morehouse Lane
Red Hook, NY 12571 USA
Phone: 845-758-0400
Fax: 845-758-2633
Email: curran@proceedings.com
Web: www.proceedings.com

TABLE OF CONTENTS

457913 Fundamental Studies and Applications of Nano-Structured Catalysts in Domestic Fuel Production

Abstract:

Fundamental studies and Applications of Nano-structured Catalysts in Domestic Fuel Production

Cun Wen, University of South Carolina, Columbia, SC

Liquid fuels are the largest commodity on the market in terms of total volume of sales, but the **U.S. is lagging behind the curve of novel technologies for domestic fuel production and upgrading**. For instance, U.S. is projected to be a fuel exporting country in the next few years because of the booming of shale gas. However, exporting shale gas has the lowest market value per energy content. Upgrading shale gas into gasoline, on the other side, can boost the price by more than 230%. Unfortunately, this technology is yet to appear on the horizon for the U.S. shale-gas producer, while South Africa and China have already built and operated such units for decades. Besides civilian fuel production, the situation of military fuel generation is not optimistic either. U.S. still can't fly a High-hypersonic jet with speed exceeding Mach 10, while other countries have already tested their high-hypersonic vehicles for several years. Lagging behind in powering high-hypersonic jet will potentially expose our homeland to other country's high speed jets and missiles, and therefore is a serious national-security concern. One of the main problems for powering high-hypersonic jet is to provide high energy-density fuels to the engine. Another example of domestic fuel production is the biofuels. In 2013, the US Navy and Airforce were still paying more than $15 dollars/gallon for biofuels. The high production cost was a result of energy-intensive processes that were needed at the time. Domestic production of transportation fuel is one of the United States' top priorities, as underscored by President Obama in his 2012 State of the Union

Address: *"This country needs an all-out, all-of-the-above strategy that develops every available source of American energy...... and nowhere is the promise of innovation greater than in American-made energy."*

To produce fuel domestically, catalyst development resides in the heart of the solution, because a good catalyst can lead to fuels with higher through-put, lower price, and higher energy-density. **Herein, my group will be focused on developing novel catalysts for domestic fuel production and upgrading.** Through my enriched experience in energy catalysis, I have studied reactions involved in almost every aspect of domestic fuel production, such as methane partial oxidation, water gas shift, Fischer-Tropsch, and hydrodeoxygenation for biodiesel production.[1-11] One uniqueness in my problem-solving technique is design catalyst's nanostructure based on fundamental understanding of reaction mechanism and kinetics. A recent example of how I apply such strategy to solve long-standing problems is designing the first-of-its-kind self-healing catalysts for Fischer-Tropsch synthesis. Cobalt catalysts have been considered the most attractive catalyst among all candidate metals, but a bottle-neck issue with cobalt catalysts is their tendency to deactivation through oxidation by water produced during the process. Through a comprehensive study of the reaction mechanism and kinetics, my idea is to design a cobalt catalyst that can be easily reduced with hydrogen in the feedstock so that the oxidation/deactivation process can be reversed. This idea, in the end, led to the first self-healing catalysts and attracted great amount of interests from academia and industry.[11,12] On the other hand, instead of solely relying on fundamental understanding in literature, my group will interrogate the reaction of interest with a unique approach of combining spectroscopic and kinetic studies. Often people study either reaction mechanism or kinetic one at a time. However, reaction mechanism is the reaction pathway that has the highest reaction rate among all routes. Therefore, a combination of spectroscopic and kinetic studies can lead to more comprehensive understanding of the reaction, and this approach has led to our new understanding in century-old reactions, such as methane partial-oxidation.[1,8,10] With my strength in catalyst design and fundamental understanding, breakthroughs in domestic fuel production and upgrading are expected to benefit our society and prosperity.

Research Interests: Catalysis, nanomaterial, renewable energy

Teaching Interests: thermos, kinetics

Read Me More:

1. Wen C, Liu Y, Guo Y, Wang Y, Lu G. Strategy to eliminate catalyst hot-spots in the partial oxidation of methane: enhancing its activity for direct hydrogen production by reducing the reactivity of lattice oxygen. *Chem Commun.* 2010;46(6):880-882.

2. Wen C, Dunbar D, Zhang X, Lauterbach J, Hattrick-Simpers J. Self-healing catalysts: Co3O4 nanorods for Fischer-Tropsch synthesis. *Chem Commun.* 2014;50:4575-4578.

3. Wen C, Zhu Y, Ye Y, et al. Water–Gas Shift Reaction on Metal Nanoclusters Encapsulated in Mesoporous Ceria Studied with Ambient-Pressure X-ray Photoelectron Spectroscopy. *ACS Nano.* 2012;6:9305–9313.

4. Wen C, Liu Y, Tao F. Integration of surface science, nanoscience, and catalysis. *Pure Appl Chem.* Jan 2011;83(1):243-252.

5. Liu Y, Wen C, Guo Y, Lu G, Wang Y. Modulated CO Oxidation Activity of M-Doped Ceria (M = Cu, Ti, Zr, and Tb): Role of the Pauling Electronegativity of M. *J Phys Chem C.* 2010;114(21):9889-9897.

6. Liu Y, Wen C, Guo Y, Lu G, Wang Y. Effects of surface area and oxygen vacancies on ceria in CO oxidation: Differences and relationships. *J Mol Catal a-Chem.* Feb 2010;316(1-2):59-64.

7. Liu Y, Wen C, Guo Y, et al. Mechanism of CO Disproportionation on Reduced Ceria. *ChemCatChem.* 2010;2(3):336-341.

8. Pichon A. Catalysis: Lower reactivity means more hydrogen. 2010; http://www.nature.com/nchina/2010/100120/full/nchina.2010.7.html.

9. Saxton C. Enhancing catalytic activity. 2010; http://www.rsc.org/Publishing/ChemTech/Volume/2010/02/enhancing_catalytic.asp.

10. Chinese New Year. *Chem Commun.* 2010.

11. Jacoby M. Custom Catalyst Resists Oxidation. *C&EN News.* 2014;92(13):25.

12. Catalysts C. Self-healing F-T catalysts. 2014; http://www.caicatalysts.com/news-a-links/news/69-self-healing-f-t-catalysts.

-

460715 Catalytic Hydrotreatment for the Development of Alternative Transportation Fuels

Abstract:

Research Interests:

Biologically-derived feedstocks are a highly desirable source of alternative transportation fuel. They can be grown renewably and can produce fuels similar in composition to conventional fossil fuels. They are also versatile and wide-ranging. Plant oils can produce renewable diesel and wood-based pyrolysis oils can be made into renewable gasoline. Catalytic hydrotreatment can be used to reduce the oxygen content of the oils and increase their viability as a drop-in transportation fuel, since they can then easily be blended with existing petroleum-based fuels. However, product distribution depends strongly on feedstock composition and processing parameters, especially temperature, hydrogen/oil flow rates and catalyst. Current literature contains little relevant information for predicting process-level data in a way that can be used for proper life cycle or techno-economic assessment. For pyrolysis oil, the associated reaction pathways have been explored via experimental studies on model compounds in a bench scale hydrotreatment reactor. The reaction kinetics of each compound were studied as a function of temperature and catalyst. This experimental data is used to determine rate constants for a hybrid, lumped-parameter kinetic model of paradigm compounds and pyrolysis oil, which can be used to scale-up this process to simulate larger, pilot-scale reactors. For plant oils, enough appropriate data was found in the literature and adapted for a preliminary model. With a systematic collection of kinetic data, hydrotreatment models can be developed that can predict important LCA inputs, such as hydrogen consumption, energy consumption and GHG production, which are necessary for regulatory and assessment purposes.

Teaching Interests:

While my research is focused on reaction engineering, my teaching experience has been focused in thermodynamics. As a TA, I lectured, held office hours, ran review sessions, graded both undergraduate and graduate level work, and helped write problems for homework and exams. I also have experience with teaching Unit Operations, and would be confident running an experimental laboratory class as well. I received the Outstanding Teaching Award from Michigan Technological University in 2015.

450820 First-Principles Modeling of Catalysts: Novel Algorithms and Reaction Mechanisms

Abstract:

Research Interests:

My research employs first-principles (ab initio) electronic-structure theory and molecular simulation to understand heterogeneous and homogeneous catalytic phenomena under realistic conditions, and to generate a platform for rational catalyst design and use in chemical synthesis and pollution abatement. Specific areas of interest include the evaluation and design of amorphous catalytic materials; the kinetic characterization of heterogeneous catalysts, with an emphasis on the use of supported nanoclusters and metals atomically-dispersed on metal-oxides; theory and simulation of catalyst dynamics under reaction conditions; Metal-organic frameworks as catalysts and catalyst supports; and the development of efficient computational methodologies for predicting accurate thermodynamics and kinetics of homogeneous reactions (e.g., C-H activation of alkanes and chiral catalysis).

Teaching Interests:

I look forward to developing and teaching courses at both the graduate and undergraduate level. Teaching has been an interest of mine since a young age, and I see classroom instruction as a natural counterpart to the mentoring that happens with students and post-docs in a research group. My teaching philosophy is to encourage students to reflect on their own understanding of the course material and to scrutinize the possibilities that arise when example problems are modified. My ultimate goal is to equip students with the engineering tools and mindset to tackle the pressing challenges facing modern society. I am enthusiastic to develop Catalysis and Molecular Modeling electives at both the graduate and undergraduate level.

452811 Development of Catalysts for Energy and Environmental Applications

Abstract:
Research Interests:

Over the next 30 years, the global energy demand is expected to increase by 60% to maintain pace with current projections of population and economic growth.[1] Fossil fuels will continue to provide a significant fraction of the energy supply; however, cleaner technologies, including solar, wind and nuclear will become prominent industries to meet the required demand. As the burning of fossil fuels continues into the foreseeable future, it is necessary to develop more efficient processes to decrease energy usage and the associated CO_2 emissions. With the threat of climate change and ocean acidification becoming more apparent each year,[2] efforts must be put forth to not only develop new forms of clean energy, but also capture and utilize CO_2 as a commodity. These large-scale CO_2 conversion processes require the discovery of active and low-cost catalysts.

Development of catalysts for CO_2 conversion can be divided into three areas: synthesis, *in-situ* characterization, and discovery, with significant research opportunities available in each area. In catalyst synthesis, the most promising methods include: atomic layer deposition, colloidal synthesis and incipient wetness impregnation. Although innovative synthesis methods, such as core-shell nanoparticles, can be extremely effective, uniform bimetallic particles can be easily and uniformly synthesized via incipient wetness impregnation. Characterization of catalysts by *in-situ* techniques is necessary to identify critical intermediates and understand the effect of reaction environments on catalyst structure. A previous study on CO_2 conversion used *in-situ* X-ray absorption near edge spectroscopy (XANES) on Mo_2C to confirm the effect of the reaction environment on catalyst oxidation states.[3]

Catalyst discovery is the most general area of catalyst development, but perhaps the most important. Previous studies demonstrate that reducible oxide supports, such as CeO_2, exhibit high activity in CO_2 conversion experiments because of their ability to exchange oxygen with CO_2.[4] Molybdenum carbide (Mo_2C), a non-precious metal catalyst, also shows promise as an active and selective catalyst for CO_2 conversion.[3] These materials are strikingly different, but have a fundamental similarity, which is they are each able to uptake and exchange oxygen with CO_2. Because of the importance of oxygen exchange in reactions with CO_2, oxygen binding energy could be a valuable descriptor when searching for CO_2 conversion catalysts. The discovery of such descriptors is an extremely powerful tool that can be applied beyond CO_2 conversion and into other applications of environmental catalysis.

Teaching Interests: Courses that I would be interested in teaching are kinetics and an introductory course in chemical engineering principles. The introductory course could be basic heat and mass transfer or an overview of the state of the field. A professor at Johns Hopkins once told me, “Everything is chemical engineering. If you name an item in your life, I can tell you how chemical engineering is involved.” I believe this principle should be imparted into young chemical engineers as early as possible. The classic view of chemical engineers as designing large reactors and processes is still relevant, but preparing students for current challenges in energy and climate change is also important. Broadening students’ views of the field has the potential to attract bright minds that might not be set on chemical engineering. I would also like to design an elective course that prepares students for solving problems in energy. These problems naturally lend themselves to applications of thermodynamics, transport, and kinetics, while exposing students to the current state of the field. This type of course and thought process clearly exemplifies my dual responsibility as an academic.

467618 Upgrading Biomass-Derived Platform Chemicals By Electrochemical and Photoelectrochemical Catalytic Oxidation

Abstract:

Research Interests:

Conversion of biomass to chemicals and fuels is a potentially viable route to minimize our dependence on unsustainable petroleum resources. However, unlike petrochemicals biomass-derived platform chemicals such as HMF, glycerol, and propylene glycol are typically highly oxygenated and require development of new catalytic processes to tune their functionality. Electrocatalysis is an extremely versatile tool where the reaction activation barriers and product selectivity can be directly regulated by tuning the electrode potential. Additionally, electrochemical methods and analysis give unique insight into complex processes at the electrode/electrolyte interface which would be otherwise unavailable without complex ***in situ***spectroscopic techniques. Electrochemical systems also give rise to fuel cells, which transform chemical energy to electricity and can also coproduce valuable chemicals. Finally, electrocatalysis can take advantage of renewable energy sources directly using photoelectrochemical (PEC) cells or indirectly by using renewable electricity to drive reactions.

My doctoral research first investigated the cogeneration of electricity and chemicals in direct alcohol anion-exchange membrane fuel cells (AEMFC) with supported noble metal anode electrocatalysts.[1] [2] It was demonstrated that a propylene glycol-fed AEMFC successfully cogenerated electricity and valuable chemicals on carbon-supported platinum or gold nanoparticles (Pt/C, Au/C). Pt/C was very selective for primary alcohol group oxidation to lactic acid under fuel cell operating conditions, while Au/C gave significant amounts of pyruvic acid, a product that has previously eluded heterogeneous catalytic studies on Au. Product selectivity on Au/C was very sensitive to anode potential. Later, I showed that HMF can be successfully oxidized to FDCA, an important bio-based polymer precursor, on supported bimetallic Pd-Au/C electrocatalysts, and the reaction benefited from a synergistic effect between Pd and Au components.[2]More recently, we have explored the use of stable nitroxyl radicals including TEMPO as electron mediators for selective oxidation of biomass-derived substrates in electrochemical and PEC reactors. This route opens opportunities to use non-noble metal electrodes and to take advantage of solar energy to drive electrochemical conversions.

Teaching Interests:

My background highlights the diverse benefits of electrochemical systems for the upgrading of biomass-derived chemicals, with unique examples in fuel cells, electrolysis flow reactors, and PECs. Electrocatalysis is a thriving research field with blooming practical applications and importance. As an independent faculty researcher, I will lead an interdisciplinary team to engineer advanced electrochemical materials and systems to meet future needs in energy conversion, production and storage, and sustainable chemical manufacturing as well as develop new curriculum for undergraduate and graduate-level electrocatalysis education.

References:

(1) Chadderdon, D. J. et al., ACS Catal. (2015), 6926–6936.

(2) Han, X. T. et al., Int J Hydrogen Energy (2014), 19767–19779

(3) Chadderdon, D. J. et al., Green Chem. (2014), 16 3778–3786.

468204 Rational Design of Catalytic Sites for Energy Applications

Abstract:

Motivation:

The rapidly increasing global demand for energy coupled with large population growth necessitates the development of increasingly efficient systems for utilizing our finite energy and chemical resources. While no single approach will solve these complex issues, I strongly believe that catalysts have an important role to play in addressing these challenges. Our society is already heavily reliant on these materials to produce the chemicals that we need to sustain our lifestyles. Catalysts are incredible because of their ability to facilitate selective chemical transformations while converting between chemical, thermal, radiant, and electrical forms of energy. Coupling catalytic systems with renewable energy sources presents a unique opportunity to regenerate useful high value chemicals from low value feed stocks like CO_2 and H_2O.

In order to develop these technologies, it is vital to understand the pathway through which these chemical transformations proceed and how catalytic systems respond to different energetic and environmental stimuli. It is also vital to identify key characteristics of active, stable, and selective catalysts to enable the design of materials with optimal sites for a particular reaction. I have implemented this strategy in my academic pursuits as a PhD student and postdoctoral researcher.

Research Interests:

As faculty I plan to leverage my past research experience to improve catalytic performance for various chemical, electrochemical, and photochemical applications. My efforts will focus on the design of cheap, active, selective, and stable catalysts for carbon and nitrogen fixation, renewable H_2generation, and selective upgrading of low-value oxygenates into important fuel and chemical feedstocks. I anticipate closely collaborating with theoreticians and other experimental specialists to decipher the true nature of catalytic sites under reaction conditions. With this information, structure-property relationships can be develop to give further insight to the field.

Research Experience:

Over the course of my graduate studies under Prof. Suljo Linic, we have investigated how geometric and chemical manipulations of catalytic sites influence overall catalytic performance. Specifically, we focused on understanding how changing the size, shape, and composition of Ag and Pt alloy electrocatalysts affects their activity towards the oxygen reduction reaction (O_2+4H^+/e^-â†'2H_2O, ORR) in alkaline and acidic electrolytes. First, kinetics studies of the oxygen reduction mechanism on these surfaces allowed us to identify key chemical descriptors (like OOH or OH adsorption energies) that are correlated with catalytic activity. We then utilized quantum chemical calculations to determine how these descriptors could be selectively influenced in platinum based catalysts through perturbations such as alloying or inducing geometric strain. After identifying promising catalysts, we developed sophisticated synthesis strategies to prepare these materials. We then tested their activity, with rigorous experimental protocol allowing us to obtain valid kinetic rates from the observed rates, and found that improved activity was achieved. Finally, we performed extensive in situ and ex situ characterization to confirm both that the atomic structure of the electrocatalyst matches that of our model and that the observed activity enhancement is a direct result of the improved key chemical descriptors of this particular nanostructure.

As a postdoctoral researcher working in Prof. Will Medlin group, I have focused on controlling rates and product selectivity for various thermochemical processes (such as CO oxidation, epoxidation, alcohol dehydration, and hydrodeoxygenation of phenolics) by manipulating the relative abundance and accessibility of different catalytic sites. Furthermore, the addition of chemical promoters to metal, support, or interfacial sites has been another method for altering the material flux through particular chemical pathways.

Teaching Interests:

My preference would be to teach chemical kinetics and reactor design because it closely aligns with my research background and extensive teaching experience. At Michigan, I served as a graduate teaching assistant (TA) for a graduate-level electrochemistry elective course (~70 students) as well as graduate chemical kinetics (~25), developing and evaluating homework and exam questions in addition to administering the course website and hosting review sessions. During my final year at UM, I was a Graduate Student Teaching Fellow for the undergraduate chemical reactor design course (~170). I was responsible for preparing course materials, managing personnel, and lecturing a 1/3 of the classes. Additionally, I am open to developing a graduate level elective course describing the fundamentals of catalyst design, which would be of particular interest to undergraduate and graduate students. Lastly, I have much experience as a mentor to both undergraduate and graduate students during my doctoral and postdoctoral research.

469170 Theory-Guided Understanding and Design of Heterogeneous Catalysts

Abstract:

Research Interests:

-Efficient catalyst screening and design:
Adsorption energies are good predictors of catalytic performance, but calculating adsorption energies on a large number of potential catalytic surfaces remains a challenge. I have developed methods for efficiently and accurately predicting adsorption energies of important intermediates on transition metal surfaces. These methods are quite general, applying to any transition metal alloy surface, and linking any adsorbate to electronic structure or to any other adsorbate. They are also efficient and physically transparent, allowing high-throughput screening or rational design. Future work will involve applying these methods to design novel catalysts (particularly those that break traditional scaling relations), and further improving their accuracy, efficiency and applicability.

-Nonadiabatic dynamics, excitations in metal surfaces and nanoparticles:
Most ab initio studies of catalysis and other processes on metal surfaces assume that the system remains in the ground state. However, using an efficient nonadiabatic dynamics code employing density functional theory and Ehrenfest dynamics, I have tested this assumption for nitrogen dissociation on Ru. These calculations show that even in thermal catalysis, surface processes can excite electrons in metal surfaces. Future work will involve understanding the conditions necessary for these electronic excitations, and their possible role in surface chemistry, as well as studies of plasmon-driven catalysis and photocatalysis.

Teaching Interests:
-Thermodynamics
-Materials chemistry and structure
-Mathematical methods
-Fluid dynamics and transport
-Quantum and computational chemistry

469497 In silico Engineering of a Future Energy Infrastructure

Abstract:

Research Interests:

The decreasing supplies of petroleum resources coupled with an increase in demand has created a need for an alternative sources of chemical building blocks and renewable energy. While current research has identified several promising technologies, high material costs and/or low yields forces our continual dependence on non-renewable sources. As a future faculty member, I seek to utilize my prior expertise to develop and distribute computational models and software which will allow for potential catalytic materials to be identified and readily examined inexpensively. During this Session I will demonstrate how my prior experience will enable my future research into the in silico design of new electrocatalytic materials for fuel cell applications.

Fuel cells are an attractive prospect for meeting our future energy demands due to their high energy density and low emissions. Utilization of H2 generated via the photo-oxidation of water further enhances the green aspects of this technology. While prior work has focused on the application of proton exchange membranes, these devices have limited applications due the high cost of the platinum based electrocatalytic materials involved. Recent work with hydroxide exchange membrane fuel cells have demonstrated that under akaline environments, cheaper base metals (i.e., nickel) can be used; however such materials are still limited due to their low activity. The future design of electrocatalytic materials will involve the detailed understanding of the kinetics of the relevant chemical processes at the anode and cathode, as well as the influence of the solvation environment. Unfortunately, such an understanding is currently lacking. My goal is to apply new computational tools to elucidate these effects.

My graduate work at the University of Pittsburgh focused on understanding the interactions between extended -systems (i.e., graphene) and polar molecules (i.e., water),[1,2] which is critical first step in understanding how aromatic carbon supports interacts with the solvation environment. During my post-doctoral work at the University of Wisconsin examined how support materials affect the catalytic properties of metal nanoparticles,[3,4] the results of which were recently used to explore the nature of novel carbon supports for fuel cell applications.[5] My current post-doctoral work at the University of Delaware has focused on the production of value-added chemicals and fuels from biomass-derived building blocks.[6–8]

Graduate Advisor: Kenneth D. Jordan (University of Pittsburgh)

Post-doctoral Advisors: Jordan R. Schmidt (University of Wisconsin),

Dionisios G. Vlachos (University of Delaware)

Select Publications (Total citations: 269):

(1) Jenness, G. R.; Karalti, O.; Jordan, K. D. Phys. Chem. Chem. Phys. 2010, 12, 6375–81.
(2) Jenness, G. R.; Karalti, O.; Al-Saidi, W. A.; Jordan, K. D. J. Phys. Chem. A 2011, 115, 5955–64.
(3) Jenness, G. R.; Schmidt, J. R. ACS Catal. 2013, 3, 2881–2890.
(4) Hermes, E. D.; Jenness, G. R.; Schmidt, J. Mol. Simul. 2014, 41, 123–133.
(5) Zhuang, Z.; Giles, S. A.; Zheng, J.; Jenness, G. R.; Caratzoulas, S.; Vlachos, D. G.; Yan, Y. Nat. Commun. 2016, 7, 10141.
(6) Jenness, G. R.; Christiansen, M. A.; Caratzoulas, S.; Vlachos, D. G.; Gorte, R. J. J. Phys. Chem. C 2014, 118, 12899–12907.
(7) Jenness, G. R.; Vlachos, D. G. J. Phys. Chem. C 2015, 119, 5938–5945.
(8) Gilkey, M. J.; Panagiotopoulou, P.; Mironenko, A. V.; Jenness, G. R.; Vlachos, D. G.; Xu, B. ACS Catal. 2015, 5, 3988–3994.

Teaching Interests:

Based on my chemistry and research background, I am most suited to teach classes that are related to physical chemistry (quantum mechanics, thermodynamics, statistical mechanics etc.), as well as classes on surface science, molecular simulations, and catalysis. Furthermore, I have an interest in formulating electives related to the application of python programming to chemical research and in the writing, preparation, and presentation of scientific results.

My approach to teaching has always been a reductionist approach with an emphasis on hands-on learning. In this approach, the student is presented with a relevant, big picture problem that is relevant to the current state of research. As an instructor, my goal is to dissect this problem into smaller problems, and instruct the student on how to solve these smaller pieces. By engaging the student with similar problems as homework and class projects, the student then learns how to perform such a decomposition on their own.

In addition to the reductionist approach in the classroom, I will employ a homework model that will emphasize both student collaboration and public speaking. Inevitably, students assigned homework problems will seek out help and advice from their fellow students. As an educator, it is critical to encourage such collaborations. To this end, I will require my students to both list those they collaborated with and their contributions. Class projects with a group component are also key, as this will not only encourage collaboration, but also give the students experience in discussing scientific literature and results in a public setting. In fostering collaboration and promoting public speaking, the student will form connections that can persist throughout their careers, while providing opportunities for the student to enhance their communication skills.

471981 Insight into Pt-Bi Bimetallic Catalysts: An Experimental and DFT Study

Abstract:

Research Interests: Catalytic Reaction Engineering

Teaching Interests: Reaction Engineering

Owing to environmental concerns and scarcity of known reserves, biomass has been shown to be an important renewable source, which can be converted into both sustainable fuels and chemicals. Heterogeneous catalysis plays an important role in such processes, while the following two issues still remain challenging and limit their industrial applications: tuning the catalyst selectivity toward target products and improving its stability against coking/carbon deposit. Both issues can be addressed by use of bimetallic catalysts, containing a primary metal (e.g. Pt, Pd, Ru, Rh, etc.) and a promoter (e.g. Mo, Sn, Bi, etc.). In the present work, we report an example of Pt-Bi bimetallic catalyst that selectively converts biodiesel waste product glycerol to 1,3-dihydroyacetone (DHA) and improves catalyst stability for reactions involving CH_4 (guaiacol deoxygenation using CH_4 in place of H_2. Based on experimental observations and theoretical analysis, Bi locates in an adatom-like configuration Pt and a cooperative effect between Pt as the primary metal and Bi as a promoter was identified. Insight into Pt-Bi bimetallic catalysts also suggests other catalytic applications, such as short-chain alcohol oxidation and oxidative coupling of methane.

474682 Mechanisms of Redox Catalysts for the Greener Processing of Shale Gas Via Chemical Looping

Abstract:

Postdoctoral Research Scholar

Research Interests:

The North American shale gas boom has stimulated renewed interest in producing value-added products from light hydrocarbons. Unfortunately, the traditional industrial approaches to produce chemicals and liquid fuels from shale gas components are energy-intensive and have severe environmental consequences. The primary challenges include: (i) selective activation of light, saturated hydrocarbons such as methane and ethane is very difficult; (ii) the production of high-enthalpy products or intermediates, such as hydrogen, requires significant external energy inputs. The push for improved, less polluting, more efficient energy conversion technologies has spurred increased interest in catalyst development. In my career, I hope to further the use of catalyst characterization techniques to applied catalysis to expand both fundamental knowledge, as well as tools for the rational design of energy-conversion catalysts. I have significant research experience in the fields of: redox catalyst for chemical looping; selective oxidation of light hydrocarbons; heterogeneous catalysis (Synthesis, Reaction testing, Characterization); platinum group catalyst formulation; surface/catalyst characterization techniques; ultra-high vacuum (UHV) system operation and maintenance; research scale reactor design, optimization, and operation.

Teaching Interests:

I have a strong interest in teaching fundamental chemical engineering subjects in addition to graduate and undergraduate courses closely related to my field (e.g. kinetics, catalysis/surface science). I also greatly enjoy training, mentoring, and collaborating with all levels of researchers down to undergraduates. In my postdoctoral work I have:

- Mentored Ph.D. students that, to date, has led to one peer review publication, two articles under review, and two nearing submission;
- Overseen interactions with facilities and environmental health and safety, and trained graduate students in best practices for laboratory safety and hazardous waste handling;
- Assisted collaboration between graduate students in the Chemical Engineering and Mechanical and Aerospace Engineering Departments;
- Successfully collaborated with graduate researchers in the Chemistry Department, leading to two high quality publications;
- Directly trained and supervised multiple undergraduates in the lab and supervised experimental plans for three master's students.

Research and Funding Successes: I have 14 authored and co-authored refereed articles across a variety of catalytic reactions. I have supervised multiple research projects, mentoring over 12 graduate and undergraduate researchers, and have participated in writing proposals, SOPO's, and reports leading to project successes. I co-invented a novel chemical looping ODH technology, which was awarded $3.8 million overall ARPA-E funding and licensed to a commercial partner (EcoCatalytic LLC).

PhD Dissertation: ***"Nano-Particle Supported Catalyst"*** Advised by Dr. Helena Hagelin-Weaver in the Department of Chemical Engineering at the University of Florida

Research Experience:

Chemical Looping Redox Catalyst: Formulated and characterized redox catalysts for the oxidative coupling of methane; Formulated and characterized a novel core-shell redox catalyst for partial oxidation of methane to syn-gas; Co-invented a licensed catalytic/chemical-looping technology for oxidative dehydrogenation of ethane.

Direct Methanol Fuel Cells: Developed test procedures for electrode characterization, including quantification of water flux and oxygen starvation, to facilitate control in a novel membrane electrode assembly that allows passive water recovery by forcing water back into the fuel stream; Performed multiple surface and electrochemical characterizations of electrodes to identify the causes of cell degradation.

Platinum Group Metal Catalyst: Prepared and characterized Pd on nano-particle oxide supports for selective oxidation; Prepared and characterized high dispersion Pt/TiO_2 catalyst for fundamental hydrogenation studies. Prepared Cu/Zn on nano-particle alumina for catalytic reforming of methanol; Designed and built gas phase reactor for catalyst activity testing; Operated and maintained UHV (Ultra High Vacuum) XPS system.

Selected Publications:

Luke M. Neal, Seif Yusuf, Fanxing Li. "Oxidative Dehydrogenation of Ethane: A Chemical Looping Approach." ***Energy Technology, June 2016 (in press).***

Ronghui Zhou, Evan W Zhao, Wei Cheng, **Luke M. Neal**, Haibin Zheng, Ryan E. Quiñones, Helena E. Hagelin-Weaver, Clifford R. Bowers. "Parahydrogen-induced polarization by pairwise replacement catalysis on Pt and Ir nanoparticles." ***Journal of the American Chemical Society, September 2015, Volume 137, Pages 1938–1946.***

Luke M Neal, Arya Shafiefarhood, Fanxing Li. "Dynamic Methane Partial Oxidation Using a Fe_2O_3@ $La_{0.8}Sr_{0.2}FeO_{3-\delta}$ Core–Shell Redox Catalyst in the Absence of Gaseous Oxygen." ***ACS Catalysis, September 2014, 4 (10), Pages 3560–3569.***

Luke M. Neal, Michael L. Everett, Gar B. Hoflund, Helena E. Hagelin-Weaver. "Characterization of Palladium Oxide Catalysts Supported on Nanoparticle Metal Oxides for the Oxidative Coupling of 4-Methlypyridine." ***Journal of Molecular Catalysis A: Chemical*, *February 2011, Volume 335, Issues 1-2, Pages 210-22.***

Samuel D. Jones**, Luke M. Neal**, Helena E. Hagelin-Weaver. "Steam Reforming of Methanol Using Cu-ZnO Catalysts Supported on Nanoparticle Alumina." ***Applied Catalysis B: Environmental*, *December 2008, Volume 84, Issues 3-4, Pages 631-642.***

Luke M. Neal, Helena E. Hagelin-Weaver. "C–H Activation and C–C Coupling of 4-Methylpyridine Using Palladium Supported on Nanoparticle Alumina." ***Journal of Molecular Catalysis A: Chemical*, *April 2008*, *Volume 284, Issues 1-2*, *Pages 141-148.***

474687 Nanostructured Transition Metal Dichalcogenide Catalysts for Electrochemical Energy Systems

Abstract:

Transition Metal Dichalcogenide Catalysts for Elecrochemical Energy Systems

Mohammad Asadi

Department of Mechanical and Industrial Engineering, University of Illinois at Chicago, Chicago, IL, 60607, USA.

Abstract:

Electrochemical energy systems have shown a tremendous potential to store energy in the form of chemical bonds using renewable energy sources. In general, energy can be converted or stored into chemical bonds through catalytic electrochemical processes such as fuel cells, oxygen reduction reaction (ORR) and carbon dioxide reduction reaction (CO_2RR). However, in practice existing catalysts are too inefficient: either weak binding interactions between the reaction intermediates and the catalyst give rise to high overpotentials, or slow electron transfer kinetics result in low exchange current densities. Both of these metrics depend not only on the intrinsic electronic properties of the catalyst, but also on the solvent and the catalyst morphology.

Here, I present the nanoarchitectures two-dimensional (2D) of molybdenum disulfide (MoS_2) and other transition metal dichalcogenides (TMDs) that manifest much higher performance for electrocatalytic ORR and CO_2RR reactions in the ionic liquid 1-ethyl-3-methylimidazolium tetrafluoroborate (EMIM-BF_4). Atomic level study of TMD nanoflakes before and after electrochemical experiments using scanning transmission electron microscopy (STEM) indicate that; (i) mainly metallic edge atoms are responsible for this remarkable activity and (ii) the edge atoms remain stable during the reactions. Moreover, I have successfully tested MoS_2 nanoflakes that work with the ionic liquid EMIM-BF_4 electrolyte as an effective co-catalyst for discharge and charge in a Li-O_2 battery.

The presented TMDs/IL co-catalyst system with remarkable electrocatalytic performance could open a new route to the advancement of energy conversion and storage systems.

Research Interests:

1- Synthesis and characterization of catalysts

2- Energy conversion systems (i.e., CO_2reduction reaction, water splitting, hydrogen evolution reaction, oxygen reduction reaction, oxygen evolution reaction)

3- Energy storage systems (i.e., metal-ion and metal-air batteries)

4- Electrochemisty and photoelectrochemistry

Teaching Interests:

1- Fluid Mechanics

2- Thermodynamics

3- Heat Transfer

4- Transport Phenomena

5- Reaction Engineering

6- Mass Transfer

store energy in into the chemical bonds in the form of electricity or fuel.

461299 Exploring Structure-Function Correlations of Nanomaterials in Heterogeneous Catalysis

Abstract:

Research Interests: Catalysis is the science and technology of influencing the rates of chemical reactions, which are at the core of chemical industry. Practical success of catalysis needs the interdisciplinary intersection of chemistry, chemical engineering, and material science. My research interests lie on understanding the mechanism of chemical reactions and the role of catalysts under reaction conditions in order to transform qualitative observations and knowledge into predictive science. My research program will focus on developing novel structure-reactivity relationships combining *operando* observation, materials synthesis and characterizations guided by theoretical calculations as applied to the field of heterogeneous catalysis inculding light alkanes activation, biomass conversion and energy conversion & storage.

Teaching Interests: I am confident that my academic experience makes me well prepared to teach both core and complementary courses. Given my research background, I would particularly enjoy teaching undergraduate level courses in reaction engineering, kinetics and electron microscopy. As for the graduate level, courses in interdisciplinary methods in heterogeneous catalysis is a natural fit for my background.

475338 Mechanistic, Spectroscopic and Theoretical Assessment of Porous Catalytic Materials

Abstract:

Research Interests:

Porous crystalline materials such as zeolites or porous organic polymers provide a large and diverse pool of heterogeneous catalysts and catalyst supports. Thorough mechanistic investigations that employ characterization of material properties and surface structure, kinetic and isotopic studies, and computational modeling are needed in order to elucidate how these materials work as catalysts regardless of reaction or application. Further, this approach can be used in conjunction with specialized synthetic strategies in order to produce well-defined catalysts, imperative in understanding catalytic function.

One class of these materials, Brønsted acidic zeolites, catalyze a plethora of hydrocarbon and oxygenate reactions, which are mediated by ion-pair transition states. My graduate research with Prof. Enrique Iglesia investigated the factors that affect reactivity, specifically the size, shape and charge of the organic cation (derived from reactants) and anion (formed after deprotonation of the inorganic catalyst). We combined state-of-the-art theoretical methods (density functional theory) with rigorous kinetic measurements from a range of reactants and catalysts that led to a copious array of transition states and precursors. Key properties of these systems include how the anion (i) accepts the negative charge (acid strength), which results in an electrostatic interaction with the organic cation, and (ii) provides van der Waals interactions, which are dependent on the size and shape of the organic component and the surrounding pore environment (confinement). The synergy between theory and experiment exposed unprecedented mechanistic detail, such as the flexibility of zeolites to distort locally to increase van der Waals interactions with the cation. These distortions cost energy, however, and can be expensive if the transition state is too large. Thorough studies incorporating both experiment and theory can provide clarity into complex relationships, such as this one between organic moieties and inorganic zeolite catalysts.

Future plans will utilize these results and methods, along with incorporating synthetic techniques, as a framework for studying porous catalytic materials from different catalyst classes and for other reaction pathways/applications. This framework can be extended to understand effects of important catalyst properties on reactivity and selectivity in order to advance rational catalyst design.

Teaching Interests:

Teaching, in the classroom and in the laboratory, is an outlet to share my zeal for science and a major driving force in my pursuit of an academic career. One benefit of research in catalysis is that multiple pillars of chemical engineering (kinetics, transport and thermodynamics) must be mastered, though I most enjoy teaching chemical reaction engineering courses. I was a graduate student instructor at UC Berkeley for the undergraduate and graduate level as well as designed and taught a new course titled â€œEnergy and Atom Efficient Catalytic Conversions,â€ which covered both current and green(er), alternative industrial catalytic reactions for energy, fuel and chemical production.

475552 In-Situ Drifts Studies on CuNi Catalyst for Ethanol Hydrogen Production

Abstract:

In-situ DRIFTS Studies on CuNi Catalyst for Ethanol Hydrogen Production

Anand Kumar

Department of Chemical Engineering, Qatar University, P O Box 2713, Doha, Qatar

Ethanol dehydrogenation mechanism was investigated over catalysts containing Cu and Ni metals prepared by solution combustion synthesis. In situ DRIFTS studies were conducted on individual metals (Cu and Ni) as well as on CuNi alloy to understand the metal-metal interaction in bimetallic systems and their implications on reaction pathway. The catalyst synthesis parameters (such as fuel to oxidizer ratio in combustion synthesis and combustion temperature etc.) were monitored and correlated with the synthesized materials properties. The amount of fuel content in the combustion solution was found to greatly affect the phase and microstructure of the synthesized catalysts. Along with the microstructure, surface phase distribution was also studied for various fuel to oxidizer ratio using XPS and changes on surface composition were monitored after catalyst activation (reduction) and after the ethanol dehydrogenation reaction. A change in reaction pathway was clearly observed over Cu and Ni catalysts. Cu mainly favors synthesis of acetaldehyde and acetate whereas Ni and CuNi both showed more tendencies towards methane and carbon monoxide formation.

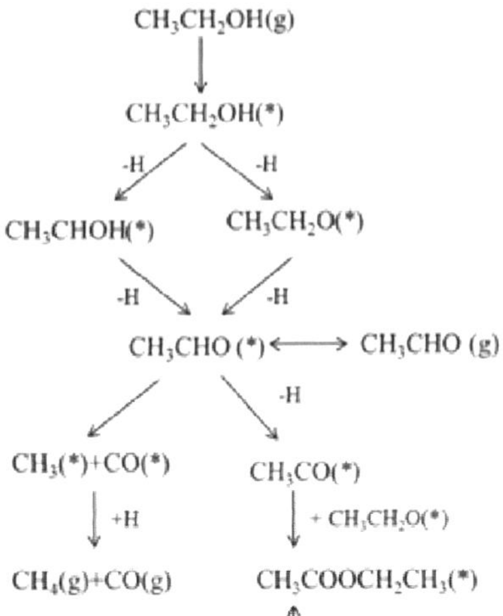

Figure 1: Postulated ethanol dehydrogenation reaction pathways over Cu and Ni based catalysts

Research Interests: Heterogeneous catalysis, combustion synthesis, microfluidic synthesis, nanomaterials, ethanol hydrogen production, methane reforming, carbon dioxide conversion, fuel cell reactions, solar fuels, electro-catalysis.

Teaching Interests: Chemical Reaction Engineering, Transport Phenomena, Chemical Process Control, Heterogeneous Catalysis, Mass Transfer, Heat Transfer, Advanced Engineering Mathematics, Numerical Methods, Natural Gas Processing, Engineering Economics

475627 Semi-Permeable Membrane Reactor for Catalysis, Hydrocarbon Processing and CO2 Reuse

Abstract:

PhD expected in January 2017

PhD Dissertation: "***Syngas Production via Water and Carbon Dioxide Thermolysis on Mixed Ionic-Electronic Conductive Membranes.***"

Under supervision of Professor Ahmed F. Ghoniem, Ronald C. Crane (1972) Professor; Director, Center for Energy and Propulsion Research; Director, Reacting Gas Dynamics Laboratory, Massachusetts Institute of Technology

Research Interests:

Inorganic membrane reactor finds numerous industrial and scientific applications such as hydrogen production, hydrocarbon processing, CO2 splitting and kinetic studies. By using a semi-permeable membrane, reactants can be doped or products can be separated. Therefore, the thermodynamic limit can be overcome and the selectivity can be enhanced. For example, a mixed ionic-electronic conductive membrane facilitates H2O/CO2 splitting at elevated temperatures and increases the selectivity to syngas in partial oxidation of methane. The implementation of such an oxygen permeable membrane also reveals the kinetics of partial oxidation of methane. Experimental and numerical studies are used to understand the reaction kinetics and limiting steps and to optimize the reactor design. Ab initio studies can reveal the fundamentals of charged species transport and transition states in reactions. These information can all lead to the optimization of reactor in industrial scale.

My ultimate goal for all my potential researches is to utilize the fundamental understanding in science to guide and optimize the operation in engineering. Therefore, in the future, I foresee myself collaborating closely with scientists and industrial partners to make the findings in my own laboratory into reality.

Teaching Interests:

Thermal science and energy related subjects such as heat transfer, fluid mechanics, transport phenomenon, thermodynamics, energy conversion

Chemistry and materials related subjects, such as electronic materials, inorganic materials, chemical kinetics

Awards:

1. MIT Golden Beaver Award – Institute award for excellence in leadership (Board of MIT European Club), 2016

2. Best Presentation Award in the 2015 AIChE Annual Meeting, 2015

3. Audience Award in MIT Mechanical Engineering Research Exhibition Poster Session, 2015

4. MISTI-MITEI fellow for Internship in Germany, MIT, 2015

Research Experience:

I have been working on several fields on chemical engineering: heat/mass transfer, electrochemical reaction/diffusion, reactor analysis and chemical kinetics. My previous work in Zhejiang University was focused on experimental studies on different types of heat transfer enhancement techniques: fin tubes for falling film evaporation and nanofluids for supercritical heat transfer. Both projects were based on industry applications. Now I am working on an interdisciplinary research as my PhD thesis, which involves chemical kinetics (surface reaction) and electrochemical diffusion (ion transport membrane). Tools in various physical scales are used in my thesis: femto (Density Functional Theory), nano (Microscopy Characterization), macro (System Scale Thermodynamics). This means both experiments and simulations are used to understand the chemical process. Additionally, my work experience in Wolverine Tube, INC and Robert Bosch LLC granted me the insights in researches in industry, which will benefit my collaboration with industrial partners.

Teaching Experience:

At MIT, I took a graduate course "Instruction in Teaching Engineering " and applied the knowledge on teaching when I TA'ed two courses at MIT: Fundamental of Advanced Energy Conversion (both undergrad and grad) and Introduction to Heat Transfer (undergrad). I got evaluation from students with 6.5/7 and 6.6/7 for these two course, respectively. Besides, I have mentored three undergraduates at MIT under my thesis work: one from MIT-Egypt Exchange Fellowship Program (3 months) and two from MIT Undergraduate Research Opportunities Program (UROP) (5 and 8 months).

Selected Publications:

1. **X.Y. Wu,** M. Uddi, A.F. Ghoniem, "Enhancing co-production of H_2 and syngas via water splitting and POM on surface-modified oxygen permeable membranes", invited to submit to AIChE Journal as "Best paper initiative", 2016, under review

2. **X.Y. Wu**, L. Chang, M. Uddi, P. Kirchen, A.F. Ghoniem, "Toward enhanced hydrogen generation from water using oxygen permeating LCF membranes", ***Physical Chemistry Chemical Physics***, 17 2015 10093-10107

3. D. Huang, **X. Y. Wu**, Z. Wu, W. Li, H. T. Zhu, B. Sunden, "Experimental Study on Heat Transfer of Nanofluids in a Vertical tube at Supercritical Pressures", ***International Communications in Heat and Mass Transfer***, 63 2015 54-61

4. W. Li, **X.Y. Wu**, Z. Luo, R.L. Webb, "Falling Water Film Evaporation on Newly-Designed Enhanced Tube Bundles", ***International Journal of Heat and Mass Transfer***, 54 (13-14) 2011 2990-2997

5. W. Li, **X.Y. Wu**, Z. Luo, S.C. Yao, J.L. Xu, "Heat Transfer Characteristics of Falling Film Evaporation on Horizontal Tube Arrays", ***International Journal of Heat and Mass Transfer***, 54 (9-10) 2011 1986-1993

Personal website:

http://xywu.scripts.mit.edu/xywu/

475657 A Practical Way to Separate Uncondensed Lignin during Biomass Pretreatment and Quantitatively Depolymerize It at a Low Temperature of 120 °C with a Cheap Ni/Al_2O_3-SiO_2 Catalyst

Abstract:

High-yield lignin monomer production from separated lignin in practical ways could greatly increase biorefinery productivity and profitability. However, lignin condensation caused by the formation of inter-unit carbon-carbon linkages during biomass pretreatment dramatically reduces lignin monomer yields; besides, high temperatures and noble-metal catalysts used in lignin hydrogenolysis depolymerization further increase the impracticality.

Recently, we have two significant findings which can largely drive the valorization of lignin. First, adding formaldehyde during biomass pretreatment produces a soluble lignin liquor that can be converted to guaiacyl and syringyl monomers at near-theoretical yields during subsequent hydrogenolysis; second, separated lignin can be readily depolymerized at a low temperature of 120°C with a cheap Ni/Al_2O_3-SiO_2 catalyst. We achieved a yield of 45% for beech wood, and a yield of 78% for transgenic poplar wood with low levels of carbon-carbon linkages in native lignin (yields were on the basis of native lignin). These yields were 3-7 times those obtained without formaldehyde. Formaldehyde reacted with lignin side-chain alpha- and gamma- hydroxyl groups to form stable 1,3-dioxane structures and with lignin aromatic rings to block the active positions, which completely quenched the condensation of lignin and facilitate its depolymerization under mild catalytic conditions.

Research Interests: Catalytic conversion of biomass into fuels, chemicals and materials.
Teaching Interests: Biomass conversion

475803 ***Combined Quantum and Classical Computational Approaches for Investigating Complex Surface Interactions Impacting Heterogeneous Catalysis***

Abstract:

2nd Year Post-Doctoral Associate

Post-Doctoral Project:Â *â€œAb initio design of semiconductor materials for the heterogeneous photo-electrocatalytic reduction of CO_2â€Â* Under the supervision of Prof. Emily A. Carter, Department of Mechanical and Aerospace Engineering, Princeton University

Ph.D. Dissertation:Â *â€œDevelopment of multi-scale computational techniques for modeling phase formation in Pd-based oxide catalystsâ€* Under the joint supervision of Prof. Michael J. Janik and Prof. Adri C. T. van Duin, Department of Chemical Engineering, Pennsylvania State University

Research Interests:

I am interested in research that employs atomistic simulation techniques at both electronic and classical levels of theory, with a particular focus on applying such methods to address issues related to energy conversion, storage, and utilization. I began academic research as an undergraduate under the supervision of Prof. William F. Schneider at the University of Notre Dame. There I was introduced toÂ *ab initioÂ* simulation techniques while completing an undergraduate thesis entitled â€œ*Ab initio*Â insights into ionic-liquid/CO_2Â binding trends for CO_2Â capture applications.â€ My training in atomistic simulation continued at the Pennsylvania State University, where I completed my Ph.D. in Chemical Engineering under the advisement of Prof. Michael J. Janik and Prof. Adri C. T. van Duin. My graduate research included the joint application of quantum and classical simulation techniques for investigating phase formation affecting the overall activity of oxide-supported metal catalysts, with a particular emphasis on identifying design principles for methane conversion catalysts. I am currently a post-doctoral associate under the advisement of Prof. Emily A. Carter at Princeton University, where I am conductingÂ *ab initio*Â investigations of heterogeneous catalytic mechanisms for CO_2Â reduction over semiconductor photo-electrodes. The aforementioned projects have led to 15 total publications (7 first-authored, 1 joint first-authored, 5 second-authored, and 2 co-authored), many of which were written with experimental collaborators.

Future Research Directions:

Through my previous research experiences, I have become well versed in both quantum and classical simulation techniques, which I will continue to apply in research aiding the development of sustainable energy technologies. I intend to integrate computational techniques I have learned from the Carter, Janik, and van Duin groups into my independent research program. I foresee that my group will collaborate with experimental researchers in the fields of catalysis, materials science, and surface characterization. In particular, my research plan consists of three principal focus areas:

(1) *Extending multi-physical simulation methodologies to the emerging field of heterogeneous catalysis over photo-electrode surfaces.* Combined quantum/classical simulation techniques will be used to assess the structural evolution of catalytically active photo-electrode surfaces that are in contact with a reactive aqueous environment. Full consideration of the reactive environment is essential, as the nature of active sites on the surface will be affected by structural changes induced by interaction with the solvent, and a reactive force field approach will be indispensable for assessing the complexities of the surface-solvent interface. Once the overall surface morphology is understood, *ab initio* methods will be used to elucidate the electronic structure of excited states that are accessible to photo-excited electrons involved in the catalytic mechanism. These studies will be relevant for H_2O splitting and CO_2 reduction applications.

(2) *Applying quantum and classical simulation techniques in tandem to assess strong metal/support interactions impacting the catalytic behavior of metal oxides.* I will continue to use combined quantum and classical simulation methods to characterize metal/support interactions leading to emergent catalytic behavior. My previous research determined the origin of highly active sites derived from unique interactions between Pd nanoclusters and a CeO_2 support. Preliminary results, obtained by an undergraduate student under my supervision, indicate that similar interaction trends occur between other metal/support pairs, thus affecting sintering rates and cluster size distributions. My group will further explore these trends, with the objective of developing novel and concrete optimization principles informing the rational design of oxide-supported metal catalysts.

(3) *Developing accelerated simulation methodologies for modeling kinetically limited processes with the ReaxFF potential.* I have developed various methods tailored to the classical ReaxFF potential for assessing the thermodynamic stability of active sites on catalytic surfaces. These methods, however, are not able to identify kinetic limitations present in the system, as processes involving significant kinetic barriers require prohibitively long simulation timescales to model. However, various well-known computational â€œtricksâ€ are available for accelerating kinetic sampling during a simulation. Some have been implemented in the literature for ReaxFF already, such as parallel replica dynamics, meta-dynamics, and bond-boost methods. My group will continue to develop such methods for ReaxFF, and in particular will implement an â€œon-the-flyâ€ kinetic Monte Carlo technique (OF-kMC) for simulating phase formation on catalyst surfaces. During OF-kMC simulations, reaction barriers are calculated directly as the state of the system evolves, thus allowing the simulation to explore phase space without the restriction of having pre-defined reaction paths. ReaxFF is uniquely suited for this, as the required barrier calculations, which typically are prohibitively expensive at the quantum level, become computationally tractable at the ReaxFF level.

Teaching Interests:

My previous teaching experiences have prepared me to serve as the instructor for both undergraduate and graduate courses traditional to the Chemical Engineering core; namely, thermodynamics, transport phenomena, and reaction engineering. While a graduate student at Penn State, I served as a Teaching Fellow forÂ *ChE 220 Introduction to Chemical Engineering Thermodynamics*. As a Teaching Fellow, I co-instructed â€" with a ChE faculty member (Prof. Scott Milner) â€" a core undergraduate thermodynamics course, where I was responsible for developing and implementing a significant portion of the class material. I served as a Teaching Assistant forÂ *ChE 544 Transport Phenomena*, which is a core graduate transport course covering momentum, heat, and mass transport. Although I have never formally taught reaction kinetics, I am confident that I can do so given my primary background in catalysis research. In addition to teaching traditional ChE courses, I assisted my advisors, Prof. Mike Janik and Prof. Adri van Duin, in teaching their co-instructed graduate elective course, titledÂ *Atomistic Simulations for Engineers*, for two semesters. I was responsible for designing and delivering lectures covering quantum mechanical simulation techniques. As a ChE faculty member, I hope to design a similar graduate course for introducing students to atomistic simulation techniques covering both quantum and classical methodologies.

Selected Publications:

Review Article:

Senftle, T. P.; Hong, S.; Islam, M. M.; Kylasa, S. B.; Zheng, Y.; Shin, Y. K.; Junkermeier, C.; Engel-Herbert, R.; Janik, M. J.; Aktulga, H. M.; Verstraelen, T.; Grama, A.; van Duin, A. C. T., â€œThe ReaxFF reactive force-field: development, applications and future directions.â€Â *npj Computational Materials*Â 2016, 2, 15011.

Research Articles:

Senftle, T. P.; van Duin, A. C. T.; Janik, M. J., â€œThe role of site stability in describing methane activation on $Pd_xCe_{1-x}O$Î´.â€Â *ACS Catalysis*Â 2015, 5 (10), 6187-6199.

Strayer, M. E.;Â **Senftle, T. P.**; Winterstein, J. P.; Vargas-Barbosa, N. M.; Sharma, R.; Janik, M. J.; Mallouk, T. E., â€œStrong adsorption of late transition metal nanoparticles correlates with sintering resistance on niobium and silicon oxides.â€Â *Journal of the American Chemical Society*Â 2015, 137 (51), 16216-16224.

Addou, R.*;Â **Senftle, T. P.***; Oâ€™Connor, N.; Janik, M. J.; van Duin, A. C. T.; Batzill, M., "Influence of hydroxyls on Pd atom mobility and clustering on rutile TiO_2(011)-(2Ã—1).â€Â *ACS Nano*Â 2014, 8(6), 6321-6333. *Equal author contribution.

Senftle, T. P.; Meyer, R. J.; Janik, M. J.; van Duin, A. C. T., â€œDevelopment of a ReaxFF potential for Pd/O and application to palladium oxide formation.â€Â *Journal of Chemical PhysicsÂ* 2013,Â 139 (4), 044109-15.

Book Chapter:

Senftle, T. P.; van Duin, A. C. T.; Janik, M. J., â€œApplication of computational methods to supported metal-oxide catalysis.â€Â *Computational CatalysisÂ* 2014, Royal Society of Chemistry, 157-191.

475819 Electrochemical Energy Transformation Processes: An Atomistic Perspective

Abstract:

Research Interests:

The upward trend in worldwide energy demand, combined with an increasingly urgent need to attenuate climate change, drives the search for renewable sources of energy. To this end, electrochemical energy conversion is a powerful technique that connects chemical energy (which serves as storage) to electrical energy, currency we can spend on a desired process. As an example of energy up-conversion, electrochemical CO_2 reduction is attractive because it potentially decreases atmospheric CO_2 concentrations while providing a carbon-neutral means to generate liquid fuels and fine chemicals. On the side of energy down-conversion and to prevent further large-scale CO_2 emissions, metal-air batteries are promising for electrifying transportation.

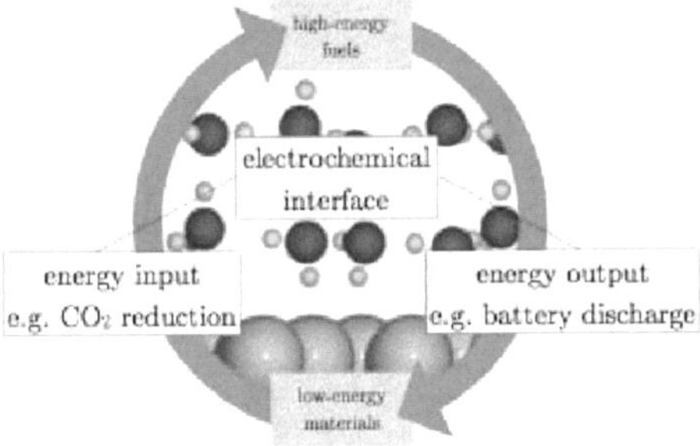

Yet even today, these energy transformation processes generally have low efficiencies. Pinpointing the source of these inefficiencies requires a detailed understanding of the stepwise mechanisms, which is where density functional theory (DFT) has been instrumental in providing insight. As illustrated by the graphic, my PhD work has three distinct yet interconnected topics with a central theme on studying these **electrochemical energy transformation processes from an atomistic perspective**:

(1) Anodic dissolution in metal-air batteries where electrical energy is the output: with DFT, I was able to elucidate the fundamental limitations on total energy output in the anodic dissolution of Al and Mg in aqueous electrolytes.

(2) Electrochemical CO_2 reduction where electrical energy is the input: I demonstrated how including the effects of electric field (in addition to potential alone) in the DFT model can significantly alter the free energy landscape of a catalytic process.

(3) The electrochemical interface, an indispensable component of these energy transformation processes: I have made progress in describing charged species at the outer Helmholtz plane and their implications on reaction energetics with DFT.

Teaching Interests:

Sharing knowledge and communicating effectively is very important to me as a scholar. In my graduate career, I have served as a Teaching Assistant (TA) for three different undergraduate courses: Organic Chemistry Laboratory, Organic Mechanisms Lecture, and Biochemistry. The most valuable undergraduate teaching experience for me came from the Organic Mechanisms Lecture course, for which I led two-hour discussions twice a week. Preparation was key for these sessions to engage my students in group problem-solving and fruitful discussions. I read up on the course material as thoroughly as possible to answer probing questions from my students.

I also served as a TA in a graduate course on Molecular Modeling. Here, the challenge came in the form of creating problem sets with the appropriate level of scope and difficulty for students who only had one fewer year of experience. This required some independent work on my part searching through textbooks and online resources for inspiration to problems that would be both in-line with the course content and interesting.

Proposal Experience:

Multidisciplinary University Research Initiatives (MURI) Program 2015

475913 Nanostructured Hybrid Materials: Directing Catalytic Activity and Selectivity By Design

Abstract:

Research Interests: and Teaching Interests:

Nanostructured Hybrid Materials: Directing Catalytic Activity and Selectivity by Design

In the field of catalysis, there is a need for highly active, selective, and stable catalysts for more efficient and cleaner use of existing and alternative energy sources. In this respect, hybrid materials with multifunctional properties show great promise and offer several important advantages over traditional catalysts. A rational design of hybrid systems provides a platform (i) to combine heterogeneous and homogeneous catalysis, which have been explored rather separately, and employ the best of the both fields, (ii) to produce extraordinary or synergystic properties at the interface of different types of materials, (iii) to carry out consecutive reactions in tandem, and (iv) to exploit the flexibility and diversity of organic, inorganic, and metallic materials. In addition, recent breakthrough in the synthesis of nanostructures and discovery of exotic materials renders the investigation of hybrid systems an attractive and promising area of research.

In this poster, I will present a brief overview of my graduate and postdoctoral studies, and my research interests and future plans on the topic. My graduate research primarily focused on the synthesis of TiO_2 nanocrystals with well-defined particle size and characterization of particle growth and phase transformation mechanisms. First, I experimentally demonstrated that the particle size and particle interface produced during nanocrystal growth play a crucial role in the thermodynamic stability of nanocrystals. Then, I developed a mathematical model to describe the types of mechanisms and characterize the kinetics of phase transformation.

With the expertise and experience in the synthesis and characterization of nanostructured materials acquired during my graduate years, as a postdoctoral researcher I work towards designing multicomponent catalysts and testing them on scientifically and industrially important chemical reactions. I have synthesized highly selective bifunctional catalyst comprising platinum nanoparticles with finely tuned particle size and mesoporous silica post-synthetically modified with aluminum, and investigated its catalytic properties on n-hexane (n-C_6H_{14}) and n-hexadecane (n-$C_{16}H_{34}$) conversions, model reactions for industrial reforming of naphtha and heavy-gas oil, respectively. In addition, metal-organic framework (MOF) was demonstrated to encapsulate and disperse a large amount of phosphotungstic acid ($H_3PW_{12}O_{40}$), a solid acid with high catalytic activity and strong acidity, but with extremely low surface area. Combined with platinum nanoparticles, the MOF-based material was catalytically active towards gas-phase isomerization of n-hexane.

In the future, I aim to establish a research program on the fundamental investigation of molecular-level electronic interaction and structure-property relationship between organic and inorganic materials and the design of hybrid catalysts for environmentally and industrially important chemical transformations. I will utilize the unique structural, chemical, and compositional features of metal-organic and covalent organic frameworks, chelating ligands and capping agents, and electronically and catalytically active nanostructures to achieve the objectives mentioned in the introduction.

475963 Towards the Computational Design of Heterogeneous Electrocatalysts

Abstract:

Research Interests:

Design of cost-effective electrocatalysts with enhanced activity and stability is of paramount importance for the next generation of energy conversion systems, including fuel cells and electrolyzers. Ultrathin (hydroxy)oxide films on transition metal substrates possess impressive activities for these reactions. These hybrid systems with novel properties are often prepared and characterized under very specific conditions, such as ultrahigh vacuum. When subsequently used in various applications, however, significant structural evolution, which depends strongly on the reaction conditions in-situ, is expected but remains largely unexplored. In order to identify structure-property relationships and, ultimately design/screen new materials with improved performance, the development of such understanding is essential.

The complex interfaces have long been a challenge for modeling, because of the complications introduced by strongly-correlated transition metal oxides and mysterious interface structures. As a consequence, even many ancient questions are still unanswered, for example the origin of the strong metal support interaction (SMSI) that plays a central role on the stability of the interfaces.

Recently, I have resolved the major challenge toward self-consistent and highly accurate description of strongly-correlated transition metal oxides, hydroxides and oxyhydroxides, through synergistic use of a Hubbard U correction, a state-of-the-art dispersion correction, and a water-based bulk reference state for the calculations. The strong performance is illustrated on a series of bulk transition metal (Mn, Fe, Co and Ni) hydroxides, oxyhydroxides, binary, and ternary oxides, where the corresponding thermodynamics of redox and (de)hydration are described with standard errors of 0.04 eV per (reaction) formula unit.

I have applied these methodologies to investigate the growth of monolayer Ti- and Fe-(hydroxy)oxide films on VIIB group and IB group noble transitional metal surfaces, and have obtained mechanistic understanding of the origin of strong metal support interaction. This understanding has been extended to predict the structure evolution of the other monolayer transitional metal (Mn, Co, and Ni) (hydroxy)oxide films on a variety of substrates. It has been demonstrated that the structure and oxidation state of the films can be systematically tuned by changing the applied electrode potential, the nature of substrates and/or the size of film cluster, which is further confirmed by synchrotron experiment on the monolayer Ni-(hydroxy)oxide/Pt, a hydrogen evolution (HER) electrocatalyst with improved performance in comparison with traditional Pt catalyst in alkaline conditions.

The identified interface under hydrogen evolution HER condition, NiOH/Pt, is subsequently used for a detailed structural and electrocatalytic analysis at three-phase boundaries of NiOH films, Pt substrates, and the surrounding water. The kinetic analysis shows that water can be dissociated to adsorbed OH* and H* groups in a bifunctional mechanism that significantly lowers the activation barriers for alkaline HER and provides a clearly measureable acceleration of the catalyst process as compared to mono-functional Pt catalysts. The mechanistic understanding and kinetic analysis also suggest a new catalyst, NiOH/Pt_3Ni, with improved performance in comparison with NiOH/Pt, which is further confirmed by electrochemical experiment.

The radical development on highly accurate description of strongly-correlated transition metal oxides, the mechanistic understanding of interface structure evolution and structure-properties relationships, and the successful prediction of new electrocatalyst has provided the foundation towards rational design of novel catalysts with increased activity, improved selectivity and/or under-controlled stability. In my future research, by leveraging and combining my experience, we will open up fundamentally new research thrusts, providing fundamental scientific insight into the interfacial properties of the relevant systems, and ultimately identifying new electrocatalysts of fuel cells, electrolyzers, CO oxidation and CO2 reduction, and heterogeneous catalyst of CO oxidation and water gas shift reactions.

Teaching Interests:

Albert Einstein once said "Most of the fundamental ideas of science are essentially simple, and may, as a rule, be expressed in a language comprehensible to everyone". I would like to say it is also a rule for my teaching. I find that most courses are essentially made up of concepts, not simply equations or mathematics. The concepts may be expressed in many specific ways and used in different fields, but the essence does not change, which is exactly like the second law in thermodynamics. The essential simplicity and generality make the concepts can be retained in students' memories for a long time period, and can be smoothly used to different fields later in life. Thus, good teaching should focus on the concepts; this is what I have learned from my study and research experience. For example, there are so many equations and derivations I learned but never use in my research; however, I still clearly remember conceptual mechanisms behind and can use them correctly when necessary. I have found this strategy useful when I teach "Advanced Modeling for Catalysis Studies" class and when I supervised graduate and undergraduate students. It has been highly effective when I first introduced the essential concept, then explained the role of each individual equation, and left the details for the students themselves to deepen the understanding. Thus my future teaching will mainly focus on the ideas and concepts, and using mathematics as a tool to express them more simply.

475985 Engineering the Electrochemical Interface for Sustainable Energy Conversion and Storage

Abstract:

Center for Individual Nanoparticle Functionality, Technical University of Denmark, Maria.Escudero@fysik.dtu.dk

Department of Chemical Engineering, Stanford University, mariaesc@stanford.edu

Research Interests:

One of the most important challenges facing humanity is the switch from fossil fuels to sustainable energy. Electrochemical energy conversion and storage devices such as fuel cells, electrolyzers and batteries should play a key role in this switch. My research combines electrochemical methods with surface science X-ray based characterization techniques, in situ spectroscopy and scanning probe microscopy, in order to investigate problems pertaining to interfacial electrochemistry and electrocatalysis. Reactions of interest include water splitting, oxygen reduction, hydrogen oxidation, oxidation of carbon monoxide and methanol, carbon dioxide reduction and hydrogen peroxide production. My scientific interests can be summarized as follows:

(a) Studying the descriptors determining the activity, stability and selectivity of electrocatalysts for energy conversion and storage devices [1-3].

(b) Determining and designing the catalytically active site [1-5].

(c) Investigating the role of the geometric structure (atomic ensemble effects [1]) and electronic properties of the surface atoms (electronic effects [2,3]) in electrocatalysis.

(d) Engineering the electrode-electrolyte interface [1,6].

Teaching Interests:

Not only have I a strong dedication to my scientific research, but also to teaching. Since 2013 I have taught a graduate course on Materials for hydrogen production and storage at the Technical University of Denmark. As a Chemical Engineer with a PhD in Physical Chemistry, I am interested in teaching both undergraduate and graduate courses in Sustainable Chemistry, Physical Chemistry, Catalysis, Spectroscopy and Chemical Engineering.

References:

[1] D. Strmcnik, **M. Escudero-Escribano**, K. Kodama, V. Stamenkovic, A. Cuesta, *Nature Chem*. **2010**, *2*, 880.

[2] **M. Escudero-Escribano**, et al., *J. Am. Chem. Soc*. **2012**, *134*, 16476.

[3] **M. Escudero-Escribano**, et al., *Science* **2016**, *352*, 73.

[4] A.F. Pedersen, et al., *Nano Energy* **2016**, doi:10.1016/j.nanoen.2016.05.026

[5] P. Malacrida, **M. Escudero-Escribano**, A. Verdaguer-Casadevall, I.E.L. Stephens, I. Chorkendorff, *J. Mater. Chem. A* **2014**, *2*, 4234.

[6] **M. Escudero-Escribano**, et al., *ChemPhysChem* **2011**, *12*, 2230.

476000 Controlled Synthesis and in-Situ Spectroscopic Study of Highly Efficient Ptfe Bimetallic and Ptrufe Trimetallic Nanocatalysts

Abstract:

Research Interests: Heterogeneous catalysis

Teaching Interests: Heterogeneous catalysis

Controlled synthesis of highly efficient nanocatalysts and studying their structure-activity relationship are of significant importance in both fundamental research and industrial applications. In this study, we developed several well defined Pt-based nanocatalysts with quasi-core-shell structure or nanoalloy structure for the preferential oxidation of CO in the presence of H_2 (PROX), which is a key reaction for the production of clean H_2. By means of various ex- or in-situ characterization method, such as HR-TEM, quasi-in-situ XPS and in-situ DRIFTS, the structure-activity relationships of these catalysts have been extensively studied. It was concluded that ferrous iron species could work as oxygen activation sites, and Ru could lower the CO adsorption strength. Furthermore, the utilization of Pt was much higher on nanoalloy catalysts as compared with quasi-core-shell catalysts. Based on these findings, highly efficient PtFe and PtRuFe nanoalloy catalysts with ultra-low Pt loading were designed, on which CO can be completely removed even under -50 ºC. We have also developed a novel strategy to in-situ monitoring the adsorbed species on the nanocatalysts by shell-isolated nanoparticles-enhanced Raman spectroscopy (SHINERS), which was then proved to be a general strategy for the in-situ study of heterogeneous nanocatalytic process. By this method, the interactions between CO/O_2 and metal can be studied, and for the first time, direct spectroscopic evidence for the activation of O_2 on ferrous center has been observed.

476052 Developing Enhanced Catalysts for Renewable Fuels through Spectroscopic Insights

Abstract:

Research Interests:

My research interests focus on the pursuit of fundamental understanding of catalytic reactions towards production of renewable fuels using insights from *in situ* and *operando* spectroscopy techniques uniquely adapted to investigate these catalytic materials. Increased global energy consumption along with heightened awareness of environmental, health, and political issues with fossil fuels are driving the need for efficient and economic conversion of renewable energy sources. Wind and sun provide more than enough energy to meet the growing demand, provided that challenges with intermittency, scale, and cost effectiveness can be overcome. These obstacles can be mitigated through development of highly active catalysts using abundant and inexpensive materials to convert solar and wind energy into fuels and chemicals.

My Ph.D. dissertation with Professor Thomas Jaramillo at Stanford University in the Department of Chemical Engineering focused on development of enhanced catalysts for electrochemical water splitting. Using a combination of electrochemical and spectroscopic techniques I was able to draw correlations between the structure, oxidation state, and catalytic activity of various transition metal oxide systems to gain a better understanding of their fundamental operation. For example, using x-ray absorption spectroscopy, I was able to track the changing oxidation state of a manganese-based catalyst in the presence and absence of a small amount of gold to understand why the addition of gold causes an order of magnitude increase in catalytic activity. During my Ph.D. I also developed a model to quantify loss mechanisms in photoelectrochemical water splitting based on the current state of materials research and to calculate maximum solar-to-hydrogen conversion efficiencies. Knowledge of practical limits for this process can help researchers assess their technology and guide future directions.

Currently, I am in my first year of postdoctoral research with Professor Clemens Heske and Dr. Lothar Weinhardt at the Karlsruhe Institute for Technology in the Institute for Photon Science and Synchrotron Radiation. I was awarded a Fellowship by the Helmholtz Association Postdoc Programme to pursue my proposed project of using advanced spectroscopic techniques to investigate the chemical and electronic structure of catalysts under operating conditions. During my time here I plan to assist in the development of advanced experimental set-ups for *in situ* and *operando* soft x-ray spectroscopy measurements and expand the application of these techniques to study catalyst materials for electrochemical and thermochemical CO/CO_2 reduction as well as NO_x selective catalytic reduction. *In situ* hard and soft x-ray experiments for these reactions are just beginning to be realized and because these techniques are such a powerful probe to study electronic structure, there is great potential to expand work and knowledge in this area.

At the poster session I will present a more thorough overview of my past and present research, as well as my plan for future research activities which will continue to seek fundamental insight towards improving efficiency for the conversion of renewable energy sources to fuels and chemicals. By studying, understanding, and improving the various components of these technologies, it is my goal to help drive the field forward and provide an efficient and economical solution that can shift our global energy dependence away from fossil fuels and towards renewable energy.

Teaching Interests:

In addition to serving as a teaching assistant during my undergraduate and graduate studies for computer programming and reaction kinetics courses, I have been an invited guest lecturer for various courses at Stanford University including graduate level Fundamentals and Applications of Spectroscopy, Electrochemical Energy Conversion, undergraduate Chemical Engineering Lab, and a Solar Fuels Master Class. I have also enjoyed mentoring graduate rotation students and a visiting masterâ€™s student as well as volunteering at local schools for science demonstrations.

476055 Computational Design of Hetero-Structured Catalysts for Energy

Abstract:

Research Interests:

Given the crucial role of catalysis science in our society, understanding how to design catalysts with the proper structure for desired catalytic activity and selectivity is a grand challenge. Hetero-structured materials such as core shell nanoparticles provide an extra degrees of freedom for activity tuning and control.

My expertise is in computational materials and catalyst design in close collaboration with experimental collaborators. My PhD research was dedicated to developing a fundamental and detailed understanding of the relationship between the structure of nanoscopic metallic catalysts and their function using first participial simulations. Instead of quantifying absolute catalytic activity; rather, I showed in a quantitative way how reactivity descriptors (which are known to correlate with the activity) vary with the particle structure and composition. During my post-doc at Stanford University and University of Pennsylvania, I have extended my research to a broader pool of materials with a higher complexity, such as metal oxides, metal carbides, and two-dimensional materials, and diversified my computational tool set.

My future research directions will focus, but not be limited to:

1) Development of a general picture for structure-reactivity correlation of different classes of hetero-structures including core shell nanoparticles.

2) Development of an understanding of dynamic catalytic processes on nanoparticles. The descriptor-based screening approach only provides a static and mean-field description of the catalysis on a nanoparticle. In reality, the morphology of the nanoparticle is varying during reaction, where a dynamic understanding is crucial to establish an accurate access of the catalytic function of the nanoparticle.

3) Development of a long-time scale simulation method suitable for distributed computing environments and its application to catalytic processes of complex materials.

Post-doctoral Experiences

University of Pennsylvania, Chemical and Biomolecular Engineering Supervisor: Aleksandra Vojvodic

Projects (a. Hetero-structured materials)

Stanford University, Chemical Engineering

Supervisor: Jens. K. Nørskov

Projects (a. Tailoring the Catalytic Activity of Perovskites b. CO activation under confinement)

PhD Thesis:

University of Texas at Austin, Chemistry

Supervisor: Graeme A. Henkelman

Dissertation: "Theoretical study of correlation between structure and function for nanoparticle catalysts"

Teaching Interests:

I have two years teaching experiences as a research educator of the Freshman Research Initiative (FRI) program at UT-Austin, where undergraduate students are exposed to authentic theoretical chemistry research. In addition to lecturing classes, I have TA:ed Physical Chemistry lab classes at UTAustin. In addition, I have also mentored graduate students in my group.

Selected Publications:

1. **L. Zhang** and G. Henkelman, Computational design of alloy-core@shell metal nanoparticle catalysts, *ACS Catal.* 5, 655-660 (2015).
2. **L. Zhang**, S. Chill, and G. Henkelman, Distributed Replica Dynamics *J. Chem. Phys.* 143, 174112 (2015).
3. S. García, **L. Zhang**, G. W. Piburn, G. Henkelman, and S. M. Humphrey, Microwave Synthesis of Classically Immiscible Rhodium-Silver and Rhodium-Gold Alloy Nanoparticles: Highly Active Hydrogenation Catalysts, *ACS Nano* 8, 11512-11521 (2014).
4. G. M. Mullen, **L. Zhang**, E. J. Evans Jr., T. Yan, G. Henkelman, and C. B. Mullins, Oxygen and hydroxyl species induce multiple reaction pathways for the partial oxidation of allyl alcohol over Au(111), *J. Am. Chem. Soc.* 136, 6489-6498 (2014)
5. **L. Zhang**, R. Iyyamperumal, D. F. Yancey, R. M. Crooks, and G. Henkelman, Design of Pt-shell nanoparticles with alloy cores for the oxygen reduction reaction, *ACS Nano* 7, 9168-9172 (2013).
6. **L. Zhang**, H.-Y. Kim, and G. Henkelman, CO oxidation at the A...
7. R. Iyyamperumal, **L. Zhang**, G. Henkelman, and R. M. Crooks, Efficient electrocatalytic oxidation of formic acid using Au@Pt dendrimer-encapsulated nanoparticles, *J. Am. Chem. Soc.* 135, 5521-5524 (2013).

*****Formatting Issue - Best Available Paper/Graphic*****

476074 Efficient Catalytic Pathways for Carbon Utilization

Abstract:

Research Interests:

Development of emission control technologies for remediation of toxic gases from stationary and mobile combustion sources is of great interest due to the stricter regulations and changes in the emissions as a result of the improvements in burner and engine technologies. Within my previous research efforts, one of my main objectives was to understand pollutant formation mechanisms to develop better emission remediation technologies. In particular, I have worked on removal of trace metals from coal-fired power plants, and elucidated the heterogeneous reaction mechanism of mercury over activated carbon. Additionally, I have developed a low temperature hydrocarbon oxidation catalyst for diesel exhaust applications. Aside from emission control technologies, my research focused on spectroscopic investigation of CO adsorption and oxidation over single crystal surfaces at different pressures, developing a catalyst for cracking of hydrocarbons to valuable low hydrocarbons and designing analytical instruments for high-throughput experimentation. The wide range of research topics that I have studied has provided me with vast experience in catalysis, spectroscopy, and instrument design and development.

As a tenure-track faculty, I would like to continue my research on developing nano materials and novel processes in the area of low temperature exhaust remediation, CO_2 utilization, tri-reforming of methane and alternative synthesis routes for the production of C_2-C_4 hydrocarbons. Currently, in most of the states, coal and natural gas power plants still remain the largest energy source for electricity production and play a crucial role to compensate the base loads in the electricity grid. Therefore, there is an urgent need for developing breakthrough technologies for CO_2 utilization and capture, and advanced combustion systems that can be implemented to the existing fossil fuel-fired power plants. I will initially develop my research program focusing on the aforementioned areas in carbon utilization and exhaust remediation and expand my research to create alternative energy pathways. My goal is to create a collaborative research environment interacting with national laboratories and industrial partners, and ultimately bridge the gap between fundamental science and industrial applications. In my poster presentation, I will discuss different projects to demonstrate my expertise in the area of catalysis and spectroscopy.

Teaching Interests:

Teaching has been my passion throughout my education and academic career. I gave lectures in various chemical engineering courses: Chemical Engineering Thermodynamics I, Chemical Kinetics and Modeling, Chemical Engineering Design, Chemical Plant Design Project, Mass Transfer, Kinetics and Reactor Design and Molecular Modeling. I am very interested in teaching core chemical engineering courses such as thermodynamics, heat transfer and chemical kinetics. I also plan to develop a special topics course on in-situ spectroscopy for surface science applications. I will discuss my teaching vision during the poster presentation.

References

1. Sasmaz, E.; Mingle, K.; Lauterbach, J. â€œHigh-throughput screening using Fourier transform infrared imagingâ€ *Engineering,* 2015, 1, 234.
2. Wang, C.; Sasmaz, E.; Wen, C.; Lauterbach, J. â€œPd supported on SnO_2-MnO_x-CeO_2 catalysts for low temperature CO oxidationâ€ *Catalysis Today*, doi:10.1016/j.cattod.2015.02.021.
3. Kim, S.; Mayeda, K.; Sasmaz, E.; Lauterbach, J. â€œOne step process for the production of BTEX and LPG-like fuel from pentanediolâ€ *ACS Sustainable Chemical Engineering*, 2015, 3, 381.
4. Kim, S.; Sasmaz, E.; Lauterbach, J. â€œEffect of Pt and Gd on coke formation and regeneration during JP-8 cracking over ZSM-5 catalystâ€ *Applied Catalysis B: Environmental,* 2015, 168, 212.
5. Galloway, B.D.; Sasmaz, E.; Padak, B. â€œBinding of SO_3 to Fly Ash Components: CaO, MgO, Na_2O, and K_2Oâ€ *Fuel*, 2015, 145, 179.
6. Lauterbach, J.; Sasmaz, E.; Bedenbaugh,J.; Kim, S.; Hattrick-Simpers, J. â€œDiscovery and optimization of coking and sulfur resistant bi-metallic catalyst for cracking JP-8: from thin film libraries to single powdersâ€, Modern Applications of High Throughput R&D in Heterogeneous Catalysis, 2014, 84-112
7. Bedenbaugh,J.; Kim, S.; Sasmaz, E.; Lauterbach, J. â€œHigh-throughput investigation of zeolite catalysts for military aviation fuel cracking to liquefied petroleum gasâ€ *ACS Combinatorial Science*, 2013, 15, 491.
8. Sasmaz, E.; Kirchofer, A.; Saha, A.; Abram, D.; Jaramillo, T. F. Wilcox, J. â€œMercury chemistry on brominated activated carbonâ€ *Fuel,* 2012, 99,189.
9. Sasmaz, E.; Aboud, S.; Wilcox, J. â€œMercury binding on Pd alloys and overlaysâ€ *J. Phys. Chem. C,* 2009, 113, 7813.
10. Sasmaz, E.; Wilcox, J. â€œMercury species and SO_2 adsorption on CaO(100)â€ *J. Phys. Chem. C*, 2008, 112, 16484.

476116 New Routes and Heterogeneous Catalysts Development for Biomass Conversion

Abstract:

Research Interests: Heterogeneous Catalysis, Biomass Conversion, Materials Science, Platform Technology

Teaching Interests: Inorganic Chemistry and Catalysis, Applied Chemical Technology

Crude oil market volatility, together with pressing concerns about greenhouse gas effects offer opportunities for the industrial-scale production of renewable fuels and chemical building blocks from biomass. In this light, the development of active, selective and stable solid catalysts, and bio-based approaches is necessary to achieve a more sustainable and economically viable catalytic conversion of such resource. My doctoral research was about the design and development of multifunctional materials, catalysts and technologies for energy-related and environmental applications, conducted in the University of Connecticut (USA) under the supervision of Prof. Steven L. Suib and supported by the US DOE. After obtaining my PhD in Chemistry in 2013, I moved to the Netherlands and worked with Dr. Pieter C. A. Bruijnincx and Prof. Bert M. Weckhuysen as a postdoctoral research fellow in Utrecht University (Inorganic Chemistry and Catalysis). For my first project, I explored new avenues in developing strategies and heterogeneous catalysts such as bifunctional solid acids for the efficient (tandem-catalytic) Diels-Alder aromatization of biomass-derived furanics in the liquid and the solid phase. This project was in collaboration with Wageningen University and Research Center - Food and Biobased Research and our industrial partner, Avantium Chemicals B.V. in Amsterdam. For my second and current project, I have been investigating the influence of synthesis, feed impurities, and process parameters on the stability of atomically dispersed metal catalysts anchored to transition metal oxide supports for levulinic acid hydrogenation. This project is in collaboration with University of Groningen (Engineering and Technology Institute) within the framework of the CatchBio Program and supported by BASF and DSM companies. Exciting recent results from these two postdoctoral research projects will be presented in this poster.

In the near future, my research work will continue to focus on the development of new approaches and high-performing heterogeneous catalysts for the valorization of bio-based platform molecules. It will aim to gain fundamental understanding of reaction engineering and catalytic processes across length scales, including the elucidation of structure-function relationships and expert systems, the investigation of inorganic catalyst stability, deactivation and mitigation, and the application of advanced in situ and operando techniques to understand reaction mechanism. I strongly believe that biomass catalysis and technology research will aid the transition of our chemical industry to a more sustainable one, as well as promote collaboration among various fields and sectors at both national and international levels.

476170 Transition Metal-Oxides for Sustainable Energy Conversion and Storage: The Computational Catalysis Perspective

Abstract:

Research Interests: The main goal of my research is to use the state-of-the-art computational methods in tandem with experiment to achieve understanding of the structure and the reaction mechanism of the most promising heterogeneous catalysts [1,2]. The largest breakthrough is needed in the catalysis of the electro-chemical water splitting, CO_2 reduction and ammonia synthesis, where the efficient, stable and earth-abundant catalysts are still to be discovered. I am especially interested in the catalytic functionality of the novel nanostructured first-row transition-metal oxides [3], which can be precisely synthesized and exhibit high level of activity and tunability.

Teaching Interests: Core level courses Chem. E, Chemistry Physics and Material Science. Ideally envision designing my own course in either Computational heterogeneous catalysis or in the Modern approaches to computational materials design.

[1] Bajdich, Sargent,el.al., Homogeneously-Dispersed Multi-Metal Oxygen-Evolving Catalysts, Science10.1126/science.aaf1525 (2016).

[2] Bajdich, Jaramillo, el.al., Gold-supported cerium-doped NiOx catalysts for water oxida- tion, Nature Energy10.1038/NENERGY.2016.53 (2016)

[3] Bajdich, Lauritsen, ACS Nano10.1021/acsnano.5b00158 (2015); Special Issue in Topics in Catalysis (2016); Nat. Materials (under review).

476301 Impacts of Metal-Adsorbate Bonds on Photon Activation Mechanism and Nanoparticle Reconstruction in Heterogeneous Catalysis

Abstract:

More than 80% of modern industrial chemical processes utilize heterogeneous catalysts, including the production and processing of fuel, pollution mitigation, and commodity chemical production. The control of selectivity in complex chemical reaction networks has become the focus of heterogeneous catalysis in the 21st century, in order to decrease unwanted byproducts and to increase energy efficiency and sustainability. My work at UC-Riverside has focused on several approaches to determine how molecules interact with catalytic metal surfaces, and ways to specifically activate targeted molecules on the surface, with the goal of achieving high catalytic selectivity. These approaches include: utilizing photocatalysis in order to activate targeted strong metal-adsorbate bonds by matching photon wavelength with bond energies of hybridized metal-adsorbate bonds, modifying metal-support interfaces to efficiently remove atmospheric contaminants in confined space pollution control applications, and exploiting *in-situ* spectroscopy techniques with a correlated analysis of reactivity in order to reveal catalytic active sites. My future work will include utilizing *in-situ*probes to reveal fundamental insights of catalytic processes for sustainable useful chemical production.

Research Interests:

Heterogeneous catalysis, *In-situ* spectroscopy, selective chemical production, conversion of solar to chemical energy

Teaching Interests:

Preferences: Chemical Kinetics, Catalysis & Reaction Engineering, Physical Chemistry, Thermodynamics, Transport Phenomena, but would be open to teaching most undergraduate courses.

Selected Publications:

[1] **Kale, M. J.**; Christopher, P. Plasmons at the interface. *Science* **2015,** *349*, 587-588.

[2] **Kale, M. J.**; Avanesian, T.; Xin, H.; Yan, J.; Christopher, P. Controlling Catalytic Selectivity on Metal Nanoparticles by Direct Photoexcitation of Adsorbate-Metal Bonds. *Nano Letters* **2014,** *14*, 5405-5412.

[3] **Kale, M. J.**; Avanesian, T.; Christopher, P. Direct Photocatalysis by Plasmonic Nanostructures. *ACS Catalysis* **2014**, *4*, 116â€“128.

[4] **Kale, M. J.**; Christopher, P. Utilizing quantitative in-situ FTIR spectroscopy to identify well-coordinated Pt atoms as the active site for CO oxidation on Al_2O_3 supported Pt catalysts. *In Review.*

[5] **Kale, M. J.**; Gidcumb, D.; Christopher, P. Evaluation of Precious Metal Catalyst Stability for Confined Space Pollution Control. *In Preparation*.

475756 Electrocatalysis for Sustainable Energy Technologies

Abstract:

ELECTROCATALYSIS FOR SUSTAINABLE ENERGY TECHNOLOGIES

Drew Higgins, Postdoctoral Fellow, Stanford University

RESEARCH INTERESTS

Achieving a sustainable energy economy is among the biggest challenges of the 21st century. Electrochemical energy storage and conversion devices (***i.e***., batteries, fuel cells, electrolyzers) will be integral components of clean energy infrastructures, as they boast zero emission operation and high theoretical efficiencies. These systems, however, are not currently techno-economically viable at scale, primarily due to electrocatalyst limitations. My research interests lie in the development and device integration of advanced, high performance electrocatalysts. These materials have the potential to address the performance, durability and cost challenges facing many emerging clean electrochemical technologies. The ultimate goal is to transform the landscape of the energy sector by providing cost competitive devices that can be applied in the transportation sector, grid scale energy storage, or industrial applications.

FUTURE PLANS

My career goal is to manage a research group at an academic institution that works on the cutting edge of clean energy technologies, particularly with respect to energy transformations and electrocatalysis. In the short term, my research will be focused on two thrust areas: (1) ***Electrochemical Carbon Dioxide Reduction***, an integral process of artificial photosynthesis that converts CO_2 directly into fuels or industrial chemicals; and (2) ***Oxygen Electrocatalysis,***the most pertinent and performance limiting processes occurring in fuel cell and electrolyzer devices. I envision early work focusing on the development of new electrocatalyst materials that will be characterized and electrochemically evaluated to understand performance capabilities. The best in class electrocatalysts will be engineered into electrode structures for device performance testing and optimization.

In the long term, my vision is to lead a research program that spans: (1) ***Fundamental Electrocatlaysis/Surface Science***; (2) ***Electrocatalyst Design and Synthesis***; and (3) ***Electrochemical Device Design, Integration and Prototype Demonstration***. These three core areas will be strategically integrated to operate in a feedback loop, whereby new knowledge generated in one area will be applied to achieve advances in other areas. Over time, this research program will be expanded to encompass different thrust areas that are synergistic with those listed above, including hydrogen electrochemistry, new battery chemistries (***i.e.***, metal-air or sodium-ion) and electrochemical purification techniques. The overarching goals of this research program will be two-fold. First, is to develop innovative material solutions and technologies that address the pertinent energy and environmental challenges facing the modern world. Second, is to educate and train the next generation of scientists and engineers that will be lifelong contributors in this sector of increasing global importance. Academic, industrial and government collaborations will be fostered and leveraged to establish an interdisciplinary research environment, and to provide targeted training and avenues for research personnel to enter the workforce.

RESEARCH EXPERIENCE

Banting Postdoctoral Fellow – Chemical Engineering (start date, Sep 2015)

Stanford University, Stanford, California

Supervisor: Dr. Thomas Jaramillo

- Electrochemical reduction of carbon dioxide to produce fuels and chemical products
- Physical vapour deposition of binary metallic thin films to serve as model electrocatalyst systems
- Integration of electrodes into carbon dioxide electrolysis cell for performance testing and product detection

Visiting Scholar – Materials Physics and Applications 11 (Jan 2014 – Dec 2014)

Los Alamos National Laboratory, Los Alamos, New Mexico

Supervisor: Dr. Piotr Zelenay

- Synthesis and integration of non-precious metal catalysts into polymer electrolyte fuel cell membrane electrode assemblies for performance validation

PhD and MASc – Department of Chemical Engineering(2009 –2015)

University of Waterloo, Waterloo, Ontario

Supervisor: Dr. Zhongwei Chen

- Synthesis, characterization and electrochemical evaluation of nanostructured oxygen reduction and oxygen evolution catalysts

Undergraduate Researcher – Department of Chemical and Biochemical Engineering (2007, 2008)

University of Western Ontario, London, Ontario

Supervisor: Dr. Jingxu (Jesse) Zhu

SELECTED SUCCESSFUL PROPOSAL WRITING

- Banting Postdoctoral Fellowship
- Natural Sciences and Engineering Research Council of Canada (NSERC) Postdoctoral Fellowship
- NSERC Discovery Grants Program
- 2 x NSERC Strategic Project Grants
- 3 x NSERC Collaborative Research and Development Grants
- Ontario Centres of Excellence: Collaborative Research Program
- Ontario Centres of Excellence: Voucher for Commercialization
- Ontario Research Fund: Small Infrastructure Fund
- NSERC Postgraduate Fellowship
- Ontario Graduate Scholarship

TEACHING EXPERIENCE

- **Graduate Student Mentor**, Stanford University (2015 – present), University of Waterloo (2012 - 2015)
- **Guest Lecturer** for ChE 331, ChE 102 and NE 491, University of Waterloo (2012 - 2015)
- **Certificate of University Teaching Program,** University of Waterloo (2012 – 2013)
- **Fundamentals of University Teaching Program,** University of Waterloo (2011)
- **Teaching Assistant** for ChE 331 and ChE 290, University of Waterloo (2010 – 2011)

SELECTED PUBLICATIONS

1. **D. Higgins**, P. Zamani, A. Yu, Z. Chen, "The application of graphene and its composites in oxygen reduction electrocatalysis: a perspective and review of recent progress", ***Energy & Environmental Science***, 9 (2016) 357-390.
2. **D. Higgins**, F. Hassan, M. Seo, J. Choi, A. Hoque, D. Lee, Z. Chen, "Shape-controlled octahedral cobalt disulfide nanoparticles supported on nitrogen and sulfur-doped graphene/carbon nanotube composites for oxygen reduction in acidic electrolyte", ***Journal of Materials Chemistry A***, 3 (2015) 6340-6350. Featured on the back cover of the associated issue.
3. A. Hoque, F. Hassan, **D. Higgins**, J. Choi, M. Pritzker, S. Knights, S. Ye, Z. Chen, "Multigrain platinum nanowires consisting of oriented nanoparticles anchored on sulfur-doped graphene as a highly active and durable oxygen reduction electrocatalyst", ***Advanced Materials***, 27 (2015) 1229-1234.
4. **D. Higgins**, R. Wang, A. Hoque, P. Zamani, S. Abureden, Z. Chen, "Morphology and composition controlled platinum-cobalt alloy nanowires prepared by electrospinning as oxygen reduction catalyst", ***Nano Energy***, 10 (2014) 135-143.
5. **D. Higgins**, A. Hoque, M. Seo, R. Wang, F. Hassan, J. Choi, M. Pritzker, A. Yu, J. Zhang, Z. Chen, "Development and Simulation of Sulfur-doped Graphene Supported Platinum with Exemplary Stability and Activity Towards Oxygen Reduction", ***Advanced Functional Materials***, 24 (2014) 4325-4336. Accepted as VIP article and featured on front-piece of the associated issue.
6. **D. Higgins**, A. Hoque, F. Hassan, J. Choi, B. Kim, Z. Chen, "Oxygen Reduction on Graphene-Carbon Nanotube Composites Doped Sequentially with Nitrogen and Sulfur", ***ACS Catalysis***, 4 (2014) 2734-2740.
7. V. Chabot, **D. Higgins**, A. Yu, X. Xiao, Z. Chen, J. Zhang, "A Review of Graphene and Graphene

Oxide Sponge: Material Synthesis and Applications", ***Energy & Environmental Science***, 7 (2014) 1564-1596.

8. **D. Higgins**, Z. Chen, D. Lee, Z. Chen, "Activated and nitrogen-doped exfoliated graphene as air electrodes for metal-air battery applications", ***Journal of Materials Chemistry A***, 1 (2013) 2639-2645.
9. **D. Higgins**, Z. Chen, "Recent progress in non-precious metal catalysts for PEM fuel cell applications", ***Canadian Journal of Chemical Engineering***, 91 (2013) 1881-1895.
10. J. Liao, **D. Higgins**, G. Liu, V. Chabot, X. Xiao, Z. Chen, "Multifunctional TiO2@C/MnO2 Core@Double-shell Nanowire Array as a High Performance 3D Electrode for Li-ion batteries", ***Nano Letters***, 13 (2013) 5467-5473.
11. **D. Higgins**, J. Choi, J. Wu, A. Lopez, Z. Chen, "Titanium nitride carbon nanotube core-shell composites as effective electrocatalyst supports for low temperature fuel cells", ***Journal of Materials Chemistry***, 22 (2012) 3727-3732.
12. **D. Higgins**, J. Wu, W. Li, Z. Chen, "Cyanamide derived thin film on carbon nanotubes as metal free oxygen reduction reaction electrocatalyst", ***Electrochimica Acta***, 59 (2012) 8-13.
13. **D. Higgins**, Z. Chen, S. Ye, S. Knights, "Highly Durable Platinum-Cobalt Nanowires by Microwave Irradiation as Oxygen Reduction Catalyst for PEM Fuel Cell", ***Electrochemical and Solid-State Letters,*** 6 (2012) B83-B85.
14. Z. Chen, A. Yu, **D. Higgins**, H. Wang, H. Liu, Z. Chen, "Highly Active and Durable Core-Corona Structured Bifunctional Catalyst for Rechargeable Metal-Air Battery Application", ***Nano Letters***, 69 (2012) 1946-1952.
15. **D. Higgins**, Z. Chen, Z. Chen, "Nitrogen doped carbon nanotubes synthesized from aliphatic diamines for the oxygen reduction reaction", ***Electrochimica Acta***, 56 (2011) 1570-1575.
16. ***16.*** Z. Chen, **D. Higgins**, A. Yu, L. Zhang, J. Zhang, "A Review on Non-precious Metal Electrocatalysts for PEM Fuel Cells", ***Energy & Environmental Science***, 4 (2011) 3167-3192.
17. **D. Higgins**, D. Meza, Z. Chen, "Nitrogen Doped Carbon Nanotubes as Platinum Catalyst Supports for Oxygen Reduction Reaction in PEM Fuel Cells", ***The Journal of Physical Chemistry C***, 114 (2010) 21981-21988.
18. W. Li, A. Yu, **D. Higgins**, B. Llanos, Z. Chen, "Biologically Inspired Highly Durable Iron Phthalocyanine Catalysts for Oxygen Reduction Reaction in PEM Fuel Cells", ***Journal of the American Chemical Society***, 132 (2010) 17056-17058.

Research Interests: (see above)

Teaching Interests: (see above)

476174 Rational Catalyst Design for Renewable Energy Technologies

Abstract:

Research Interests:

Today, there is an extensive interest in replacing fossil fuels to supply mankind energy needs sustainably on a long-term basis. In recent years several renewable energy technologies have been developed such as fuel cells, metal-air batteries and electrolyzers. There are, however, severe shortcomings of the present technologies, which need to be overcome to make these processes more economically attractive. One of the most important problems is the lack of active, selective and inexpensive catalysts for these processes. My research focuses on designing cost-effective and efficient heterogeneous catalysts for the application in energy conversion devices such as fuel cells, batteries, and chemical synthesis processes. The design of catalysts will be guided by computational modeling in particular DFT with an emphasis on understanding the electronic structure of the catalyst and adsorbed surface species. In particular I am interested in 1) developing efficient catalysts for oxygen reduction reaction (ORR) in fuel cells and oxygen evolution reaction (OER) in metal-air batteries and water electrolyzers. 2) Additionally I am interested in synthesizing fuels using biomass, methane oxidation and electrochemical conversion of $CO_2$to hydrocarbons.

Recent theoretical understandings indicate that there are universal correlations (scaling relations) between surface bond energies of different adsorbed species that bind to the surface through similar atom for specific reaction.[1,2] These correlations pose limitations on many of the known catalysts for reactions such as ORR, OER, CO_2 reduction and N_2 reduction reactions.[3] Therefore, catalyst design requires development of new, effective and non-precious catalysts with active site motifs, which do not follow the similar correlation. ***My work utilizes theoretical tools in particular DFT to model two-dimensional materials with different modification strategies including doping and defect design to describe the surface reactions. These strategies have not been explored so far. I seek to find a breakthrough for scaling relations and my goal is to discover novel catalysts for the following applications: 1) improving the efficiency of the devices that are related to sustainable energy applications, i.e. fuel cells[4], electrolyzers[5], and metal-air batteries[6], and 2) to discover/design new, effective and non-precious catalysts for ORR, OER, and CO_2 reduction reaction, fuel synthesis from biomass, and methane oxidation.***

Another important issue facing humanity is related to the potable water. Millions of people in the developing world are lacking access to drinking water due to its contamination with urban, industrial and agricultural waste. These pollutants can in principle be removed by their total oxidation by hydrogen peroxide (H_2O_2) as a strong oxidizing agent without leaving any residues behind. Production and transportation of H_2O_2 to the developing world, however, is challenging due to high costs and safety issues, making this potentially game-changing chemical largely inaccessible to the massive number of people who need it the most. ***My research combines the expertise from fundamental understanding of the reaction and DFT with versatile synthetic methods, electrochemical evaluation and catalyst characterization to develop an electrochemical route by designing an efficient and cost effective electrocatalyst for selectively reducing O_2 to H_2O_2 to enable continuous, small-scale and decentralized production of H_2O_2.***[7,8]

References:

(1) Viswanathan, V.; Hansen, H. A.; Rossmeisl, J.; Nørskov, J. K. ***ACS Catal.* 2012**, ***2***(8), 1654–1660.

(2) Man, I. C.; Su, H.-Y.; Calle-Vallejo, F.; Hansen, H. A.; Martínez, J. I.; Inoglu, N. G.; Kitchin, J.; Jaramillo, T. F.; Nørskov, J. K.; Rossmeisl, J. ***ChemCatChem* 2011**, ***3***(7), 1159–1165.

(3) Vojvodic, A.; Nørskov, J. K. ***Natl. Sci. Rev.* 2015**, ***2***(2), 140–149.

(4) Siahrostami, S.; Björketun, M. E.; Strasser, P.; Greeley, J.; Rossmeisl, J. ***Phys. Chem. Chem. Phys.* 2013**, ***15***(23), 9326–9334.

(5) Siahrostami, S.; Vojvodic, A. ***J. Phys. Chem. C* 2015**, ***119***(2), 1032–1037.

(6) Siahrostami, S.; Tripković, V.; Lundgaard, K. T.; Jensen, K. E.; Hansen, H. a; Hummelshøj, J. S.; Mýrdal, J. S. G.; Vegge, T.; Nørskov, J. K.; Rossmeisl, J. ***Phys. Chem. Chem. Phys.* 2013**, ***15***(17), 6416–6421.

(7) Siahrostami, S.; Verdaguer-Casadevall, A.; Karamad, M.; Deiana, D.; Malacrida, P.; Wickman, B.; Escudero-Escribano, M.; Paoli, E. a; Frydendal, R.; Hansen, T. W.; Chorkendorff, I.; Stephens, I. E. L. S.; Stephens, I. E.; Rossmeisl, J. ***Nat. Mater.* 2013**, ***12***(12), 1137–1143.

(8) Verdaguer-Casadevall, A.; Deiana, D.; Karamad, M.; Siahrostami, S.; Malacrida, P.; Hansen, T. W.; Rossmeisl, J.; Chorkendorff, I.; Stephens, I. E. L. ***Nano Lett.* 2014**, ***14*** (3), 1603–1608.

Teaching Interests:

I would like to teach courses such as Physical Chemistry, Thermodynamics, Quantum Chemistry, Molecular Spectroscopy and Chemical Kinetics. Moreover, I would be interested in developing and offering new courses in accordance to my area of specialty and research experiences.

476259 Engineering Visible-Light Organic Photocatalysis for Polymers in Biomaterials, Biosensing, and Photomedicine

Abstract:

Research Interests:

Visible-light organic photocatalysis is one of the most promising methods to synthesize and modify polymers of medical and biomedical relevance. These reactions permit spatiotemporal control at biologically benign irradiation levels with low-cost biocompatible organic photocatalysts. However, we have poor control of these reactions. This stems from the general lack of knowledge of the complex reaction mechanisms around proteins. Therefore, discovery of the right photocatalyst for every biomedical application has advanced slowly through trial-and-error. Molecular engineering of photocatalysts promises to accelerate the implementation and expand the utility of visible-light photocatalysis in biomedicine. My goal is to promote specific reaction pathways that yield desired outcomes, e.g. radical initiation, polymer/protein grafting, reversible crosslinking, or protein crosslinking, by engineering organic photocatalysts.

I have discovered that the photocatalysis of methylene blue mimics photosynthesis in that it involves a $2e^-/H^+$ photocatalytic cycle, as opposed to the concerted e^-/H^+ transfer that is generally assumed for photoinduced electron transfer reactions. This allows synthesis of centimeter thick optically absorptive polymers for dental and orthopedic restorative medicine. Then, I investigated the photocatalysis of Eosin to initiate radical polymerization of hydrogels against a thousand times excess oxygen around proteins, useful for biosensing in medical diagnostics. The next step is elucidating the mechanism for the crosslinking of collagen proteins with Rose Bengal useful for surgical wounds closure, corneal sealing, nerve, tendon and vessel repair, and cartilage regeneration. Applying chemical engineering principles to photochemical reactions will allow me to design photocatalysts that crosslink collagen fibers with minimal tissue inflammation.

Teaching Interests:

A startled classroom of junior and senior Chemical Engineering students looked at me as I opened a lecture with a simulation of the explosion of a chemical reactor. As TA of the Reaction Engineering and Chemical Kinetics course at the University of Colorado I wanted to do something different to get students out of our classical equation-heavy class routine. I selected the best video I could find on Runaway Reactions to engage students in the topic of incorporating heat considerations into reactor design. Office Hours were well spent observing their progress and discussing specific questions in small groups. Most questions were directly related to problem sets and midterms, but some served as discussion points to chat about chemical kinetics, numerical methods, and reaction mechanisms. I saw our students making their schematics and postulating new questions as the learning process sparked their curiosity. I believe this classical learning/teaching interaction is invaluable to stimulate active learning. I enjoyed designing online assessments to assess studentsâ€™ knowledge, critical thinking, and inventiveness as TA of the Statistics for Engineering course at the University of Colorado. A combination of Backwards Design and Constructive Alignment will allow me to create an engaging environment where my students achieve the intended learning outcomes. I work to engage all students in their own learning experience. I observed this first hand in the students I mentored as the founding president of the AIChE student Chapter at my undergraduate University. All this coalesced in my Teaching Philosophy developed during the Kaufman Teaching Certificate Program at MIT.

476641 Catalysts for Emission Control and Energy Conversion: Computational Study Based on DFT Calculations

Abstract:

Research Interests:

Catalysts for emission control and energy conversion

- Emission control: Metal-exchanged zeolite for NO_x reduction; "single-atom" catalyst for CO oxidation
- Energy conversion: bimetallic catalyst for biomass conversion; photocatalyst for using solar energy; "single-atom" catalyst for fuel cell

Research Experiences:

(1) Computational K-edge X-ray absorption near edge structure (XANES) and X-ray emission spectroscopy (XES) studies on properties of Cu ions in Cu-SSZ-13 and Cu-SAPO-34, collaborating with experimental results;

(2) Characterized NO chemisorption on the catalyst of Cu/SSZ-13 by comparing computational and experimental infrared (IR) spectra;

(3) Characterized benzene adsorption on Pt step surface by integrating computational and experimental X-ray photoelectron spectroscopy (XPS);

(4) Discovered the accurate model of "29"-structure Cu oxide by performing computational scanning tunneling microscope (STM), and comparing with experimental STM;

(5) Innovatively created another single-atom catalyst of Pt/TiN suitable for reactions in proton exchange membrane fuel cells;

(6) Through a method of surface modification, to functionalize silicon namowires as photocatalysts for water splitting by using solar energy;

(7) Investigated effect of electric field and strain field on properties of silicon nanowires, such as band structure and band gap.

Teaching Interests:

Theoretical surface and interface science

Reaction at solid surface

Selected Publications:

Zhang, R.; Szanyi, J.; Gao, F.; and McEwen, J.-S. "The Interaction of Reactants, Intermediates and Products with Cu Ions in Cu-SSZ-13 NH_3 SCR Catalysts: An Energetic and ab initio X-ray Absorption Modeling Study", *Catal. Sci. Technol.*, 6 (15), 5812-5829 (2016)

Therrien, A.J.; Zhang, R.; Lucci, F.R.; Marcinkowski, M.D.; Hensley, A.J.; McEwen, J.-S.; and Sykes, E.C.H., "Structurally Accurate Model for the "29"-structure of Cu_xO/Cu(111): A DFT and STM Study", *J. Phys. Chem. C*, 120 (20), 10879-10886 (2016)

Zhang, R.; McEwen, J.-S.; Kollár, M.; Gao, F.; Wang, Y.; Szanyi, J.; and Peden, C. "NO Chemisorption on Cu/SSZ-13: A Comparative Study from Infrared Spectroscopy and DFT Calculations", *ACS Catal.*, 4 (11), 4093-4105 (2014)

Hensley, A.J.; Hong, Y.; Zhang, R.; Zhang, H.; Sun, J.; Wang, Y.; and McEwen, J.-S., "Enhanced Fe_2O_3 Reducibility via Surface Modification with Pd: Characterizing the Synergy within Pd/Fe Catalysts for Hydrodeoxygenation Reactions", *ACS Catal.*, 4 (10), 3381-3392 (2014)

Zhang, R.; Hensley, A.J.; McEwen, J.-S.; Wickert, S.; Darlatt, E.; Fischer, K.; Schoppke, M.; Denecke, R.; Streber, R.; Lorenz, M.; Papp, C.; and Steinruck, H.-P., "Integrated X-ray Photoelectron Spectroscopy and DFT Characterization of Benzene Adsorption on Pt(111), Pt(355) and Pt(322) Surfaces", *Phys. Chem. Chem. Phys.*, 15 (47), 20662-20671 (2013)

Zhang, R.; Lee, T.H.; Yu, B.D.; Stampfl, C.; and Soon, A., "The Role of Non-conventional Supports for Single-atom Platinum-Based Catalysts in Fuel cell Technology", *Phys. Chem. Chem. Phys.*, 14 (48), 16552-16557 (2012)

Zhang, R.; Liu, X.M.; Wen, Z.; and Jiang, Q., "Prediction of Silicon Nanowires as Photocatalysts for Water Splitting: Band Structure Calculated Using Density Functional Theory", *J. Phys. Chem. C*, 115 (8), 3425-3428 (2011)

Ao, Z.M.; Jiang, Q.; Zhang, R.; Tan, T.T.; and Li, S., "Al Doped Graphene: A Promising Material for Hydrogen Storage at Room Temperature", *J. Appl. Phys.*, 105 (7), 074307 (1-6) (2009)

476917 Development of Heterogeneous Catalysis towards a Sustainable Future

Abstract:

Research Interests:

The prosperity of a growing population has placed significant stress on the environment and society, and led to many grand challenges in climate, energy, water, and food, etc. Due to its transforming power, catalysis is definitely a key part of the solutions to these problems. My research interests are to understand and develop heterogeneous catalysis for a sustainable future.

In my Ph.D. research with Prof. Yuriy Román-Leshkov in the Department of Chemical Engineering at MIT, I have developed effective catalytic strategies to convert biomass-derived molecules into mono- and dicarboxylic acids and esters. In particular, by exploiting the dual-function heteroatom-containing zeolites, we enabled the activation of carbonyl groups for carbon-carbon coupling to upgrade biomass-derived oxygenates into renewable monomers, and we also developed stable reducible metal oxides supported gold catalysts for base-free aqueous aerobic oxidation of glucose by maximizing the anchoring and activating benefits from the oxygen vacancies of metal oxides. In my current postdoctoral research with Prof. William H. Green in the Department of Chemical Engineering at MIT, my research interest is on the prediction of reaction kinetics in liquid phase.

Motivated by those grand challenges, I would like to extend and expand the two types of chemistry, i.e. activation of carbonyl groups and aqueous aerobic oxidation of organic molecules, to contribute to the solutions. Specifically, I am interested in developing heterogeneous catalysis for and understanding the mechanism and kinetics of 1) conversion of carbon dioxide into chemicals and fuels, and 2) remediation of pollutants in water.

Teaching Interests:

My preference includes kinetics and reaction engineering courses for undergraduate and graduate students, which naturally fit my research experience. I would also like to develop electives that focus on heterogeneous catalysis and/or kinetics. During my Ph.D., I have been the teaching assistant for undergraduate kinetics courses. I have also mentored undergraduate students and master students in my Ph.D. and current postdoctoral research.

478099 Understanding and Improving Electrocatalysts for Energy Conversion and Waste Remediation

Abstract:

Research Interests:

The challenge of sustainably providing the increasing demand for energy and resources relies on improving existing renewable energy processes or developing new processes. Major issues include the storage of energy, conversion of renewable energy to storable transportation fuels, and the remediation of waste streams. Catalysis and electrocatalysis can play a major role in providing sustainable energy and remediating waste streams, but improvements are needed in electrocatalyst activity, selectivity and stability. My interest is in synthesizing and testing new classes of materials (such as doped and undoped metal sulfides) that can achieve activities, selectivities and stabilities that are unmet by the existing metal electrocatalysts. I am also interesting in characterization of electrocatalysts under operating conditions through the use of spectroscopy and kinetic analysis. To evaluate potential processes, I will also use technoeconomic analysis and sensitivity analysis to identify which factors have the largest impact on the final product. As a faculty member I will focus on the following research projects:

Project 1. Understand and improve the electrocatalytic activity of flow batteries. A focus of my graduate work was on new electrocatalysts for H_2/Br_2 flow batteries, whose activity and stability have a major impact on the feasibility of flow batteries. In this project, I hope to extend my previous work to other flow batteries including the H_2/Cerium battery (where competing oxygen evolution and changing oxidation states of the catalyst are a challenge)[1] or vanadium batteries (where the competition of hydrogen evolution is an issue). *The specific aims of this project are to test new electrocatalyst materials, and characterize them under operating conditions using techniques such as X-ray absorption spectroscopy (XAFS) to understand the effects of applied potential on oxidation state or adsorption of reactant molecules.*

Project 2. New methods to convert solar energy to transportation fuels. Research on solar fuels has focused on water splitting (to H_2 and O_2) and carbon dioxide reduction, but other processes are also achievable (and possibly more viable) if new electrocatalysts are developed. A sample reaction is using electrochemical potential provided from sunlight to hydrogenate waste biomass, which enables easier deoxygenation to fuel-grade hydrocarbons. Aqueous electrochemical hydrogenation of biomass has a low overpotential compared to CO_2 electro-reduction (~0.1 V compared to 1 V), important for solar driven processes where the voltage provided by photons is limited. However, electrocatalyst activities rates must be improved, and photoelectrochemical hydrogenation of biooil has not been demonstrated using sunlight. *The specific aims of this project are to evaluate new electrocatalysts (such as metal sulfides) for the conversion of lignin-based biooils to hydrocarbons, and coupling the reaction with light absorbers powered by sunlight.* The characterization techniques such as XAFS used to evaluate flow batteries in Project 1 can also be used here to understand reactant coverages and electronic properties of the catalyst.

Project 3. Remediation of wastewater through electrochemical treatment. One of the major concerns for a sustainable future is treatment of waste streams. Nitrate is a biological poison that is present in waste as varied as fertilizer run-off, electrochemical industrial waste and low-level nuclear waste, and can potentially be treated electrochemically through nitrate reduction. Yet, the mechanism of nitrate remediation is not well-understood, and active catalysts that can maintain activity in the presence of metal ions or other poisons do not exist. *The specific aim of this project is to determine the properties of electrocatalysts that control the activity for nitrate reduction and develop catalysts with high activity and resistance to poisoning.* Similar experimental techniques from the other two proposed projects can be applied to develop waste remediation electrocatalysts.

PhD Research:

My graduate research with Professor Eric McFarland and Professor Horia Metiu at the University of California, Santa Barbara focused on understanding new types of electrocatalysts for the conversion of energy. This consisted of identifying metal sulfides and doped metal sulfides for active and stable H_2/Br_2 flow battery electrocatalysts,[2–5] and use of catalysts for the conversion of solar energy to hydrogen.[6,7] We coupled experiments with theory to understand molecular descriptors for catalyst activity (such as hydrogen binding energy on metal sulfides) and to identify active sites.[8,9] Through selective synthesis of particular phases of rhodium sulfide we were able to improve electrocatalytic activity and stability. I also completed a technoeconomic analysis of the H_2/Br_2 flow battery to identify key components in the levelized cost of electricity.[10]

Postdoctoral Research:

My postdoctoral research with Prof. Charles T. Campbell and Prof. Johannes Lercher at the University of Washington and Pacific Northwest National Laboratory consists of the interrogation of electrocatalysts for electrocatalytic hydrogenation (ECH) of phenol. This includes *in situ* characterization of the catalysts by XAFS to identify phenol and hydrogen coverages on the surface as well as electronic and structural properties of the catalyst. In addition, we identified the cause for deactivation at elevated temperatures by kinetic analysis and modified reaction conditions (e.g. hydrogen pressure) to prevent formation of poison species, and enable unprecedented rates at temperatures relevant for ECH (<100 °C). I am currently a Washington Research Foundation Innovation Fellow with the Clean Energy Institute at UW, and successfully presented for a PNNL Initiative as well as co-wrote a proposal for Laboratory Directed Research and Development (internal funding for PNNL).

Teaching Interests:

During my time as a student, the professors of my core Chemical Engineering courses helped me develop my interest in Catalysis and Reaction Engineering, and I hope to be able to do the same for future chemical engineers. I am comfortable with the core Chemical Engineering course structure, but feel best suited towards teaching Reaction Engineering and in developing elective courses in Electrochemistry or Energy. As a graduate student I was a Teaching Assistant for Undergraduate Chemical Reaction Engineering and an Energy elective. In terms of teaching in a research environment, I supervised six undergraduate students during my graduate work, including education on the experimental aspects of catalysis, electrochemistry and photoelectrochemistry.

References:

1. Tucker, M. C., Weiss, A. & Weber, A. Z. Improvement and analysis of the hydrogen-cerium redox flow cell. *J. Power Sources* **327,** 591–598 (2016).

2. Ivanovskaya, A., **Singh N.**, Liu, R., Kreutzer, H., Baltrusaitis, J., Nguyen, T. V., Metiu, H. & McFarland, E. Transition Metal Sulfide Hydrogen Evolution Catalysts for Hydrobromic Acid Electrolysis. *Langmuir* **29,** 480–492 (2013).

3. Masud, J., Nguyen, T. V., **Singh, N.**, McFarland, E., Ikenberry, M., Hohn, K., Pan, C.-J. & Hwang, B.-J. A RhxSy/C Catalyst for the Hydrogen Oxidation and Hydrogen Evolution Reactions in HBr. *J. Electrochem. Soc.* **162,** F455–F462 (2015).

4. Nguyen, T. Van, Kreutzer, H., Yarlagadda, V., McFarland, E. & **Singh, N.** HER/HOR Catalysts for the H2-Br2 Fuel Cell System. *ECS Trans.* **53,** 75–81 (2013).

5. Masud, J. Walter, J., Nguyen, T. V., Lin, G., **Singh, N.**, McFarland, E., Metiu, H., Ikenberry, M., Hohn, K., Pan, C.-J. & Hwang, B.-J. Synthesis and Characterization of RhxSy/C Catalysts for HOR/HER in HBr. *ECS Trans.* **58,** 37–43 (2014).

6. **Singh, N.**, Mubeen, S., Lee, J., Metiu, H., Moskovits, M. & McFarland, E. W. Stable electrocatalysts for autonomous photoelectrolysis of hydrobromic acid using single-junction solar cells. *Energy Environ. Sci.* 1–4 (2014). doi:10.1039/c3ee43709d

7. Mubeen, S., **Singh, N.**, Lee, J., Stucky, G. D., Moskovits, M. & McFarland, E. W. Synthesis of chemicals using solar energy with stable photoelectrochemically active heterostructures. *Nano Lett.* **13,** 2110–5 (2013).

8. **Singh, N.**, Upham, D. C., Liu, R., Burk, J., Economou, N., Buratto, S., Metiu, H. & McFarland, E. Investigation of the active sites of rhodium sulfide for hydrogen evolution/oxidation using carbon monoxide as a probe. *Langmuir* **30,** 5662–8 (2014).

9. **Singh, N.**, Upham, D. C., Metiu, H. & McFarland, E. W. Gas-Phase Chemistry to Understand Electrochemical Hydrogen Evolution and Oxidation on Doped Transition Metal Sulfides. *J. Electrochem. Soc.* **160,** A1902–A1906 (2013).

10. **Singh, N.** & McFarland, E. W. Levelized cost of energy and sensitivity analysis for the hydrogen–bromine flow battery. *J. Power Sources* **288,** 187–198 (2015).

478752 Development of Novel Single-Site and Isolated Bimetallic Alloy Catalysts for C-H Bond Activation

Abstract:

Research Interests:

My research interests are primarily focused on the development of novel single-site and isolated bimetallic alloy catalysts and their application in heterogeneous catalysis, particular in the activation of C-H bonds of alkanes. In an era of cheap and abundant alkanes, there is a growing interest in the conversion of alkanes, to value-added products, in which C-H activation is the elementary first step. Despite extensive studies, C-H activation under mild conditions continues to be the "holy-grail" of catalysis. The difficulty in C-H activation arises not only from the relatively inert nature of C-H bond, but also from the poor selectivity, which leads to unattractive products, carbon dioxide and water.

One particular important application of C-H activation is the methane activation and conversion to more valuable products, such as methanol and acetic acid. Many studies over the past several decades have failed to show promise for practical applications of the direct methane to desired liquid products. Examples include the approaches using Pd and Pt homogeneous catalysts, the conversion over CuFe/ZSM-5 catalysts using expensive H_2O_2 as the oxidant in aqueous media, and the non-catalytic two-step approaches focusing on Cu/MOR catalysts, which involves first the activation and conversion of methane at high temperatures, and second an *ex-situ* extraction process at room temperature. This stoichiometric reaction produces only trace amount of methanol; therefore is economically impractical. In my previous research, I have found the first truly catalytic cycle of the direct conversion of methane to practical yields of methanol and acetic acid with using molecular oxygen as the oxidant at low temperatures. The reactions are performed in aqueous media, under mild pressures of methane, oxygen and carbon monoxide, the latter used as catalyst promoter and reactant if the desired product is acetic acid. The catalysts are Rh/ZSM-5 with tunable high selectivity to acetic acid, and Cu promoted Ir/ZSM-5 with tunable high selectivity to methanol. The active sites are likely isolated Rh and Ir cations anchored on the internal walls of zeolite. This work opens a new door for the solving of long-standing problem of the direct catalytic conversion of methane to desirable liquid products using heterogeneous catalysts under mild operating conditions. Nevertheless, further investigations that avoid using precious metals and carbon monoxide are warranted.

My near future research will continue on the developing of novel heterogeneous catalysts for the activation of C-H bonds. In particular, I would like to using my obtained knowledge and experience to design novel single-site and isolated bimetallic alloy catalysts using non-precious metal to directly convert methane to value-added products.

Research Experience

My academic career path has been a blend of two main fields in heterogeneous catalysis. The first is the UHV study of the interaction of H_2O, CH_4, H_2, CO, CO_2 with single crystal surfaces, as well as with supported nanoparticles. The second involves synthesis of heterogeneous catalysts, performing catalytic measurements under realistic conditions, and in-situ characterizations with APXPS, XANES, EXAFS, DRIFTS, and various other techniques. In the meantime, I had trained and supervised new graduate students to start their respective projects. These experiences are very helpful to maintain a productive leading-edge research group when I start my own academic career.

Teaching Interests:

In addition to doing research, I also had a strong interest in teaching undergraduate and graduate students. I spent a semester teaching high-school students basic science when I was an undergraduate student. As a graduate student, I taught courses in physical chemistry and general chemistry. Given my background and experience in chemistry and catalysis, I am most interested in teaching classes related to physical chemistry and interdisciplinary techniques used in heterogeneous catalysis.

Selected publications

25. J. Shan, M. Li, L. F. Allard, S. Lee, M. Flytzani-Stephanopoulos

Direct conversion of methane to methanol or acetic acid with molecular oxygen on Rh- or Ir-ZSM-5 catalysts

Submitted, under the revision.

24. J. Shan, N. Janvelyan, H. Li, J. Liu, T. M. Egle, J. Ye, M. M. Biener, J. Biener, C. M. Friend, M. Flytzani-Stephanopoulos

Selective non-oxidative dehydrogenation of ethanol to acetaldehyde and hydrogen on highly dilute NiCu alloys

Submitted, under the revision.

23. J. Shan, S. Zhang, T. Choksi, L. Nguyen, C. S. Bonifacio, Y. Li, W. Zhu, Y. Tang, Y. Zhang, J. C. Yang, J. Greeley, A. I. Frenkel, F. Tao

Tuning catalytic performance through a single or sequential post-synthesis reaction(s) in a gas phase

Submitted, under the revision.

22. J. Shan, F. R. Lucci, J. Liu, M. El-Soda, M. D. Marcinkowski, L. F. Allard, E. C. H. Sykes, M. Flytzani-Stephanopoulos

Water co-catalyzed selective dehydrogenation of methanol to formaldehyde and hydrogen

Surface Science, 2016, 650, 121-129.

21. S. Zhang, **J. Shan**, L. Nie, L. Nguyen, Z. Wu, F. Tao

In situ studies of surface of $NiFe_2O_4$ *catalyst during complete oxidation of methane*

Surface Science, 2016, 648, 156-162.

20. C. S. Bonifacio, **J. Shan**, F. Tao, J. C. Yang

Atomic-scale characterization of restructured PtCu nanocubes

Microscopy and Microanalysis, 2015, 21, 1067-1068.

19. F. Tao, **J. Shan,** L. Nguyen, Z. Wang, S. Zhang, L. Zhang, Z. Wu, W. Huang, S. Zeng, P. Hu

Understanding of complete oxidation of methane on spinel oxides on a molecular level

Nature Communications, 2015, 6, 7798.

18. J. Shan, L. Nguyen, S. Zhang, F. Tao

Water–gas shift on $Pd/\alpha\text{-}MnO_2$ *and* $Pt/\alpha\text{-}MnO_2$

Catalysis Letters, 2015, 145, 1571-1580.

17. J. Shan, W. Huang, L. Nguyen, Y. Yu, S. Zhang, Y. Li, A. I. Frenkel, F. Tao

Conversion of methane to methanol with a bent mono(μ-oxo)dinickel anchored on internal surface of micro-pores

Langmuir, 2014, 30, 8558-8569.

16. S. Zhang, **J. Shan**, Y. Zhu, L. Nguyen, W. Huang, H. Yoshida, S. Takeda, F. Tao

Restructuring transition metal oxide nanorods for 100% selectivity in reduction of nitric oxide with carbon monoxide

Nano letters, 2013, 13, 3310-3314.

15. S. Zhang, **J. Shan**, Y. Zhu, A.I. Frenkel, A. Patlolla, W. Huang, S. Yoon, L. Wang, H. Yoshida, S.

Takeda, F. Tao

WGS catalysis and in-situ studies of CoO_{1-x}, $PtCo_n/Co_3O_4$, and $Pt_mCo_{m'}/CoO_{1-x}$ nanorod catalysts

Journal of the American Chemical Society, 2013, 135, 8283-8293.

14. J. Shan, Y. Zhu, S. Zhang, T. Zhu, R. Seigei, F. Tao

Catalytic performance and in-situ surface chemistry of pure α-MnO_2 nanorods in selective reduction of NO and N_2O with CO

Journal of Physical Chemistry C, 2013, 117,8329-8335

13. J. Shan, A. Chakradhar, K. Andeson, J. Schmidt, S. Dhuey, U. Burghaus

Butane adsorption on silica supported MoO_x clusters nanofabricated by electron beam lithography

Nanotechnology for Sustainable Energy, American Chemical Society, Washington DC, USA, 2013, Chapter 12, pp 295-310.

12. J. Shan, A. Chakradhar, M. Komaneni, M. Lu, U. Burghaus

Rim effects in the adsorption of CO_2 on silica supported copper oxide clusters - utilizing electron beam lithography

Journal of Physical Chemistry C, 2012, 116, 18930-18936.

11. A. Chakradhar, **J. Shan**, M. Komaneni, M. Lu, U. Burghaus

Identifying rims along nano-sized clusters as catalytically active sites – The case of CuO_x/silica model catalysts nanofabricated by electron beam lithography

Chemical Physics Letters, 2012, 544, 70-72.

10. M. Komaneni, **J. Shan**, A. Chakradhar, E. Kadossov, S. Cabrini, U. Burghaus

Adsorption dynamics of CO on silica supported CuO_x clusters: utilizing electron beam lithography to study methanol synthesis model systems

Journal of Physical Chemistry C, 2012, 116, 5792-5801.

9. C. Hahn, **J. Shan**, Y. Liu, O. Berg, A. W. Kleyn, L. B. F. Juurlink

Employing a cylindrical single crystal in gas-surface dynamics

Journal of Chemical Physics, 2012, 136, 114201.

8. J. Shan, A. Chakadhar, Z. Yu, U. Burghaus

Adsorption of water on a hydrophobic surface – the case of antimony(111)

Chemical Physics Letters, 2011, 517, 46-50.

7. J. Shan, M. Komameni, U. Burghaus

Adsorption dynamics of CO on copper and gold clusters supported on silica –how special is nanogold?

Chemical Physics Letters, 2011, 517, 59-61.

6. M. Komaneni, **J. Shan,** U. Burghaus

Adsorption dynamics of CO on silica–supported Cu clusters: a molecular beam scattering study

Journal of Physical Chemistry C, 2011, 115, 16590-16597.

5. C. Hahn, **J. Shan,** I. M. N. Groot, A. W. Kleyn, L. B. F. Juurlink

Selective poisoning of active sites for D_2 dissociation on platium

Catalysis Today, 2010, 154, 85-91.

4. J. Shan, A. W. Kleyn, L. B. F. Juurlink

Adsorption of molecular hydrogen on an ultrathin layer of Ni(111) hydride

Chemical Physics Letters, 2009, 474, 107-111.

3. J. Shan, A. W. Kleyn, L. B. F. Juurlink

Identification of hydroxyl on Ni(111)

Chemphyschem, 2009, 10, 270-275.

2. J. Shan, J. F. M. Aarts, A. W. Kleyn, L. B. F. Juurlink

Co-adsorption of water and hydrogen on Ni(111)

Physical Chemistry Chemical Physics, 2008, 10, 4994-5003.

1. J. Shan, J. F. M. Aarts, A. W. Kleyn, L. B. F. Juurlink

The interaction of water with Ni(111) and H/Ni(111) studied by TPD and HREELS

Physical Chemistry Chemical Physics, 2008, 10, 2227-2232.

479774 Smart Materials through Molecular Networking

Abstract:

Research Experience:

I have a broad experience with chemical kinetics and reaction engineering in flow systems. I developed the first synthetic enzymatic oscillator during my first postdoc at the University of Nijmegen in the Netherlands. In this work, I precisely arranged reactions in continuous stirred tank reactor (CSTR) to get oscillatory dynamics.[1] At George Whitesides group at Harvard, I applied the similar strategy to obtain bistable and oscillatory dynamics from organic reactions in CSTR.[2] I have also worked with coupled reaction-diffusion systems, microfluidics, and soft materials.[3] In my independent research, I would like to apply my experience with engineering ofreaction kinetics in open system to problems in materials and catalysis.

Research Interests:

My research program will focus on autonomous regulation in chemical systems by regulatory chemical reaction networks— that is, cascades of chemical reactions that receive a chemical signal, process this signal, and respond with a chemical output. Chemical regulation is centrally important to cellular life; it orchestrates cellular metabolism and guides the growth and movement of cells in heterogeneous environments. Although electrical and mechanical control systems are usually the choice for regulating macroscopic devices, chemical control becomes a preferable option for the regulation at the micro- and nanoscales. When applied to catalysis, self-regulation can improve efficiency of chemical reactions. My group will exploit these opportunities in designing "smart" materials and regulating catalytic transformations.
My research will have three major thrusts:

(i) Smart soft materials;

(ii) Chemical regulation at micro- and nanoscales through the surface modification.

(iii) Self-regulation in mixtures of organometallic catalysts.

Teaching Interests:

I have had a variety of teaching experiences. I voluntarily gave two full lecture courses (8 and 6 lectures) in organometallic chemistry in the Materials Science Department of Moscow State University in 2011 and 2012. I examined and graded student that attended these courses. I worked as a lab assistant for the Inorganic chemistry laboratory course 1 at the University of Zurich. I also worked as a chemistry teacher in the School for young chemists (12-14 years old pupils), which is associated with the Moscow State University. As a prospective faculty member, I am interested in teaching thermodynamics, chemical kinetics, and reaction engineering.

Selected Publications:

1. **Sergey N. Semenov,** Albert S.Y. Wong, Martijn van der Made, Sjoerd Postma, Rik van Roekel, Tom F. A. De Greef, Wilhelm T. S. Huck "Rational design of functional and tunable oscillating enzymatic networks" // *Nature Chemistry*, (2015**)**, *7*, 160 – 165.
2. **Sergey N. Semenov**, Lewis J. Kraft, Alar Ainla, Mengxia Zhao, Mostafa Baghbanzadeh, Victoria E. Campbell, Kyungtae Kang, Jerome M. Fox, George M. Whitesides "Autocatalytic, Bistable, Oscillatory Networks of Biologically Relevant Organic Reaction" // *Nature*, (2016), *537*, 656 – 660.
3. **Sergey N. Semenov**, Albert J. Markvoort, Tom F. A. de Greef, Wilhelm T. S. Huck "Threshold Sensing Through a Synthetic Enzymatic Reaction-Diffusion Network" // *Angew. Chem. Int. Ed.* (2014), 53, 8066 – 8069.

476305 Condition and Support Dependent Development of Computational Methods for the Engineering of Materials

Abstract:

Research Background:

My Ph.D. work under the supervision of Dr. William F. Schneider is an example of a system where I integrated density functional theory calculations with my own Monte Carlo code to accurately predict the distribution of active sites in a metal-exchanged zeolite post-synthesis. I then used a combination of ab initio molecular dynamics and analytical statistical mechanical models to describe the dynamic response of the active sites to varying gas exposure conditions. These findings brought the insight of solvated and mobile active sites to this system, a more malleable view of these active sites than had previously been assumed in the literature and more closely resembled a homogenous catalyst under catalytic conditions than a heterogeneous one which was critical to describing the catalytic mechanism. The thermal motion of these active sites also had direct consequences to their electronic structure, which required the integration of time-dependent density functional theory and ab initio molecular dynamics to accurately model. These â€œoperandoâ€ type computational models were validated through in situ and operando spectroscopic experiments that I played a major role in while a visiting student at Purdue University for one semester under the supervision of Dr. Fabio H. Ribeiro, Dr. W. Nicholas Delgass, and Dr. Rajamani Gounder, and at Argonne National Laboratory under the supervision of Dr. Jeffrey T. Miller.

Research Interests:Â

Development of computational models that account for the dynamic response of active sites to their contacting support and fluid phase is a critical step for being able to predict new materials that are not only active, but thermally and chemically stable and also capable of being synthesized. These are concepts often neglected in current computational models where the focus has historically been activity, in large part due to the difficulties involved with taking quantum calculations at 0K performed on small systems and scaling these results up to macroscopically significant length and time scales. Computational methods integrating interactions between active sites, supports, and the contacting environment are also important for providing a direct integration between theory and the bleeding edge of experimental methodology in catalysis which is highly focused on in situ and operando characterization of catalysts.

I provide several examples below for models informed by quantum chemical calculations and ab initio molecular simulations that can be coarse-grained to macroscopic length and time scales and a wide range of catalyst exposure conditions through the use of analytical and modern numerical statistical mechanics based theories using novel approaches that either combine computational methods that work efficiently on different energy scales, or those that work akin to classical molecular simulations but integrate quantum based energetics and structures into the model which is a new approach in this field:

I) The restructuring of supported nanoparticles at high surface coverages and temperatures is important for many catalysts, including those used for hydrocarbon chemistry and electrocatalysis, and can lead to dynamical behavior such as oscillatory rates or deactivation. A major challenge is the coupling of nanoparticle reconstruction, which involves high energy barriers and longer time scales, with surface coverage, which will influence the nanoparticle structure and is tied to the chemical potential at the interface and that in the fluid phase and must be equated. Iterative numerical schemes coupling a genetic algorithm with density functional theory computed molecular binding energies allow for rapid convergence of structure and coverage, creating accurate condition-dependent based models of these materials that naturally lend themselves to parallel computing environments such as GPU computing. Accurately modeling these systems informs support engineering at the atomistic scale and allows for mediation of stability challenges through approaches such as 3D printing of supports and chemical tethering of nanoparticles.

II) In microporous media modeling of percolation informs both the birth of the catalyst in support synthesis, metal exchange, or nanoparticle encapsulation, and its death in deactivation, from coking for example. Quantum chemical calculations describe how and where blockages occur, which are then integrated into statistical mechanical models that simulate the zoning off of portions of the support at macroscopic length scales and these geometries can be easily transferred into computational fluid dynamics software. Once these models are validated, new support geometries, for instance including mesoporous and microporous domains, can then be simulated to predict materials with advantageous synthesis properties or enhanced resistance to deactivation.

III) Computational models that incorporate fluid phase and support effects have use beyond just catalysis in the design of materials where thermodynamic properties are critical. Examples of this include supports designed for storage of gaseous NOx or COx species through chemisorption, and Lithium batteries where the formation of dendrites and ultimately shortage is obviously undesirable. Nucleation phenomena like dendrite formation is propagated by the exposure conditions of the contacting fluid and solid interface and take place over long timescales. The dendrite propagation occurs in a way that can be described by analytical fractal models once initiated, but only quantum chemical calculations can describe the initial nucleation accurately, which creates a separation of time and length scales. This can be resolved by tying these analytical models to condition-dependent quantum ones that accurately describe initiation and then use the analytical models to extend to deactivation timescales that will be macroscopically observable.

All of these processes involve systems where quantum-based methods are critical to describe the chemisorption of molecules or covalent interactions between active sites and a support, but must be integrated with analytical and numerical modeling to extend to phenomena that occur on the length and time scales of catalyst synthesis, restructuring, and deactivation.

Teaching Interests:

I have a strong background in analytical and numerical mathematical methods including:

I)analytical: solution to different classes of ODEs, PDEs, asymptotic methods, integral transforms, perturbation theory, and non-linear dynamics

II)numerical: classes of numerical PDE solvers and stability, finite element and volume methods, numerical solutions to the Navier-Stokes equations and higher order expansions of them

While many of these methods are not appropriate for an undergraduate curriculum, what my exposure to a diversity of analytical and numerical methods has taught me is that typically one methodology works well when the other does not and due to this complementary nature both analytical and numerical techniques should be emphasized in an undergraduate curriculum. In particular a strong theoretical foundation is critical to any undergraduate education since an experiment or computational simulation performed without an underlying theory behind it is worthless.

My experience performing fundamental experiments both as an undergraduate at Notre Dame and as a Graduate Student visiting at Purdue also provides me with the background necessary to teach any undergraduate chemical engineering course, including those involving experiments such as unit operations, phase equilibria, process control, and catalytic systems.

A priority of mine as an educator will be on exposing undergraduates to programming languages, with emphasis on python/matlab which are very similar and have a small learning curve, as well as interfacing with unix based operating systems. Many of the best catalysis experimental laboratories currently incorporate automated reactor design, "big" data analysis that relies on codes to facilitate the interpretation of large amounts of collected data, and working with unix based systems that rely on terminal commands, macros, and skilled use of on-site software when performing experiments at government funded laboratories. Contemporary approaches to the incorporation of coding into classical chemical engineering courses will be easy (for instance, automation of steam tables or UNIFAC) to integrate naturally into the core chemical engineering curriciulum. In the modern era some computational knowledge is key not only for theoretical models and simulation, but also for the design and analysis of experiments.

477292 Conversion of Waste Oil to Biofuels in a Single Catalytic Process over Bifunctional Catalysts

Abstract:

Research Interests:

My research interest relates to two patents I filed on biofuel field. It would be a great idea to convert the waste oil to biofuels in a single catalytic process. According to data I have found zeolites are the nominate catalyst for this purpose because of the high percentage of decarboxylation and high selectivity they have for branched and aromatic compounds. The other plan I have for my future research is to utilize the nanoscience and computational fluid dynamics knowledge together to cure cancer with magnetic nanoparticles. The concept of this work is cancer hyperthermia.

Teaching Interests:

I have worked in a variety of academic institutions in a wide range of teaching areas. Teaching has always been a rewarding experience for me, mainly because of the great satisfaction that comes with sharing my knowledge with others. My teaching interest includes kinetics and reaction engineering, transport phenomena and fluid dynamics courses for undergraduate and graduate students, which naturally fit my research and teaching experience. Because of the fact that my research includes kinetics and reaction engineering, heat and mass transfer and fluid mechanics concepts, so I could give students team working projects and provide them the opportunity to improve their knowledge about the fundamental principles and technical details of basics of Chemical Engineering. I also enjoy teaching hands-on laboratory classes, like unit operation laboratory.

477582 Rational Design of Catalytic Materials for Advancing the Use of Alternative Energy Sources

Abstract:

Rational Design of Catalytic Materials for Advancing the Use of Alternative Energy Sources

Research Interests:

Due to the depletion of fossil fuels and the presence of greenhouse gases in the atmosphere, alternative sources of energy which do not emit greenhouse gases will become essential in the near future. Both challenges of reducing greenhouse gas emissions and utilizing alternative energy sources will be accomplished through catalysts. My research directions focus on storage of energy which potentially comes from alternative sources. Specifically, my first research direction is determining catalytic materials for using electricity to split water for the production of hydrogen fuel. Hydrogen may be used in fuel cell vehicles for example and does not emit greenhouse gases when consumed. Solid oxide fuel cell catalysts are considered for this electrolysis of water reaction. At present, the most economical approach to the generation of hydrogen is through steam reforming of hydrocarbons followed by addition of water to allow water-gas shift ($CO + H_2O \leftrightarrow CO_2 + H_2$). However, electrolysis of water[1] ($2H_2O \leftrightarrow O_2 + 4H^+ + 4e^-$) is an alternative hydrogen production reaction which has the advantage that no hydrocarbons, i.e. fossil fuels, are required. The electricity required for electrolysis may be drawn from alternative energy sources such as wind and solar.

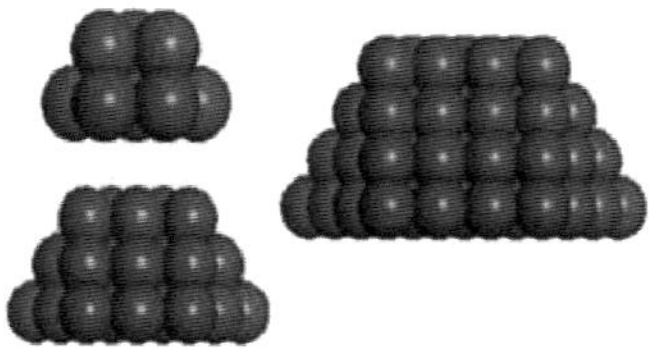

Figure 1. Different size metal nanoparticles have different ratio of terrace, edge and corner sites. Determining which site is active for a reaction can guide the rational design of catalysts.

Figure 1. Different size metal nanoparticles have different ratio of terrace, edge and corner sites. Determining which site is active for a reaction can guide the rational design of catalysts.

My second research direction is proton exchange membrane fuel cells (PEMFC) for generating electricity from hydrogen. Current challenges for increasing PEMFC efficiency lie in improving the catalyst. In order to rationally design an improved catalyst, a knowledge of the active site of reaction at the atomistic scale is necessary. The benefit of this knowledge can be seen from Figure 1 which shows the size of a Pt nanoparticle catalyst affects the fractions of edge, corner and terrace sites present. Knowledge of which type of site is active suggests the optimal size of Pt nanoparticles. Quantum chemistry and in particular Density Functional Theory (DFT) has shown great success in predicting the activity of catalysts and is becoming an essential tool for the rational design of catalysts. However, when conducting DFT investigations it is necessary to include a quantification of uncertainty of DFT with the reason of not being misleading in making conclusions.[2] A quantification of DFT uncertainty will be conducted in order to generate reliable results.

[1] Smith, R. D. L.; Prévot, M. S.; Fagan, R. D.; Zhang, Z.; Sedach, P. A.; Siu, M. K. J.; Trudel, S.; Berlinguette, C. P. Photochemical Route for Accessing Amorphous Metal Oxide Materials for Water Oxidation Catalysis. *Science* **2013**, 340, 60-63.

[2] Deshpande, S.; Kitchin, J. R.; Viswanathan, V. Quantifying Uncertainty in Activity Volcano Relationships for Oxygen Reduction Reaction. *ACS Catal.* **2016**, 6, 5251-5259.

Teaching Interests:

In the undergraduate curriculum, I am particularly interested in and have the background for teaching courses such as: Transport Phenomena, Statistical and Numerical Methods for Chemical Engineering, and Chemical Engineering Kinetics. In the graduate curriculum, I am interested to teach courses such as Chemical Reactor Design and Chemical Process Principles. In any course I teach, I believe the inclusion of computer programming will ensure that students learn attention to detail and how to formulate engineering problems in an explicit manner. I am interested in developing a course entitled Quantum Chemistry of Materials for Catalysis to be offered as a graduate elective. This course will begin with principles of quantum chemistry starting with the Schroedinger equation, followed by Kohn-Sham density functional theory as an approximate solution to the Schroedinger equation. Topics of entropy contributions to free energy and sources of uncertainty in quantum chemistry will be discussed. A project will be conducted at the end of the semester in which students will simulate materials using quantum chemistry for catalyzing chemical reactions of industrial interest.

480133 Catalysis Reactions towards Advanced Energy Applications

Abstract:

Research Experience:

Optimizing reactive efficiencies of particles that possess catalytic properties proves to be a source of major interest from the perspective of generating energy. Energetic materials can be tailored for intended applications through such enhanced surface characteristics. These enhancements can be achieved through simple techniques such as improving the hydration layer around reactant materials and thereby adding to their polarities. This change in polarity is known to boost their thermal properties adding to the energy released during the process of combustion. Simulation models for such structured crystals proves that bonds are affected by the polar nature of the environment particles are subjected to by altering their bond distances, atomic charges and stretching frequencies due to polar and hydrogen bonding forces. It would prove to be greatly beneficial if highly functionalized reactive materials could be developed for generating more effective energetic/ thermitic composites. Figures 1. a. and b. capture the idea of functionalized surface of actively polar molecules and the pronounced effect observed on their reactive output.

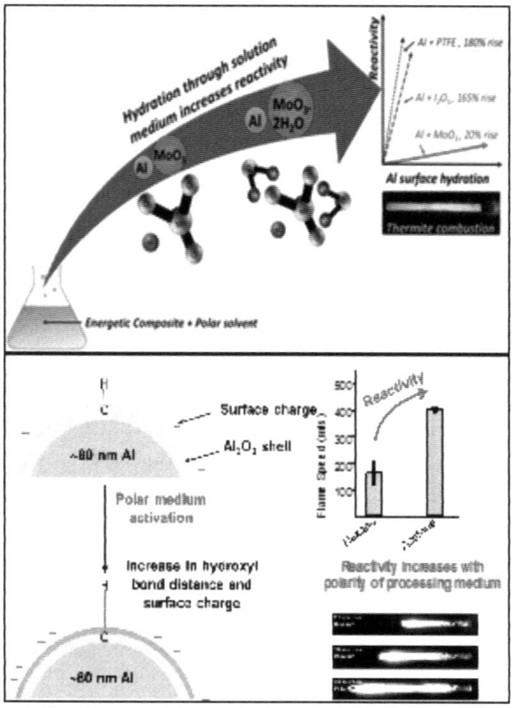

a.

THE GRAPHIC ON THIS PAGE WAS NOT AVAILABLE ON THE E-MEDIA VERSION OF THIS PAPER.

b.

Figure 1: a. Surface functionalization of alumina shell surrounding aluminum particles. b. Oxidizing agents can also be catalyzed similar to Al particles.

Research Interests:

Despite having a background in Mechanical Engineering, my research entails a lot of inputs from Chemical Engineering and Chemistry. I work with energetic compositions, with a focus on aluminum as a fuel. My dissertation has been inspired by the curious finding that the mere presence of polar media around polarized compounds like the passivating alumina shell in the fuel crystal can do wonders to improve its thermal behavior under fluorination. Through the understanding of results obtained off the Differential Scanning Calorimetry, I observed that the enthalpy of reaction could be increased significantly when water from the solution affects the reactant particle surface as against when a non-polar medium containing no trace of water was used for suspending these particles. Fourier Transform Infrared spectroscopy enabled me to detect the actually augmented hydration layer at the particle interface. I am also trained to handle a Hitachi S4300 Scanning Electron Microscope – to look at the homogeneity of dispersion matrix. Even with this hand-on experience using such highly precise instrumentation; it is extremely difficult to gauge interactions taking place at the molecular level. This brings the need for me to collaborate with

Chemists who specialize in performing ab initio calculations. Exposure to this area of research added depth to my understanding of chemical kinetics from an entirely unique perspective.

Through my knowledge of catalysis and catalysts used in various processes, I would want to focus on its applicability towards understanding basic molecular reactions and surface properties towards producing greener and cleaner fuels by maximizing conversion efficiency. Engineered materials that could be used as adsorbents and catalysts in fuel cells could effectively be used to improve their performance efficacy. Zeolites that are used in ion – exchange beds for water purification and softening can be improved for their efficiency in attracting dirt and minerals by polarizing catalysts to act as electromagnets. Understanding of catalysts through ab initio calculations helps me predict the nature of bonds affected by the hydrogen bonding forces to alter crystal structures when exposed to various environments. This information leads to the ability to tailor materials for suited applications.

In other collaborative simulation efforts, I would like to focus my attention on deriving intermediate reactions taking place which are studied in calorimetric techniques such as the DSC. From this data generated, it would be easier to predict which pathways any reaction mechanism could possibly take enabling us to better concentrate on the most desirable results from these reactions.

Teaching Experience:

I have experience teaching the undergraduate Heat Transfer and Combustion courses for Mechanical Engineering juniors and seniors respectively. I have been actively involved in mentoring masters' students in our research group where we evolve in an extremely symbiotic atmosphere.

I have found that it always important to keep students engaged through interactions so they develop more interest in what they are learning and in return take back much more than they would have. I like getting feedback from them as to how they were following what I am trying to convey in my lectures. It is really necessary for me to make classes increasingly interesting so students have the motivation to attend each class with equal zeal.

I prefer to start my classes with a short review of the previous lecture and put those in the form of points on the board so they retain what we talked about and then gradually introduce the day's topics. I think it is very important to grab the students' attention in the initial few minutes. I try to include a very relatable example for every concept so students understand its applicability. I try to involve the class in a discussion of the concept involved.

Proposal:

Submitted a grant proposal to the Army Research Office on the topic: 'Energetic Materials' (PI: Dr. Michelle L. Pantoya).

Ph.D. Dissertation:

'Altering surface properties toward enhanced aluminum reactivity'

Supervised by Dr. Michelle L. Pantoya, J. W. Wright Regents Endowed Chair Professor, Department of Mechanical Engineering, Texas Tech University, Lubbock, TX – 79409, USA

-

Publications:

Examining Hydroxyl – Alumina Bonding Toward Al Nanoparticle Reactivity; Padhye, R.; McCollum, J.; Korzeniewski, C. L.; Pantoya, M.L.; *Journal of Physical Chemistry C.* 2015, 119(47):26547-53 (DOI: 10.1021/acs.jpcc.5b08408; Publication date: November 17, 2015). Effect of Polar Environments on the Aluminum Oxide Shell Surrounding Aluminum Particles: Simulations of Surface Hydroxyl Bonding and Charge; Padhye, R.; Aquino, A. J. A.; Tunega, D.; Pantoya, M. L.; *ACS Applied Materials and Interfaces.* 2016, 8 (22), pp 13926–13933 (DOI: 10.1021/acsami.6b02665; Publication date: May 13, 2016). The Influence of Environmental Processing Conditions on the Reactivity of Aluminum with Various Oxidizing Agents; Padhye, R; Smith, D; Pantoya, M. L. (submitted to *Applied Surface Science*).

480715 Synthesis and Spectroscopic Characterization of Heterogeneous Catalysts for Energy Production

Abstract:

Research Interests:

- *In situ/operando* spectroscopic (DRIFTS, Raman, X-ray adsorption) characterization of catalysts
- Tuning surface acid-base properties of metal oxides for C-C coupling of small oxygenates to olefins
- Functional control of (bi)metallic catalysts for hydrodeoxygenation and steam reforming reactions
- Syntheses of multifunctional nano-catalysts for C1 (i.e., CO_x) activation

Successful Proposals: Washington State University new faculty seed grant

Research Experience:

My research career path has been following two major directions: materials science and catalysis. My graduate research was devoted to the structure control of nanoporous materials via self-assembly and sol-gel chemistry, as well as further synthesis of uniform metal nanoparticles catalysts confined within the pores of nanoporous materials via host-guest chemistry. During my PhD studies, I also spent three months (July, 2005-September, 2005) in the department of inorganic chemistry, Fritz Haber Institut der MPG, Germany where I received a full training on using high-resolution scanning/transmission electron microscope for materials characterizations. During my postdoc studies, I grasped the Schlenk line synthesis and *in situ* X-ray absorption spectroscope (i.e, XANES and EXAFS) characterization of structural simple supported metal clusters/mononuclear. My current research topics include the fundamental understanding and rational design of highly active and selective steam reforming catalysts, the development of bimetallic hydrogenation catalysts for removing oxygen from biomass derived phenolics, and the balance of acid-base chemistry for a cascade aldolization and self-deoxygenation of biomass derived small oxygenates. State-of-the-art *in situ/operando*characterization techniques (e.g., IR, Raman, and X-ray absorption etc.), reaction kinetics, and theoretical calculations have been used to elucidate the reaction mechanism of the catalysts. **Teaching Interests:**

During my graduate studies, I TA'd in the undergraduate physical chemistry labs taught by my advisor, and supervised a few undergraduate students to successfully finish their research projects at Dalian Institute of Chemical Physics. Since then, I have been presenting scientific results periodically in different forums, seminars and international/national conferences. As an Assistant Research Professor in the Gene and Linda Voiland School of Chemical Engineering and Bioengineering (VSCEB), I am also serving as a thesis committee member to supervise both graduate and undergraduate students independently. During the five years of working in VSCEB, I have gained valuable experiences in leading the group meetings and providing mentorship for students who are working under both government and industrial funded projects.

Selected Publications:

1. Rebecca A.L. Baylon, Junming Sun,* Kevin J. Martin, Padmesh Venkitasubramanian and Yong Wang* "Beyond ketonization: selective conversion of carboxylic acids to olefins over balanced Lewis acid–base pairs", *Chemical Communications*, 2016, 52, 4975-4978.
2. Junming Sun, Rebecca A.L. Baylon, Changjun Liu, Donghai Mei, Kevin J. Martin, Padmesh Venkitasubramanian and Yong Wang "Key roles of Lewis Acid-Base Pairs on $Zn_xZr_yO_z$ in Direct Ethanol/Acetone to Isobutene Conversion" *Journal of the American Chemical Society*, 2016, 2, 507-517.
3. He Zhang, Junming Sun,* Changjun Liu, Yong Wang* "Distinct water activation on polar/non-polar facets of ZnO nanoparticles" *Journal of Catalysis*, 2015, 331, 57.
4. Junming Sun, Ayman M. Karim, Donghai Mei, mark Engelhard, Xinhe Bao, Yong Wang "New insight into reaction mechanism of ethanol steam reforming on Co-ZrO2" *Applied Catalysis B: Environmental*, 2015, 162, 141-148.
5. Junming Sun, Ayman M. Karim, He Zhang, Libor Kovarik, Xiaohong Shari Li, Alyssa J. Hensley, Jean-Sabin McEwen, Yong Wang, "Carbon Supported Metal Catalysts for Vapor-phase hydrodeoxygenation of Guaiacol: A comparative Study" *Journal of Catalysis*, 2013, 306, 47-57.
6. Zhehao Wei, Junming Sun*, Yan Li, Abhaya K. Datye, Yong Wang* "Bimetallic catalysts for hydrogen generation" *Chemical Society Reviews*, 2012, 24, 7994-8008.
7. Kake Zhu, Junming Sun,* He Zhang, Jun Liu, Yong Wang* "Carbon as a Hard Template for Nano Materials Catalysts" *Journal of Natural Gas Chemistry*, 2012, 21, 215-232.
8. Junming Sun, Kake Zhu, Feng Gao, Chongmin Wang, Jun Liu, Charles H.F. Peden, Yong Wang "Direct Conversion of Bio-ethanol to Isobutene on Nanosized $Zn_xZr_yO_z$ Mixed Oxides with Balanced Acid-Base Sites", *Journal of the American Chemical Society*, 2011, 133, 11096.
9. Junming Sun, Miaofang Chi, Lobo L. Rodrigo, Mehraeen Shareghe, Browning Nigel, Gates C. Bruce "Tantalum Clusters Supported on Silica-Alumina: Influence of Support Composition and Chemistry on Cluster Structure", *Langmuir*, 2009, **25**, 10754-10763.
10. Junming Sun, Ding Ma, He Zhang, Xiumei Liu, Xiuwen Han, Xinhe Bao, Gisela Weinberg, Norbert Pfänder, and Dangsheng Su "Toward monodispersed silver nanoparticles with unusual thermal stability", *Journal of the American Chemical Society*, 2006, 49, 15756-15764.

461800 Accelerating Materials Design: Computer Simulations and QSAR Modeling

Abstract:

Research Interests:

My future research group will focus on designing soft materials and processes that can help us face critical health, energy and environment challenges. Computer simulations and QSAR modeling have shown their ability to accelerate materials discovery and process design by revealing critical molecular-level mechanisms, providing rational design principles, and identifying key molecular architectures. My three research projects are to design (1) zwitterionic electrolytes for Li^+batteries and solar cells, (2) peptide materials for high-performance medical diagnostics and (3) amyloid-forming peptides for industrial catalysts using computer simulations and QSAR modeling.

My PhD and postdoctoral research experiences equip me with the expertise to apply/develop computer simulations/QSAR modeling in cooperation with experimentalists. My PhD research in Prof. Shaoyi Jiangâ€™s group at the University of Washington focused on understanding and designing zwitterionic anti-biofouling materials using computer simulations and experiment. Zwitterionic materials possess cationic and anionic groups in one moiety. We developed force field parameters for zwitterionic moieties, investigated hydration, protein interaction, ion interaction, and self-association of zwitterionic materials, and prepared and characterized zwitterionic polymers. This work explained the anti-biofouling mechanisms of zwitterionic materials and developed principles for designing new zwitterionic moieties. One moiety designed *in silico*by us was verified to resist biofouling in experiment.

My postdoctoral research in Prof. Carol Hallâ€™s group focuses on understanding and predicting nanoparticle toxicity. Excitement about the applications envisioned for nanomaterials in energy, medicine and electronics is tempered by concerns about their toxicity. As the first response of a biological fluid to the intrusion of nanoparticles is

the formation of a protein corona that determines the nanoparticleâ€™s toxicity, we take a computational approach to learn the mechanisms that govern the formation of a protein corona around nanoparticles. A coarse-grained model was developed to describe protein-protein and protein-nanoparticle interactions. Combining this coarse-grained model, atomistic simulations and QSAR modeling, we (1) explore how binding to the nanoparticle affects the structure and dynamics of a key blood protein, albumin, (2) decode the common features in the binding configurations of 2315 human proteins and gold nanoparticles, and (3) develop models that can predict protein-nanoparticle interaction energies. This work deepens our understanding of protein-nanoparticle interactions, paving the way to predict and assess nanoparticle toxicity.

Teaching Interests:

I earned my bachelor and PhD degrees in chemical engineering and have been conducting research in chemical engineering for more than 10 years. I am confident of my abilities to teach all chemical engineering courses at the undergraduate and graduate levels, and I am eager to do so. I am particularly interested in teaching thermodynamics, transport processes, process design and applied mathematics.

465464 Computationally Assisted Discovery of Well-Designed Materials for Applications to Energy, Environment, and Catalysis

Abstract:

Research Interests:

The worldâ€™s dependence on fossil fuels has led to the need for alternate sourcesÂ of energy as supplies dwindle, as well as a growing need to removeÂ harmful compounds from the air. Hydrogen energy and lithium-ion batteries are promising candidates for supplanting fossil fuels for automobile applications while novel adsorbents like metal-organic frameworks (MOFs) are promising materials for removing harmful gases. To date, my researches have been concerned with both aspects of the fossil fuel problem in keeping with my interest in achieving better performance on polymer solar cells and understanding catalysis on alkylamine functionalized silica surfaces. Regarding the lithium-ion batteries, my researches are focused on fundamentally understanding the thermodynamics and redox properties of well-designed electrode material candidates which would directly affect the battery capacity. It is emphasized that the redox properties of carbon materials such as quinone derivatives and graphenes can be strongly correlated with their structural and electronic properties. One of the highlighted observations is that the cathodic activity of the quinone derivatives during the discharging process relies strongly on the number of carbonyl groups available for further Li binding. On the other hand, my research focus in the field of the hydrogen energy is on understanding the thermodynamics of metal hydride reactions for hydrogen storage applications. Specifically, the goal is to identify thermodynamically promising metal hydride reactions from a full database of metal hydride mixtures using first-principles calculations. The large-scale screening approach ultimately provides a number of promising single-step or multi-step metal hydride reactions. My researches on MOFs are related to investigating promising MOFs for the selective capture of harmful gases as well as gas separations. Quantum mechanical methods are used to screen and assess functional groups that would be incorporated into MOF ligands to preferentially adsorb harmful gases under humid conditions. In the last research topic, my molecular dynamics simulation approach is employed to understand the aldol condensation between acetone and 4-nitrobenzaldehyde on alkylamine-functionalized silica surfaces. In conclusion, my research goal is to contribute to the community for the transition to environmentally friendly ecosystems.

Inspired by my research experience in the field of the lithium-ion battery, my future research interests are aimed at the development of methodologies for systematically designing promising electrode materials with the high redox properties as well as the high energy and power densities. A new protocol of designing hypothetical electrode materials followed by the large-scale screening approach will be employed to expedite the material-search process. The newly designed materials are expected to provide exceptional performance in four aspects: (1) structural stability, (2) desired redox property, (3) high charge capacity, and (4) high ionic diffusivity. I believe that this research will significantly contribute to the battery community for discovering completely new electrode materials.

Teaching Interests:

I describe the topics I would like to share with my students in a future position. First, my education on quantum chemistry and molecular simulation makes it possible for me to teach an introduction to theoretical calculations and simulations. I am highly interested in sharing and discussing a number of thermodynamic and kinetic problems, which should sometimes be solved using density functional theory calculations. Second, based on my experience in gas adsorption experiments on porous materials, I would like to teach topics related to an introduction to adsorption and specific techniques for measuring adsorption isotherms at the thermodynamic equilibrium state. Third, the topics arising from my research background, namely energy materials and science, alternative energies, hydrogen storage technology, metal hydride thermodynamics, lithium-ion battery technology, adsorption thermodynamics on porous materials, carbon capture and storage, solar cell technology, and metal-organic frameworks provide interesting material to teach on both undergraduate and graduate levels. In addition, the basic knowledge I gained as an undergraduate student on general chemistry, physical chemistry, thermodynamics, fluid mechanics, heat transfer, mass transfer, differential equations, numerical analysis, and transport phenomena would prove helpful for teaching basic undergraduate classes.

472231 Computationally Driven Discovery of Novel Materials for Separation and Catalysis

Abstract:

Research Interests: Theory and molecular modeling have seen tremendous growth in recent years; capitalizing on theoretical and algorithmic developments and advancements in computer hardware, applications of computational methods have begun to penetrate into all fields of the chemical sciences. In this poster presentation, I will highlight separation and catalysis applications using crystalline nanoporous materials, primarily zeolites, and show how molecular modeling can be used both to *gain a microscopic understanding into phenomena difficult to probe experimentally* and to *predict accurate data at a scale beyond the reach of the traditional trial-and-error approach*. I developed a new, transferable force field for all-silica zeolites, TraPPE-zeo, that is applicable to any type of sorbate molecule and framework structure. With the improved intermolecular potentials as well as better methodologies, it was then possible to study several complex systems involving articulated molecules and liquid mixtures, such as the multi-component adsorption of aqueous alcohol, polyol, glucose, and furfural solutions. These studies demonstrate the important effects of hydrogen-bonding, which render both the bulk and adsorbed phases highly non-ideal and are for certain molecules so strong that adsorption becomes entropically rather than enthalpically driven. In another application, the unusual adsorption and diffusion behaviors of a hierarchical zeolite were investigated, and our simulations clearly indicate regimes where the introduction of mesoporous "highways" impedes, rather than enhances, molecular transport in a material without structural defects, and provide insights into the nature of surface resistance and its roles in the practical application of hierarchical materials and ultrathin membranes. More recently, the effect of reactant chain lengths and architectures and the degrees of framework confinement on the reactivity and selectivity of zeolites as solid acid catalysts were carefully examined using first-principles calculations. Information obtained from this work provide the basis for an effort in which a large number of materials were screened with the purpose of finding better candidates for two energy-related applications, ethanol/water separation and hydrocarbon iso-dewaxing, and some of our predictions were subsequently verified by experiments. This strategy was recently extended to account for transition-state selectivity [1], and several advancements were made to allow for highly parallel ab-initio Monte Carlo simulations, which together bring closer the possibility of using theory and computation in a predictive mode to expedite the pace of materials discovery for complicated, realistic, and industrially relevant applications.

Teaching Interests: Thermodynamics and statistical mechanics, catalysis and chemical kinetics, heat and mass transfer, numerical methods and linear algebra, computational chemistry

[1] "A method to screen nanoporous catalysts for transition-state selectivity," to be presented at 2016 AIChE Annual Meeting, Symposium on 'Catalysis with Microporous and Mesoporous Materials,' San Francisco, CA.

476096 Towards Accurate Atomistic Description of Reactive Interfaces for *in silico* design of Novel Functional Materials

Abstract:

Research Interests:

The complex interplay of chemical reactions, defect-chemistry, transport phenomena, and structural evolution at reactive interfaces underpin much of energy capture, conversion and storage by materials. A fundamental understanding of these dynamical processes at angstrom-to-mesoscopic length scales over several nanoseconds is crucial, and urgently needed to enable development of novel functional materials for various energy applications. However, development of accurate atomistic models of reactive interfaces remains a grand challenge, mainly due to the non-systematic and intuition-driven methods employed. Here, I will first present a robust, efficient, and automatic computational framework to develop reactive interatomic potentials (force-fields; FFs) that significantly advances the state-of-the-art by completely obviating the need for human intuition. Using supervised machine learning empowered by genetic algorithms, and large datasets obtained from high-throughput *ab initio* calculations, we sample the parameter landscape efficiently to obtain a highly transferable set of FF parameters that accurately capture structural, dynamical, energetic, and other physical/chemical properties. We have successfully employed this framework for a wide range of material systems, including gold, ZrN_x, CoC_x, IrO_2, and 2D tin.[1,2] Subsequently, I will demonstrate how a synergistic integration of large-scale reactive molecular dynamics (RMD), electronic structure calculations, and *ab initio* MD (AIMD) can pave path towards an accurate description of reactive interfaces, and consequently enable computational discovery of novel materials. Specifically, I will highlight two representative examples: (1) using AIMD, and million-atom RMD simulations, we discovered a novel catalytic phenomena possible only under the extreme local conditions afforded by friction at sliding interfaces; this enabled design of novel catalytically active coatings (e.g., MoN_x-5%Cu) with exceptional wear resistance (~25% lower friction than best blended oils). We found that such coatings transform base lubricating oils (long-chain alkenes) into low-friction diamond-like carbon (DLC) film via dehydrogenation, backbone scission of long chain alkenes to form short fragments, and their subsequent polymerization. Our atomistic simulations also indicate that the formation of DLC-like tribofilm is crucially linked to the propensity of the transition metal catalyst to form carbides; carbide-forming metals, e.g., V preclude the formation of DLC tribofilm.[3] (2) RMD simulations revealed a new transfer-free route to synthesize wafer-scale high quality single/multi-layer graphene directly on diamond.[4]We found that a Ni thin film placed on ultra-nanocrystalline diamond (UNCD) causes amorphization of diamond lattice via diffusion of Ni into the UNCD grain boundaries (GBs); concurrently, C from UNCD GBs diffuse through the Ni film, and form a graphene layer on the surface. This opens up avenues to explore novel synthesis pathways, and atomistic mechanisms underlying exotic phenomena in other 2D materials, e.g., stanene, phosphorene etc.

Research Experience

My research experiences lie in the broad area of theoretical/computational materials science, which have made me well-versed with developing accurate models at multiple length scales (angstrom to meso) and employing them to gain novel physical insights into exotic materials behavior. My expertise includes density functional theory calculations, *ab initio*molecular dynamics, large-scale classical molecular dynamics simulations, multi-parameter optimization for empirical force-field development, machine learning, advanced sampling and mesoscopic length-scale models. Specifically, I have employed these tools to investigate (i) structural evolution, ionic diffusion, and solvation dynamics near electrochemical interfaces during aqueous corrosion, (ii) atomic-scale diffusion mechanisms in ionic liquid based electrolytes for Al-batteries, (iii) mechanisms underlying friction and chemical reactions at sliding interfaces, (iv) growth mechanisms for 2D materials on transition metal surfaces, and (v) self-assembly of biomolecules (like collagen, peptoids) on inorganic surfaces. Furthermore, as a part of the user program at Center for Nanoscale Materials (DOE nanoscience user facility at Argonne National laboratory), I have fostered successful collaborations with several experimental research groups at US universities; Through these user-collaborations, I have worked on materials science problems at various length/timescales including investigations on large area synthesis of 2D materials via electrochemical exfoliation, chemical reactions under tribological conditions, and self-assembly of ligand decorated nanoparticles.

Postdoctoral Project

*"Bridging the electronic and atomistic scales: force field development for reactive interfaces from first principles"*Under supervision of Dr. Subramanian Sankaranarayanan, Center for Nanoscale Materials, Argonne National Laboratory

PhD dissertation

*"Understanding structure-property relationships in b-eucryptite through atomistic simulations"*Under supervision of Prof. Cristian V. Ciobanu (Mechanical Engineering) and Prof. Ivar E. Reimanis, (Metallurgical and Materials Engineering), Colorado School of Mines.

Future Direction

Despite recent strides, the field of computation-driven materials design remains plagued by the dearth of accurate models (at nano-to-meso length scales), as well as lack of efficient strategies to design new realistic functional materials (i.e, with defects, grain boundaries etc.) and identify routes to synthesize them. Motivated by this challenge, as a faculty, I seek to establish a successful research program titled "Multi-scale *in silico* materials design", which focuses on devising new strategies/tools to address these challenges, and applying these tools to investigate various technologically relevant materials systems. Specifically, I intend to develop robust computational frameworks to (1) use advanced sampling methods (e.g., evolutionary algorithms, metadynamics) to search for new metastable/stable structures for low-dimension materials (i.e., clusters, 2D sheets), (2) couple high throughput-DFT calculations with advanced global (e.g., genetic algorithms) and local (e.g., Simplex) optimization routines to develop new reactive interatomic force field (FF) models, which capture bond formation/dissociation events and transition states accurately; such FFs are crucial to advance fundamental understanding of interfacial reactions, mechanistic origins of materials phenomena, and structure-property relationships at length scales spanning tens of nanometers, (3) develop machine learning strategies to rapidly design new materials for superlubricity, battery electrolytes and anti-corrosion coatings, as well as (3) develop accurate coarse-grained models to understand self-assembly of hierarchical architectures at *m*m sizes, including superlattices of ligated nanoparticles.

Teaching Interests:

I possess a strong interest in teaching and mentoring students. During my postdoctoral appointment at Argonne National Laboratory (ANL), I pursed and obtained the opportunity to mentor several graduate students who actively collaborate with my group on various computational materials science research problems. While working towards my PhD dissertation, I delivered several guest lectures for the course "Ceramic Engineering" hosted by the Department of Materials Science and Engineering in Colorado School of Mines (CSM); additionally, I have also served as a teaching assistant for various materials science courses at CSM [e.g., Electronic Properties of Materials, Mechanical Properties of Materials, Nanomechanical Engineering etc.]. These opportunities have helped me develop the qualities necessary for effective teaching, and for being a good mentor.

Selected Publications

1. **B. Narayanan**, A. Kinaci, F. Sen, M. J. Davis, S. Gray, M. K Y. Chan, and S. Sankaranarayanan, "*Describing the diverse geometries of gold from nanoclusters to bulk – a first-principles based hybrid bond order potential*", Journal of Physical Chemistry C 120, 13787 (2016).
2. **B. Narayanan**, K. Sasikumar, A. Kinaci, F. Sen, M. J. Davis, S. Gray, M. K Y. Chan, and S. Sankaranarayanan, "*Development of a modified embedded atom force field for zirconium nitride using multi-objective evolutionary optimization*", *Accepted for publication*, Journal of Physical Chemistry C (2016).
3. A. Eridemir, G. Ramirez, O. Eryilmaz, **B. Narayanan**, Y. Liao, G. Kamath, and S. Sankaranarayanan, "*Cabon-based protective tribofilms from lubricating oils*", *In Press*, Nature (2016).
4. D. Berman, S. Deshmukh, **B. Narayanan**, S. Sankaranarayanan, Z. Yan, A. Baladin, A. Zinovev, D. Rosenmann, and A. Sumant, "*Metal induced ultra-fast transformation of diamond into single domain graphene on wafer scale*", Nature Communications 7, 12099 (2016).
5. **B. Narayanan**, S. A. Deshmukh, L. K. Shreshta, K. Ariga, V. Pol, and S. K. R. S. Sankaranarayanan, "*Cavitation and radicals drive the sonochemical synthesis of functional polymer spheres*", *Accepted for publication*, Applied Physics Letters (2016)
6. **(Invited) B. Narayanan**, S. Deshmukh, S. Sankaranarayanan, and S. Ramanathan, "*Strong correlations between structural order and passive state at water–copper oxide interfaces*", Electrochimica Acta 179, 386 (2015)
7. **B. Narayanan**, G. Gilmer, J. Tao, J. DeYoreo, and C. Ciobanu, "*Self-assembly of collagen on surfaces: the interplay of collagen-collagen and collagen-substrate interactions*", Langmuir 30, 1343 (2014)
8. **B. Narayanan**, A. van Duin, B. Kappes, I. Reimanis, and C. Ciobanu, "*A reactive force field for lithium–aluminum silicates with applications to eucryptite phases*", Modelling and Simulation in Materials Science and Engineering 20, 015002 (2012)

WAGENINGEN UR
For quality of life

Solubility and thermodynamic properties of α-amino acids in a model system of industrial residues

Nathan A. Bowden, Johan P.M. Sanders, Marieke E. Bruins

Background

- Amino acids have the potential to play an important part in a Bio-based Economy,
- This potential is expressed in the possibility of extracting amino acids out of agro-industrial residues for feedstock or bio-based products,
- Extracting and crystallizing single amino acids from a solution of mixtures of amino acids will facilitate reaching their economic potential in the bio-based pyramid (figure 1)

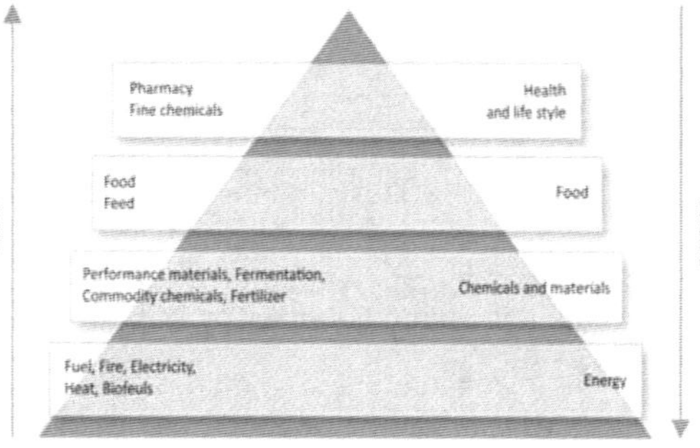

Figure 1: The bio-based pyramid

Methodology and Objectives

1. Extract individual α-amino acids from a mixture of 20 α-amino acids by crystallization,
2. Measure the solubility of α-amino acids in a model system of all 20 α-amino acids from 277.15 K to 335.15 K
3. Compare the solubility of these α-amino acids in in a model mixture to themselves in single solution

Results

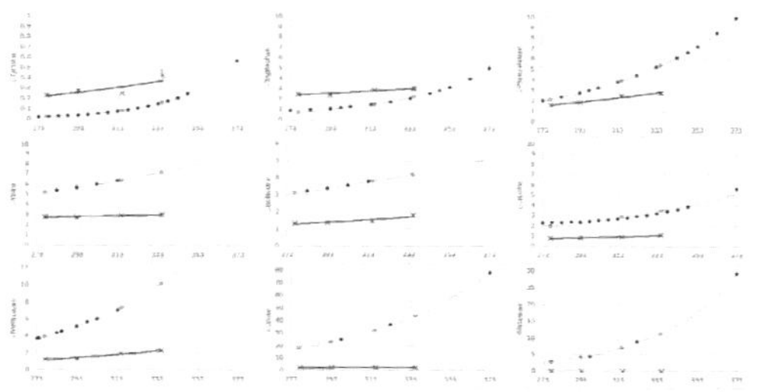

Figure 2: Solubility (m_i) of α-amino acids in g/100 g water at various temperatures (K). Solid circle (single amino acid from literature), empty circle (single amino acid calculated from literature values), solubility as a function of temperature (dotted line) crosses with error bars (measured in the model system of multiple amino acids). The solid line represents the function of the solubility of the amino acid in the model system of 20 α-amino acids.

Results (Cont.)

- L-tyrosine and L-tryptophan dissolved in higher concentrations in the model system of 20 α-amino acids than in aqueous solution by themselves,
- The solubility of the aliphatic α-amino acids (L-phenylalanine, L-valine, L-isoleucine, L-leucine and L-methionine) were lower in the model system of 20 α-amino acids,
- L-lysine and L-glutamine showed the most decrease when in the model system of 20 α-amino acids.

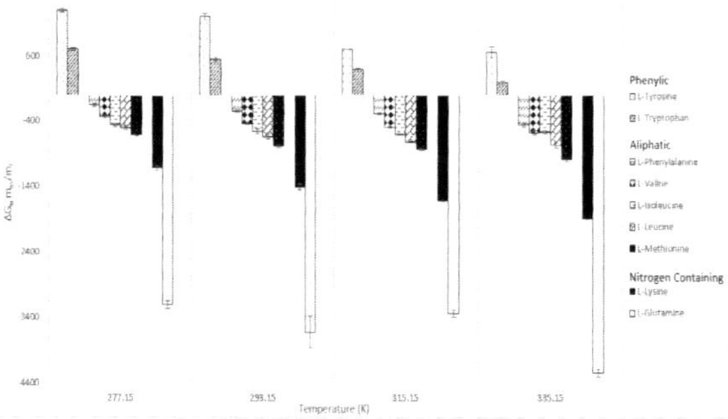

Figure 3: The standard free energy of transfers, ΔG^{o}_{tr}, for the α-amino acids grouped by their side chain physical characteristics

Conclusions

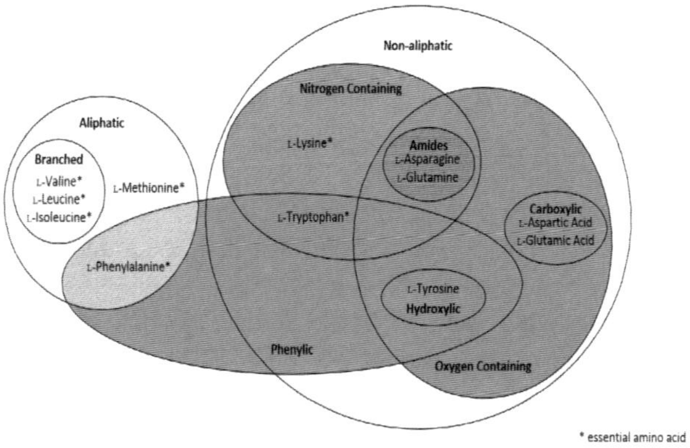

Figure 4: A Venn diagram of the physical characteristics of the side chains of 12 α-amino acids found in agricultural residue in this study. Solubility when in a mixture: increased (blue), slightly reduced (yellow) and greatly reduced (red).

- Our results demonstrate that α-amino acids behave differently in mixtures than in aqueous solution alone and since structurally similar α-amino acids behave similarly in the model system, we assert that the form of an α-amino acid predicts its solubility.

Wageningen UR, Agrotechnology and Food Sciences Group,
Sub-department of Environmental Technology
and Biobased Chemistry and Technology Group
Droevendaalsesteeg 4, 6708 PB Wageningen
Contact: elvira.bozileva@wur.nl
T + 31 (0)317 482943
www.wageningenUR.nl

avans hogeschool

453432 Dispersion-Corrected Density Functional Tighiting Binding Modeling of the Natural Metastable Twin Boundary of Organic Energetic Materials: Beta-Octahydro-1,3,5,7-Tetranitro-1,3,5,7-Tetrazocine

Abstract:

A comparative study of the geometries and electronic properties of twin boundary (TB), as a typical planar defect, in crystalline Î²-boctahydro-1,3,5,7-tetranitro-1,3,5,7-tetrazocine (Î²-HMX) was conducted by using self-consistent charge density functional tight binding with dispersion corrections (SCC-DFTB-D). The possibleÂ twinned structures were predicted and compared with many experiments. The calculated electronic properties of the proposed TBs showed that the interface of twin boundary were more active than the bulk region. The TBs grown along the (001) plane was shown to be more energetically stable than other TBs along (101) and (10-1) plane. The most stable TB structure was compared with available experimental results. The formed TB in the ideal bulkÂ increases internal instability and thus increases the sensitivities of the real materials to the external stimuli.

Research Interests: theoretical study of crystal defects of organic materials.

Teaching Interests: molecular simulation methods.

467626 Quantitatively Reliable Molecular Modeling and Simulation of Vapor-Liquid Equilibria

Abstract:

To reach the quantitative precision required in engineering applications, molecular force-field models need to be adjusted to experimental data for diverse thermodynamic properties. This implies the presence of multiple conflicting objectives in the optimization task, making it difficult to achieve an optimal agreement for the different properties with a single model parameterization. In the present work, this problem is solved by computing and analysing the Pareto set, i.e. the set of all rational compromises between objectives, from which a user can choose the one which fits best to a particular application scenario.

For a variety of low-molecular fluids, two-center Lennard-Jones plus point-quadrupole (2CLJQ) and two-center Lennard-Jones plus point-dipole (2CLJD) models are applied to describe bulk and interfacial properties of pure-component vapor-liquid equilibria. Computing the surface tension at a high accuracy by molecular dynamics simulation requires a massively-parallel simulation environment with a numerically efficient long-range correction for planar interfaces [1, 2]. For this purpose, a highly scalable molecular dynamics code is used: ls1 mardyn [3]. Literature models, which were adjusted to bulk properties, but not to interfacial properties, are validated against the surface tension of real fluids. These models, on average, overestimate the surface tension by 15 to 20% [4, 5].

On this basis, the following optimization task is studied: For ten different pure fluids, the 2CLJQ model class is explored as a whole, and by computing the Pareto set, it is evaluated to what extent the vapor pressure, the liquid density, and the surface tension can be accurately modelled by one and the same model. In various cases, it is found that any pair of two out of the three properties can be described by a 2CLJQ model with an accuracy close to the accumulated error from experiment and simulation, but it is impossible to reconcile all three properties to that level of accuracy [6].

It is computationally expensive to determine the Pareto set. Brute force methods usually cannot be applied even for moderate dimensionalities of the objective and parameter spaces, whereas Monte Carlo methods often yield poor results. For the present study, two algorithms are combined: Sandwiching and hyperboxing. This is computationally efficient and enables a specification of the numerical error, illustrating that multicriteria optimization provides a versatile framework for developing thermodynamic models. User-friendly visualization techniques for the Pareto set are discussed, including a novel approach based on self-organizing patch plots [6].

[1] S. Werth, G. Rutkai, J. Vrabec, M. Horsch, and H. Hasse, Molecular Physics 112 (2014) 2227-2234.
[2] S. Werth, M. Horsch, and H. Hasse, Molecular Physics 113 (2015) 3750-3756.
[3] A. Heinecke, W. Eckhardt, M. Horsch, and H.-J. Bungartz, Supercomputing for Molecular Dynamics Simulations, Springer, Heidelberg (2015).
[4] S. Werth, K. Stöbener, P. Klein, K.-H. Küfer, M. Horsch, and H. Hasse, Chemical Engineering Science 121 (2015) 110-117.
[5] S. Werth, M. Horsch, and H. Hasse, Journal of Chemical Physics 144 (2016) 054702.
[6] K. Stöbener, P. Klein, M. Horsch, K. Küfer, and H. Hasse, Fluid Phase Equilibria 411 (2016) 33-42.

Research Interests:

Computational Molecular Engineering (see above).

Teaching Interests:

Courses presently taught by Martin Horsch at the University of Kaiserslautern, Germany:

Molecular Thermodynamics (in German): Lecture covering statistical mechanics (ensemble, phase space, partition function, analytically tractable problems) as well as molecular modeling and simulation (Monte Carlo and molecular dynamics, computation of thermodynamic properties).

Molecular Simulation (in German): In this "computer laboratory" course, the students implement and apply molecular simulation methods under individual supervision. This is accompanied by a lecture concerning the simulation of phase equilibria (entropic quantities, Grand Equilibrium, interfacial properties) and non-equilibria (fluctuation and dissipation, transition states, work done by small systems), extending the knowledge from Molecular Thermodynamics.

Molecular Modeling (in English): Compact course within the international graduate school IRTG 2057 "Physical Modeling for Virtual Manufacturing Systems and Processes" (University of Kaiserslautern, UC Berkeley, and UC Davis), covering basic molecular modeling and simulation approaches and relating them to other methods in computational engineering.

Martin Horsch was positively evaluated as a Junior Professor in 2014, and following a training in didactics, he obtained the Rhineland-Palatinate Certificate for University Didactics in 2015. Preceding his faculty appointment, he served as a teaching assistant in Mechanical Engineering (Molecular Thermodynamics) and Computer Science (Theoretical Computer Science II and III) at the University of Stuttgart, and in Mechanical Engineering at the University of Paderborn (Thermodynamics II and Refrigeration Technology).

476178 Modeling Chemical Reactivity for Nanoscale Design

Abstract:

Research Interests: The overarching goal of my research is to understand the connection between molecular interactions and macroscopic dynamic behavior. I employ computer simulations at both classical and quantum levels of detail to probe molecular interactions, and use reaction rate theories and statistic mechanics to evaluate the related macroscopic kinetic and transport parameters. My research underscores the need for non-empirical and transferrable predictive tools for the design of new solvents and catalysts. Specific areas of interest include identification of mechanisms for solvated and heterogeneous reactions; design of solvents and solid surfaces with unique interactions that increase selectivity; development of efficient simulation methods for calculating transport properties in amorphous condensed phases with long relaxation times.

Postdoctoral Project: Reaction equilibria in ionic liquids using Monte Carlo simulations
In collaboration with Ed Maginn & Steve Corcelli, Chemical and Biomolecular Engineering, Notre Dame

PhD Dissertation: "Mechanisms of Rare Events in Condensed Phases"
Under supervision of Baron Peters & Joan-Emma Shea, Chemical Engineering, University of California Santa Barbara

Teaching Interests: I am prepared to teach any of the core chemical engineering undergraduate courses, particularly mass transfer, chemical reaction engineering, and thermodynamics. I am additionally interested in teaching introductory chemical engineering courses to first and second year students, a course on analytical mathematical methods, or a course on numerical methods / computer programming. I am eager to develop elective courses on molecular simulation at both the undergraduate and graduate levels, and graduate courses in reaction rate theory and mass transfer. I find that illustrating fundamental chemical engineering processes (e.g. diffusion, phase separation, reaction equilibrium) using visuals at both the macroscopic and molecular levels better prepares students to reason through new scenarios. When feasible, I plan to incorporate course projects with hands-on demonstrations or molecular-level visualization in order to build an online repository of teaching aids.

476189 Enhanced Molecular Simulations for Applications in Protein Stabilization, Crystallization, and Structural Determination

Abstract:

1st Year Postdoctoral Fellow

Research Interests:

Molecular modeling and simulation can provide researchers with a more detailed understanding of the interactions that underlie important biological and chemical processes. With ever-advancing computer power, researchers can now simulate the molecular-level interactions of proteins, DNA, large polymers, and other organic macromolecules on experimentally relevant timescales. One of the major shortcomings of standard molecular simulations techniques is that they sample molecular conformations that are near local free energy minima. Thus transitions between various states of a system and the barriers associated with those states are rarely visited during a simulation. To overcome these limitations, several enhanced sampling techniques have been developed that allow a researcher to exhaustively sample states of the system and escape local free energy minima. Some of the rarely visited states are often crucial to completely understand thermophysical and transport properties, and such information can not only help us to understand chemical phenomena but also to predict the behavior of chemical systems *in silico*. I an expert in the application of enhanced and non-equilibrium sampling methods that allow us to answer pertinent scientific questions.

Successful Proposals: Alexander von Humboldt Postdoctoral Fellow. Assisted in writing and gathering preliminary results for NSF and XSEDE/Teragrid applications.

Postdoctoral Project: *Enhanced Molecular Simulation and Thermodynamic Characterization of* Protein Mutations for Improved Crystallization by Surface Entropy Reduction

Externally funded through the Alexander von Humboldt Foundation. Under the supervision of Prof. Bert de Groot at the Max Planck Institute for Biophysical Chemistry. Göttingen, Germany.

PhD Dissertation: *Developing a Simulation Framework for Modeling Biomolecules in Ionic Liquids*

Under the supervision of Prof. Jim Pfaendtner at the University of Washington. Seattle, WA.

Research Experience:

The focus of my research has been the computational modeling of protein and ionic liquid systems. As such, I have become an expert in the field of molecular dynamics simulations with an emphasis on enhanced sampling methods that allow us to characterize the thermodynamics of rare events. Specifically, I employ metadynamics and alchemical mutations to determine the relative free energies of multiple states of biomolecular systems and rates of transitions between states as in protein unfolding and ligand unbinding. When I initiate a simulation study, I always look to better explain the molecular interactions that drive experimental results or to predict results of future experiments. Other times, experimental collaborators request my assistance in performing molecular simulations to determine the feasibility of a hypothesis. As such, I have worked closely with several collaborators and even collected experimental data myself. Consequently, I have experience in performing simulations that are relevant to IR, Raman, and SFG spectroscopy as well as microcalorimetry, reaction kinetics, AFM, electron microscopy, XPS, SAXS, and SANS. Therefore, I am confident that I can design and implement simulations that (a) answer important questions about biological and chemical phenomena, (b) support the development of hypothesis in collaboration with other scientists, and (c) efficiently and quickly predict the behavior of biomolecular and organic systems when no experimental data is yet available.

Teaching Interests:

During my time as a PhD student, I had the privilege of training several undergraduate, masters, and fellow young PhD students in the fundamentals of molecular simulation. I find teaching and training of students to be one of the most rewarding parts of my academic career. As a TA, I had the opportunity to teach both core chemical engineering courses such as unit operations, mass transport, and separations as well as electives such as introductory courses to molecular and nanoscale principles. During these courses I gained experience in developing lectures and in the hands-on training of students for the operation of scientific equipment. As a professor, I would feel comfortable in teaching any core chemical engineering course, and I would like to develop a course on molecular simulation to help students better understand the fundamental physics that underlie the thermophysical phenomena we observe as scientists and engineers.

Future Direction:

As faculty, I would like to move my research forward in three distinct fields:

First, I would like to expand my work on thermostabilizing and entropy-reducing protein mutations to include alchemical thermodynamic calculations on mutations that affect the stability of proteins in the presence of chemical denaturants and of enzymes in confinement. With such simulations, we can design proteins that better tolerate a wide range of conditions. The engineered proteins are industrially relevant for a wide variety of applications where enzymes are subjected to stress as in biomass processing, food production, and long-term therapeutic protein storage.

Second, I would like to study peptides that precipitate inorganic nanoparticles. Our previous work indicates that coarse-grained metadynamics simulations can give us a general idea of the shapes of nanostructures formed by a core of small charged peptide with only 1-2 days of simulation time. In nature, many organisms use proteins to transform simple soluable molecules into complex inorganic structures such as diatom exostructures, mollusk shells, and vertebrate bones. Using natural proteins as a starting point, we can run a campaign of simulations to screen thousands of potential nanostructure-forming mutant peptides per year. This work offers the possibility to make well-defined inorganic nanostructures at mild conditions for applications such as catalyst supports or drug delivery.

Third, I would like to develop a protocol for accurately predicting the properties of mixtures of ionic liquids using polarizable force fields. Ionic liquids are liquid organic salts that have many unique properties making them interesting for industrial applications such as biomass processing, batteries, and carbon dioxide capture. Our previous work on ionic liquids has demonstrated that the thermophysical properties of many pure ionic liquids can be predicted using molecular simulation with a generalized non-polarizable force field. We can improve our results by using established polarizable force fields, and we can begin to model mixtures of multiple ionic liquids. With the right computational tools, we can develop highly-tuned ionic liquid recipes to fit the properties required for a wide range of applications.

Selected Publications:

V.W. Jaeger and J. Pfaendtner. S*tructure, Dynamics, and Activity of Xylanase Solvated in Binary Mixtures of Ionic Liquid and Water. ACS Chemical Biology.* 2013. *8* (6), 1179-1186.

V. Jaeger, P.R. Burney, and J. Pfaendtner. *Comparison of Three Ionic Liquid-Tolerant Cellulases by Molecular Dynamics. Biophysical Journal.* 2015. *108 (4)*, 880 â€“ 892.

J.E. Baio, A. Zane, **V. Jaeger**, A.M. Roehrich, H. Lutz, J. Pfaendtner, G.P. Drobny, and T. Weidner. *Diatom Mimics: Directing the Formation of Biosilica Nanoparticles by Controlled Folding of Lysine-Leucine Peptides. Journal of the American Chemical Society.* 2014. *136* (43), 15134-15137.

K.G. Sprenger, **V.W. Jaeger**, and J. Pfaendtner. *The General AMBER Force Field (GAFF) Can Accurately Predict Thermodynamic and Transport Properties of Many Ionic Liquids. The Journal of Physical Chemistry B.* 2015. *119* (18), 5882-5895.

V. Jaeger, W. Wilson, and V.R. Subramanian, *Photodegradation of Methyl Orange and 2,3-Butanedione on Titanium-Dioxide Nanotube Arrays Efficiently Synthesized on Titanium Coils. Applied Catalysis B: Environmental.* 2011. 110, 6-13.

476191 Advanced Materials Design Using Molecular Simulation, Evolutionary Computing and Machine Learning

Abstract:

Research Interests: Polymer and Soft Matter; Glass Transiton; Self-Assemly, Phase Transition and Free Energy of Materials; High Throughput Materials Design; Machine Learning

Teaching Interests: Thermodynamics, Statistical Mechanics, Theory of Liquids, Polymer Physics, Numerical Methods, Molecular Modeling

Abstract: Molecular simulation has emerged as a remarkable technique for investigating structure-property relationships in a wide variety of materials. Combining it with evolution based optimization methods such as genetic algorithm can provide tremendous opportunity for discovering new materials of extremal properties. However, the major roadblock in this direction is that the determination of a molecular property via simulations is very slow. Alternatively, machine learning tools have been progressively adopted by the materials science community to accelerate design of materials with targeted properties. Nevertheless, in the search for new materials exhibiting properties and performance beyond that previously achieved, machine learning approaches are frequently limited by two major shortcomings. First, they are intrinsically interpolative. They are therefore better suited to the prediction of properties within the known range of accessible behavior than to the discovery of new materials with extremal behavior. Second, they require the availability of large datasets, which in some fields are not available and would be prohibitively expensive to produce. Here we describe a new strategy for combining genetic algorithms, machine learning, and molecular simulation to discover materials with extremal properties in the absence of pre-existing data. Predictions from progressively constructed machine learning tools are employed to bias the evolution of a genetic algorithm, with fitness evaluations performed via direct molecular dynamics simulation. We survey several materials design problems, in particular, polymer compatabilizer, porous polymeric materials, and polymer glasses, with in this framework. Further development of this computational framework and its implication to future material design will be discussed in this presentaion. In additon, we employ molecular simulation to provide useful guidelines for designing polyelectrolytes and polymer nanocomposites.

476138 Exploring Fundamentals of Zeolite Catalysis â€“ a Theoretical Perspective

Abstract:

Research Interests:

Zeolites are among the most important industrial catalysts today. Their diverse structure leads to a large variety of different possible active sites combined with a high selectivity due to their microporous structure. However, chemical reactions take place inside the material, which makes a full understanding of them highly challenging. This is one of the reasons, why many fundamental aspects of zeolite catalysis are still little understood. In this contribution I describe my efforts in the past years to elucidate several of the underlying principles in context of several timely problems within zeolite catalysis.

One of the main challenges in automotive catalysis using lean burn engines is the removal of nitrous oxides from exhaust gases. A system that has been found to be a particularly efficient catalyst in this context is copper exchanged SSZ-13 (Cu-SSZ-13). Even though SSZ-13 is a zeolite in the chabazite structure, the zeolite structure with the most simple primitive unit cell, the true nature of the active centers for this reaction is still under discussion. In this contribution I describe my efforts to combine ab-initio molecular dynamics simulations, post Hartree-Fock calculations and thermodynamic modeling to arrive at an unambiguous assignment of infrared spectra of adsorbed probe molecules [1,2] and UV-vis spectra of reduced Cu-centers [3]. Due to the thermal motion at finite temperature, the Cu cation is highly mobile within the framework. This mobility changes with the oxidation state. While Cu^{II} oscillates around its local minimum, Cu^{I} changes its coordination on a pico-second timescale [1,2,4] and Cu^{0} even moves from its favored position in the six-ring of the structure to silanol-defect sites [3]. However, as soon as water is present in the system, the cation is solvated and moves away from the framework. While it is possible to follow this mobility by comparing calculated spectra to experimentally measured ones, it is still exciting to see in what way these fundamental insights on the nature of the active sites will affect our understanding of the reaction cycle and whether they can be transferred to other zeolite catalyzed reactions.

Another reaction that has recently drawn large amounts of attention is the conversion of methane-to-methanol over transition metal exchanged zeolites. First of all these systems show a very low activation energy and at the same time the active centers are very similar to those in enzymatic catalysts, responsible for catalyzing this reaction in nature. Here I will discuss two different aspects of this problem. It is safe to assume that the confining environment is a key factor for the activity of these materials. Confinement effects are rather complicated to understand, since entropy and enthalpy affect adsorption and conversion of reactants and products. A first step is understanding the impact of confinement on the activation enthalpy of the methane-to-methanol conversion over Fe-oxo centers in SSZ-13 [5]. I conclusively show that confinement stabilizes the reaction intermediates more than the initially adsorbed methane molecule, which in turn leads to a significant lowering of the activation energies. This allows to extrapolate generally valid statements about the impact of confinement on reaction enthalpies, which will most likely impact our understanding of similar processes in the future. However, more recently Cu-exchanged zeolites have drawn larger amounts of attention in the scientific community for this reaction. I will present a thermodynamic model for $Cu_xO_yH_z$ sites in the zeolite SSZ-13 [6]. Interestingly not only one, but several different types of active sites exist in this material and depending on the exact conditions, different ones are thermodynamically most stable. The theoretical model is also supported by Raman- and UV-vis measurements, which show excellent agreement with the modeled results.

The problems discussed in this contribution show, how of state of the art electronic structure theory combined with experimental measurements can help elucidating fundamental concepts in zeolite catalysis. In the future I am looking forward to continue my work along this path to improve the conceptional understanding of zeolite catalysis, to generalize these ideas and apply them to catalysts of industrial interest and ideally arrive at an approach that does not only help to understand, but also predict the activity of a large number of zeolite catalysts. Furthermore I belive that collaboration is one of the key aspects of succesful scientific work and I am looking forward to establish joint projects with other faculty members.

Teaching Interests:

Besides my achievements in science, I have always had a passion for learning and teaching, which has been reflected in mentoring graduate and undergraduate students as well as teaching selected classes during my current post-doc. In my time as a faculty member I plan to develop a course describing how state of the art electronic structure methods can be applied to gain insights into heterogeneous catalysis, either by understanding reaction pathways or by predicting observed spectra. It will provide the necessary theoretical background as well as practical examples and hands-on experience for the students. Among the undergraduate core classes I ready to teach thermodynamics and, depending on the departmental needs, am looking forward to develop classes close to my core competencies in fundamental physics, quantum mechanics, catalysis or materials science.

476139

Abstract:

Reverse Engineering of Molecular Structure

Dr. Farhad Gharagheizi

Email: fghara@gmail.com ; farhad.gharagheizi@ttu.edu ; Tel: (+806) 252 8892

Post-Doctoral Researcher (Advisor: Ronald Hedden)

Department of Chemical Engineering, Texas Tech University,

Box 43121, Lubbock, Texas 79409-3121, United States

Advanced computational techniques are invaluable for investigating how the molecular structure of pure organic/inorganic compounds, ionic liquids, polymers, and their mixtures affect their physical, chemical and biological properties. Molecular structure-based knowledge gained from computational screening can be harnessed to reverse-engineer new chemical compounds with desired properties. Molecular structures of chemical compounds possess encoded details that govern their thermophysical properties. Decoding these structural characteristics through data-driven computational screening enables reverse engineering of molecular structure to obtain molecules with desired properties. Reverse engineering methodology is relevant to research in the fields of Petroleum Engineering, Natural Gas Science and Engineering, Pharmaceutical Science, Environmental Science, and Energy.

In the petroleum field, reverse engineering of molecular structure could be used to design new materials that have attractive transport properties (such as viscosity, thermal conductivity, heat capacity, density, and vapor pressure). As an example, one could target the design of a "perfect" lubricant whose viscosity has minimal temperature dependency within a desired range of temperature and pressure. In natural gas science and engineering, reverse engineering of molecular structure could also be used to design safer separation media. For instance, the technique could be used to find a new chemical compound that promotes the storage capacity of gas hydrates and improves process performance.

In the energy area, electrolytes play a key role in energy storage systems (i.e. batteries, super-capacitors, actuators, dye sensitized solar cells, and thermo-electrochemical cells). Also, the tunable phase behavior of ionic liquids also permits the creation of phase-change thermal energy storage materials having melting points tuned to any desired application. Reverse engineering of molecular structure could be used to design new ionic liquids that exhibit attractive combinations of viscosity and conductivity.

In the pharmaceutical industry, "Drug Discovery" generally refers to a high-throughput screening process in which the main goal is to find a chemical compound that meets predefined performance criteria. Such criteria are generally defined based on the values of a series of properties/activities (i.e. aqueous solubility (blood solubility), skin permeation, organ diffusivity, etc.). Therefore, molecular reverse engineering could play a key role in identifying undiscovered drug molecules with structural features that induce the desired performance characteristics.

As an assistant professor, I plan to establish a mainly computational research group in molecular engineering, develop strong collaborations with industry, and serve as a key element of interdisciplinary research teams with experimentalists and theoreticians.

- Abstract Text

476146 Developing Molecular Theories/Simulations to Understand and Optimize Soft Matter Systems: From Ions to Polymers to Gels

Abstract:

Research Interests:

Modern chemical engineering becomes increasingly molecular-based. The development of theoretical frameworks that capture the essential physics, address the molecular feature and quantify the systems across multiple time and length scales is vitally important. My research focuses on using statistical mechanics, which bridges the molecular information and macroscopic phase/interfacial properties, to study a variety of soft matter systems including electrolyte solutions, liquid crystals, polymers, polyelectrolytes, polymer-protein conjugates and polymer networks/gels.

I did my PhD work at Caltech with Prof. Zhen-Gang Wang, where we developed renormalized Gaussian fluctuation theory to study the effect of self energy of ions on the double layer structure and interfacial properties. Our theory incorporates different components of self energy in a unified framework and involves only intrinsic parameters. We show the consequence of the self energy by revealing three essential effects: the image charge effect, inhomogeneous screening effect and specific ion effect. We are able to quantitatively explain a couple of long-standing puzzles in physical chemistry and soft matter physics, such as the salt concentration dependence of surface tension, like-charge attraction and charge inversion, Hofmeister series in the interfacial affinity and strong interfacial adsorption of hydrophobic ions.

My current postdoctoral research in MIT with Professors Bradley Olsen, Alfredo Alexander-Katz and Jeremiah Johnson is to use theory and simulation to quantify the topology and elasticity of polymer networks by addressing cyclic defects (loops). I develop a kinetic graph theory which demonstrates a universal dependence of loop fractions on the preparation condition and reveals the intrinsic relation between different cyclic topologies. I also develop a real network theory which quantifies the individual negative impacts of different loops on the elasticity. These theories shows excellent agreement with experimental data, providing, for the first time, a quantitative understanding of gel elasticity based on molecular details. By bridging the topology and elasticity, we are now seeking to optimize the mechanical property of the polymer networks through controlling the molecular connectivity.

In the future, I would like to carry theoretical and simulational research to solve fundamental challenges in physical chemistry and soft matter physics. I will build simple models that capture the essential molecular characteristics (constitution, interaction and connectivity), which are then incorporated into the framework of statistical mechanics and further solved by using or developing modern analytical, numerical and computational methods This methodology facilitates the study of inhomogeneous systems, fluctuation effects and nonequilibrium processes, where most of the challenges exist. On the other hand, I will seek for extensive collaboration with experimentalists to use this fundamental knowledge for rational design of new materials and processes. I will initially focus on the following topics:

1) Complex interfacial and kinetic phenomena driven by electrostatics;

2) Conformation, association and interfacial behavior of polyelectrolyte solutions;

3) Topology of polymer/polyelectrolyte networks vs. phase/interfacial behaviors.

Teaching Interests:

Teaching is extremely important for the professional development of my future students, and also benefits my own research work. In my future course, I would like my students to obtain clear basic concepts, the ability of critical thinking, the training of solving real problems and the knowledge about current challenges in this field. To achieve this goal, I will organize the course materials and develop the teaching methods by addressing the following aspects: 1) encourage the students to think about one problem from different perspectives; 2) guide the students to find the underlying physics that explains different phenomena; 3) design problem sets and term projects from recent research progresses.

As a graduate student at Caltech, I had the opportunity to work as a teaching assistant in graduate level thermodynamics, statistical mechanics and polymer physics. The evaluation from the students shows that I was one of the best teaching assistants in recent years. Besides these courses, my education and research experiences also make me well quantified to teach kinetics, transport process, interfacial phenomena, mathematics in chemical engineering and computer modeling and simulation. Currently as a postdoc in MIT, I am also training for mentoring undergraduate and graduate students, which is extremely helpful for my teaching skills. The course about soft materials right now is limited due to the rapid progress in this field and the lack of systematic textbooks. I plan to develop an in-depth course that convers a broad spectrum of soft matter systems in the future.

476165 Simulation of Selectively Permeable Novel Polymeric Membranes

Abstract:

My research focuses on the use of computational modeling to understand systems important in energy and materials sciences. Access to cheap, reliable energy is a grand development challenge, while simultaneously the transition to clean energy sources is a grand sustainability challenge. Often, understanding processes in energy generation requires knowledge of the physical chemistry of reactions, which occur at the atomic level on the picosecond time scale, yet also being able to track energy inputs and outputs for the total macroscopic device over the course of days or even years. Having experience with a variety of modeling methodologies that address different time and length scales (density functional theory, molecular dynamics, and stochastic kinetic algorithms) uniquely positions me to address these types of systems.

Presented here, my current project at the Joint Center for Artificial Photosynthesis (JCAP, a DOE Energy Innovation Hub supported under Award Number DE-SC0004993) is aimed at improving design and understanding of novel polymeric separations membranes. The overall goal of JCAP is to develop and demonstrate a device that can convert carbon dioxide and water into useful hydrocarbon fuels using solar energy. A key determinant of performance for an artificial photosynthesis device will be the ability to block transport of products of the carbon dioxide reduction reaction (e.g., methanol, formate) while allowing ion transport (via the bicarbonate ion). Current commercially available membranes are not designed for this purpose, and the key structural features that would enable high function for carbon dioxide reduction systems have not been identified. I am performing mesoscale simulations to develop a better understanding of key membrane characteristics. Using stochastic kinetic algorithms, I model how solutes in a complex solution move into and through polymeric materials. I work closely with experimentalists to explain experimental observations and elucidate guiding principles for design of novel membrane materials. Improvements in membrane design could have impacts in a wide range of technologies including reverse osmosis and fuel cells.

Research Interests:

My overarching vision is to improve computational modeling in order to address problems in development and sustainability. Specifically, I would (1) build on my graduate work in development of advanced force fields to expand the types of systems which can be modeled accurately with molecular dynamics, (2) extend my modeling of permeation of polymeric material to better design the controlled release of drug molecules in the body, and (3) take advantage of the long time scales that can be simulated using stochastic algorithms to understand aging of materials. Connection with experiment is and will continue to be a vital part of my research.

Teaching Interests:

As a secondary school teacher in Tanzania and then a mentor to undergraduates during my doctoral and post-doctoral work, I have found that being able to relate in-classroom knowledge to the world at large is the best way to engage students with material and have them retain the information. As such, my preferred teaching style is project-based learning, especially for upper-level classes in physical chemistry. While most class time should be spent on ensuring that students have a firm grounding in the fundamentals, they would also work on a semester-long overarching project. Also related to my interest in project-based learning, a key component of my teaching would be to involve undergraduate students in research.

476299 Towards More Rational Design of Electrocatalyst for Carbon Dioxide Reduction

Abstract:

Research Interests:

The energy appetite from our global society to satisfy the requirements of mankind's development is insatiable. However, most of the present fossil fuels (coal, oil, and natural gas) are not renewable. For the sustainable development of our society, we thus have to increasingly tap into alternative energy resources, such as wave, geothermal, solar, and wind sources, to satisfy our energy needs in the near future. In addition, the emission of carbon dioxide is unavoidable in the case of fossil fuel combustion, which results in global warming and other environmental issues. Therefore, discovering alternative energy sources and solving the environmental issues of carbon dioxide emission become more and more important. This work aims to design highly efficient catalysts for energy storage and conversion based on an innovative paradigm, i.e. theoretical descriptors based catalyst design, to solve the energy problems mentioned above. We find the binding energy of CO on electrode is an important descriptor for the activity of CO_2electroreduction.

Teaching Interests:

1. theoretical descriptor of catalysis,

2. energetic scaling relationship,

3. potential-dependent electrochemistry

476192 Computational Soft Matter

Abstract:

Research Interests:

My research aims to answer the questions: What are the molecular foundations of soft materials and how can simulations guide the design of such materials? How can we predict the assembly of novel forms of colloidal matter, based on entropy and shape? How can we design novel, biological materials from protein building blocks using simulations, and unravel the role of anisotropic shape and specificity of interaction? How can we simulate and contribute to the understanding of molecular recognition processes of proteins that are essential to the design of drugs?

I will address these questions in a vibrant research group by combining high-performance simulation on Graphics Processing Units (GPUs), theory, and coarse-grain modeling. We will perform Molecular Dynamics (MD) and Monte Carlo (MC) simulations of colloidal and biological matter, predict the assembly of proteins into aggregates and crystals, and collaborate with experimental groups on the design of protein-based materials.

Teaching Interests:

I want to convey to students the essential skills they need to participate in scientific discussion by encouraging active problem solving, presentation of research papers in small learning circles, and individual small research projects, as well as foster an inclusive learning environment. Moreover, by motivating them to arrive at solutions to problems that have multiple possible correct answers, my goal is to prepare the students for problems in research and the professional world they have not been directly prepared for. As it relates to soft matter theory, students should be able to appreciate the intellectual adventure that is science and engineering, by having a solid foundation both in classical thermodynamics as well as modern computer simulations. Ultimately, I aim to incite curiosity in the same questions that inspired the science that we teach, which I view as the best foundation for research and professional success.

476210 Design and Discovery of Multifunctional Nanoporous Materials

Abstract:

Rapid developments in design and synthesis of nanoporous materials have revolutionized both established and emerging fields including heterogeneous catalysis, gas separations, energy transformations and chemical sensing. In particular, zeolites, metal-organic frameworks (MOFs) and porous organic polymers (POPs) have received significant attention.

Given the diversity of available structures and their possible variations, atomistic simulations have proven to be an invaluable tool for understanding and predicting various properties in nanoporous materials. In this work, I will summarize several classical and quantum chemical modeling strategies for predicting reactivity, dynamics and adsorption characteristics of zeolites and MOFs. Further, I will show that these multi-scale approaches enable efficient screening of catalysts and adsorbents for industrially-relevant methane upgradation, CO_2 adsorption and olefin/paraffin separation applications.

Ultimately, as the viability of a functional material depends on a number of factors such as ease of synthesis and functionalization, long-term stability, degree of disorder and reactivity; computational techniques that transcend different time and length-scales are necessary. This work highlights the growing importance of molecular simulations, which will ultimately lead to rational design and discovery of complex, multifunctional porous materials vital for a sustainable future.

Research Interests: Nanoporous materials, multiscale molecular simulations, computational catalysis

Teaching Interests: Chemical engineering fundamentals, catalysis and reaction engineering, computational chemistry, adsorption and separations

476228 Hydrodynamic Model of Complex Liquids with Microstructure

Abstract:

Complex liquids, such as liquid crystals (LCs) and colloidal suspensions, have unique properties that find wide applications in mechanical, chemical and material engineering areas. How the underlying microstructure of such material determines its mechanical, optical and electromagnetic characteristics is not fully elucidated. The inverse problem, that what specific microstructure is needed to satisfy certain material properties, is however fundamentally important for the design of next-generation functional, active and smart materials. The goal of my proposed research is to develop and generalize continuum and molecular models to understand the physics of known experiments on structured liquids and propose the recipe of materials that can be useful for the future energy and bio-inspired material applications.

Specifically, I plan to model and simulate the following systems:

1) LC smectic phase. Highly periodic self-assembling materials are important for optoelectronics and nanotechnology applications. Focal conic domains (FCD), a typical periodic structure formed from smectic A phase when strongly confined, is the candidate for such materials. I plan to use a Q-tensor model of Landau-de Gennes free energy funtional to simulate the 3D structure of FCD. We can control the spatial configurations of FCD by introducing patterned surface. As a template, FCDs may selectively trap nanoparticles, which then show highly ordered self-assembling structure. My work should help design new periodic patterns for the future 3D electronic devices.

2) Biopolymer networks. Active nematic liquids, e. g. microtubule-kinesin-ATP film exhibits many peculiar non-equilibrium behaviors, such as active turbulence at very low Reynolds number, oscillating defect structure when

confined within a shell, and long-range orientational order among +1/2 and -1/2 defects. The underlying physics is poorly understood. I plan to develop both continuum and molecular models to explain the above phenomena at different length scales. Validated by reproducing known phenomena, my model can help understand such active materials and may guide the design of new bio-inspired materials.

3) Dense suspensions. They are important for petroleum industry, food manufacture and functional material design. Current research is focused on steady state rheology, spherical particle shape and single length scale. I plan to generalize the current simulation method, Stokesian dynamics, to account for the transient dynamics as well as multi length scale particles. My model should help understand the underlying physics of dynamic jamming, and enable the simulation of impact mitigation applications.

During my Ph. D study, I worked with Prof. Joel Koplik at Levich Institute in CCNY, on multi scale simulations of microfluidic phenomena. During my postdoc experience with Prof. Juan de Pablo at the University of Chicago, I developed the hybrid lattice Boltzmann method to simulate flowing LC system in the continuum level, and applied to the microtubules system to understand the active nematics phenomena.

Research Interests:

Theory and modelling of soft matter physics:

- Liquid crystals and its hydrodynamics

- Active nematics

- Colloidal self-assembly

- Dense suspensions

Simulation expertise:

- Molecular Dynamics

- Lattice Boltzmann method

- Stokesian dynamics

Teaching Interests:

- (Computational) Fluid Mechanics

- Statistical physics

- Molecular modelling

- Transport phenomena

- General physics

476321 New Generation of Polarizable Reactive Force Fields for Multiscale Simulations of Complex Materials

Abstract:

Research Interests:

Critical to the development and manufacturing of new generations of functional materials is the design of the non-equilibrium dynamical processing to achieve desired microstructures and transport properties. Here, we need methods for *in silico* designing and simulating of such processes to produce novel functional materials and to facilitate technological breakthroughs. These methods must be able to describe: **I**) non-equilibrium transport phenomena (e.g. of charges, ions, atoms, mass, and defects) in bulk and across interfaces over multiple length and time scales, **II**) the dynamics of phase changes during material growth, processing and operation, and **III**) the microstructures of complex heterogeneous structures that provide the desired functional properties.

Non-equilibrium phenomena determine the behavior of materials for practical applications, and understanding them is critical to optimize material synthesizability, characterization, and design trade-offs. Specific examples of non-equilibrium in material systems include ionic and electronic conductance in semiconductors, synthesis of thin films and nanostructures (e.g. crystallization), and functional self-organizing soft materials (e.g. adsorption polymers, catalyst support materials, soft magnets), condensed phase reaction kinetics. In all these materials, non-equilibrium dynamics determines all structural characteristics and their properties.

We need methods that provide the accuracy of Quantum Mechanics (QM) for reactive non-equilibrium dynamics simulations of extremely large spatial scales (1 billion atoms to describe a system of 100 nm on a side) and time scales (milliseconds to seconds) to characterize the non-equilibrium dynamical processes required to synthesize complex chemical and material systems. At this level we need to match to the continuum simulations required for developing manufacturing processes.

Successful Proposals: Department of Energy, National Science Foundation, Office of Naval Research

Postdoctoral Projects:

1. "New Generation Reactive Force Fields based on Valence Bond Concepts with Polarized Charge Distributions (VaPoX), Fundamental Concepts and Applications"

2. "Quantum Optimized Polarizable Charge Equilibration (PQEq) For Predicting Accurate Charges in Molecules and Solids

3. "Prediction of Structures and Properties of Green Energetic Materials from Density Functional Theory (DFT) and Reactive Molecular Dynamics"

Advisor: Prof. William A. Goddard III, Division of Chemistry and Chemical Engineering, California Institute of Technology (Caltech)

PhD Dissertation:

"A process-based molecular model of silicon carbide (SiC) nano-porous membranes"

Advisors: Profs. Muhammad Sahimi and Theodore T. Tsotsis, Mork Family Department of Chemical Engineering and Materials Science, University of Southern California (USC)

Research Experience:

My research focuses on first principles-based multiparadigm, multiscale strategies for simulating complex materials processes. It contains, 1) development of advanced computational algorithms for atomistic modeling, and 2) their applications to current problems in chemistry and chemical engineering. For example, I have developed VaPoX, a new generation of polarizable reactive force field. VaPoX allows an accurate description of the local polarization of atoms needed to describe dynamic dielectric properties. In addition, it uses valance bond concept to accurately describe the bonding. VaPoX is a great deal of progress toward *in silico* design and simulation of non-equilibrium dynamical processes required to manufacture complex functional materials to achieve novel functional and transport properties.

These research projects bring together concepts in physics, chemistry, chemical engineering, numerical methods and algorithms analysis that I have developed during my doctoral and post-doctoral work studying chemical phenomena. I have received several awards for excellence in my research including the University PhD Achievement Award (USC 2013), the Material Engineering and Science (MESD) Award (AIChE 2012), the Soft Matter Journal Award (AIChE 2012), the Mork Family Department of Chemical Engineering & Materials Science Best PhD Thesis Award (USC 2013), and Project Navigator Ltd. Environmental Engineering Awards (USC 2011 and 2012).

Future Plans:

There is an economic and environmental imperative to develop new and alternative energy technologies. As a professor, I would like to continue developing polarizable reactive force fields to design new functional materials and to facilitate technological breakthroughs.

I propose to build a new formulation of polarizable reactive force fields based on my previous successes (e.g. VaPoX) but with the goal of matching the highest level QM while making it more systematic so that the transferable reactive force fields will be attainable. This will enable not only my group but also a broad range of scientists to study nonequilibrium, adiabatic and non-adiabatic, reactive dynamics simulations of large-scale functional material systems. I would like to prove that it is possible to construct a generic force field capable of describing complex reaction dynamics at nearly the accuracy of QM on systems that are many orders of magnitude larger than QM size and time-scales.

My research direction is motivated by a diverse set of applications including energy technologies, nanotechnology, biomaterials, catalysts, and organic materials where experimental approaches are inaccessible and/or inefficient. In particular, I would like to develop new polarizable reactive force fields to design and study high energy density materials under extreme conditions, nonporous membranes for gas separation, energy storage and conversion, hydrogen storage and supercapacitors, gas purification, fuel cells, solar cells, and Li–ion batteries.

As a faculty, I hope to educate new engineers and establish an externally funded and internationally renowned research group that develops energy technologies. My research group will focus on socially important problems and the translation of scientific studies into practical technologies. I will foster a learning environment and will actively maintain collaborations with colleagues from within the engineering community and from diverse fields. Scientific advancement is best accomplished through efficient communication of results and analyses; as such, I will strive to disseminate my works at the highest scientific levels as well as to the broader public. I am confident in my abilities to manage a research group and I am highly motivated to begin working on the development of new functional materials and energy technology.

Teaching Interests:

I have experience teaching and mentoring students in the classroom and in the laboratory. As a teaching assistant I have given lectures and tutorial sessions in Transport Phenomena (i.e. Mass Transfer, Heat Transfer, and Viscous Flow), Reactor Design and Kinetics, and Advanced Engineering Mathematics. My teaching efforts and strategies received commendable reviews and ratings at USC. I received university outstanding teaching assistant award in the category of chemical engineering (2012) and Mork Family Department of Chemical Engineering and Materials Science best teaching assistant award (2013). In addition, I have received several offers from prestigious institutions such as Harvey Mudd College to teach undergraduate chemical engineering courses. I am enthusiastic about the opportunity to teach undergraduate and graduate courses including Quantum Mechanics, Computer Simulation and Modeling, Transport Phenomena, Chemical Reaction Engineering, Engineering Thermodynamics, and Advanced Engineering Mathematics.

477817 Flexible and Dynamic Porous Crystals

Abstract:

Research Interests:

1. Fundamentals of the thermodynamics of porous solids with dynamic constituents

Some porous materials possess dynamic constituents that respond to guest molecules and external stimuli. Such enzyme-like complexity can enhance the selective recognition and capture of molecules. I propose to construct and analyze simple, yet illuminating statistical mechanical models to study porous crystals with dynamic constituents to harness them for gas separations, chemical sensing, drug delivery, and catalysis.

2. Computationally-accelerated discovery of flexible materials for chemical sensing

Using a Materials Genome approach, I propose to explore the chemical space of gate-opening materials by constructing structure models from judiciously chosen molecular building blocks and carrying out high-throughput computational screenings to identify candidates for sensitive and selective chemical sensors.

3. Mechanisms of bistability and pattern formation in dynamical systems

I propose to develop mathematical models to elucidate the mechanism of the shifts in the microbial composition in the mucus layer of coral that are involved in bleaching, the water-biomass feedback involved in the self-organization of vegetation in arid landscapes, and the organization of proteins that mediate the formation of invadopodia in invasive cancer cells. These projects stem from my background in applied mathematics, but involve striking parallels with phase transitions in chemistry and reaction modeling and fluid dynamics in chemical engineering.

Teaching Interests:

Statistical Mechanics, Transport Phenomena, Reaction Engineering, Thermodynamics, generally any course on mathematical or numerical methods, data analysis, data visualization, or computer programming

476231 Computational and Experimental Investigation of Membrane Biomechanics

Abstract:

Research Interests:

Potentially toxic drugs and compounds from endogenous or exogenous sources constantly interact with the human body. Many of these compounds are small molecules that alter the physicochemical properties of physiological membranes, including changing permeability, fluidity, packing, electrostatic potential, rheology, and barrier function. In many instances, this may lead to the development of various pathophysiological states. The molecular mechanisms underlying these changes and their effects on proteins and other membrane structures are not entirely clear and depend strongly on the chemical identities of the offending compounds. My research interests focus on elucidating these molecular mechanisms through a combination of computational and experimental techniques and revealing their role in pathogenesis.

As a faculty member, I will build upon the experience and skills I have gained during my postdoctoral and graduate school training. In my lab, I look forward to applying a combination of: (i) coarse-grained molecular dynamics simulations, (ii) atomistic simulations, (iii) spectroscopic methods, and (iv) cell biology techniques. By utilizing these methods I aim to elucidate the mechanisms underlying the biophysical and biomechanical consequences of molecules involved in various pathophysiological states, leading to cardiovascular, metabolic, and immunological diseases. My research expertise, experience authoring scientific publications and writing grant proposals, as well as my experience mentoring both graduate and undergraduate students will enable me to establish a competitive research program that focuses on topics at the interface of Chemical Engineering and Biomedicine.

In my current position as a National Institutes of Health (NIH) Pulmonary T32 Postdoctoral Research Fellow at the University of Illinois College of Medicine, under the supervision of Dr. Irena Levitan, I am investigating the influence of bioactive oxidized species on model bilayer and cell membrane biomechanics. These oxidized species have been implicated in the development of atherosclerosis. In this role, by utilizing a combination of experimental and computational methods, I aim to uncover the molecular mechanisms underlying the paradoxical impact of pro-atherogenic compounds on endothelial cell stiffening and membrane biomechanics.

As a PhD student in the Chemical Engineering Department at the University of Illinois, Chicago, I undertook coarse-grained molecular dynamics simulations to probe the effects of small molecules on the structural and dynamic properties of phospholipid layers. Specifically, I investigated the membrane altering capability of toxins as applied to: (i) the differential cytotoxicity of anticoagulant species, (ii) the role played by endogenous metabolites in causing consciousness impairment, and (iii) the association of various drugs with lipid emulsions acting as detoxification agents. Using my simulation approach, I was able to obtain detailed molecular-level information on the absorption, transport, and partitioning of toxin molecules into model membranes, which provided useful insight and predictions.

Teaching Interests:

In addition to my research experience, I have extensive teaching experience, from elementary through college level. My teaching interests are Numerical and Computational Methods in Chemical Engineering, Biochemical Engineering, Material & Energy Balances, Transport Phenomena, Thermodynamics, and any other special topics related to my research background. Overall, my expertise in interdisciplinary science will serve as a framework for innovative study using both computational and experimental techniques.

476244 Machine Learning and Molecular Dynamics Map Conformational Landscape of μ-Opioid Receptor

Abstract:

Research Interests:

The μ Opioid Receptor (μOR) is a key mediator of pain and critical target for analgesic medications. The current generation of opiates – drugs that bind to μOR – are beset by dangerous side effects such as respiratory depression and addiction. In this work, molecular dynamics (MD) simulations are deployed to generate a high-resolution map of the conformational landscape of μOR. By seeding such simulations from the two X-Ray crystal structures of μOR, we not only identify key intermediates between these two structures, but also reveal new metastable states that are unforeseeable from crystallography alone. The application of Markov State Models with machine learning demonstrates tight coupling between the orientations of opiates in the binding pocket with downstream changes in the receptor's overall conformation. This clearer model of μOR dynamics may be useful to medicinal chemists in developing opiates with a more desirable therapeutic profile.

Teaching Interests:

466153 Smart City, Smart Energy, Smart Decision Making

Abstract:

Research Interests:

Integrated system thinking and analyses are more and more important for sustainable development by addressing challenges in natural resources, economic sectors, and human behaviours in a systematic manner. Process systems engineering has abundant modelling, simulation, and optimization tools to solve these complex problems through a comprehensive 'system of systems' approach. My research focuses on the demand side management of the constantly growing and developing populations, and sustainable provisioning of energy, water and material resources to satisfy needs, as well as the subsequent processing and management of wastes and pollutions. The nexus among energy-water-food is a representative example to adopt system engineering tools for overall economic, environmental and social benefits improvements.

I am developing a decision-making platform that combines comprehensive database, agent based simulation and resource technology network optimization, to deal with the challenges of energy, water, transportation and many interlinked systems in urban development. The open-source, data-driven, integrated systematic tool can gather social, environmental and economic data in a bottom-up approach to inform urban planning, investment and policymaking for city regions globally. Long-term socio-economic scenarios with temporal and spatial feedback loops are addressed to form smart city development strategies.

From macro-level down to detailed systems, sustainable energy and water treatment technologies are studied as fundamental to system analyses. I am researching smart energy management strategies for both generation and demand side with applications to residential and industrial users, as well as emerging technologies. For example, hybrid renewable energy systems (HRES) that consist of photovoltaic thermal, wind, fuel cells, and diesel generator as a backup resource as well as storage components turn to be a promising solution to overcome the drawbacks of single-resource systems. Compared with traditional operation of energy systems, the on-line process I am developing that combines sensors, optimizers and controllers to realize real-time prediction, optimization and demand-responsive scheme will improve system efficiency and take better advantage of all available resources.

Postdoctoral Project: "Water and Energy Systems in Sustainable City Development - ***Resilience.IO*** Resource Economic Human Ecosystem Modeling Platform"

Team under supervision of Professor Nilay Shah, Centre of Process Systems Engineering, Imperial College London.

PhD Dissertation: "***Modeling, Operation and Optimization of Energy Systems Integrating Demand Management***"

Co-advised by Professor Ahmet Palazoglu and Professor Nael H. El-Farra, Department of Chemical Engineering and Materials Science

Teaching Interests:

The philosophy of Knowing, Acting, and Being in education constitute my teaching career to encourage students' engagement, practice, and confidence, during instructing a course or even as simple as teaching a concept, method or software. I look upon a deeper connection between theory and practice when leading a class and creating impressive moments by challenging students and challenging myself in the meantime. Whether for teaching the classic mathematic theories and methods that remain the same, or engineering skills that advance with time, I would put the topics in context, evolve my teaching style and optimize class materials timely to always make the learning and teaching experience satisfactory and pleasant from the feedback of my students. It is the best reward to a teacher to witness how the "WOW" moments positively impact the students' career and life.

I had more than five-year teaching experience giving lectures with excellent ratings, for courses such as computational methods, process design and analysis, separations and unit operations. I was also the instructor for a series of software and tools training such as Aspen Plus and Matlab application. I supervised more than 10 undergraduate and master students' research projects. As a junior faculty member, I bear the mission to foster cooperation and ambition to bridge interdisciplinary researchers and build up a cooperative environment. I would organize series of webinars by inviting prestigious experts in the specific fields to discuss the trending topics by sharing knowledge and insights. With my broad exposure to the research environment of the U.S., UK, Asia, and Africa, I can take the advantage to unite researchers for global collaboration. It is also vital to use research models and results to get involved in government policy making internationally and engage the public to build collaborative communities that expand beyond the academic campus. I had some good practice and will transform the research output through more channels.

472947 Solution Approaches for Large Scale Multistage Stochastic Programs with Endogenous and Exogenous Uncertainty

Abstract:

Research Interests:

Multistage stochastic programs (MSSPs) with endogenous and exogenous uncertainty arise from problems where there is uncertainty in one or more of the problem parameters. In chemical engineering these problems can include the pharmaceutical clinical trial planning problem (Colvin and Maravelias, 2008), the oil field infrastructure planning problem (Gupta & Grossmann, 2011), and the process synthesis problem (Goel & Grossmann, 2006). Multistage stochastic programming formulations are limited computationally by problem size. The growth of the problem is related to the number of uncertain parameters and the number of realizations for each uncertain parameter. Linear increases in the number of uncertain parameters cause the number of scenarios in a MSSP to grow exponentially. Similarly, non-linear growth in problem size is also seen when the number of realizations for each uncertain parameter is increased. Increasing the number of scenarios in a MSSP increases the size of the decision variables and the number of constraints. As a result, most MSSPs considered in literature are limited to a few uncertain parameters with a few realizations for each parameter.

Often real-world size problems have many uncertain parameters with multiple realizations. Therefore, the MSSPs of these problems quickly become computationally intractable. One approach to addressing the computation complexity of MSSPs is to develop improvements for existing solution procedures. Sample Average Approximation (SAA) is a commonly applied approach to determine upper and lower bounds for a MSSPs. One challenge faced in generating the SAA form of the problem is the generation of the constraints. During my PhD I have been able to use my of graph theory to develop a systematic approach to generate

constraints for the SAA problem.

One drawback to the SAA approach is that the approach can only provide bounds on the true solution. In my graduate studies, I have worked on the development of a solution procedure to address the computation challenges presented by the MSSP. The solution procedure uses a branch and bound approach to gradually solve the MSSP formulation. The first step in the development of this procedure was to develop a way to quick generate a feasible solution. To do this I developed two heuristic approaches. The first heuristic approach decomposed the original problem into a series of two-stage stochastic problems (Christian & Cremaschi, 2015).The benefit of this approach is that the number of constraints in each of the problems solved was reduced, however the number of scenarios in the problem caused the approach to quickly become computationally intractable. The second approach decomposed the problem into a series of knapsack problems (Christian & Cremaschi, 2015). Unlike the two-stage decomposition the knapsack did not need the full set of scenarios and was able to generate solutions for very large MSSPs. To complement the feasible solution it was important to develop an approach for solving a relaxed version of the problem. Currently my research focuses on the development of a progressive hedging based approach to solve a relaxed version of the MSSP. The benefit of using progressive hedging is that approach is highly parallelizable.

Continued growth in the computational abilities of computers has opened the door for the use of large datasets in the development of chemical engineering models and optimization. Going forward, an increased understanding of how datasets can be incorporated into existing optimization frameworks will become increasingly important. In my future work I plan to be use my knowledge of statistics, programming, optimization, and graph theory to facilitate the advancement and development of new approaches to incorporating large data sets into chemical engineering applications.

Christian, B., & Cremaschi, S. (2015). Heuristic solution approaches to the pharmaceutical R&D pipeline management problem. *Computers and Chemical Engineering*, *74*, 34–47.

Colvin, M., & Maravelias, C. T. (2008). A stochastic programming approach for clinical trial planning in new drug development. *Computers & Chemical Engineering*, *32*, 2626–2642. http://doi.org/10.1016/j.compchemeng.2007.11.010

Goel, V., & Grossmann, I. E. (2006). A Class of Stochastic Programs with Decision Dependent Uncertainty. *Mathematical Programming*, *108*(2), 355–394.

Gupta, V., & Grossmann, I. E. (2011). Solution strategies for multistage stochastic programming with endogenous uncertainties. *Computers and Chemical Engineering*, *35*(11), 2235–2247. http://doi.org/10.1016/j.compchemeng.2010.11.013

Teaching Interests:

During my graduate studies I have had the opportunity to be a teaching assistant. As a teaching assisant I worked under a faculty advisor to develop assignments and lesson plans to engage students. My goal as a teacher will be to develop a program that promotes curiousity and a desire for life long learning. To do this I plan on incorporating real life examples and hands on learning to develop student passion for chemical engineering.

475545 Multi-Scale Process Systems Engineering

Abstract:

Multi-Scale Process Systems Engineering

Bruno A. Calfa

bruno.calfa@wisc.edu

In this poster session, I will discuss my Ph.D. work in the area of Process Systems Engineering (PSE) at Carnegie Mellon University, my Postdoc work at UW-Madison, the research topics and projects I am proposing to pursue, and my teaching interests and experience. For more information, please visit my website: http://bacalfa.com/.

Research Interests:

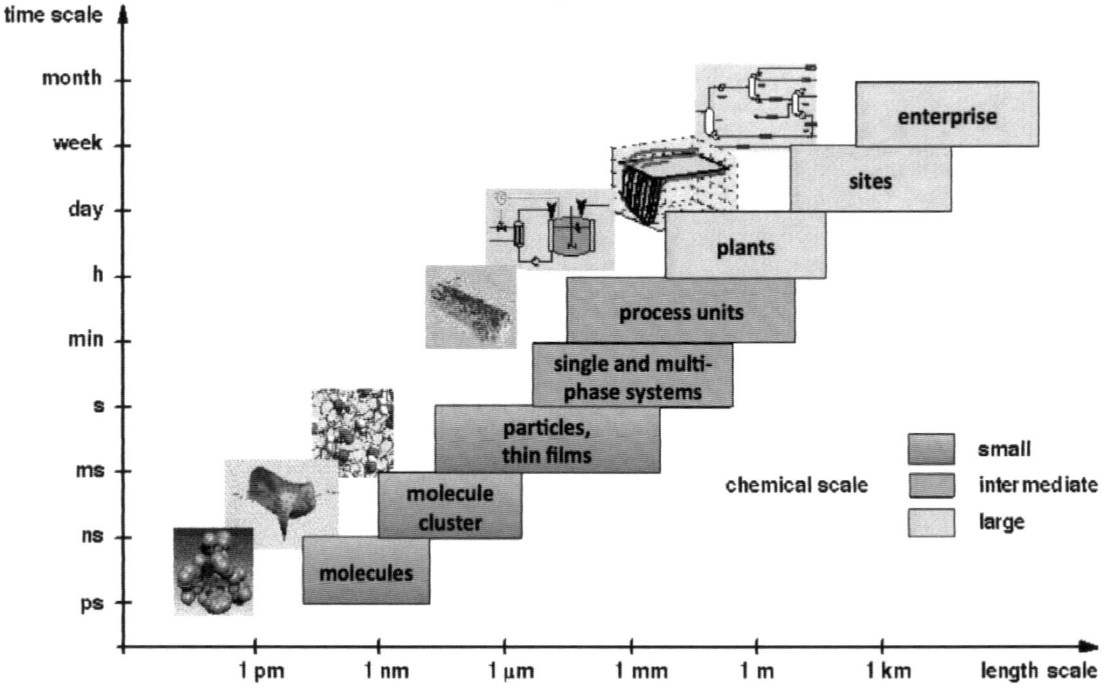

Figure 1. Multiple scales in Process Systems Engineering (PSE) research.

My Ph.D. research focused on the large scale top blocks in Figure 1. I **integrated** and **efficiently solved** planning and scheduling models for a network of batch plants, and developed **data-driven**approaches for modeling uncertainty in Enterprise-wide Optimization (EWO) problems [4-11]. I am interested in developing PSE methods (modeling, simulation, optimization, and control) to solve multi-scale problems of practical importance.

- **Large Scale:** novel *data-driven* models for uncertainty in sales and operations planning; multilevel optimization with contracts and pricing.
- **Intermediate Scale:** reduced-order modeling and optimization; conceptual/operational analysis of *sustainable* technologies, e.g., solar fuels [1].
- **Small Scale:** *property prediction* [3] and computer-aided *materials design* (e.g., crystals) via optimal inverse problems; microkinetics and optimal catalyst design.

I am the lead solver of a winning solution-a systematic crystal design framework-to the Materials Science and Engineering Data Challenge put forward by the Air Force Research Lab. I received the **2015 Ken Meyer Award for Excellence in Graduate Research**, which is given by the Department of Chemical Engineering at CMU in which the faculty base their selection of the student on research quality, productivity, recognition, and impact.

Teaching Interests:

I had extensive teaching experience as a TA at CMU. I prepared several teaching materials, and gave guest lectures and tutorials. I received the **2012 Mark Dennis Karl Outstanding Graduate Teaching Award**, which is given by the Department of Chemical Engineering at CMU to a student judged by the faculty to have done an outstanding job as a teaching assistant.

At UW-Madison, I have developed case studies that are incorporated into a *Chemical Business Simulator* (http://uwchembussim.che.wisc.edu/), which is an education tool that integrates engineering and business best practices [2]. I have also volunteered as a tutor at the Undergraduate Learning Center for freshman and sophomore Chemical Engineering courses.

References

[1] B. A. Calfa and C. T. Maravelias. Conceptual Analysis of Process Alternatives for Solar Thermochemical Methanol Production: The Role of Chemical Storage. In Preparation.

[2] B. A. Calfa and W. F. Banholzer. Web-Based Simulation Games for the Integration of Engineering and Business Best Practices. In Preparation.

[3] B. A. Calfa and J. R. Kitchin. Property Prediction of Crystalline Solids from Composition and Crystal Structure. In: AIChE Journal. (2016). DOI: http://dx.doi.org/10.1002/aic.15251.

[4] I. E. Grossmann, R. M. Apap, B. A. Calfa, P. Garcia-Herreros, and Q. Zhang. Recent Advances in Mathematical Programming Techniques for the Optimization of Process Systems under Uncertainty. In: Computers & Chemical Engineering. (2016). DOI: http://dx.doi.org/10.1016/j.compchemeng.2016.03.002.

[5] B. A. Calfa, A. Agarwal, S. J. Bury, J. M. Wassick, and I. E. Grossmann. Data-Driven Simulation and Optimization Approaches to Incorporate Production Variability in Sales and Operations Planning. In: Industrial & Engineering Chemistry Research. 54.29 (2015), pp. 7261-7272.

[6] B. A. Calfa and I. E. Grossmann. Optimal Procurement Contract Selection with Price Optimization under Uncertainty for Process Networks. In: Computers & Chemical Engineering. 82.1 (2015), pp. 330-343.

[7] B. A. Calfa, I. E. Grossmann, A. Agarwal, S. J. Bury, and J. M. Wassick. Data-Driven Individual and Joint

Chance-Constrained Optimization via Kernel Smoothing. In: Computers & Chemical Engineering. 78.1 (2015), pp. 51-69.

[8] B. A. Calfa. A Memory-Efficient Implementation of Multi-Period Two- and Multi-Stage Stochastic Programming Models. Carnegie Mellon University. Technical Report, 2014. URL: http://repository.cmu.edu/cheme/246/.

[9] B. A. Calfa, A. Agarwal, I. E. Grossmann, and J. M. Wassick. Data-Driven Multi-Stage Scenario Tree Generation via Statistical Property and Distribution Matching. In: Computers & Chemical Engineering. 68.1 (2014), pp. 7-23.

[10] I. E. Grossmann, B. A. Calfa, and P. Garcia-Herreros. Evolution of Concepts and Models for Quantifying Resiliency and Flexibility of Chemical Processes. In: Computers & Chemical Engineering. 70 (2014), pp. 22-34.

[11] B. A. Calfa, A. Agarwal, I. E. Grossmann, and J. M. Wassick. Hybrid Bilevel-Lagrangean Decomposition Scheme for the Integration of Planning and Scheduling of a Network of Batch Plants. In: Industrial & Engineering Chemistry Research. 52.5 (2013), pp. 2152-2167.

[12] B. A. Calfa and M. L. Torem. Bioreagents: their use in the removal of heavy metals from liquid streams by biosorption/bioflotation. (in Portuguese). In: Revista Escola de Minas. 60.3 (2007), pp. 537-542.

[13] B. A. Calfa and M. L. Torem. On the Fundamentals of Cr(III) Removal from Liquid Streams by a Bacterial Strain. In: Minerals Engineering. 21.1 (2007), pp. 48-54.

Formatting Issue - Best Available Paper/Graphic

475589 Petroleum Coke Morphology Mapping: A Mechanistic Approach Using Machine Learning

Abstract:

Petroleum Coke Morphology Mapping: A Mechanistic Approach Using Machine Learning

Amorim, P.H., University of Tulsa

2nd Year PhD Student

Research Interests:

Machine learning is an established and important technique for introducing intelligent systems to chemical engineering and augment the knowledge of a subject-matter. In the field of big data analytics, a full comprehension of data is easily achievable using this powerful tool, which can analyze a gamma of parameters simultaneously, and tell how they correlate with the outputs that one may be interested in. This feature, which is not found in other modeling techniques, gives an advantage to a machine learning user, since he will get the correct interpretation of variables that have complex association with each other. While machine learning can surely be applied to innumerous research fields (such as process safety, process optimization, or any field that contains big data processing), the application within this particular research, which is being advised by Dr. Keith Wisecarver, is to classify and map the morphology of petroleum coke, according to different chemical and physical parameters. Moreover, the mechanistic approach will give an explanation on why each type of coke morphology is formed. It does so by analyzing suspension stability and rheological properties of the feedstock (which can be considered a suspension) and observing and extracting information out of the liquid crystal carbonaceous mesophase generated by thermolysis and polymerization reactions. Further modeling techniques involving large eddy simulations will be used, as an attempt to simulate the different coke morphologies in flow conditions.

Research Experience:

My research background is a blend of diverse experimental methods and computational work. Initially, I focused in microemulsion systems applied to environmental remediation, and process design in the oil & gas industry, where I worked in collaboration with Petrobras and the Brazilian Association of Oil, Gas, and Renewable Fuels (ANP). Currently, I am focused on the structure and rheology of complex fluids, big data analytics using machine learning, with applications in process design and process monitoring. My current research group, the University of Tulsa Delayed Coking Project (TUDCP), gives me the opportunity to conduct research funded by several national and international oil companies.

Teaching Experience:

As an undergraduate student, I was the teaching assistant of Unit Operation Involving Solids and Fluids at the Federal University of Rio Grande do Norte for one and a half years, during which I had the opportunity to be the lecturer and conduct process design projects. Moreover, I created and was the teacher of an English class for low income students for 6 months, also at the same university.

Teaching Interests:

My teaching interests are in Fluid Mechanics, Process Design (HYSYS, Comsol, Ansys), Computational Fluid Dynamics Applied to Chemical Engineering, Heat Transfer, Numerical Methods for Chemical Engineering, and Process Analysis and Optimization Using Machine Learning.

Future Direction:

As a faculty member, I intend to continue applying machine learning in different research projects which have complex association of variables and also conduct research associated with industry. More specifically, these would be conducted in two main areas: Process analysis & optimization, and Process safety. The former will include two tasks, including analyzing variables extracted from energy applications involving solar thermal energy and biomass and applying algorithms, such as Neural Networks, Support Vector Machines, and Naive Bayes to get a better understanding of the association of operating and chemical parameters; and to optimize the design and performance of the process at the same time. The latter will involve troubleshooting an already existing process using the same techniques previously listed, as a way to improve the process safety and prevent hazardous situations in industrial plants, a topic that can attract a variety of fields. Machine learning has the potential to be one of the main tools for data processing, in a world where computers and chemical processes are getting increasingly interconnected.

Additionally, I will conduct studies of turbulent flow using large eddy simulations simulation models, especially with applications involving simulations of the human body systems (respiratory treat simulations to study allergens deposition, drug delivery simulations), particle agglomeration and deposition in suspensions, and even studies of air and water pollution. These studies will establish research on the fields of medicine, fluid suspensions, and environment. Besides these topics, research involving oil & gas industry will still be in my list of interests. Therefore, because of the different research topics, I see myself establishing an interdisciplinary research network with other faculty members, in a way to increase the scientific collaboration among a variety of fields.

Publications:

Amorim ValenÃ§a, P.H., Waturuocha, A., Wisecarver, K. â€œ2D Axisymmetric Temperature Profile Modelling of a Delayed Coking Drum During Pre-Run Warm Up.â€ Proceedings from Comsol Conference Boston, 2015.

475776 Simulation and Optimization of Chemical Processes for CO2 Sequestration and New Clean Energy: Cyclic Adsorption Process, Membrane, and Direct Methanol Fuel Cell

Abstract:

Research Interests:

Because the goal of my research is to solve environmental and energy issues, the focus is the simulation and optimization of CO_2 sequestration processes and a fuel cell system to prepare against global warming. As the methods to reduce the CO_2 emission, gas separation processes such as cyclic adsorption processes (CAPs) and membrane processes are considered in this study.

At first, this study presents the optimization approach (Ko et al., 2005) and the developed simulation model (Ko, 2016) of CAPs to sequester CO_2. The strength of the developed CAP model is to be able to accurately calculate an interstitial gas velocity within adsorption beds resulting in successful scale up designs from a laboratory scale (0.5Nm^3/h) to a large commercial scale (5,000Nm^3/h).

Then, a dynamic membrane simulation model for the removal of CO_2 is introduced of which mass and energy balances are derived. Optimizations are performed to conveniently perceive the scale up design conditions up to commercial scale (500Nm^3/h) by solving mixed integer nonlinear programming (MINLP) problems.

Finally, this work also introduces a non-isothermal dynamic optimization model of DMFC which can generate clean energy (Ko et al., 2008). The optimal operating strategy of DMFC systems was attained through the dynamic simulation and optimization.

In summary, my research includes 1) the development of the scale up design technologies based on the mathematical approaches and 2) dynamic simulation and optimization works for various chemical processes such as the gas treating and energy systems

My research will focus on the optimization technology for the process design and operation. The target processes for the optimization studies will mainly be 1) the gas separation/purification processes to sequester and capture CO_2 and 2) energy systems such as hydrogen plants, fuel cell system, or lithium ion batteries.

Teaching Interests:

What students learn should be practical as well as academic. In other words, I would like to say "to be practical" is same as "to be academic" and the courses of all the chemical engineering area must be helpful for the students to apply to industries. Thus, I would like to teach the core chemical engineering courses including chemical engineering thermodynamics, fluid dynamics, heat transfer, mass transfer, reaction engineering, and process control for the undergraduate level. At the graduate level, I would like to select specialized fields of chemical engineering processes such as process dynamic simulation, gas separation technology, and energy systems. These specialized courses are based on my industrial and academic experiences. The results of my research and teaching must be able to be applied to real industries.

475836 Simulation of the Oil-Treatment Process in the Oil Sands Plant

Abstract:

Research Interests: Process Simulation

Teaching Interests: Process Simulation

The purpose of this paper is to define criteria to be used in the engineering design of an oil sands plant equipped with the steam assisted gravity drainage process. In this effort, the oil treatment process of the oil sands plant on a pilot scale was focused for detailed investigation. The thermodynamic properties of the process fluid, which is mainly composed of bitumen and water, were estimated with the CPA model. The commercial aspen HYSYS was implemented in the analysis throughout this work along with the provided input data and some necessary assumptions. From the simulation results, the heat and mass balances of the process for a 300 BPD oil sands plant were established in order to define standard data for its modular design. In particular, the designing basis for equipment size, heat transfer areas, capital cost and operation cost was extensively discussed.

475841 Effects of the Mixed Refrigerant Composition on the Peformance of the Rankine Cycle Driven By LNG Cold Energy

Abstract:

Research Interests: Process Simulation

Teaching Interests: Process Simulation

The power generation system empowered by the LNG cold energy, which evolves in a large amount during the vaporization process of the liquefied natural gas, was designed in favor of the Rankine cycle with a mixed refrigerant as the working fluid. In this study it is intended to identify the limits of the working fluid composition in respect of equipment safety in the Rankine cycle-type power generation system driven by the cold energy. The thermodynamic properties of the working fluid, which is a hydrocarbon mixture, were calculated with the Peng-Robinson equation. In the steady state simulation of the power generation system by using the commercial simulator Aspen HYSYS, the feed conditions of LNG from Test Bed Train No.1 along with some necessary assumptions were incorporated. The results indicated that deterioration of the mechanical performance of the equipment as well as its safety would be brought about if contents of C_2H_6 and C_3H_8 in the mixture become, respectively, too high or too low.

476056 Process System Engineering for Advanced Modular Continuous Pharmaceutical Manufacturing Platform

Abstract:

Research Interests: My main research area is process system engineering with focus on continuous pharmaceutical manufacturing. I have contributed significantly in the field of process monitoring and control, process modelling and simulation, optimization, PAT, QbD, control hardware/software integration, new methodology and software tools development.

Process system engineering (PSE) is a very advanced field and has been very successful in solving complex problems of chemical based product manufacturing. However, the particulate systems are still poorly understood and there is different level of complexities involved, therefore new techniques and tools need to be developed for pharmaceutical manufacturing through extensive fundamental and applied research.

Currently, major pharmaceutical industries are shifting toward continuous manufacturing technology in which PSE can play a very significant role. Furthermore, there is a need of advanced modular continuous manufacturing platform technology for fast and robust development of solid oral products. Master formulations, process models, standardized sensing and control architecture, superior automation and novel methods and tools need to be developed to enable platform technology. Author's main research area is pharmaceutical process system engineering. This area of research and education are highly encouraged and supported by pharmaceutical companies, academic institutions, regulatory authority (FDA) and national science foundation (NSF). For example, national science foundation of USA has established a 100 million dollar research center for structured organic particulate systems (C-SOPS) at Rutgers University New Jersey. Furthermore, more than 40 companies are supporting this center economically and providing scientific inputs so that the research outcomes can be directly applied to the companies.

In this work, a scientific foundation for the optimal design, optimization and control of the modular continuous process involving structured organic particulate systems has been developed. The mathematical models for continuous manufacturing of active pharmaceutical ingredients and final product tablet, has been developed for virtual experimentation and applied for plant design, optimization and control. Advanced modelling techniques such as PBM, DEM and CFD have been applied for pharmaceutical process modeling. The developed continuous tablet manufacturing plant and process model has been adapted by several pharmaceutical companies. A flexible monitoring and control system incorporating different PAT techniques and tools and control strategies for continuous pharmaceutical manufacturing process has been designed and implemented to Rutgers plant. Different control strategies such as PID, PID coupled with dead time compensator, and MPC has been compared and evaluated practically for pharmaceutical process control in terms of set point tracking and disturbances rejection ability, robustness, and stability/instability. A new hybrid MPC-PID control strategy has been developed and implemented into the pharmaceutical tablet manufacturing plant and its performance has been found better than the other control strategies. For the first time, the closed-loop operation of the tablet manufacturing plant using NIR sensor has been demonstrated, which is a significant advancement in pharmaceutical manufacturing as endorsed by the Food and Drug Administration (FDA) and other regulatory authorities as part of the Quality by Design (QbD) paradigm. For the first time NIR has been applied for real time monitoring of powder bulk density. Feedforward control system has been integrated with feedback control system to take proactive mitigation actions on raw materials and process variability. A systematic generic framework including the methods and tools through which the cost and profit can be estimated and the Dynamic Real Time Optimization (DRTO) can be integrated with hybrid control system has been also developed. The integrated DRTO provides the optimal operational set points for the plant control system in real time optimizing an objective function. DRTO integrated with hybrid control strategy ensures the maximum possible profit irrespective of the market demand fluctuations.

Presently, I am working as an assistant research professor at Department of Chemical and Biochemical Engineering of Rutgers University and currently managing and leading ten projects funded by NSF, industries and regulator with total grants exceeding 18 million USD. I was involved in proposal writing of these projects and I am PI/Co-PI of some of these projects. Prior to this, I worked in a large scale European pharmaceutical project (F^3 Factory project), which involves 25 partners from industry and Academia; where my responsibility was to manage and lead the DTU's parts of the F^3 Factory project and also to coordinate with the other partners associated with it. My doctoral research work, done at Department of Chemical and Biochemical Engineering, Technical University of Denmark, has been awarded for ***prestigious "EFCE Excellence Award 2010", given in Recognition of an Outstanding PhD Thesis, from European Federation of Chemical Engineering***. I have exemplary academic records, previous teaching experiences, and an established trail of valued research accreditations'. I have published several peer reviewed scientific international Journal articles, popular reputed magazine articles, book chapters and presented in more than 80 international conferences. I am actively serving as a conference session chair, guest editor of Journal, manuscripts reviewer and bachelor, master and PhD projects supervisor.

The objective of this presentation is two-fold. First to give the overview of my research and teaching experience, and second to demonstrate the simulation, design, optimization, monitoring and control of advanced modular continuous pharmaceutical manufacturing plant.

Teaching Interests:

My teaching experience that I acquired serving as a teaching assistant, lecturer, and assistant research professor makes me confidant to develop and teach any courses of chemical and biochemical engineering. The industrial invited crash courses that I have given provide me opportunity to extend my teaching skills to cover more applied courses and to interact with highly experienced scientist and operators from manufacturing companies. I demonstrated my skills to instruct hands-on exercise through several workshops that I conducted. Several training courses that I conducted for industrial and academic participants make me instrumented with additional capability of building highly skilled workforce. In my department, I am a lead organizer of postdoctoral training program with the aim of building highly skilled workforce for industry and academia. I developed two online course modules where I learn how to develop and give online courses. My unique research area- applied process system engineering- provides me dual ability to teach theoretical as well as practical courses. I am open to teach any courses that your university will assign me. I am especially interested to teach courses related to process control, process modeling and simulation, process monitoring, chemical reaction engineering, process and tool integration and pharmaceuticals. If needed then I can develop new courses based on my research outcomes.

Grants:

Extensive experience in funding proposals writing. Wrote several grants proposals for national science foundation (NSF), FDA, and pharmaceutical industries. 11 projects with total grant more than 18 million USD got funded in which I have contributed significantly.

Publications:

Scientific articles (43); Book chapters (4); International conference presentations (80); Invited industrial/academic lectures (8); Workshops (5); Thesis (3); Reports (>100).

Activities in AIChE 2015 conference:

Session co-chair

1. 10B08 Process Control Applications I: 32339
2. 10B12 Process Control Applications II: 34890
3. 10C03 Dynamic Simulation and Optimization: 32320
4. 10D01 Advances in Computational Methods and Numerical Analysis: 32306
5. 10E02 Big Data Analytics in Chemical Engineering: 32315

Presenter (Oral presentations)

1. **Singh, R.,** Muzzio, F. J, Ramachandran, R. Ierapetritou, M. **(2016)**. Advanced flexible control system implementation into direct compaction continuous pharmaceutical manufacturing pilot-plant. Oral presentation at AIChE annual meeting, San Francisco, CA, USA, 13 – 18 November. ***Session: Product Quality Attribute Monitoring and Control in Drug Product Manufacturing. Schedule: Thursday, November 17, 3: 59 PM.***
2. **Singh, R.,** Cao, H., Mushnoori, S., Higgins, B., Kolipara, C., Fermier, A., Hausner, D., Jha, S., Ierapetritou, M., Ramachandran, R. **(2016)**. Data management and integration for continuous pharmaceutical manufacturing. Oral presentation at AIChE annual meeting, San Francisco, CA, USA, 13 – 18 November. ***Session: Industrial Applications of Data Analysis, Information Management, and Intelligent Systems. Schedule: Wednesday, November 16, 9.46 AM.***
3. **Singh, R.,** Pereira, G. C., Soni, N., Román-Ospino, A. D., Ierapetritou, M., Ramachandran, R. **(2016)**. Feedforward control of continuous pharmaceutical manufacturing process. Oral presentation at AIChE annual meeting, San Francisco, CA, USA, 13 – 18 November. ***Session: Quality By Design in Drug Product Formulation, Design, & Process Development II. Schedule: Friday, November 18, 8.30 AM.***

Co-author

1. Wang, Z., Escotet-Espinoza, M. S., **Singh, R.,** Muzzio, F. J., Ierapetritou, M. G. **(2016)**. Surrogate-Based Optimization Methodology for Pharmaceutical Tablet Manufacturing Processes. Oral presentation at AIChE annual meeting, San Francisco, CA, USA, 13 – 18 November. ***Session: Design and Operations Under Uncertainty I. Schedule: Thursday, November 17, 8:30 AM.***
2. Wang, Z., Escotet-Espinoza, M. S., **Singh, R.,** Muzzio, F. J., Ierapetritou, M. G. (2016). Feasibility Analysis of Flowsheet Models in Continuous Pharmaceutical Manufacturing Processes Considering the Effects of Noise. Poster presentation at AIChE annual meeting, San Francisco, CA, USA, 13 – 18 November. ***Poster Session: Pharmaceutical. Schedule: Monday, November 14, 3:15 - 5:45 PM.***

Selected publications:

1. International scientific articles:

Year 2016

1. Wu, S., Panikar, S. S., **Singh; R.**, Zhang, J., Donepudi, A., Glasser, B., Ramachandran, R. **(2016)**. Systematic framework to monitor mulling processes using Near Infrared spectroscopy. Advanced Powder Technology. doi:10.1016/j.apt.2016.03.022.
2. **Singh, R.,** Muzzio, F., Ierapetritou, M., Ramachandran, R. **(2016)**. Systematic framework for design and implementation of plant-wide multilayer, sensing and control architecture into continuous pharmaceutical manufacturing plant. Computer Aided Chemical Engineering. Accepted.
3. Román-Ospino, A. D., **Singh, R.**, Ierapetritou, M., Ramachandran, R. Méndez, R., Ortega, C., Muzzio, F. J., Romañach, R. J., **(2016)**. Near Infrared Spectroscopic Calibration Models For Real Time Monitoring Of Powder Density. International Journal of Pharmaceutics. Accepted.

Year 2015

1. **Singh, R.,** Román-Ospino, A. D., Romañach, R. J., Ierapetritou, M., Ramachandran, R. **(2015)**. Real time monitoring of powder blend bulk density for coupled feed-forward/feed-back control of a continuous direct compaction tablet manufacturing process. International Journal of Pharmaceutics, 495, 612-625.
2. **Singh, R.,** Muzzio, F., Ierapetritou, M., Ramachandran, R. **(2015)**. A combined feed-forward/feed-back control system for a QbD based continuous tablet manufacturing process. PROCESSES Journal, 3, 339-356.
3. **Singh, R.,** Sen, M., Ierapetritou, M., Ramachandran, R. **(2015)**. Integrated moving horizon based dynamic real time optimization and hybrid MPC-PID control of a direct compaction continuous tablet manufacturing process. Journal of Pharmaceutical Innovation, 10 (3), 233-253.
4. Simon, L. L., .. **Singh, R.,** et al. **(2015)**. Assessment of Recent Process Analytical Technology (PAT) Trends: A Multi-author Review. Organic Process Research & Development, 19, 3-62.
5. **Singh, R.,**Zhang, J., Ierapetritou, M., Ramachandran, R. (2015). Designing a novel continuous manufacturing plant with superior monitoring and control. European Pharmaceutical Review, 20(6), 37-41.
6. Sen, M., **Singh, R.,** Ramachandran, R. **(2015)**. Model Manufacturing. The Medicine Maker, February, 05, 42-45.
7. Escotet-Espinoza, M. S., **Singh, R.,** Sen, M., O'Connor, T., Lee, S., Chatterjee, S., Ramachandran, R., Ierapetritou, M., Muzzio, F. **(2015)**. Flowsheet Models Modernize Pharmaceutical Manufacturing Design and Risk Assessment, Pharmaceutical Technology 39 (4), 34-42.
8. **Singh, R.,** Muzzio, F., Ierapetritou, M., Ramachandran, R. **(2015)**. Plant-wide control of a continuous tablet manufacturing for Quality-by-Design based pharmaceutical manufacturing. Computer Aided Chemical Engineering, 37, 2183 - 2188.
9. Karry, K. M., **Singh, R.,**Muzzio, F. J. (2015). Fit-for-Purpose Miniature NIR Spectroscopy for Solid Dosage Continuous Manufacturing. American Pharmaceutical Review, 18(4), 64 - 67.
10. **Singh, R.,**Ierapetritou, M., Ramachandran, R. (2015). The scopes of PAT in real-time advanced control of tablet quality. European Pharmaceutical Review 20(2), 76-80.

Year 2014

1. **Singh, R.,** Sahay, A., Karry, K. M., Muzzio, F., Ierapetritou, M., Ramachandran, R. (**2014**). Implementation of a hybrid MPC-PID control strategy using PAT tools into a direct compaction continuous pharmaceutical tablet manufacturing pilot-plant. International Journal of Pharmaceutics, 473, 38–54.
2. **Singh, R.,** Sahay, A., Fernando Muzzio, Ierapetritou, M., Ramachandran, R. (**2014**). Systematic framework for onsite design and implementation of the control system in continuous tablet manufacturing process. Computers & Chemical Engineering Journal, 66, 186-200.
3. **Singh, R.,** Barrasso, D., Chaudhury, A., Maitraye Sen, Ierapetritou, M., Ramachandran, R. (**2014**). Closed-Loop Feedback Control of a Continuous Pharmaceutical Tablet Manufacturing Process via Wet Granulation. Journal of Pharmaceutical Innovation, 9, 16-37.
4. Sen, M., Barrasso, D., **Singh, R.**, Ramachandran, R. (**2014**). A Multi-Scale Hybrid CFD-DEM-PBM

Description of a Fluid-Bed Granulation Process. Processes Journal, 2(1), 89-111.
5. Sen, M., **Singh, R.,** Ramachandran, R. (**2014**). Simulation based design of an efficient control system for the continuous purification and processing of active pharmaceutical ingredients. Journal of Pharmaceutical Innovation, 9, 65-81.
6. Sen, M., **Singh, R.,** Ramachandran, R. **(2014)**. A hybrid MPC-PID control system design for the continuous purification and processing of active pharmaceutical ingredients. PROCESSES Journal, 2, 392-418; doi:10.3390/pr2020392.
7. Sen, M., Chaudhury, A., **Singh, R.,** Ramachandran, R. **(2014)**. Two-dimensional population balance model development and validation of pharmaceutical crystallization processes. American Journal of Modern Chemical Engineering, 1, 13-29.

Year 2013

1. **Singh, R.,** Ierapetritou, M., Ramachandran, R. (**2013**). System-wide hybrid model predictive control of a continuous pharmaceutical tablet manufacturing process via direct compaction. European Journal of Pharmaceutics and Biopharmaceutics, 85(3), Part B, 1164-1182.
2. **Singh, R.,** Godfrey, A., Gregertsen, B., Muller, F., Gernaey, K. V., Gani, R., Woodley, J. M. (**2013**). Systematic substrate adoption methodology (SAM) for future flexible, generic pharmaceutical production processes. Computers & Chemical Engineering Journal, 58, 344 - 368.
3. Sen, M., Dubey, A., **Singh, R.,** Ramachandran, R. (**2013**). Mathematical Development and Comparison of a Hybrid PBM-DEM description of a Continuous Powder Mixing Process. Journal of Powder Technology, http://dx.doi.org/10.1155/2013/843784.
4. Sen, M., Chaudhury, A., **Singh, R.,** John, J., Ramachandran, R. (**2013**). Multi-scale flowsheet simulation of an integrated continuous purification–downstream pharmaceutical manufacturing process. International Journal of Pharmaceutics, 445 (1-2), 29-38.
5. Sen, M., Rogers, A., **Singh, R.,** Chaudhury, A., John, J., Ierapetritou, M., Ramachandran, R. (**2013**). Flowsheet optimization of an integrated continuous purification-processing pharmaceutical manufacturing operation. Chemical Engineering Science, 102, 56 – 66.
6. **Singh, R.,** Ierapetritou, M., Ramachandran, R (**2013**). Hybrid advanced control of a flexible multipurpose continuous pharmaceutical tablet manufacturing process via direct compaction. Computer Aided Chemical Engineering, 32, 757-762.
7. Muzzio, F., **Singh, R.,** Chaudhury, A., Rogers, A., Ramachandran, R. Ierapetritou, M. (**2013**). Model-predictive design, control and optimization of pharmaceutical process. Pharmaceutical Technology magazine, 37(6), 40-41, 77.
8. **Singh, R.,** Sahay, A., Oka, S., Liu, X., Ramachandran, R., Ierapetritou, M., Muzzio, F. (**2013**). Online monitoring, advanced control and operation of robust continuous pharmaceutical tablet manufacturing process. BioPharma magazine Asia, 2(5), 18-25.

Year 2012

1. **Singh, R.,** Ierapetritou, M., Ramachandran, R. (**2012**). An engineering study on the enhanced control and operation of continuous manufacturing of pharmaceutical tablets via roller compaction. International Journal of Pharmaceutics, 438 (1-2), 307-326.
2. Sen, M., **Singh, R.,** Vanarase, A., John, J., Ramachandran, R. (**2012**). Multi-dimensional population balance modeling and experimental validation of continuous powder mixing processes. Chemical Engineering Science, 18, 349-360.
3. **Singh, R.**, Raquel Rozada-Sanchez, R., Dean, W., Perkins, J., Muller, F., Godfrey, A., Gernaey, K. V., Gani, R., Woodley, J. M. (**2012**). A generic process template for continuous pharmaceutical production. Computer Aided Chemical Engineering, 31, 715-719.
4. **Singh, R.,** Boukouvala, F., Jayjock, E., Ramachandran, R. Ierapetritou, M., Muzzio, F. (**2012**). Flexible Multipurpose Continuous Processing: integration of process flow modeling for continuous processing of pharmaceutical solid dosage forms. PharmPro Magazine, Pharmaceutical Processing, 27(6), 22-25.
5. **Singh, R.,** Boukouvala, F., Jayjock, E., Ramachandran, R. Ierapetritou, M., Muzzio, F. (**2012**). Flexible Multipurpose Continuous Processing of Pharmaceutical Tablet Manufacturing Process. GMP news, European Compliance Academic (ECE), http://www.gmp-compliance.org/ecanl_503_0_news_3268_7248_n.html.

Years 2011-2008

1. **Singh, R.,** Gernaey, K. V., Gani, R. **(2010)**. ICAS-PAT: A Software for Design, Analysis & Validation of PAT Systems. Computers & Chemical Engineering Journal, 34(7), 1108-1136.
2. **Singh, R**., Gernaey, K. V., Gani, R. **(2010)**. An ontological knowledge based system for selection of process monitoring and analysis tools. Computers & Chemical Engineering Journal, 34(7), 1137-1154.
3. **Singh, R.,** Gernaey, K. V., Gani, R. **(2009)**. Model-based computer-aided framework for design of process monitoring and analysis systems. Computers & Chemical Engineering Journal, 33(1), 22-42.
4. Samad, N. A. F. A., **Singh, R.**, Sin, G., Gernaey, K. V., Gani, R. **(2011)**. A generic multi-dimensional model-based system for batch cooling crystallization processes. Computers & Chemical Engineering Journal, 35(5), 828-843.
5. Samad, N. A. F. A., **Singh, R.,** Sin, G., Gernaey, K. V., Gani, R. **(2011)**. Systematic Procedure for Generating Operational Policies to Achieve Target Crystal Size Distribution (CSD) in Batch Cooling Crystallization. IEEE, 1-6. Print ISBN: 978-1-4577-0003-3. DOI: 10.1109/ICMSAO.2011.5775588.
6. Samad, N. A. F. A., **Singh, R.,** Sin, G., Gernaey, K. V., Gani, R. **(2011)**. Integration of Generic Multi-dimensional Model and Operational Policies for Batch Cooling Crystallization. Computer Aided Chemical Engineering, 29, 86-90.
7. Samad, N. A. F. A., **Singh, R.**, Sin, G., Gernaey, K. V., Gani, R. **(2010)**. Control of process operation and monitoring of product qualities through generic model-based in batch cooling crystallization. Computer Aided Chemical Engineering, 28, 613-618.
8. **Singh, R.**, Rozada-Sanchez, R., Wrate, T., Muller, F., Gernaey, K. V., Gani, R., Woodley, J. M. **(2010)**. A retrofit strategy to achieve "Fast, Flexible, Future (F^3)" pharmaceutical production processes. Computer Aided Chemical Engineering, 29, 291-295.
9. **Singh, R.,** Gernaey, K. V., Gani, R. **(2009)**. A software tool for design of process monitoring and analysis systems. Computer Aided Chemical Engineering, 26, 321-326.
10. **Singh, R.,** Gernaey, K. V., Gani, R. **(2008)**. Off-line design of PAT systems for on-line applications. Computer Aided Chemical Engineering, 25, 423-428.

2. Book Chapters

1. **Singh, R.,** Figueroa, CV, Sahay, A., Karry, KM, Fernando Muzzio, F., Ierapetritou, M., Ramachandran, R. **(2014)**. Chapter 7: **Advanced Control of continuous pharmaceutical tablet manufacturing processes**. Book title: Process Simulation and Data Modeling in Solid Oral Drug Development and Manufacture. Publisher: Humana Press, ISBN: 978-1-4939-2995-5, 191 – 223.
2. Ierapetritou, M., Escotet-Espinoza, M. S., **Singh, R. (2016)**. **Process Simulation and control for continuous pharmaceutical manufacturing of solid drug products**. A chapter of book "Continuous manufacturing of pharmaceuticals" edited by Peter Kleinebudde, Johannes Khinast and Jukka Rantanen. Publisher: Wiley-VCH. Accepted.
3. Oka, S, Escotet-Espinoza, M. S., **Singh, R.,** Scicolone, J., Hausner, D., Ierapetritou, M., Muzzio, F. **(2016)**. **Design of an integrated continuous manufacturing system**. A chapter of book "Continuous manufacturing of pharmaceuticals" edited by Peter Kleinebudde, Johannes Khinast and Jukka Rantanen. Publisher: Wiley-VCH. Accepted.
4. **A Book chapter in "Product and Process Modelling: A case study approach":**

- **Singh, R. (2011). Fermentation process modeling**. Chapter 12.2 of book "Product and Process Modelling: A case study approach" edited by I Cameron & R. Gani. Publisher: Elsevier, pp 380-396.
- **Singh, R. (2011). Milling process model.** Chapter 12.4 of book "Product and Process Modelling: A case study approach" edited by I Cameron & R. Gani. Publisher: Elsevier, pp 407-413.
- **Singh, R. (2011). Granulation process model.** Chapter 12.5 of book "Product and Process Modelling: A case study approach" edited by I Cameron & R. Gani. Publisher: Elsevier, pp 413-422.
- **Singh, R. (2011). Pharmaceutical tablet pressing process model.** Chapter 12.6 of book "Product and Process Modelling: A case study approach" edited by I Cameron & R. Gani. Publisher: Elsevier, pp 422-430.
- **Singh, R. (2011). Milk pasteurization process modeling**. Chapter 12.3 of book "Product and Process Modelling: A case study approach" edited by I Cameron & R. Gani. Publisher: Elsevier, pp 396-406.

476073 Optimization-Based Quantification of Performance Limits for Process Networks

Abstract:
Research Interests: Process systems engineering, global optimization, energy systems, process synthesis and design, modeling and simulation, computational fluid dynamics.

Teaching Interests: Transport phenomena, mass transfer, heat transfer, thermodynamics, chemical process computer-aided design and analysis, numerical and mathematical methods, process dynamics and control.

The goal of any synthesis task is to design a feasible network that can deliver desired products from a set of inlet raw material using one or more chemical processes. Computational tools are usually employed to perform this task, aiming to find an optimal solution to a given design problem. Nevertheless, existing tools cannot ensure that the identified solution is a global rather than local optimum, keeping open the possibility of further performance improvements.

In order to find a global optimal solution for the process network synthesis problem, a methodology that employs the Infinite DimEnsionAl State-Space (IDEAS) conceptual framework is presented in this work. The IDEAS framework is based on the techniques of optimization by vector spaces, convex analysis and chemical engineering fundamentals, which guarantee global optimality.

This work presents several examples of globally optimal synthesis of process networks involving reactive/non-reactive distillation of azeotropic mixtures.

476079 Metabolic Modeling for Improved Bioprocess Efficiency

Abstract:

Research Interests:

Microbial fermentation plays a critical role in efforts to enable sustainable manufacturing of commodity chemicals, and the economic feasibility of the integrated biorefinery concept hinges sharply on the productivity, yield, and titers that can be achieved by metabolic hosts. Flux balance modeling has emerged as a fundamental tool in pushing biological transformations towards profitability, as it enables model-based design of gene knockout and overexpression efforts. My interests center on the unbiased search of intervention strategies for models of central carbon metabolism, together with the consideration of the dynamic systems which govern metabolic flows within the network.

Postdoctoral Projects: Metabolic modeling of industrially relevant organisms for improved yield and productivity

Under supervision of Yannick Bomble and Gregg Beckham, Biosciences Center, National Renewable Energy Laboratory, Golden CO

PhD Dissertation: ÒMathematical Approaches to Understanding Mammalian Circadian RhythmsÓ

Under supervision of Francis J. Doyle III, Department of Chemical Engineering, University of California, Santa Barbara.

Future Direction:

As a faculty I would like to work towards standardizing and accelerating model-based design of metabolic interventions, particularly in non-model organisms. Industrial bioprocesses predominantly leverage specialized

organisms to achieve the desired operating conditions, and while these microbial platforms offer unique advantages over traditional hosts, their relative obscurity slows the design/build/test cycle for determining optimum metabolic manipulations. To speed this process I aim to develop standardized workflows and new computational techniques that exploit the overlap in metabolic functionality between classes of microbes. By extracting general trends in gene regulation and enzyme kinetics and combining this knowledge with organism-specific results, searches for the optimum overexpression or gene knockout could explicitly consider the underlying uncertainty in the gene regulatory network.

In addition to metabolic engineering, my interests extend to bio- and cheminformatics, where machine-learning algorithms can be used to extend patterns and trends to new, unmeasured points. I hope to apply these techniques throughout the pipeline from feedstocks to commodity chemicals, as well as in applications in synthetic biology and pharmaceuticals.

Teaching Interests:

My primary teaching interests lie in computational and analytical methods in dynamic systems, statistics, and process control. As a teaching assistant I have given several guest lectures in introductory statistics, and have instructed undergraduate and graduate students on how to perform computational analyses in numerous programming languages.

Proposal Experience: Mitsubishi Chemical Fellowship recipient

NSF GRFP Honorable Mention

Selected Publications: 312 Citations, h-index 4.

Abel, J.H., Meeker, K., Granados-Fuentes, D., **St. John, P.C.**, Wang, T.J., Bales, B.B., Doyle F.J. III, Herzog, E.D., and L.R. Petzold. Functional network inference of the suprachiasmatic nucleus (2016) *PNAS*, 113 (16) pp. 4512-4517

St. John, P.C. and F.J. Doyle III. Quantifying stochastic noise in cultured circadian reporter cells (2015), *PLoS Computational Biology* 11(11): e1004451.

St. John, P.C., Taylor, S.R., Abel, J.H., and F.J. Doyle III. Amplitude metrics for cellular circadian bioluminescence reporters (2014) *Biophysical Journal*, 107 (11) pp. 2712-2722

St. John, P.C., Hirota, T., Kay, S.A. and F.J. Doyle III. Spatiotemporal separation of PER and CRY posttranslational regulation in the mammalian circadian clock (2014) *PNAS*, 111 (5) pp. 2040-2045.

St. John, P.C. and F.J. Doyle III. Estimating confidence intervals in predicted responses for oscilla- tory biological models (2013) *BMC Systems Biology* 7:71.

Hirota, T., Lee, J.W., **St. John, P.C.**, Sawa, M., Iwaisako, K., Noguchi, T., Pongsawakul, P.Y., Son- ntag, T., Welsh, D.K., Brenner, D.A., Doyle, F.J. III, Schultz, P.G., Kay, S.A., Identification of small molecule activators of cryptochrome (2012) *Science*, 337 (6098) pp. 1094-1097.

476087 Dynamics of Discrete Systems: At the Interface of Engineering and Medicine

Abstract:

Dynamics of Discrete Systems: At the Interface of Engineering and Medicine

Anwesha Chaudhury, PhD,

Center for Systems Biology, Massachusetts General Hospital; Pathology, Massachusetts General Hospital; Systems Biology, Harvard Medical School.

Email: achaudhury@mgh.harvard.edu, anwesha_chaudhury@hms.harvard.edu

Education

Doctor of Philosophy

December 2014

Department of Chemical and Biochemical Engineering

Rutgers-The State University of New Jersey, NJ, USA

Masters of Technology

August 2010

Department of Chemical Engineering

Indian Institute of Technology, Kharagpur, India

Bachelors of Technology

June 2008

Department of Chemical Engineering

West Bengal University of Technology, India

Research Interests:

Research experience:

I am a Chemical Engineer by training and have worked on a wide variety of research projects in various realms of basic sciences and engineering. The common denominator across all my research experiences is mathematical modeling and systems engineering. The multifaceted areas of my expertise include working on non-linear system dynamics pertaining to reaction engineering, developing mechanistic mathematical models for improving processes operation in pharmaceutical drug manufacturing, and clinical data analyses related to human pathophysiology.

During the course of my research expedition I have been interested in diverse fields in which I feel I may be able to make a significant contribution using the core skill sets of systems modeling. While working as a theoretician during my PhD, I worked closely with my experimental colleagues for developing realistic mathematical models and to better understand the underlying assumptions required for a specific model. Furthermore, I have had opportunities to become adept in various hands-on laboratory techniques including the use of a fluidized bed granulator. Towards the end of my PhD, I got drawn into the enigma of systems biology, which led me to switch fields and broaden the applications of my mathematical modeling skills.

Currently as a postdoctoral research fellow I work in an extremely multi-disciplinary environment with clinicians, biologists, chemists, engineers, and statisticians. I am leveraging my engineering training to develop models that can improve disease prognosis and alleviate the shortcomings of a delayed therapeutic intervention. Moreover, I am collaborating with experts in the field of microfluidic device development with the aim of developing a point-of-care system to detect biomarkers for sickle cell disease using single-cell measurements.

Throughout my PhD and postdoctoral journey I have received several awards and grants that exhibit my ability to secure external funding and inculcate others' interests in my research.

Research Vision:

I plant to integrate my training in various fields of engineering and medicine, with a special focus on systems engineering, by developing tools that describe hematological systems and particulate processes. In the short-term, I aim to develop data-driven population dynamics models for blood cells to predict the occurrence of diseases. Furthermore, by using this approach I intend to develop hypotheses for the underlying mechanisms of diseases. These hypotheses in turn will be investigated by performing laboratory bench experiments. I plan to integrate these models for various lineages of white blood cells in order to investigate the crosstalk within the immune system. Additionally, I plan to develop models (macroscopic, microscopic and mesoscopic) for quantitative systems pharmacology and then consolidate the dynamic behavior of the various components to study the overall pharmacodynamic behavior of the human body.

From a process systems perspective, I aspire to continue developing mathematical models for particulate systems, which can then be utilized for process control and optimization. I have significant experience with developing numerical techniques to solve complex partial differential equations, which I would like to continue incorporating into my research. One of my primary goals is to develop realistic models, which can actually be used for practical applications. With this intent in mind, I plan to continue collaborating closely with experimentalists in order to use systems engineering with a practical application.

Teaching Interests:

In addition to research, I have a keen interest in mentoring students and teaching core as well as advanced engineering topics. During the tenure of my PhD I was a teaching assistant (TA) for several courses including: transport phenomenon-I, Junior design and economics, and process control. I have derived immense pleasure and a sense of accomplishment from teaching and helping undergraduate students learn the chemical engineering curriculum. I have mentored several undergraduate and masters students, who have successfully performed research and have co-authored publications in leading peer-reviewed journals. I have persistently encouraged undergraduate students to pursue research in addition to their coursework. I enjoy teaching the core chemical engineering subjects and am extremely excited to design and teach new courses that combine the knowledge of chemical engineering, mathematics and statistics.

Research appointments:

Research fellow(February 2015-Present), Center for Systems Biology, Massachusetts General Hospital, MA, USA;

Pathology, Massachusetts General Hospital, MA, USA;

Systems Biology, Harvard Medical School, MA, USA.

Advisor: John Higgins, MD.

Project: Mathematical modeling of pathophysiological processes for disease prognosis.

Graduate Student(September 2010-December 2014), Department of Chemical and Biochemical Engineering, Rutgers University, NJ, USA.

Advisor: Rohit Ramachandran, PhD.

PhD dissertation: Mathematical modeling, simulation and optimization of pharmaceutical manufacturing processes.

Exchange research student(June 2014-August 2014), Department of Mathematics and Statistics, University of Konstanz, Germany.

Advisor: Stefan Volkwein, PhD.

Project: Reduced order modeling and optimal control of granulation processes.

Exchange research student(July 2012-October 2012), Research Center Pharmaceutical Engineering, Graz, Austria

Advisor: Johannes Khinast, PhD

Project: Investigations for particulate processes relevant to pharmaceutical drug manufacturing.

Masters student(January 2008-May 2010), Chemical Engineering, Indian Institute of Technology, Kharagpur, India

Advisor: Saikat Chakraborty, PhD

MS thesis: Pattern formation in autocatalytic reactions

Awards and Honors:

February 2016, Finalist for the Life Science Research Foundation (LSRF) postdoctoral fellowship (final result pending).

May 2015, Awarded the prestigious Outstanding Graduate Student Award by the School of Engineering, Rutgers University in recognition of the achievements as a graduate student.

June 2014, Awarded the Baden-Wuerttemberg Scholarship to pursue research at

University of Konstanz, Germany.

July 2012, Awarded the Marshall Plan Fellowship to pursue research at TU Graz, Austria.

May 2012, Awarded the best poster prize at the New Jersey Pharmaceutical Association for Science and Technology graduate student poster competition.

September 2010, School of Engineering Fellowship, Rutgers University, USA.

Selected publications: (total 22, 9 first author publications, 2 book chapters)

A. Chaudhury, A. Tamrakar, M. Sch ngut, D. Smrčka, F. Štěpánek, R. Ramachandran, Multidimensional population balance model development and validation of a reactive granulation process, Industrial Engineering & Chemistry Research, *2015*, 54(3)

A. Chaudhury, A. Armenante, R. Ramachandran, Compartment based population balance modeling of a high shear wet granulation process using data analytics, Chemical Engineering Research and Design, *2015*, 95

A. Chaudhury, H. Wu, M. Khan, R. Ramachandran. A mechanistic population balance model for granulation processes: Effect of process and formulation parameters. Chemical Engineering Science, *2014,*107

A. Chaudhury, I. Oseledets, R. Ramachandran. A computationally efficient technique for the solution of multi-dimensional PBMs of granulation via tensor decomposition. Computers and Chemical Engineering *2014*, 61

M. Sen, **A. Chaudhury**, R. Singh, J. John, R. Ramachandran, Multi-scale flowsheet simulation of an integrated continuous purification-downstream pharmaceutical manufacturing process, International Journal of Pharmaceutics, *2013*, 445(1-2)

A. Chaudhury, A. Kapadia, A.V. Prakash, D. Barrasso, R. Ramachandran. An Extended Cell-average Technique for Multi-Dimensional Population Balance Models describing Aggregation and Breakage. Advanced Powder Technology, *2013*, 24 (6)

A. Chaudhury, A. Niziolek, R. Ramachandran. Multidimensional Mechanistic Modeling of Fluid Bed Granulation Process: An Integrated Approach. Advanced Powder Technology, *2012*, 24

A. Chaudhury, S. Chakraborty, Dynamics of Mixing-Limited Pattern Formation in Non-Isothermal Homogeneous Autocatalytic Reactors: a Low-Dimensional Computational Analysis, Industrial Engineering and Chemistry Research, *2011*, 50 (8)

Book chapters:

A. Chaudhury, D. Barrasso, Daniel A. Pohlman, James D. Litster, R. Ramachandran, Mechanistic modeling of high shear and twin screw mixer granulation processes, Modeling of pharmaceutical unit operations in solid dosage forms, Elsevier, *2016*, In Press

A. Chaudhury, M. Sen, D. Barrasso, R. Ramachandran, Population balance models for pharmaceutical processes, Handbook in Process Simulation and Data Modeling in Solid Oral Drug Development and Manufacture, Humana Press, *2016*, 43-83

Teaching and research mentoring experience:

September 2011-June 2012 and September 2014-December 2014: *Teaching assistant*

Department of Chemical and Biochemical Engineering, Rutgers University

Courses taught:

Transport phenomenon I, Junior design and economics, Process control

Personnel mentored during PhD:

6 Undergraduate students, 2 Masters students.

476137 Application of Modeling and Optimization Methods in Biomedicine and Biorefineries

Abstract:

Application of modeling and optimization methods in biomedicine and biorefineries

Kirti M. Yenkie, University of Wisconsin-Madison Current Position: Second year Postdoctoral Researcher **Research Interests:** My research focuses on development of methodologies for predicting policies or forecasting decisions which can enhance the performance of existing processes as well as act as guidelines for developing novel processes in chemical and biological systems. I plan to incorporate methods such as mathematical modeling, simulation and numerical techniques for the process analysis and then apply principles from optimization and control for prediction of better operating policies and decisions. Policy prediction is particularly advantageous for processes in pharmaceutical industries where the product specifications are stringent and it is essential to maintain the product quality and purity. Also, medical treatments can be benefitted by prediction of drug dosing policies for chronic problems such as diabetes or treatments with adverse side effects such as cancer chemotherapy. These methods are also applicable to selection of best alternatives in a process; for example, selection of the best option among the available alternatives for wastewater treatment, based on initial conditions and final water quality requirements. To ensure the robustness of the results predicted, I intend to apply methods for uncertainty characterization and incorporation into process models through stochastic modeling and control methods. **Postdoctoral Project: Modeling, simulation, analysis and optimization of biorefineries Advisor:** Dr. Christos T. Maravelias, Chemical and Biological Engineering, University of Wisconsin-Madison. - Assessment of separation technologies for efficient recovery of bio-based chemicals from microbial cultures - Collaboration with experimentalists from Dr. Brian Pfleger's group for information about microbial strains, substrates, culture conditions, titers, productivity and yield. - Developing guidelines for selection of best possible separation scheme based on physical properties of selected chemicals by application of mathematical modeling, mixed-integer non-linear programming optimization and high performance computing methods. PhD Dissertation: Stochastic processes from batch crystallization to in-vitro fertilization (IVF) **Advisor:** Dr. Urmila M. Diwekar, Bioengineering, University of Illinois at Chicago (UIC). - Studied effects of uncertainties in batch crystallization, developed stochastic models using Ito processes and applied optimal control methods for predicting robust operating policies for enhanced product quality. - Analyzed clinical data for superovulation (first stage in IVF) cycles. Developed mathematical model for multiple ovulation using differential equations and applied optimization and control for predicting optimal hormone dosing policies. - Developed a new approach for parameter estimation in stochastic differential equations and applied it to develop stochastic models for multiple ovulation and follicle growth. Other Research Projects:

(i) Developed models for cellular systems in terms of stochastic differential equations as an alternative to discrete event simulation methods such as the Gillespie algorithm.

(ii) Application of optimal control methods to research on sustainability by using representative models for earth's ecological-economic interactions in terms of differential-algebraic equations and regulatory policies. Future Research Plans: ***Optimizing cancer chemotherapy*** - I plan to work on modeling the cancer propagation by using deterministic and stochastic models. Using these models, treatment therapies in terms of drug dosing regimens as well the time-gaps between dosing periods will be designed. I would also like to explore the possibility of treatment by combination of multiple drugs or multiple therapies such as chemotherapy, radiotherapy and gene therapy. ***Optimizing wastewater treatment methods*** - Processes such as activated sludge, advanced oxidation, ion-exchange and anaerobic digestion are used to treat wastewater for compliance with the environmental norms. I plan to use optimal control and MINLP optimization methods for prediction of treatment conditions as well as selection of best alternatives. **Teaching Interests:** My B. Tech. and M. Tech. degrees in Chemical Engineering and Ph.D. in Bioengineering have provided me with the opportunity to learn core chemical engineering subjects, like thermodynamics, mass transfer, reaction engineering, transport phenomena, etc. as well as some interdisciplinary topics like metabolic engineering, biochemical engineering and computational biology. Also, my interdisciplinary research facilitated me to develop skills in applied optimization, biomedicine, crystallization, separation processes, sustainability and advanced mathematics. I have been a teaching assistant for several courses such as chemical engineering thermodynamics, plant utility, reaction engineering laboratory, senior process design, and cell and tissue engineering during my graduate studies. I have graded assignments, conducted problem solving sessions and supervised experiments. I have also conducted tutorials for programming tools like MATLAB and GAMS. I have been a teaching associate for the introduction to optimization course at UIC and have taken lectures on genetic algorithms and simulated annealing. I have mentored undergraduate and graduate students for their research projects. I believe that I am capable of teaching most of the chemical engineering subjects, particularly the ones requiring computer programming and numerical methods such as transport phenomena, process design and optimization, bioprocess and biomedical engineering. I am also interested in developing and teaching innovative courses like modeling and control in biomedicine and stochastic processes in chemical engineering. Selected Publications

1) **Yenkie, K. M.**; Diwekar, U. 2013. Stochastic optimal control of seeded batch crystallizer applying Ito process. *Ind. Eng. Chem. Res.*, 52:108-122.

2) **Yenkie, K. M.**; Diwekar, U.; Bhalerao, V. 2013. Modeling the superovulation stage in in-vitro fertilization. *IEEE Trans. Biomed. Eng.*, 60(11): 3003-3008.

3) **Yenkie, K. M.**; Diwekar, U. 2014. Optimal control for predicting customized drug dosage for superovulation stage of in-vitro fertilization. *Journal of Theoretical Biology*, 355: 219-228.

4) **Yenkie, K. M.**; Diwekar, U.; Bhalerao, V. 2014. Modeling and prediction of outcome for the superovulation stage in in-vitro fertilization. *JFIV Reprod.Med.Genet.* 2(2):1000122(1-8).

5) **Yenkie, K. M.**; Diwekar, U. 2014. Comparison of optimal control methods for customized drug dosage prediction in superovulation stage of in-vitro fertilization. *Computers and Chemical Engineering*, 71: 708-714.

6) **Yenkie, K. M.**; Diwekar, U. 2014. Uncertainty in clinical data and stochastic model for in-vitro fertilization.

Journal of Theoretical Biology, 367: 76-85.

7) **Yenkie, K. M.**; Diwekar, U.; Linninger, A. 2016. Simulation-free estimation of reaction propensities in cellular reactions and gene signaling networks. *Computers and Chemical Engineering*, 87: 154-163.

476169 Advanced Adaptive Control Approaches for Complex Batch or Semi-Batch Operations

Abstract:

Research Plan:

Batch/semi-batch processes are an important set of unit operations in industries such as biofuel synthesis, synthesis of active pharmaceutical ingredients (APIs), specialty chemicals, etc. Such applications, in general, have complex dynamics and are operated in conjunction with other unit operations such as distillation, separation, etc. The products of these applications need to meet stringent quality requirements. Given the complex dynamics of such systems, it becomes essential to develop advanced control schemes that can utilize its multivariable dynamics in order achieve the desired control objectives and optimal performance of the process. Developing an all-encompassing control algorithm in order to achieve these goals involves addressing the following main challenges-

• **Adaptive identification of the pdf of the disturbances for optimal estimation of states and parameters**. While the states and uncertain parameters can be estimated using Bayesian estimation algorithms, these algorithms rely on the fact that the system model and pdfs of the unmeasured disturbances and measurement noise are perfectly known. However, this is not the case since there are uncertainties in the model parameters and the pdfs of the unmeasured disturbances and measurement noise are not known. Given the continuously time-varying nature of the batch/semi-batch processes I would like to work on developing online MLE-based approaches for estimation of the pdfs of the disturbances and noise. This would help maintain the optimal performance of the Bayesian state and parameter estimators, which will aid in improved state-feedback control and monitoring.

• **Stochastic iterative learning control and parameter learning for batch/semi-batch operations**. The repeated nature of batch/semi-batch operations means that knowledge of the control moves and parameters used in the previous batches can be used to improve the controller performance and parameter estimates for the current batch. Given the stochastic nature of batch/semi-batch operations and uncertainties in the system model arising from the time-varying nature of the processes, I would like to work on developing iterative learning control and iterative learning parameter estimation approaches that account for the probabilistic nature of these stochastic uncertainties in computing the control moves and parameter estimates respectively. This would also enable the formulation of a probabilistic control objective, thereby allowing for greater control over the final product quality.

Research Experience:

My research experience has been primarily in the areas of nonlinear Bayesian state estimation, system identification and model predictive control. During my PhD, I developed a maximum likelihood estimates (MLE)-based approach for identification of the covariance matrices associated with the Gaussian process and measurement noise from input-output process data for state and parameter estimation using extended Kalman filter (EKF)[1]. I have also developed an MLE-based approach for identification of grey-box models for state and paramter estimation, when the system under consideration is affected by time-correlated noise in the inputs or disturbances[2]. We have validated these approaches on an experimental setup as well. As a post-doctoral fellow at University of Alberta, I have worked on data-based approaches for modeling and classification for welding processes. We used approaches such as principal component analysis (PCA) and partial least squares (PLS) for classification of gas-metal arch welding jobs[3]. This work has led to an industry-academia collaboration and is being taken forward by the concerned PIs. As a post-doctoral fellow at UC Berkeley, my work has focussed on control-relevant model identification for stochastic model predictive control with a focus on cold atmospheric plasma (CAP) system for biomedical applications[4]. As a result I have research experience in multiple fields of Process Systems Engineering, both in computational and experimental studies.

References:

1. **V. A. Bavdekar**, A. P. Deshpande and S. C. Patwardhan, "Identification of process and measurement noise covariance for state and parameter estimation using extended Kalman filter", Journal of Process Control 21 (4) (2011) 585-601.
2. **V. A. Bavdekar** and S. C. Patwardhan, "Development of grey box state estimators for systems subjected to time correlated unmeasured disturbances", Journal of Process Control 22 (9) (2012) 1543-1558.
3. R. Ranjan, A. Talati, M. Ho, H. Bharmal, **V. A. Bavdekar**, V. Prasad, P. F. Mendez, "Multivariate data analysis of gas-metal arc welding process", Proceedings of the 9th International Symposium on Advanced Control of Chemical Processes (AdCHEM 2015).
4. **V. A. Bavdekar** and A. Mesbah, "Model predictive control with integrated input design for nonlinear systems with probabilistic uncertainties", accepted for the 11th International Symposium on Dynamics and Control of Process Systems, 2016.

Select Publications:

1. **V. A. Bavdekar** and A. Mesbah; *Stochastic nonlinear model predictive control with joint chance constraints*, accepted for the 10th IFAC Symposium on Nonlinear Control Systems, 2016.
2. **V. A. Bavdekar** and A. Mesbah; *Model predictive control with integrated input design for nonlinear systems with probabilistic uncertainties*, accepted for the 11th International Symposium on Dynamics and Control of Process Systems, 2016.
3. **V. A. Bavdekar** and A. Mesbah; *A polynomial chaos-based nonlinear Bayesian approach for estimating state and parameter probability density functions*, accepted for the 2016 American Control Conference.
4. **V. A. Bavdekar**, N. Nandola and S. C. Patwardhan; *Estimation of noise covariance matrices for state estimation of autonomous hybrid systems*, under review, Computers and Chemical Engineering.
5. **V. A. Bavdekar**, J. Prakash, S. C. Patwardhan and S. L. Shah; *A Moving window formulation for state estimation of systems with irregular and delayed measurements*, Industrial and Engineering Chemistry Research

53(35) (2014) 13750-13763.
6. **V. A. Bavdekar** and S. L. Shah; *Computing point estimates from a non-Gaussian posterior distribution using a probabilistic k-means clustering approach*, Journal of Process Control 24 (2) (2014) 487-497.
7. **V. A. Bavdekar** and S. C. Patwardhan; *Development of grey box state estimators for systems subjected to time correlated unmeasured disturbances*, Journal of Process Control 22 (9) (2012) 1543-1558.
8. **V. A. Bavdekar**, A. P. Deshpande and S. C. Patwardhan; *Identification of process and measurement noise covariance for state and parameter estimation using extended Kalman filter*, Journal of Process Control 21 (4) (2011) 585-601.

Research Interests:

My research interests are in the areas of nonlinear Bayesian state and parameter estimation, system identification, control-relevant model identification, stochastic MPC, data analysis uses methods such as PCA, PLS, etc. I'm interested in applications involving batch/semi-batch operations such as ethanol synthesis, algal biofuels, batch polymerizations, active pharmaceutical ingredient (API) synthesis, etc.

As a faculty member, I will use my expertise in the above areas of Process Systems Engineering to form a group that conducts leading research at the interface of systems' theory, applied mathematics and process engineering towards the development of advanced control algorithms for batch/semi-batch operations.

Teaching Interests:

At the undegraduate level, I would be interested in teaching the following courses (or their equivalents) and the laboratory/computational hours associated with them: Process Dynamics and Control, Computer-aided Process Design, Numerical Methods in Chemical Engineering and Process Optimization. I would also like to explore the option of introducing a course on Data Analysis and Statistics for Chemical Engineers, which will introduce students to basic experiment design and analyis techniques, with focus on the multivariate nature of chemical engineering data.

At the graduate level, I would be interested in teaching the following courses (or its equivalents): Digital Control and Advanced Numerical Methods for Chemical Engineers. I am interested in offering a course on Bayesian state estimation, given my research experience in this area and a course on System Identification, which can be a nice complement to the course on Digital Control.

476180 New Modeling and Decision-Making Paradigms in Systems Engineering

Abstract:

Research Interests: energy systems, sustainability, mathematical modeling, optimization

Systems engineering is an interdisciplinary field concentrated on the design and management of complex systems. The tools and perspectives of systems engineers are increasingly necessary to establish new links between research fields and drive inter-disciplinary innovation. This poster discusses my research vision in the context of three new paradigms in systems engineering.

Multi-Stakeholder Decision-Making

The design and control of engineering systems involves complex decision-making processes with multiple conflicting objectives of social, economic and environmental natures. Multi-objective optimization techniques have been applied to many chemical engineering domains, including separations, crystallization, chromatography, control systems, and chemical reactors, and have also been used to co-optimize economic and environmental metrics, quantified through life-cycle analysis. Most literature, however, assumes the perspective of a single decision-maker and considers only two or three competing objectives.

We recently proposed a framework for multi-stakeholder decision-making that addresses these two fundamental issues. The conflicting opinions from multiple stakeholders are transformed into individual stakeholder dissatisfactions that are minimized. We generalize previous approaches using the conditional-value-at-risk norm, which provides statistical interpretations to the compromise solutions and guarantees Pareto efficiency. Most notably, the method is computationally scalable. This is critical, as many sustainability studies consider hundreds of conflicting metrics. As a demonstration, we determine locations for waste-to-biogas conversion facilities while considering safety, transportation, water quality and cost priorities. Future work includes developing alternate solution metrics to explicitly address *fairness* and explore the theoretical connections with stochastic programming. I envision novel design methods for *sustainable chemical processes* based on the proposed framework, where the ambiguity of prioritizing different suitability metrics and life-cycle analysis standards is addressed through a multi-stakeholder perspective.

Stochastic Programming and High Performance Computing

In many engineering systems, uncertainty is addressed through safety margins centered on deterministic analysis. Stochastic analysis helps identify overly conservative designs and operating policies, thus further improving the realistic performance. In stochastic programming, the goal is to optimize a statistic (e.g., expected value) of the objective function while considering the probabilistic nature of input data (e.g., weather, future prices) propagated through constraints. Parallelized algorithms tailored for high performance computing platforms are of increased interest, as complex uncertainty structures can result in stochastic optimization problems with *millions* of variables and constraints.

Many energy systems benefit from stochastic analysis, as their performance depends both on design decisions and the operating policy. Consider, for example, sizing an energy storage system to be coupled with a solar or wind farm. If only average annual weather is considered, the system will likely be under-designed for extreme events. We are exploring the use of new stochastic programming algorithms to design concentrated solar power systems with thermal storage and other energy systems, while considering large historical datasets for weather and energy prices. This analysis is essential to properly value the flexibility provided by storage.

Energy Markets

We are studying the incentives embedded in multi-scale price signals from wholesale electricity markets for a diverse set of energy and manufacturing systems. We have determined that 70% of the economic opportunities are only available at fast timescales (seconds to minutes). Most literature, however, focuses on slower timescales, such as hourly prices. These findings inform the design of energy and manufacturing systems, and emphasize the importance of dynamic flexibility. This presents enormous new opportunities for energy intensive industrial systems or facilities with onsite utility plants. For example, providing regulation energy services to the market with a combined heat and power system may introduce low amplitude, high frequency (seconds) variations in steam availability for the coupled chemical process. Large separation systems take hours to days to reach steady-state, however, and may be resilient to these additional disturbances. Such a system has inherent dynamic flexibility that is not well understood, valued or utilized. Future research goals focus on characterizing the dynamic flexibility of industrial systems and developing multi-scale optimization architectures to exploit such flexibility. Long-term goals focus on inter-disciplinary research to redesign electricity markets to more effectively incentivize industrial participation and leverage dynamic flexibility from existing infrastructure to cost effectively support wide-scale adoption of wind and solar intermittent generation.

Additional details and publication record are available at www.alexdowling.net.

Teaching Interests:

I am qualified to teach all undergraduate chemical engineering core courses, and especially interested in teaching control, design, thermodynamics, separations/unit operations, mathematics, engineering problem solving, and statistics. I am also excited to teach graduate courses in optimization, numerical methods and process design, and look forward to developing new courses related to energy systems and sustainability.

Presentations at AIChE Annual Meeting:

A Framework for Multi-Stakeholder Decision-Making and Conflict Resolution
Session: Advances in Optimization I

Economics and Dynamic Flexibility of Concentrated Solar Power Technologies
Session: Sustainable Electricity: Generation and Storage

Exploiting Dynamic Flexibility to Enable Participation in Multi-Scale Electricity Markets
Session: Energy Systems Design and Operations II

476215 Management of Energy Supply Chains Under Uncertainty

Abstract:

Research Interests:

Energy demand, including electricity demand, is directly linked to factors such as economic development, population growth, and urbanization. However, meeting energy demands entails a critical challenge to use the Earth's energy resources in a sustainable fashion, enhancing quality of life, and facilitating human development. Specifically, a number of environmental impacts have been identified associated with the development of energy systems. In particular, emissions of Greenhouse Gases (GHG), the depletion and degradation of water sources, as well as the potential for underground water contamination, are major concerns associated with these systems. Thus, the design and planning of energy systems is a challenging problem where economic and environmental aspects need to be considered at both the process and supply chain levels.

Even though both levels have received significant attention from the engineering research community, there has been comparatively less research on the development of decision-making support tools for the investigation of integrated energy supply chains. For instance, only a few studies have been devoted to the integration of water management and shale gas supply chains design as well as to the integrated planning of generation and transmission expansion in power systems, which are systems that have been playing and will continue to play important roles in supplying our energy needs. I am especially interested in attacking this gap, through the development of optimization tools for supporting the decision-making processes in these two aforementioned energy systems at both national and regional scales. Specifically, I have targeted the development of decision-support optimization frameworks for the strategic design and planning of shale gas supply chains integrated with water management as well as generation and transmission expansion in power systems. In [1] we present a methodology for the simulation and evaluation of different well-pads designs for the exploitation of shale gas resources. Then, in [2] we present the mathematical formulation of a optimization framework for the integration of well-pad design and water management with shale gas supply chain design. The mathematical formulation consists of a novel large-scale Mixed Integer Linear Programming (MILP) model and a Mixed Integer Nonlinear Programming (MINLP) model. Furthermore, the analysis of the results reveals strong interactions between water management and the design of the shale gas supply chain. For instance, it was found that both the Total Dissolve Solid (TDS) concentration in wastewater and the fresh water availability are key factors not only for the selection of the water management strategy but also for the drilling scheme that should be implemented. Additionally, we demonstrated that a more accurate formulation, which results in a MINLP model, could lead to improved designs which increase by about 71% the economic benefit of the exploitation of shale gas plays in comparison with the MILP formulation. While the MILP is much more efficient from a computational viewpoint it is an approximation of the real problem and thus fails to capture some important synergies.

In addition to the shale gas project, I have been working on the development of an optimization framework for the integrated planning of power generation and transmission in interconnected power systems. In [3] we presented the mathematical formulation, MILP models, of the optimization framework as well as some revealing applications including "business as usual" and CO_2 mitigation policy scenarios. The novelty of this framework stems from the integration of power generation and transmission planning along with spinning and non-spinning reserve constraints as well as CO_2 emission constraints and mitigation options. In the case of CO_2 emission mitigation options, we consider the penetration of renewable energies, the integration of Carbon Capture and Sequestration (CCS) technologies, and the implementation of Demand Side Management (DSM) strategies. The results from the "business as usual" scenario revealed that fuel prices have a direct impact not only on the allocation of electricity generation resources but also on the selection of the fossil power plant technologies. Additionally, the analysis of the results reveals an interdependency between the transmission constraints and the decisions regarding the investment schedule and installation of new power plants. Moreover, the results from the CO_2 mitigation policy scenario demonstrated that the reconfiguration of existing power generation technologies is the most cost-effective alternative for the abatement of CO_2emissions. Furthermore, the integration of renewable energy sources was demonstrated to be preferred over the retrofitting of coal power plants with CCS technologies.

My ongoing research is concentrated on the characterization and modeling of the uncertainties inherent to the shale gas supply chain and the power generation and transmission planning problems, as well as on the development of stochastic optimization models. Specifically, global sensitivity analysis is being carried out in order to identify the most impactful among the many parameters in each model. This analysis will allow us to focused on parameters with the most impact on the optimal solution, thus increasing the computational effectiveness when modeling the uncertainty space. Moreover, two-stage stochastic models will be developed in order to solve the optimization problems under uncertainty.

Teaching Interests:

My academic and industrial experiences have offered me the opportunity to not only strengthen my knowledge of mathematical modeling, simulation, and optimization of processes and systems but also to develop deep understanding on the applicability of such tools to deal with real world design and planning problems in different domains, i.e. refinery operations, shale gas supply chains, and power systems. Therefore, my teaching interest pertains to decision making theory and applications, i.e. courses on computer aided process design and planning or optimization theory with applications to industrial processes and systems. However, I am well prepared to teach courses in the chemical engineering core, e.g., material and energy balances, separation processes, and process economics, design, and evaluation, as well as courses on mathematical and numerical methods.

References:

[1] Calderón AJ, Guerra OJ, Papageorgiou LG, Siirola JJ, Reklaitis G V. Preliminary Evaluation of Shale Gas

Reservoirs: Appraisal of Different Well-Pad Designs via Performance Metrics. *Ind Eng Chem Res*2015;54:10334–49. doi:10.1021/acs.iecr.5b01590.

[2] Guerra OJ, Calderón AJ, Papageorgiou LG, Siirola JJ, Reklaitis G V. An Optimization Framework for the Integration of Water Management and Shale Gas Supply Chain Design. *Comput Chem Eng* 2016; 92:230–255. doi:10.1016/j.compchemeng.2016.03.025

[3] Guerra OJ, Tejada DA, Reklaitis G V. An optimization framework for the integrated planning of generation and transmission expansion in interconnected power systems. *Appl Energy* 2016;170:1–21. doi:10.1016/j.apenergy.2016.02.014.

476225 Energy Systems Analysis to Enable a Sustainable Economy

Abstract:

Research Interests:

Current energy and chemical sectors predominantly depend on fossil resources. Limited fossil resources and significant increase in atmospheric greenhouse gases urge to develop and implement renewable energy conversion processes for a more sustainable economy.

The overarching goal of my research is to meet essential human needs including transportation fuel, chemicals, electricity and water, primarily from renewable energy sources, and ensure a smooth transition to a sustainable economy. My research focuses on multiscale systems analysis of energy conversion technologies for synthesis of energy efficient, synergistic processes incorporating process intensification for optimal utilization of resources.

During my PhD, I have focused on determining the optimal synergistic process design to produce liquid fuel from biomass and natural gas, developing a methodology to synthesize efficient solar thermal hydrogen production processes, identifying and optimizing efficient solar thermal hydrogen production process designs based on the developed methodology, synthesis of integrated thermal desalination processes, creating efficient and intensified processes to convert solar thermal energy into electricity and to continuously supply to the grid by integrated energy storage solutions. Moreover, I have worked on policy analysis of large-scale energy storage options for uninterrupted renewable power supply, particularly on closed loop CO_2-driven energy storage solutions.

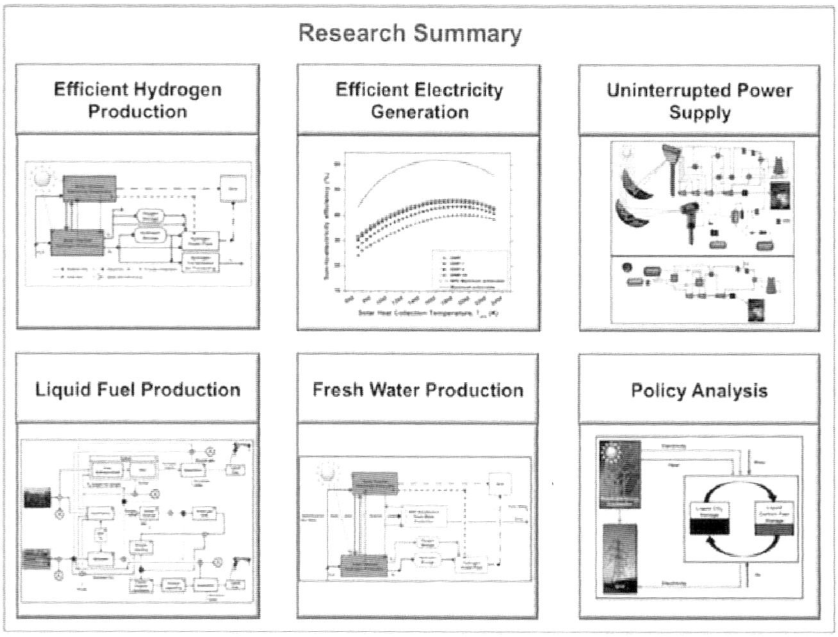

Teaching Interests:

I had extensive teaching experience as a teaching assistant at Purdue University. I gave several lectures and prepared several teaching materials including a new curriculum and a complete set of lab instructions for the newest version of Aspen Plus design class. I am passionate about teaching and I am particularly interested in enhancing students' understanding on energy conversion technologies, process design and advanced mathematical modeling. I am also interested in teaching thermodynamics, separations, heat and mass transfer.

I received several teaching awards including 2016 Graduate School's Excellence in Teaching Award, Purdue University's highest award in recognition of graduate student teachers, Excellence for Undergraduate Teaching Assistant Award as voted by the graduating seniors for Teaching Excellence in Chemical Engineering and 2013 CETA (Committee Education of Teaching Assistants) Award.

References

[1] Gençer E, Mallapragada DS, Marechal F, Tawarmalani M, Agrawal R. Round-the-clock power supply and a sustainable economy via synergistic integration of solar thermal power and hydrogen processes, Proceedings of the National Academy of Sciences (PNAS), 112(52), 15821-15826, 2015.

[2] Gençer E, Al-musleh E, Mallapragada D, Agrawal R. Uninterrupted Renewable Power through Chemical Storage Cycles. Current Opinion in Chemical Engineering, 5, 29-36, 2014.

[3] Gençer E, Agrawal R. A commentary on the US policies for efficient large scale renewable energy storage systems: Focus on carbon storage cycles, Energy Policy, 88, 477-484, 2016.

[4] Gençer E, Tawarmalani M, Agrawal R, Integrated solar thermal hydrogen and power coproduction process for continuous power supply and production of chemicals, Computer Aided Chemical Engineering, 37, 2291-2296, 2015.

[5] Gençer E, Mallapragada DS, Tawarmalani M, Agrawal R. Synergistic biomass and natural gas conversion to liquid fuel with reduced CO2 emissions. Computer Aided Chemical Engineering, 34:525-530, 2014.

476271 Deterministic and Robust Model-Based Strategies for the Online Multi-Level Optimization of Batch Operations

Abstract:

Research Interests:

In the last few decades, there has been a steady growth in applying optimization techniques to the solution of problems of industrial importance, e.g. supply-chain management, operational planning/scheduling and optimal process control/dynamic optimization. This trend has been motivated by both increasing global competition and tighter environmental regulations, which are driving industry towards reducing costs as well as minimizing environmental impacts.

Although extensive process systems engineering research has been reported so far, most of it has addressed continuous processes and only limited attention has been devoted to batch operations, which are widely used to produce high added-value products (specialty chemicals, pharmaceuticals, cosmetics, etc.). Given the opportunities that this application domain provides, I have focused my research on methodologies/applications in operational planning/scheduling and optimal process control/dynamic optimization of batch processes. However, I have also addressed supply-chain problems[1] and continue having interests in such research area.

Most recently, I have focused on developing methods for the optimal online management of both single batch units and entire batch plants. I have formulated and implemented two dynamic optimization/optimal control algorithms applicable to single batch operations (BSMBO&C [2,3], RBSMBO&C [4,5]) and a further one designed for (multi-unit) batch plants (MUBSMBO&C). These strategies are very flexible and can tackle the online optimization/optimal control of both single batch cycles and entire production campaigns. Moreover, RBSMBO&C also implements a novel dynamic scenario selection concept to handle model uncertainty.

My ongoing research involves studying uncertainty propagation/Bayesian inference with the goal of further improving RBSMBO&C. Moreover, I am also planning to extend MUBSMBO&C and make it capable of handling model uncertainty. Longer-term research perspectives include addressing the problem of supply-chain management of batch manufacturing systems and developing integrated frameworks for the simultaneous online optimization of multiple decisional levels.

Teaching Interests:

My research work and graduate studies have provided me with deep knowledge of control theory, numerical methods, dynamic/steady-state modelling and coding/parallel programming (especially in C++). Therefore, I would be interested in teaching any course (broadly) related to these subjects, e.g. process control/design, numerical methods, dynamic/steady-state modelling and process optimization.

References:

1. Rossi F, Manenti F, Reklaitis G. A general modular framework for the integrated optimal management of an industrial gases supply-chain and its production systems. *Computers and Chemical Engineering*. 2015;82:84-104.
2. Rossi F, Manenti F, Buzzi-Ferraris G. A novel all-in-one real-time optimization and optimal control method for batch systems: Algorithm description, implementation issues, and comparison with the existing methodologies. *Industrial and Engineering Chemistry Research*. 2014;53:15639-15655.
3. Rossi F, Copelli S, Colombo A, Pirola C, Manenti F. Online model-based optimization and control for the combined optimal operation and runaway prediction and prevention in (fed-)batch systems. *Chemical Engineering Science*. 2015;138:760-771.
4. Rossi F, Reklaitis G, Manenti F, Buzzi-Ferraris G. Multi-scenario robust online optimization and control of fed-batch systems via dynamic model-based scenario selection. *AIChE Journal*. 2016; DOI: 10.1002/aic.15346.
5. Rossi F, Manenti F, Pirola C, Mujtaba I. A robust sustainable optimization & control strategy (RSOCS) for (fed-)batch processes towards the low-cost reduction of utilities consumption. *Journal of Cleaner Production*. 2016;111:181-192.

476776 Modeling and Control of Hybrid and Nonsmooth Process Systems

Abstract:

Research Interests: Control and systems theory; Optimization; Biological and chemical systems.

Broadly, my interests lie in the modeling, simulation, and control of hybrid and nonsmooth systems. Hybrid systems display a mixture of continuous and discrete dynamics, where the continuous portion is often modeled using differential equations. Nonsmooth systems refers here to dynamic systems whose participating functions exhibit non-differentiability, which can be problematic in conventional simulation and optimization methods. Possible sources of such behavior include thermodynamic phase changes, flow regime transitions, and valve activation, among others. Hybrid and nonsmooth modeling frameworks have become widely applied tools in a variety of real-world problems. Examples are found in pharmaceutical manufacturing, biological systems with changes in population behavior, secure communications via supervisory switching control, and multistream heat exchanger design. My pursuits and contributions are found in this area of research, with an emphasis on sensitivity, stability, and optimization techniques for use in controller and parametric design.

Research Experiences: I am currently a postdoctoral fellow working on dynamic optimization of campaign continuous manufacturing processes, under the supervision of Professor Paul I. Barton (*successful proposal: NSERC Postdoctoral Fellowship*). This collaborative work is a part of the Novartis-MIT Center for Continuous Manufacturing initiative to revolutionize drug manufacturing. I have developed new theory to characterize sensitivities of nonsmooth dynamic systems (e.g., [1, 2]), mirroring the classical results on the topic. These findings lay the theoretical groundwork upon which efficient numerical methods can be designed for large-scale problems, where standard methods can fail. This work thus extends nonsmooth equation-solving and optimization methods to problems with nonsmooth differential-algebraic equations embedded.

I obtained my Ph.D. in Applied Mathematics at the University of Waterloo, under the supervision of Professor Xinzhi Liu (*successful proposal: NSERC Alexander Graham Bell Canada Scholarship*). My Ph.D. studies related to the analysis of complex physical phenomena exhibiting time-delays, which are an inherent feature in a wide range of biological and physical problems. By uncovering underlying mechanisms, I made contributions here in the stabilization, synchronization, and robust control of nonlinear hybrid systems with time-delays (e.g., [3, 4]). I designed switching control strategies to achieve desired performance outcomes where classical methods are either unavailable or inadequate. I connected the theoretical findings to biological systems through the introduction and analysis of hybrid modeling approaches to disease transmission (e.g., [5, 6, 7]).

Future Directions: My goal as faculty would be to combine my training and research experiences in applied mathematics and engineering to uniquely approach high-impact problems in chemical and biological process control. Using my expertise in theoretical and computational mathematics, and by collaborating with experimentalists, I plan to design, develop, implement, and optimize state-of-the-art process control methods. Short-term projects include efforts in characterizing the behavior of high-index differential-algebraic equations and new approaches to nonlinear model predictive control. Long-term projects suitable for Ph.D. students are comprised of problems in stochastic modeling of biological systems, fundamental extensions in the setting of optimal control theory, and the development of global optimization theory and methods for hybrid process systems. Industrial applications in mind include ones in continuous manufacturing processes, power and energy systems, and biological population models.

Teaching Interests: Optimization; Computational methods; Advanced dynamic systems; Process control.

I was the full-time lecturer for two undergraduate engineering courses (approximate class size of 100 each) at the University of Waterloo. I have also acted as a teaching assistant in mathematics and engineering courses (11 undergraduate, 3 graduate level) on topics ranging from computational mathematics to applied functional analysis. My teaching philosophies are instructed by these opportunities, along with the completion of the Kaufman Teaching Certificate Program at MIT. For example, my course design is dictated as follows:

- My lecture and assessment design for first and second-year courses (e.g., differential equations, numerical methods) will aim to solidify the students' foundations going forward in their field of study.
- Upper-year and graduate-level courses (e.g., control theory, optimization) will be designed so that students understand the material in a way that an expert in the field would expect.
- In specialized graduate-level courses (e.g., modeling and simulation of hybrid systems, nonsmooth analysis, topics in mathematical biology), research-oriented goals will be included whenever possible.

Selected Publications (11 accepted peer-reviewed journal articles as lead author):

[1] **P.G. Stechlinski** and P.I. Barton, **In Press**. "*Generalized Derivatives of Nonsmooth Differential-Algebraic Equations.*" Journal of Optimization Theory and Applications, DOI: 10.1007/s10957-016-0988-9.

[2] **P.G. Stechlinski** and P.I. Barton, **Accepted Invited Session Paper**. "*Generalized Derivatives of Optimal Control Problems with Nonsmooth Differential-Algebraic Equations Embedded.*" 55th IEEE Conference on Decision and Control 2016.

[3] X.Z. Liu and **P.G. Stechlinski**, **2016**. "*Hybrid stabilization and synchronization of nonlinear systems with unbounded delays.*" Applied Mathematical Modelling, 280, 140-161.

[4] X.Z. Liu and **P.G. Stechlinski**, **2014**. "*Hybrid control of impulsive systems with distributed delays.*" Nonlinear Analysis: Hybrid Systems, 11, 57-70.

[5] X.Z. Liu and **P.G. Stechlinski**, **2015**. "Application of control strategies to a seasonal model of Chikungunya disease." Applied Mathematical Modelling, 39, 3194-3220.

[6] X.Z. Liu and **P.G. Stechlinski**, **2014**. "SIS models with switching and pulse control." Journal of Applied Mathematics and Computation, 232, 727-742.

[7] X.Z. Liu and **P.G. Stechlinski**, **2013**. "Transmission dynamics of a switched multi-city model with transport-related infections." Nonlinear Analysis: Real World Applications, 14, 264-279.

476958 Energy Security and Environmental Protection: Bridging the Gap Between Theory and Application

Abstract:

Research Interests:

The convergence of recent advances in Computer-aided Engineering and theoretical developments for design and operation of industrial processes has provided unprecedented opportunities to explore new solutions for energy security and environmental protection. This poster presents my recent contributions and future plans in three dimensions; (1) The theory of process design and control, (2) Novel applications for energy security and environmental protection, and (3) Solution algorithms. The justification is that delivering energy and environmental impacts necessitates in-depth theoretical understanding, and solving industrial-scale problems requires tailor-made solution algorithms.

On the theoretical dimension, I demonstrate that integration of the process design and control decisions opens up new venues to energy efficiency and environmental protection. The features of interest include economic performance, dynamic controllability, and flexibility in the presence of disturbances. On the application dimension, I have focused on (1) production of fuel and chemical through biomass pyrolysis, (2) CO_2 capture from power generation systems, and (3) Integration of solar and wind energies into smart electricity grids. The features of interest include whole-system analysis, multi-scale modelling, process integration, and optimization under uncertainty. Finally, applying the advanced theories to address complex applications requires tailored-made solution algorithms. Over the course of my academic studies, I have developed and applied a variety of computer-aided engineering tools spanning over a wide spectrum of off-the-shelf model libraries to customizable simulation tools to low-level optimization programming. Here, the features of interest include multi-scale model integration, optimization with explicit constraints, and exploiting problem logic for improving solution algorithms.

Teaching Interests:

My teaching philosophy is highly inspired by my research vision. It will be based on a balanced combination of theory and application and will employ enabling software tools. I believe the theory of process design and control is an indispensable element of the Chemical Engineering curriculum and will enable future industrial practitioners and academic researchers to close the gap between theory and application. The students should be equipped with a variety of simulation, optimization and programming software tools, that will enable quantification and optimization of economic and environmental performance of industrial processes. I believe students should be inspired by novel applications for energy security and environmental protection and be motivated to deliver impacts to societal wellbeing and economic prosperity.

Publications -The papers are published in highly-ranked journals. The corresponding author is shown by *. The co-authors with equal contribution are shown by ¥. The current bibliometric statistics of my publications are h-index 9, i10-index 7, and Citations >240, according to Google Scholar.

[1] **Sharifzadeh M***, (2013). Integration of process design and control: a review, ***Chemical Engineering Research and Design***, 91 (12), 2515–2549 (Link).

[2] **Sharifzadeh M***, Thornhill NF, (2013). Integrated design and control using a dynamic inversely controlled process model. ***Computers & Chemical Engineering,*** 48, 121–134. (Link).

[3] **Sharifzadeh M***, (2013). Implementation of an inversely controlled process model for integrated design and control of an ETBE reactive distillation, ***Chemical Engineering Science***. 92, 21–39, (Link).

[4] **Sharifzadeh M***, Thornhill NF, (2012).Optimal selection of control structures using a steady-state inversely controlled process model. ***Computers & Chemical Engineering***, 38 (5), 126–138, (Link).

[5] **Sharifzadeh M***, Rashtchian D, Pishvaie MR, Thornhill NF, (2011). Energy induced separation network synthesis of an olefin compression section: a case study. ***Industrial & Engineering Chemistry Research,*** 50 (3), 1610–1623, (Link).

[6] Wang L, **Sharifzadeh M**, Templer R, Murphy RJ*, (2012). Technology performance and economic feasibility of bioethanol production from various waste papers. ***Energy & Environmental Science***, 5, 5717-5730, (Link).

[7] Wang L, **Sharifzadeh M**, Templer R, Murphy RJ*, (2013). Bioethanol production from various waste papers: Economic feasibility and sensitivity analysis. ***Applied Energy,*** 111, 1172–1182, (Link).

[8] Chen L¥, **Sharifzadeh M¥**, MacDowell N, Welton T, Shah N, Hallett JP*, (2014). Inexpensive ionic liquids: [HSO4]−based solvent production at bulk scale. ***Green Chemistry***, 16, 3098-3106. (Link).

[9] **Sharifzadeh M*,** Richards C, Liu K, Hellgardt K, Chadwick D, Shah N. (2015). An integrated process for biomass pyrolysis oil upgrading: the synergistic approach. ***Biomass & Bioenergy***. 76, 108–117, (Link).

[10] **Sharifzadeh M*,** Wang L, Shah N, (2015). Integrated bio-refineries: CO_2 utilization for maximum biomass conversion. ***Renewable and Sustainable Energy Reviews***, 47, 151–161, (Link).

[11] **Sharifzadeh M*,** Wang L, Shah N, (2015). Decarbonisation of olefin processes using biomass pyrolysis oil. ***Applied Energy***, 149, 404–414, (Link).

[12] **Sharifzadeh M***, Cortada Garcia M, Nilay Shah, (2015). Supply chain network design and operation: Systematic decision-making for centralized, distributed, and mobile biofuel production using mixed integer

linear programming (MILP) under uncertainty. Biomass and Bioenergy, 2015, 81, 401–414, (Link).

[13] **Sharifzadeh M***, Shah N, (2015). Carbon capture from natural gas combined cycle (NGCC) power plants: solvent performance comparison at industrial scale. ***AIChE Journal***, 62 (1), 166–179, (Link), Invited paper.

[14] **Sharifzadeh M***, Shah N, (2015). Comparative studies of CO_2 capture solvents for gas-fired power plants: Integrated modelling and pilot plant assessments. ***International Journal of Greenhouse Gas Control***, 43, 124–132, (Link).

[15] **Sharifzadeh M***, Bumb P, Shah N, (2015). Carbon capture from pulverised coal power plants (PCPP): solvent performance comparison at industrial scale. ***Applied Energy***, 163, 423–435, (Link).

[17] Eini S, Shahhosseini HR, Javidi M, **Sharifzadeh M***, Rashtchian D. Inherently safe and economically optimal design using multi-objective optimization: the case of a refrigeration cycle. Accepted for publication at ***Process Safety and Environmental Protection***.

[16] **Sharifzadeh M***, Meghdari M, Rashtchian D. Integrated safe design and operation of Thermal Power System-Solid Oxide Fuel Cell (SOFC) Triple Combined-cycle System. Under review.

[18] **Sharifzadeh M***, Lubiano H, Shah N. Optimal design and operation of integrated renewable energy systems. Under review.

[19] **Sharifzadeh M***, Siirola JD, Grossmann IE. The hybrid reformulation algorithm for generalized disjunctive programs: Implementation in Pyomo. Under review.

[20] **Sharifzadeh M***, Konda M, Wang L, Guo M, Cortada Garcia M, Sadeqzadeh M, Hallet J, Shah N. Biomass pyrolysis and upgrading: the state of art and research frontiers. Under review.

481297 Automatic Exploration of Potential Energy Surfaces: Towards Reaction Mechanisms, Rate Constants, and Product Branching Ratios From First Principles

Abstract:

Research Interests:

My research focuses on computer simulations based on first principles of gas-phase reactions in combustion, atmospheric, and astrochemical systems. With modern computing power and the latest methodologies, ab initio calculations are yielding better agreement with experimental findings yet having the ability to cover a large range of parameters and conditions due to the simulations being done in silico.

In general, the scope of my research can be divided into two categories. The first category comprises of developing methodologies and strategies for automatic transition state (TS) finding and exploring of Potential Energy Surfaces (PES). The second category is applying strategies, either existing or developed by my research, to chemical systems of interests. Later, the methodologies and applications will be extended to condensed phase systems such as catalysis and organometallic systems.

Successful Proposals:

Japan Society for the Promotion of Science (JSPS) Postdoctoral Fellow, Kyoto University, Japan.

Postdoctoral Projects:

"*Sensitivity Analysis on Extinction Strain Rate Calculations of Natural Gas (NG) and Refinement to Detailed Kinetic Models*" and "*Automatic Reaction Discovery of Chemical Reactions*" under supervision of William H. Green, Department of Chemical Engineering, Massachusetts Institute of Technology

"*Automatic Exploration of the Potential Energy Surface from the Pyrolysis and Oxidation of Jet-Fuels*" under supervision of Keiji Morokuma, Fukui Institute for Fundamental Chemistry, Kyoto University

"*Global Investigation of the Pyrolysis of C_1-C_3 Hydrocarbons: Towards the Construction of Detailed Kinetic Models*" under Keiji Morokuma, Cherry L. Center for Scientific Computation, Emory University

Ph.D. Dissertation:
"*Ab Initio Quantum Chemical Studies on Neutral-Radical Reactions of Ethynyl (C2H) and Cyano (CN) with Unsaturated Hydrocarbons.*"

Research Experience:
Succinctly, my research experience has been towards the elucidation of reaction mechanisms, rate parameters, and product branching ratios based on first principle calculations of Potential Energy Surfaces (PES). In graduate school, the reaction mechanisms were constructed used *a priori*knowledge based on the chemical intuition of how molecules react. Then as a postdoctoral associate and postdoctoral fellow, I became a practitioner of automatic reaction search methods that explore the PES without any human intervention or bias of reactivity. This allowed for a systematic and comprehensive search of reaction mechanisms. In each of my postdoctoral appointments, the strategy of automatic reaction search varied, with the Global Reaction Route Mapping (GRRM) strategy employed under the supervision of Keiji Morokuma at Emory and Kyoto University and various chain-of-state methods under the supervision of William H. Green at Massachusetts Institute of Technology. In the latter position, I also did method development towards novel automatic search methods as well as analyzing detailed kinetic mechanisms reaction steps and kinetic parameters.

Teaching Interests:
In the summer and fall semester of 2014, I was an Adjunct Professor at Georgia Perimeter College, a part of the Georgia State University system. There, I taught undergraduate chemistry laboratory courses for Department of Natural Sciences. In graduate school, I was head TA for physical chemistry laboratories for junior and senior level undergraduate students majoring in chemistry. Part of the duties included devising my own curriculum based on given experiments, creating my own grading scale, and physical chemistry laboratory maintenance of equipment and ordering of supply. I would have an open-door office hour, allowing students to come anytime they wish with advanced notice, aside from having a set office hours.

Future Direction:
As a faculty, I would like to divide work in both method development and applications of methodologies. I would focus on method development of automatic search methods of PES, combining my past knowledge and skills in this area. I would also focus on the application of methodologies by elucidating reaction mechanisms and product branching ratios of important chemical reactions in combustion, atmospheric, and astrochemical systems. I would have a research intensive program that includes undergraduate students working on summer research projects towards the application of methods to chemical systems either existing in literature or developed by my group.

Selected Publications:
Ryazanstev, M.*; **Jamal, A.***; Maeda, S.; Morokuma, K. Global Investigation of Potential Energy Surfaces for the Pyrolysis of C_1-C_3 Hydrocarbons: Toward the Development of Detailed Kinetic Models from First Principles. *Physical Chemistry Chemical Physics.* **2015**, *17,* 27789-27805. [*Contributed equally]

Yang, T.; Muzangwa, L.; Kaiser, R. I.; Jamal, A.; Morokuma, K. A Combined Crossed Molecular Beam and Theoretical Investigation of the Reaction of the Meta-Tolyl Radical with Vinylacetylene – Toward the Formation of Methylnaphthalenes. *Physical Chemistry Chemical Physics.* **2015**, *17,* 21564-21575.

Parker, D.; Dangi, B.; Kaiser, R.; Jamal, A.; Ryazantsev, M.; Morokuma, K. Korte, A.; Sander, W. An Experimental and Theoretical Study on the Formation of 2-Methylnaphthalene (C11H10/C11H3D7) in the Reactions of the Para-Tolyl (C7H7) and Para-Tolyl-d7 (C7D7) with Vinylacetylene (C4H4). *Journal of Physical Chemistry A.* **2014**, *118*, 2709-2718.

Dangi, B.; Parker, D.; Kaiser, R.; **Jamal, A.**; Mebel, A. A Combined Experimental and Theoretical Study on the Gas-Phase Synthesis of Toluene under Single Collision Conditions. *Angewandte Chemie International Edition.* **2013**, *52*, 7186-7189.

Jamal, A.; Mebel, A. An Ab Initio/RRKM Study of the Reaction Mechanism and Product Branching Ratio of Ethynyl Radical with 1,2-Butadiene. *Chemical Physics Letters.* **2011**, *518*, 29-37.

Jamal, A.; Mebel, A. Reactions of C2H with 1- and 2-Butynes: An Ab Initio/RRKM Study of the Reaction Mechanism and Product Branching Ratio. *Journal of Physical Chemistry A.* **2011**, *115*, 2196-2207.

Jones, B.; Zhang, F.; Kaiser, R.; **Jamal, A.;** Mebel, A.; Cordiner, M.; Charnley, S. Formation of Benzene in the Interstellar Medium. *Proceedings of the National Academy of Sciences.* **2011**, *108*, 452-457.

Jamal, A.; Mebel, A. An Ab Initio/RRKM Study of the Reaction Mechanism and Product Branching Ratios of the Reactions of Ethynyl Radical with Allene and Methylacetylene. *Physical Chemistry Chemical Physics.* **2010**, *12*, 2606.

476234 Fostering Student Development from Novice to Expert

Abstract:

Research Interests:

Fostering students' transition from novice to expert is a key component of education. Among the factors at play in this process, I am interested in multiple representations for enhancing problem solving abilities, use of simulations for developing engineering intuition, and self-efficacy development.

Current Research and Future Direction:

1) Simulations as a tool for developing student "engineering intuition"

Simulations are a powerful supplement to the engineering classroom, allowing for otherwise infeasible experiments to be completed virtually. However, students often fail to question the simulations' assumptions, and/or fail to assess the final answer's accuracy. This project is a collaborative effort with a faculty at a small teaching institution. Currently, we are working to understand what factors may distinguish students that more often successfully address the assumptions of a simulation problem from those that do not. This information will inform future work on better optimizing the use of simulations for the benefit of all students.

2) The time-course of, and team effects on, individual self-efficacy development during senior design.

Many studies have shown that senior design has positive effects on student self-efficacy. However, few (if any) look at how student self-efficacy changes as the course progresses. In this work, I have noticed that some self-efficacy items show gains early in the semester, while others show gains only later. In future work, I will further assess the influence of team dynamics on individual self-efficacy. These results will inform approaches to team conflict resolution, and better our understanding of design self-efficacy development.

3) Evaluating disciplinary biases in representations used by faculty, and their impact on student learning.

The use of multiple representations has been linked to enhanced learning, but the connections specifically to development of problem solving and critical thinking skills across disciplines have not been thoroughly investigated. Here, comparisons will be made not only of representations used across subdisciplines in the same general area (e.g., biology vs. chemistry), but also across schools/colleges. Currently, this work is in the development stage. A multi-institution study of faculty is envisioned, and ultimately this will lead to work incorporating representations into curriculum through low-cost interventions.

PhD Dissertation: *"Learning in Style: Investigation of Factors Impacting Student Success in Chemical Engineering at Individual and Team-Levels with a Focus on Student Learning Styles."*

Under the supervision of Dr. David W. Wood, Department of Chemical and Biomolecular Engineering, The Ohio State University

Previous Research Experience: Following a variety of experiences in the fields of chemical engineering, genetics, medicine, and education, I found my way into chemical engineering education research. I began my graduate career working on developing bacterial biosensors capable of identifying nuclear hormone receptor agonists and antagonists for the malaria vector mosquito *Anopheles gambiae*. In my second year my advisor offered me the opportunity to take on the educational research component a proposal, and I began examining student factors for success in the chemical engineering curriculum with particular attention to learning styles, student self-efficacy, and team dynamics. This research became my focus, and I have since continued in engineering education. My dissertation work included three projects assessing the factors of learning styles, student self-efficacy (individual level), collective efficacy (team level), and attitudes, perceptions, and performance (individual and team levels): (i) Individual Level (introductory Mass and Energy Balances course)—evaluating the aforementioned factors with respect to specific exam problems and understanding how the variability in the learning styles engaged by these problems may correlate to self-efficacy and performance; (ii) Team Level (senior level Unit Operations course)—understanding how team composition (homogeneous vs. heterogeneous) with respect to learning styles may influence these factors; (iii) Combinatorial (Individual and Team) Level (senior level Design course)—understanding how collective efficacy may influence individual self-efficacy and again if there are any correlations with learning styles and performance.

Teaching Experience: I have taught a variety of chemical engineering courses. While a graduate student, I taught my department's graduate level Research Communication course for two years. From my work with this class, I was given the opportunity to design and develop a brand new course for undergraduate students in Technical and Professional Communication and have taught that both at Ohio State and Bucknell. In my position as a visiting faculty at Bucknell, I have taught and co-taught courses at all levels: Thermodynamics, Separations, Senior Design (co-instructor), Technical and Professional Communication, and Mass and Energy Balances (co-instructor). This variety of experience both speaks to and enhances my ability to teach across the curriculum. Most notably, teaching senior design has better allowed me to prepare my course objectives for freshmen through junior courses. Co-instruction has further enhanced my teamwork skills, and my ability to adapt to new teaching environments and adopt styles that may not necessarily be my own natural style.

Teaching Interests: I am interested in teaching across the chemical engineering curriculum, as well as continuing to teach technical/professional communication in some form. These interests align well in capstone design, and I also enjoy working with early career students in mass and energy balances. I would be happy to teach any of the courses in my repertoire again, while slowly continuing to expand that repertoire with other courses.

475721 Predictive Tools for Modelling Adsorption Phenomena

Abstract:

Predictive Tools for Modelling Adsorption Phenomena

Richard T. Cimino, Rutgers University

1st Year Postdoctoral Fellow

Research Interests:

From cell signaling to the capture and sequestration of CO_2, phenomena that occur on the nanoscale (1-100nm) have a critical impact on our "macroscale" existence. The technological advances in this broad field are felt today in many aspects of life – from the most mundane (water filters, air purifiers and cat litter!) to the most extraordinary (drug treatments, solar cells and super capacitors). In particular, my research focuses on one of the most ubiquitous classes of nanoscale phenomena – that of adsorption. Adsorption - the process by which one nanoscale object "sticks" to another - has myriad subfields and applications and though it has been studied theoretically for over a century, our understanding of the fundamentals of the adsorption process are still the subject of scientific inquiry. Using statistical mechanics and computational modelling, my research seeks to provide the scientific community with a more complete understanding of the fundamentals of this nanoscale phenomenon, while also providing practical and predictive tools that can be used by others to further their adsorption-related research.

Postdoctoral Projects: "Calculation of the Isosteric Heat of Adsorption Using Quenched Solid Density Functional

Theory"

Under supervision of Alexander V. Neimark, Distinguished Professor, Department of Chemical and Biochemical Engineering, Rutgers University

PhD Dissertation: "Molecular Modeling of Adsorption of Simple and Complex Fluids on Nanoporous Materials"

Under supervision of Alexander V. Neimark, Distinguished Professor, Department of Chemical and Biochemical Engineering, Rutgers University

Research Experience:

My doctoral research focused on the development of computational and theoretical tools to describe the adsorption of simple and complex fluids in nanoporous materials using a variety of techniques including density functional theory, Monte Carlo simulation, and analytical modelling. Nanoporous materials consisted primarily of tailored, hierarchically structured (containing micro, meso- and macropores) carbons and silicas, which are touted for their high surface area-to-volume ratio. I have created novel hybrid Density Functional Theory kernels to analyze Ar, N_2 and CO_2 sorption isotherms on molecularly rough carbon and silica materials. These kernels have been incorporated into industrial software produced by Quantachrome Instruments and used by scientists across the world for the characterization of porous materials.

In another project, I developed a model of polymer interaction chromatography based on a fundamental thermodynamic approach utilizing Monte Carlo simulations and classical mass transfer dynamics. This model is capable of describing the three modes of liquid chromatographic elution: SEC, LAC, and liquid chromatography at critical conditions (LCCC) with minimal parameterization. This model can be used by chromatographic practitioners to predict the order of elution of polymer chains, or to determine the critical conditions of adsorption for chains of similar chemistry.

Teaching Interests:

In addition to my research experience, I have an extensive teaching history in both Chemical Engineering and Mathematics. As an undergraduate, I worked as a student mentor for three years within Rutgers' Mathematics Department, working one-on-one with students during Calculus recitation sessions. I had the unique opportunity for the duration of my senior year to teach a recitation of Precalculus Mathematics, with thirty students per semester. I worked for the majority of my Chemical Engineering graduate program at Rutgers as a Teaching Assistant, teaching Kinetics and Reactor Design four times at the Graduate level and once at the undergraduate, and teaching Fluid Mechanics two semesters at the Graduate level. In the future, I would like to teach Kinetics, as well as introductory courses in molecular simulation and statistical thermodynamics, aimed at Graduate students.

Future Direction:

Of interest to me are several problems in adsorption which concern medical/biological applications of adsorption and environmental sustainability.

Adsorption Dynamics and Nanoscale Effects during Hydraulic Fracturing

The exploitation of shale gas reservoirs by hydraulic fracturing in the United States and other countries over the past 15 years has led to a boom in natural gas production, with the US currently deriving roughly 40% of its natural gas from shale. However, speculation on future production estimates is controversial due to uncertainty about shale gas availability and concerns about hydraulic fracturing's effect on the environment. There is therefore a need to improve our understanding of the hydraulic fracturing process, in order to design safer and more cost-effective drilling operations. I intend to pursue a research program to computationally study the complex interactions of alkane mixtures in the pores of shale materials in an effort to 1.) examine the effect of slickwater (the mixture of water, sand, and chemicals pumped into wells prior to liberating gas) and on adsorption and mixing of alkanes in shale pores under geological conditions and 2.) explore the transport dynamics of alkane/slickwater mixtures in the pores of shale. I have recently submitted this project as a proposal to the NRC postdoctoral competition.

Critical Adsorption and separation of biological polymers

As an extension of my work on polymer chromatography, I would like to investigate the application of critical adsorption to the separation of realistic biopolymers – namely proteins, DNA and RNA for drug applications and gene sequencing. Some specific problems in this field that I would like to contribute are 1.) the translocation of biopolymers for sequencing, 2.) The separation of biopolymers by critical properties, 3.) the adsorption and uptake of drug molecules by tumors. 4.) Application of adsorption to protein folding and solvation.

Selected Publications:

§ R. Cimino, C. Rasmussen, Y. Brun, A. V. Neimark. Mechanisms of Chain Adsorption on Porous Substrates and Critical Conditions of Polymer Chromatography: Insight from Monte Carlo Simulations (Submitted)

§ R. Cimino, C. Rasmussen, Y. Brun, A. V. Neimark. Critical conditions of polymer adsorption and chromatography on nonporous substrates, Journal of Colloid and Interface Science. 474 p. 25-33 (2016)

§ R. Cimino, C. Rasmussen and A. V. Neimark, Communication: Thermodynamic analysis of critical conditions of polymer adsorption, J. Chem. Phys., 139 (20) p.201101-1-4 (2013)

§ R. Cimino, K. Cychosz, M. Thommes and A. V. Neimark, Experimental and theoretical studies of scanning adsorption–desorption isotherms, Colloids Surf. A: Physicochem. Eng. Aspects, 437 (SI) p.76-89 (2013)

§ K. Cychosz, X. Guo, W. Fan, R. Cimino, G. Gor, M. Tsapatsis, A. V. Neimark, and M. Thommes, Characterization of the Pore Structure of Three-Dimensionally Ordered Mesoporous Carbons Using High Resolution Gas Sorption, Langmuir, 28 (34) p.12647-12654 (2012)

§ R. Cimino, C. Rasmussen, Y. Brun, A. V. Neimark. Simulation of Polymer Interaction Chromatography: A Review (in preparation for submission)

§ R. Cimino, K. Cychosz, M. Thommes and A. V. Neimark, Characterization of Micro-Mesoporous Carbons by High-Pressure CO_2 Adsorption with Hybrid QSDFT Methods (in preparation for submission)

§ R. Cimino, P. Kowalczyk, A. V. Neimark, "Calculation of the Isosteric Heat of Adsorption Using Quenched Solid Density Functional Theory" (in preparation for submission)

466046 Multiscale Design of Nanomaterial Synthesis

Abstract:

Research Interests:

- Multiscale modeling and simulation of aerosol reactor design for the synthesis of nanostructured materials (discrete models â€" Molecular Dynamics and Brownian Dynamics, continuum models â€" balance equations)
- Nanoparticle interactions with biological interfaces (adhesion, self-assembly, dynamics in confined geometries)
- Theory and simulation of polymer nanocomposites and modeling of their rheological and mechanical properties (effect of polymer architecture and molecular weight, particle size, polydispersity, temperature)

Teaching Interests:

- Molecular Simulations and Statistical Mechanics
- Physical Chemistry, Chemical Engineering Thermodynamics, Statistical Thermodynamics
- Reaction Engineering and Surface Chemistry
- Aerosol Dynamics and Growth

Abstract

Aerosol particle technology is used for manufacturing of carbon black, fumed SiO_2, pigmentary TiO_2, filamentary Ni and ZnO for rubber vulcanization as well as for advanced materials (e.g. photocatalysts, nanofluids and biomaterials) with a number of functionalities and high-performance applications (Goudeli and Pratsinis, 2016a). Even though such materials can be made by both gas or liquid processes, aerosol technology has attracted a lot of interest due to its undisputable capacity for large-scale production of nanostructured commodities (up to 25 t/h), high product particle purity, easier collection and fewer unit operations (Pratsinis, 2010). Such nanostructured commodities consist of clusters of primary particles (PPs) that are formed by chemical reactions, condensation/evaporation or surface growth and grow further by sintering and coagulation. Depending on process conditions and particle residence time, coagulation and sintering (or partial coalescence), gas-phase processes typically result in aggregates (highly-sintered particles) which are attractive in catalysis, lightguide preforms and electronics and, downstream in the process in agglomerates (PPs held together by rather weak, physical forces) that are attractive in nanocomposites, pigments and liquid suspensions.

This motivates research for understanding of the formation of such gas-phase-generated nanoparticles. Multiscale design of aerosol reactors for synthesis of nanomaterials includes continuum, mesoscale, molecular dynamics (MD) and quantum mechanics models allowing fundamental understanding of particle growth dynamics. More specifically, quantum mechanics account for the electronic structure of matter determining the interatomic potentials in MD models for accurate estimation of sintering and crystallization rates. Particle properties, like structure and crystallinity (that cannot be easily obtained experimentally) are determined from first principles by highly parallelized MD simulations (Goudeli and Pratsinis, 2016b) as they affect material performance in a score of applications (e.g. sensors, catalysts, biometerials). Furthermore, MD simulations can be used also to determine the surface structure of bimetallic nanoparticles (e.g. Ag-Au) that facilitates understanding of catalytic properties.

Such accurate MD-derived sintering rates and crystallinity dynamics can be incorporated in mesoscale models that provide the transport properties and coagulation rate of multi-particle structures and continuum models that describe the effect of process variables on product particle size and morphology at various process temperatures and residence times (Buesser and Pratsinis, 2012). Thus, Discrete Element Modeling (DEM) simulations can used to track the detailed structure and size distribution and agglomerate size of fractal-like agglomerates consisting of monodisperse (Goudeli et al., 2015a) or polydisperse primary particles undergoing Brownian coagulation. The DEM-obtained mobility size and number-based geometric standard deviation of the mobility radius are compared to real-time, online measurements (e.g. Scanning Mobility Particle Sizer and Differential Mobility Analyzer combined with Aerosol Particle Mass analyzers) as function of the number of primary particles per agglomerate (Goudeli et al., 2016). Particle morphology and structure are quantified by the fractal dimension, D_f, and mass-mobility exponent, D_{fm}, which are related to the cluster optical and transport properties, respectively, from spherical particles to fractal-like agglomerates until they attain their well-known asymptotic structure.

The above easy-to-use relations can be readily interfaced with climate dynamics, meteorological models or computational fluid dynamics describing the reactor operation and particle production. For example, the effect of varying particle structure during the formation of fractal-like crystalline TiO_2 and amorphous SiO_2 is investigated (Goudeli et al., 2015b) by employing the above DEM-derived D_f evolution in monodisperse continuum models. Neglecting the evolving particle structure overestimates the hard-agglomerate diameter by up to 25% for TiO_2 and 30% for SiO_2, especially at high maximum temperatures and cooling rates that hard-aggregates consist of only a few primary particles.

Interfacing the above models that span 10 â€“ 15 orders of magnitude in length and time, respectively, can facilitate the understanding and scale-up design of aerosol reactors for synthesis of nanoparticles whose properties can be closely controlled during scale-up from laboratory scale to commercial products. This systematic approach to study particle formation can offer significant insight into fundamental physical principles and mechanisms that may be exploited by biomedical or pharmaceutical industry and nanotechnology.

Keywords: particle technology, multiscale modeling, aerosol reactors, gas-phase synthesis

References

Buesser, B., and Pratsinis, S.E. (2012) *Annu. Rev. Chem. Biomol. Eng.*, **3**, 103-127.

Goudeli, E., Eggersdorfer, M.L., and Pratsinis, S.E. (2015a) *Langmuir* **31**, 1320-1327.

Goudeli, E., Eggersdorfer, M.L. and Pratsinis, S.E. (2015b) *J. Aerosol Sci.* **89**, 58-68.

Goudeli, E., Gröhn, A.J. and Pratsinis, S.E. (2016) *Aerosol Sci. Technol.* **50**, 591-604.

Goudeli, E. and Pratsinis, S.E. (2016a) *Particul. Sci. Technol.* in press.

Goudeli, E. and Pratsinis, S.E. (2016b) *AIChE J.* **62**, 589-598.

Pratsinis, S.E. (2010) *AIChE J.* **56**, 3028-3035.

472242 Charge Storage Mechanisms of Carbides and Nitrides Based Supercapacitors

Abstract:

Research Interests:

- Electrode-structure function relationships in chemical energy storage

- Design, synthesis and characterization of functionalized nanostructured materials

- Nanostructured materials for electrochemical biosensors

Teaching Interests:

Iâ€™m interested in teaching traditional chemical engineering courses such us thermodynamics, heat and mass transfer, and reaction engineering and design. Additionally, I intend to develop and teach non-traditional courses including electrochemical engineering, design and synthesis of nanostructured materials, and energy storage systems.

475478 Dynamic Properties of Interfaces in Soft Matter

Abstract:

Research Interests:

The dynamic properties of soft materials are central to a wide range of engineering and biomedical applications. While the synthesis and structural characterization of soft materials have come a long way, there are still many big gaps in both our fundamental understanding and practical applications of soft materials, especially in the realms of quantification of intermolecular forces, transitions between different length scales, and dynamics (non-equilibrium, time- and rate-dependent physical and engineering properties). The knowledge filling these gaps will allow rational design of functional soft matter interfaces, and has profound implications across a wide range of disciplines, including colloidal and interface science, biomaterials and bioengineering, polymer chemistry and physics, and nanotechnology.

The overall goal of my research is to understand the connection between a soft material systemâ€™s molecular architecture and interactions, and its dynamic surface properties. The research aims at:

(1) understanding the physical laws that govern the structure, dynamics, and cooperative behavior of soft materials at different time and length scales;

(2) developing and characterizing advanced soft materials and surfaces with novel physical and chemical properties through molecular design and chemical synthesis.

These goals will be pursued and achieved by establishing the physiochemical interactions of soft matter molecules at interfaces through novel chemical synthesis, sensitive surface forces measurements, state-of-the-art scattering techniques, and theoretical calculations and modeling. The proposed research is *interdisciplinary* and *collaborative*, and easily integrated with other major chemical engineering research areas, such as synthesis, rheology, interfacial/colloidal science, polymer science, and biomaterials. Specific interests include (1) bio-inspired adhesives, (2) structure and functionality of polyelectrolyte brushes, and (3) supramolecular cyclic peptides for biomedical applications;.

I received my PhD in Chemical Engineering at the University of California, Santa Barbara in December 2012, under the supervision of Prof. Jacob Israelachvili. I am currently a postdoctoral associate, working with Prof. Mathew Tirrell in the Institute for Molecular Engineering at the University of Chicago and Argonne National Laboratory.

Teaching Interests:

I look forward to teaching and developing courses at both the undergraduate and graduate level. During my PhD at UCSB, I was the teaching assistant (TA) for graduate-level Advanced Theoretical Methods in Engineering, and undergraduate-level Heat Transfer, Thermal Dynamics, and Biomaterials & Biosurfaces courses. Additionally, I mentored 4 undergraduate students in the Summer Intern programs at UCSB and Caltech. Through my TA experiences and summer intern mentorships, I have learned that the interaction between an instructor and a student is essential to keep the student motivated in the subject matter. Therefore I will create an interactive classroom where the students are encouraged to ask questions. I would also encourage students to meet with me periodically to discuss their progress in the class and the subjects they plan to further explore.

475750 Microstructure-Rheology Relationship in Complex Fluids:Towards Design of Soft Materials with Tunable Properties

Abstract:

Research Interests:

The field of complex fluids encompasses a wide class of materials, which exhibit unusual mechanical responses to an applied stress or strain. In virtually all complex fluids, this rich and unusual mechanical response originates from a microstructure that responds to different applied stress or strain in specific and varied ways. Thus understanding the microstructure â€“ macroscopic behavior relationship is a crucial step for systematically designing complex fluid materials for novel applications.Â The complex fluid landscape can be subdivided based on the particle-level interactions that govern their underlying microstructure and the resulting macro rheology:

In my graduate research work under supervision of **Prof. Joao Maia (Case Western Reserve University)** and in collaboration with **Prof. Norman Wagner (University of Delaware)**, I studied the interplay between the microstructure and mechanical response of repulsive and neutral (so-called hard spheres) concentrated suspensions, with emphasis on Shear-Thickening behavior, where viscosity of a dense suspension progressively increases with increasing the applied rate of deformation. To do so, I developed a mesoscale simulation scheme, based on Dissipative Particle Dynamics method, providing a platform that enables us to study the crucial role of multi-body long and short range hydrodynamic effects, frictional interactions between the constituents, particle deformability, and particle anisotropy.

During my postdoctoral research work with **Prof. Robert Armstrong** and **Prof. Gareth McKinley (MIT)**, I studied the role of microstructural evolutions of attractive systems in defining their macroscopic rheological response. I established a quantitative correlation between the microstructural measures of a system, to flow properties and heterogeneities observed in complex fluids. We showed that steady state properties, as well as transient phenomena at various time scales can be explained through detailed study of the microstructure. Furthermore, we proposed a Time-Peclet-Transformation (TPT) procedure with regards to their fundamental interactions and flow properties/history. This provides a phase map for the attractive systems such as nano emulsions and colloidal gels, prescribing novel design tools for complex fluids with modified and tunable properties.

By exploring the microstructure-rheology relationship for a wide range of complex fluids, and connecting this knowledge to electrochemical processes, my future research will focus on finding new routes for designing soft materials with tunable properties, with emphasis on applications in flow batteries and water desalination technologies.

Teaching Interests:

Will be presented upon request.

476062 Dynamics of Deformable Objects in Flowing Fluids: Polymers, Metamaterials, and Beyond

Abstract:

Research Interests:

My current research is focused on understanding the behavior of origami metamaterials in flowing fluids. Such metamaterials are two dimensional sheets with predefined creases that can be folded using fold patterns prevalent in traditional origami. Specifically, I have chosen a periodic herringbone crease pattern known as Miura pattern in the origami community as my model system. This is a particularly simple design that still shows a very rich set of behaviors. When subjected to shear flow, the Miura unit cell alternately deforms and tumbles and in contrast to simple closed Jeffrey orbits, it shows a wide variety of trajectories, including limit cycles and quasiperiodic orbits.

During my doctoral research with Kevin Dorfman and Satish Kumar at the University of Minnesota I investigated the role of hydrodynamic interaction in adsorption and desorption of polymer chains under shear flow. I showed that in case of weak adsorption longer chains desorb faster than shorter chains due to wall-induced migration, but for strongly adsorbed chains the opposite trend holds. This provided a physical explanation behind several long unresolved experimental observations in the literature of the molecular weight dependence on desorption.

In future I want to focus my research on problems involving the interaction of flowing fluids and elastic or piecewise elastic structures. These include, e.g. articulated bodies actuated in microscale flows and flows driven by aggregates of shape-morphing bodies. They key aspect of these kinds of problems is the coupling between flow-induced changes in geometry and material properties that in turn affect the flowfield. This constitutes a very rich but yet unexplored area and I believe my research will significantly aid in expanding the frontiers of fluid mechanics.

Teaching Interests:

I am willing to teach any of the core undergraduate level courses in chemical engineering. At the graduate level, I would prefer to teach fluid mechanics/transport courses. I would also like develop a new course on scientific computing techniques for engineers, aimed at advanced undergraduate or first year graduate students. The purpose of this course will not be to teach a programming language or numerical methods, rather it will focus on the design and implementation of scientific software, thereby training students to think beyond the conventional procedural paradigm and help them in developing flexible and reusable software.

476076 *Interfacial Properties and Field-Driven Assembly of Colloidal Nano/Micro Particles*

Abstract:

Research Interests:

Our ability to control interparticle potentials underlies many applications of colloids (e.g., paints, coatings, mineral recovery, personal care and food products). Control of interparticle potentials is essential to achieve colloidal stability, to impart directionality on the interactions, or to exploit the emerging properties of colloids through self- and directed-assembly. In this poster, I will explore how the interfacial arrangement of surfactant molecules determines the interaction potential of colloidal carbon nanotubes, their dispersion and separation processes. A combination of small-angle neutron scattering and analytical ultracentrifugation provides key quantitative information on the binding number, hydration (solvation) and conformation of surfactant molecules at the aqueous interface of carbon nanotubes. Further control of the interparticle potential can be achieved by imposing external electric fields on particle ensembles. I will show how the superposition of electric fields on particles with asymmetric surfaces results not only in assembly, but also in unprecedented electrokinetic behavior. Janus particles under direct current electric fields, for example, respond by levitating and forming clusters at high concentrations. Understanding the intricacies of the interactions and electrokinetics of anisotropic colloids is an important step towards achieving dense assemblies that can result in colloidal crystals. Colloidal crystals are of interest for applications such as structural colors, sensing, and camouflage.

Teaching Interests:

My teaching interest lies in the fundamental engineering courses at the undergraduate level. Therefore, although I can certainly teach most of the traditional subjects in the ChE curriculum, my background will best fit to teach classes at the undergraduate level such as Thermodynamics and Transport Phenomena. I am also interested in teaching and further developing a course on the physical chemistry of colloids that can be taken by advance undergraduate and graduate students. After serving as a teaching assistant for a year and mentoring four undergraduate students in research projects, I realized that teaching is an intellectual challenge that can potentially be a gratifying experience for my students and myself. To fulfill that potential, I am willing to tap into the collected knowledge of my colleagues, the findings of cognitive psychology and STEM education research.

475862 Molecular Modeling and Simulation for Carbon Capture and Sequestration

Abstract:

Carbon capture and sequestration (CCS) covers broad aspects of chemical, material, petroleum and geological engineering including phase equilibrium, gas separation, fluid transport in confinement, and reaction kinetics. The design of an energy efficient CCS process requires knowledge for complex fluid and solid mixtures involving water, carbon dioxide, heavy organic compounds, electrolytes. Novel materials, for example ionic liquids, polymeric membranes, or metal organic frameworks also have been proposed for the relevant separations. Molecular-level modeling and simulation are valuable alternatives to experimental studies for these complex systems, and their ability to access microscopic time and length scales makes it an ideal tool to understand phenomenon that happens at molecular level, for example transport of hydrocarbon in nanopores and carbon dioxide diffusion through membranes.

Research Interests:

My current research focuses on the modeling and simulation of thermophysical properties for CCS related systems. During my postdoctoral training in the group of Prof. A. Z. Panagiotopoulos at Princeton University, I applied molecular dynamics and Monte Carlo simulation techniques to study the mixture of carbon dioxide, water and electrolytes in order to understand the phase behaviors and transport properties of such system under the pressure and temperature conditions of geologic carbon dioxide storage. In particular, I have evaluated existing molecular models and developed new polarizable molecular force fields that substantially improve the accuracy of the simulations. [1,2] I have also developed molecular equation of state to characterize the mixture of carbon dioxide and geologic formation brines. [3] I did my doctoral studies under the supervision of Prof. Hertanto Adidharma at University of Wyoming. During my PhD studies, I developed statistical associating fluid theory to study the phase equilibria of gas hydrates under the influence of aqueous ionic liquid solutions, [4] and I received the outstanding dissertation award from University of Wyoming.

In the future, I plan to apply my expertise in molecular modeling and simulation to energy and CCS related problems. First, in order to improve the quality of simulations for mixtures of polar / nonpolar components, I plan to develop polarizable molecular models for components in flue and sour gas, and for liquid absorbents, such as ionic liquids. I also plan to develop molecular thermodynamic models for adsorption process using novel materials, such as molecular sieve. Second, I am interested in the modeling and simulation of thermodynamic and flow properties for water, electrolytes, carbon dioxide, and organic matters in geologic confinement with the aim of understanding important phenomenon in shale gas recovery and carbon dioxide geologic sequestration. Third, I plan to apply molecular simulation techniques to aid the design of ionic liquids as low dosage kinetic inhibitors of natural gas hydrate, which is a major flow assurance challenge for offshore oil and gas production.

Teaching Interests:

In addition to my research, I have experience in both teaching and mentoring students. At University of Wyoming, I taught "Introduction to Chemical Engineering Computing" for two semesters as temporary lecturer, and guest-lectured other undergraduate and graduate courses in both Chemical and Petroleum Engineering. At Princeton University, I mentored two undergraduate students for their research projects, and one of the project led to a journal publication. In the future, I have interests in teaching chemical thermodynamics, statistical mechanics and molecular simulation methods.

References

1. **Jiang, H.**; Moultos, O. A.; Economou, I. G.; Panagiotopoulos, A. Z. Gaussian-Charge Polarizable and Non-Polarizable Models for CO_2. *Journal of Physical Chemistry B*. 2016, 120, 984-994.

2. **Jiang, H.**; Mester, Z.; Moultos, O. A.; Economou, I. G.; Panagiotopoulos, A. Z. Thermodynamic and Transport Properties of H_2O+NaCl from Polarizable Force Fields. *Journal of Chemical Theory and Computation*. 2015, 11, 3802.

3. **Jiang, H.**; Panagiotopoulos, A. Z.; Economou, I. G. Modeling of CO_2 Solubility in Single and Mixed Electrolyte Solutions Using Statistical Associating Fluid Theory. *Geochimica et Cosmochimica Acta*. 2016, 176, 185-197.

4. **Jiang, H.**; Adidharma, H. Thermodynamic Modeling of Aqueous Ionic Liquid Solutions and Prediction of Methane Hydrate Dissociation Conditions in the Presence of Ionic Liquid. *Chemical Engineering Science*, 2013,102, 24-31.â€¨

475999 Investigation of Dynamics of Soft Materials to Design Multifunctional Materials

Abstract:

Research Interests:

My scientific interest involves extending the understanding of physics of soft materials such as colloidal suspensions, emulsions and polymer composites to design new/multifunctional materials that are more energy efficient, environmentally friendly, clinically safe and cost effective. These studies are becoming particularly challenging when inherent material properties (e.g. opaqueness, thermal stability, photobleaching, etc) limit utilization of conventional characterization techniques. Thus the nature of my research is based on simultaneous utilization of numerous characterization techniques, e.g. light scattering, optical microscopy, rheology, and microfluidics in conjunction with theoretical models. I attempt to answer fundamental questions such as how do materials' pristine properties, the processing conditions and molecular/mesoscopic interactions affect the bulk properties? What are the mechanisms of structure formation under the influence of various parameters? What is the time scale of structural evolution under different processing conditions? How to control material bulk properties by manipulating the properties at smaller scales? To what extent do different forces contribute to transport of soft materials (such as vesicles, double emulsions, colloidal suspensions, etc.)?

As a faculty member, I would like to extend my research in the field of complex fluids to design more energy efficient transport processes (applicable in different industries) and optimize on-chip devices for cell sorting and characterization mainly used in biomedical and clinical applications.

Teaching Interests:

- San Francisco State University (present)
  - Currently teaching: Heat Transfer, Engineering Cost Analysis, Materials and Manufacturing Processes + Lab.
- University of California Berkeley (2014-present)
  - Defined and supervised undergraduate students in research projects including:
    - Devising an optimal microfabrication protocol using thin film lamination.
    - Particle flow velocimetry in microfluidic devices: simulation and experiments.
- Schlumberger Oil Field Services (2000-2005)
  - Conducted intensive technical workshops (40 hrs/lecture) for diversified oil and gas professionals in the following subjects:
    - Production Data Management
    - Well Performance, Pipeline and Facilities and Network Analysis
    - Production Optimization
    - Risk and Value Management

Publications:

1) Khalkhal, F., K. Chaney, S. Muller, "Optimization and application of dry film photoresist for rapid, low cost fabrication of high aspect ratio microfluidic devices", **J. Microfluidics and Nanofluidics** (submitted).

2) Khalkhal, F.and S. Muller, "Formation of inertio-elastic vortices of a shear thinning fluid in planar microfluidic abrupt contractions", (journal article in preparation).

3) Khalkhal, F., M. Hwang, S. Muller, "Inertial and non-inertial migration of double emulsion droplets in wall-bounded shear flow", (journal article in preparation).

4) Ren, F., S. A. Kanaan, F. Khalkhal, C. Z. Loebick, G. L. Haller and L. D. Pfefferle (2013), "Controlled cutting of single-walled carbon nanotubes and low temperature annealing", **Carbon** (63), pages 61-70.

5) Khalkhal, F. and P. J. Carreau (2012) "Critical shear rates and structure build-up at rest in MWCNT suspensions", **J. Non-Newtonian Fluid Mechanics** 171-172, pages 56-66.

6) Khalkhal, F., ***Characterization of flow-induced structures in carbon nanotube suspensions***, Ph.D. thesis, Ecole Polytechnique de Montreal, Montreal, 2011.

7) Khalkhal, F. and P. J. Carreau (2011), "Scaling behavior of the elastic properties of non-dilute MWCNT-epoxy suspensions", **Rheologica Acta** 50(9), 717-728.

8) Khalkhal, F., P. J. Carreau and G. Ausias (2011), "Effect of flow history on linear viscoelastic properties and the evolution of the structure of MWCNT suspensions in an epoxy", **J. Rheology** 55(1),153-175.

Successful grant writing experience:

Postdoc Fellowship: Fonds de Recherche du Québec – Nature et Technologies (FQRNT), 2014 to 2016.

Web page:

https://www.linkedin.com/in/fatemeh-khalkhal-b2b33052

476004 Materials Development for Electrochemical Applications By Combined Experiment and Theory

Abstract:

Materials Development for Electrochemical Applications by Combined Experiment and Theory

Matthias J. Young, 2nd Year Posdoctoral Fellow
Applied Chemicals and Materials Division, Materials Measurement Laboratory,
National Institute of Standards and Technology, Boulder, CO 80305

Research Interests:

In order to rationally design improved materials for electrochemical applications, we need to understand the origins of electrochemical performance and instability in active materials. This is true for established electrochemical technologies (***e.g.*** lithium ion batteries and supercapacitors), as well as for emerging electrochemical applications (***e.g.*** electrochemical desalination and separation processes). Understanding the performance and instability of active materials during electrochemical operation is a formidable task given the complexity of the physics involved and dynamic material structures present under applied bias. In this poster, I discuss the use of experimental and theoretical techniques to understand where we have been, and help decide where we need to go to improve material performance for electrochemical applications. Specifically, I aim to leverage my background in nanomaterials synthesis and characterization, ***ab initio*** materials modelling, and ***in operando***materials characterization to improve existing battery materials and develop new materials for emerging applications including anion intercalation electrodes for electrochemical desalination and high rate pseudocapacitive electrodes for grid-level energy storage.

Successful Proposals: NSF Graduate Research Fellowship, NRC Postdoctoral Fellowship, Advanced Photon

Source Beam Time, Janus Supercomputer Time (CU-Boulder)

Postdoctoral Project: "***In-situ Measurement and Quantum Simulations of a Layered Double Hydroxide as an Anion Intercalation Electrode for Battery-Inspired Water Desalination***"

PhD Dissertation: "***Charge Storage in Thin Films of Cation-Incorporated Manganese Dioxide***"

Under supervision of Steven M. George & Charles B. Musgrave, Chemical and Biological Engineering, University of Colorado Boulder

Research Experience:

I began doing research as an undergraduate freshman in chemical engineering. I helped initiate a project on battery materials, which was a new area for my advisor. I continued this research throughout my undergraduate studies, and became motivated to develop a fundamental understanding of what materials properties dictate electrochemical performance. In graduate school, I sought out advisors who could help me to obtain both a strong theoretical understanding and experimental skill set and was co-advised by experts in computational chemistry and nanomaterial synthesis/characterization. With them, I developed a project aimed at understanding unique charge storage properties in manganese oxide leveraging both computational and experimental skill-sets, and successfully carried out this work under an NSF graduate research fellowship. Last year I received an NRC postdoctoral fellowship to develop new materials for electrochemical anion intercalation and during this postdoc have expanded my skill set to include ***in operando***material characterization using synchrotron techniques.

Teaching Interests:

My first experiences teaching scientific subject matter were as an undergraduate tutor. I had open tutoring office hours during my undergraduate studies covering all of the subject matter from the courses I had taken. The students I tutored gave me positive feedback for my calm and determined assistance, and my office hours were consistently packed with students. As a graduate student at the University of Colorado I worked as an instructor for a K-12 science outreach program. I also TAed and guest-lectured process control for two terms. This included conducting a weekly recitation with new course content and directing a weekly lab. As a postdoc, I co-instructed a graduate quantum simulations course and prepared in-class lectures and workshops. I have also mentored multiple undergraduate and graduate students in a research setting. I would be able to teach any course in the chemical engineering core curriculum, but am especially well-equipped to teach process control and computational methods. I would also be interested in developing an elective in electrochemistry fundamentals.

Future Direction:

As faculty I would like to continue to apply both computational and experimental techniques to advance the understanding of the origins of electrochemical performance and instability in active materials. In particular, I would like to (a) provide insight and help to develop solutions to the materials challenges facing existing battery technologies, (b) develop materials for electrochemical anion intercalation, and (c) develop improved materials for high-rate pseudocapacitive charge storage. For each of these areas, I will work with students to employ state-of-the-art materials synthesis, computational techniques, and ***in operando***materials characterization to provide a strong fundamental basis for analysis and interpretation.

The thread that connects all of my research aspirations is the need I see for a fundamental and quantitative understanding of the physical properties that dictate electrochemical performance. With a well-established fundamental understanding, I believe that we will be able to achieve new heights in electrochemical technology because we will be able to intelligently remedy the shortcomings of established technologies, and identify new materials with exciting properties. With optimized anion intercalation electrode materials comes the promise of anion-based batteries and solar-powered energy-efficient water desalination. With improved pseudocapacitive electrode materials comes the promise of battery-like devices with unprecedented charge rates, and capacities. These two stepping stones are only the beginning of a vast expanse of electrochemical technologies I see on the horizon.

Selected Publications:

M.J. Young, C.D. Hare, C.B. Musgrave, S.M. George, "Rapid Growth of Crystalline Mn_5O_8 by Self-Limited Multilayer Depsition using $Mn(EtCp)_2$ and O_3." ***ACS Applied Materials and Interfaces***. **2016**. [online ahead of print]

M.J. Young, M. Neuber, A.S. Cavanagh, H. Sun, C.B. Musgrave, S.M. George, "Sodium Charge Storage in Thin Films of MnO_2 Derived by Electrochemical Oxidation of MnO Atomic Layer Deposition Films" ***Journal of the Electrochemical Society***. **2015**. doi: 10.1149/2.0671514jes

M.J. Young, A.M. Holder, S.M. George, C.B. Musgrave, "Charge Storage in Cation Incorporated α-MnO_2," ***Chemistry of Materials***. **2015**. doi: 10.1021/cm503544e

D. M. Piper, J. J. Travis, **M. Young**, S. B. Son, S. C. Kim, K. H. Oh, S. M. George, C. Ban, and S. H. Lee, "Reversible high-capacity Si nanocomposite anodes for lithium-ion batteries enabled by molecular layer deposition," ***Advanced Materials***, **2014**. doi: 10.1002/adma.201304714

476007 Interfacial Dynamics of Soft Matter and Low-Cost Diagnostic Devices

Abstract:

Research Interests:

Complex Fluid Interfaces

Â When small biomolecules (lipids, proteins) and particles adsorb at fluid surfaces, they render complex microstructural and rheological properties to the interface. These molecular-thin layers provide macroscopic functionality in both nature and engineered systems. Examples in nature include the foam nests of frogs and spittle bugs, the membranes of living cells, the tear film of the eye and alveoli in mammalian lungs. Engineers emulate natureâ€™s tricks to develop stable interfaces in modern 3D printing, pharmaceutical formulations and for foams and emulsions in food products.

As a doctoral student with Prof. Gerald G. Fuller, I focused on studying the dynamic measurement biological interfacial films, such as the human tear film and lung surfactants. My work draws on concepts and tools from rheology, soft matter and fluid mechanics to address the interplay of structure, transport and biological function. In addition to fundamental questions of why the tear film and lung surfactant possess such remarkable interfacial properties, my work has been applied in industry for developing next-generation contact lenses and artificial lung surfactant therapeutics.

Low-cost diagnostic devices

Diagnosing patients quickly and accurately is one of the most important steps in fighting disease, and in developing nations, it is one of the biggest obstacles to treatment. An urgent challenge in global health is development of low-cost point-of-care tools (<$1) to ultimately reach a billion people who need it most. Keeping the cost of tools in consideration while maintaining the scientific rigor, quality, robustness and accuracy of devices requires creative ideas from soft matter, biophysics, chemistry and medical instrumentation.

As a postdoctoral fellow with Manu Prakash, I am working on developing a low-cost, electricity-free centrifugation technology that will only cost 20-cents. This tool addresses sample preparation challenges currently relying on centrifuges that are expensive, bulky and require infrastructure (electricity). More importantly, my work opens up possibilities for developing integrated diagnostic systems that exploit centrifugal microfluidics for one-step integrated nucleic acid based tests, that are crucial for early diagnosis and complete elimination of infectious diseases affecting human society.

Teaching Interests:

Transport phenomena, biophysics, fluid dynamics, soft matter.

476175 The Transition Kinetics of Bacterial Collective Motions

Abstract:

Research Interests: Active fluids, soft materials, phase transitions

Teaching Interests: Thermodynamics and statistical mechanics, fluid mechanics.

Few things seem to capture our imagination more reliably than the collective motions of flocking birds, schooling fish or swarming insects. Such collective motions represent a universal feature of active fluids—a novel class of non-equilibrium complex fluids. While the collective motions of active fluids in steady states have been extensively studied in recent years, the kinetic route to these striking motions is still largely unknown. Here, we study the transition kinetics of the collective motions of active fluids by using a new light-controlled bacterial strain, whose swimming behaviors can be reversibly switched between swimming and tumbling modes by blue light. We investigate the initiation of collective motions, particularly the formation of coherent vortex structures, in bacterial suspensions when the bacteria are switched from tumbling to swimming modes. Velocity correlations, energy spectrum and vortex statistics are measured for quantifying the kinetics of this non-equilibrium "phase transition". We map the phase diagram of active fluids by systematically varying bacterial concentrations and the ratio of tumblers and swimmers. The transition kinetics at different points of the phase diagram are explored.

476179 Exploring and Exploiting the Physical Properties of Biological Soft Matter: From Bacterial Infections to Metastatic Cancer

Abstract:

Research Interests:

The structural and mechanical properties of biological soft matter (i.e. biopolymers, cells, bacterial biofilms, tissues) are often connected to cellular function and disease state. However, there is frequently a gap in the understanding of the biomechanical contributions to systems level behavior. Determining the physical behaviors of biological soft matter contributes to addressing this knowledge gap and allows for the development of novel therapeutic tools targeting the physical properties of biological systems. Here I demonstrate the utilization of a soft matter approach to address two clinically relevant problems: a) bacterial infection prevention and control and b) minimally invasive cancer diagnosis and assessment.

Bacterial biofilms are structured communities of cells encapsulated in polysaccharides, proteins and DNA that are frequently responsible for clinical infections. In my work, I consider bacterial biofilms as a biocolloidal system, where the cells are analogous to colloidal particles and the matrix materials are analogous to a viscoelastic hydrogel. I find that unstressed staphylococcal biofilms have densely packed, disordered microstructures, while stressed biofilms contain open, tenuous, fractal-like microstructures. I create artificial staphylococcal biofilms and reveal the importance of pH and matrix phase behavior to biofilm mechanics and dispersal. I identify that multispecies staphylococcal biofilm structures and growth behaviors are strongly dependent on their growth environment. These findings on the structure and mechanics of biofilms have implications in the field of biofilm microbiology and in the practice of biofilm prevention and control.

Metastatic cancer is cancer that has spread from the original site to a more distant site within the body. Cancer metastasis is often initiated by circulating tumor cells (CTCs), cells that shed off of tumors and circulate in the blood stream. Here I demonstrate the use of microfluidic devices for the rapid and minimally invasive testing of the biophysical markers of cancer and its metastasis. Specifically, I report the utilization of a deformability cytometer for the high-throughput assessment of cancer cell mechanics and the use of label-free sorting devices to separate viable CTCs from blood using sound waves. These devices could serve as cancer diagnostics and prognosis assessments.

Through continuing to investigate the biophysical properties of cells and multicellular systems, I plan to reveal connections between the physical properties of living systems and their disease states and to utilize these findings to develop therapeutics and diagnostics.

Teaching Interests:

At the undergraduate level, I am most interested in teaching Material and Energy Balances, Mass Transfer, Heat Transfer, and Fluid Mechanics. I have previous experience teaching Heat and Mass Transfer, and concepts from Transport Phenomena and Fluid Mechanics are important to my research. I am also willing and able to teach any of the core chemical engineering courses as requested.

At the graduate level, I am most interested in teaching core courses like Transport Phenomena or Fluid Mechanics. I am also interested in developing graduate elective coursework in Biological Soft Matter, Principles and Applications of Colloidal Science, and Quantitative Imaging Methods.

Additionally, I plan to continue to conduct engineering education research. During my PhD, I researched student learning in an interdisciplinary graduate elective course.

Education:

Massachusetts Institute of Technology, Â Postdoctoral Associate, Materials Science and Engineering, 2015â €"Present

Project Title: *Microfluidic Devices for Minimally Invasive Testing of Biophysical Markers of Cancer*

Advisor: Dr. Ming Dao

University of Michigan,Â PhD, Chemical Engineering, 2015

Dissertation Title:Â *Staphylococcal Biofilms: Microstructure, Mechanics, Self-assembly, and Multispecies Communities*

Advisors: Professor Michael J. Solomon and Professor John G. Younger

University of Michigan, MS, Chemical Engineering, 2010

Worcester Polytechnic Institute, BS, Chemical Engineering, Minor in Materials Science, 2008

Keywords: Biological Soft Matter, Biomechanics, Bacterial Biofilms, Metastatic Cancer, Complex Fluids, Colloidal Science, Engineering Education, Interdisciplinary Learning

476204 Measurement and Control of Slip-Flow Boundary Conditions at Solid-Gas Interfaces

Abstract:

Measurement and Control of Slip-Flow Boundary Conditions at Solid-Gas Interfaces

Dongjin Seo

Department of Chemical Engineering

UC Santa Barbara, Santa Barbara, California

dongjin_seo@ucsb.edu In traditional Chemical Engineering applications one usually assumes that the no-slip boundary condition for flow applies at the solid-gas interface. However, it has been known since the time of Maxwell that, even for moderate Knudsen numbers, partial slip could occur, of order the mean free path of gases. With the growing number of applications for micro- and nano-scale devices and flow systems, it is interesting to quantify more precisely the flow boundary condition. In this work I describe unambiguous measurements where I show that partial slip does occur, that the slip length is a function of both the gas and solid, and that the solid can be altered in-situ to change the slip length. In-situ changes in slip length could in principal be used to control flow in a variety of important applications. The main theme of this presentation is the effect of solid surface properties on the flow boundary condition. The effect of water films, organic films, electric fields and the gas species was studied. Water films had a large, but complex effect. On bare hydrophobilic glass, the tangential momentum accommodation coefficient (TMAC) for nitrogen on hydroxyl-terminated silica changed from 0.25 to 0.88 when the humidity changed from 0 to 98 %. On hydrophobized glass, TMAC changed from 0.20 to 0.56 in the same range. The effect of humidity was interpreted with the formation of water film, verified with a quartz crystal microbalance. TMAC on octadecyltrichlorosilane-coated glass surfaces was examined for five different gases (helium, nitrogen, argon, carbon dioxide, hexafluoride sulfur). A lower TMAC occurred for greater molar mass. I also developed methods for controlling the flow boundary condition using external stimuli – temperature and electric fields. Using Atomic Force Miscroscopy (AFM), each of these stimuli altered the roughness of the film, so control of the boundary condition was attributed to surface roughness change. The temperature was used to change the roughness of an octadecyltrichlorosilane film on glass, and the electrical field was used to change the roughness of a 16-mercaptohexadecanethiol film on gold. Considering collisions of simple spheres that conserved momentum and kinetic energy, a surface with periodic triangular grooves had higher TMAC with steeper slopes. A novel method of measuring gas pressure using an AFM cantilever was presented. The frequency spectrum of the cantilever vibration in the gas was measured, which was then fit to a continuum description that depends on the density. Use of the continuum approximation restricts calculation of the pressure to the continuum regime. In the range 0.1 to 2.2 atm, the gauge deviated by less than 5 % compared to a commercial gauge.

Research Interests:

Colloids and surface science is my expertise and tool. With this, I am planning on extending my research toward energy engineering and biomedical engineering. I like to propose two research projects for the immediate years that follows. The first project, a novel gas separation technology, is intended for the submission to National Science Foundation (CBET division). This project is intended to prove a new separation principle and separate carbon dioxide from natural gas in the future. The second project, deep drug deliver to lung with aerosol surface modification, was prepared for National Institute of Health R21 funding. This simple modification to aerosol medication would deliver more active ingredient toward alveoli, reducing overdose and cost.

Separation of Gas Mixtures via Differences in Flow Boundary Conditions

The objective of the proposed work is to verify the principle of a novel gas separation method based on difference in flow boundary conditions. The principle to be proved is that different gas species intrinsically have different boundary conditions when their mean free paths are comparable to the characteristic length of the separation system. This will further extended toward the preparation of the prototype gas separation device, which is clean and easy to operate consuming low energy.

The specific aims of this project is to 1) implement simple computational simulation and calculations 2) devise a prototype device for continuous process, and 3) build a control system for a batch process. The simulation part will be done with the collaboration with graduate students studying computational fluid dynamics (CFD). The rest will be achieve with glass capillary system coated with surface molecules for passivation and protection. This system would prove the principle found in the previous work can be also applied in tubular systems. For these aims, the following activities will be implemented.

Computational Fluid Dynamics: For both methods mentioned above, I will investigate the feasibility of volumetric flow and velocity difference with computer simulation getting help from a graduate student involved in CFD. A simple model would be set up to see the difference velocity of several spheres with different boundary conditions flowing through a cylinder.

Devising adequate cylindrical geometry: An experimental equipment would be built. I propose 100 ~ 1000 nm glass capillaries. The interior of the capillaries would be coated with octadecyltrichlorosilane for protection and smoothness. The outlet would be connected to gas chromatography to detect the effluent of different gases

Continuous separation experiment: To test this method, a chamber would be built for a binary mixture with oxygen and carbon monoxide. I chose these species because detection equipment is very precise and requires low concentration. Then their concentration would be measure at the outlet.

The success of this project would provide clean separation technology with low energy consumption and high throughput without having severe operation conditions. The process will integrate the science and engineering in different fields, surface science, process control, fluid mechanics, and computational calculation, into one field. This will lead to several patents further utilized in commercial separation units.

Roughening of Aerosol for Deep Lung Delivery

1. Background

Chronic Obstructive Pulmonary Diseases (COPDs) are the third leading cause of mortality in the United States and about 15 million adults have reported they have COPD. Another lung disease, asthma, is even more common for both adults and children; in the United States: one out of 12 adults and one out of 11 children have asthma. Delivery of drugs to treat COPD, asthma, and other pulmonary diseases is achieved through aerosolized (small liquid droplet) medications, primarily delivered via inhalers. These inhalers are used to spray drugs into the lungs via the mouth, which naturally results in unnecessary deposition close to the entrance of lung, rather than deeper into the lung where delivery is required. A longer average travel distance of aerosol particles is therefore desired for better treatment.

2. Specific Aims:

The aim of this work is to enhance the travel distance of inhalant aerosol medication inside the lungs by increasing the resistance to droplet motion towards airway surfaces (See Fig. 1). By opposing motion toward the airway surfaces, deposition on the walls is delayed, and the droplets are able to travel to the distal portions of the lungs.

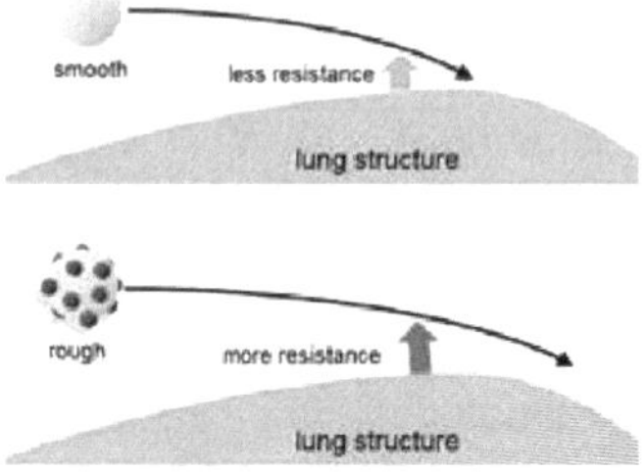

Fig. 1. A stronger lubrication force on rough particles delays approach of particles to lung surfaces, thereby extending travel into the deep lung.

Until recently, many chemical engineers assumed the boundary condition for the flow of gases was fixed. But recent research by our group has shown that very slight (nanoscale) roughening of surfaces can be used to increase the resistance to a particle approaching a solid in air. The required increase in resistance to motion towards airway surfaces will be achieved through deliberate roughening of the surface of the aerosol droplet through the addition of surfactant. Whether the aerosol particles attempt to make contact with lung passage walls due to gravity, or due to airway curvature, the additional resistance will always push the droplets away from the wall. This allows each droplet to utilize momentum parallel to lung surface, and thereby travel deeper into the lungs.

The specific aims are to test the following hypotheses: 1)The addition of surfactants increases the lubrication force acting on particles approaching a wall. 2) These surfactants adsorb to the aerosol – gas (inhaled air) interface, and increase its roughness. 3) An increased lubrication force increases the distance that a droplet travels in a numerical model.4) An increased lubrication force increases the distance that a droplet travels in an experimental model system. 5) Optimum concentrations of surfactants exist and vary by the kind of surfactant and medication.

Successful completion of this project will provide a relatively simple, inexpensive, and safe way to enhance delivery of aerosol-based drugs into the entire lung. Most importantly, the increased delivery distance will help to obtain a more even distribution of the drug, which may allow lower doses, and fewer side effects. Since this enhancement would come by simply adding appropriate FDA-approved surfactants to existing aerosol delivery systems, this additive technology is more likely to be adopted. The major impact of adoption would be better therapeutic outcomes for both sufferers of pulmonary diseases, and recipients of systematic drugs.

Teaching Interests:

I am very interested in interaction with chemical engineering students and helping them become more prepared engineers after graduation. With such interest, I am willing and able to teach general chemical engineering topics in any level, as well as special topics such as surface and colloid chemistry, atomic force microscope (AFM), refinery and chemical plant processes. My diverse research experiences scope through different areas of chemical engineering, physics, mechanical engineering, and material science. In addition to this research activities, my practical experience in a downstream oil company I worked for four years provided me with not only the basic principles of chemical engineering, but also the practical integration of knowledge that is needed for chemical engineers.

Teaching Philosophy

My basic teaching philosophy consist of 1) teaching students correct principles and knowledge, 2) encouraging them to utilize the knowledge, 3) getting them involved in teaching, and 4) following up. I believe these principle are needed for students in any fields to learn. As with other teachers, I will lecture with much preparation focusing on how to have them involved during lectures. I will make sure they understand through homework and projects they have to work together and teach together. In my belief, they have to at least attempt to speak or write out what they have learned to their colleagues in their own language. Otherwise, very little can be left with them. Therefore problems and projects will be oriented toward teaching each other. My teaching method will be also tested with exams that require not only calculations but also require interpretation of the calculated results. I have seen from much experience that the students like to produce results. I will help them make one step further by letting them think for physical meaning for the result and the application thereof. This application of knowledge is also important. I will assign a group project with which they can think of the solution to a problem with the knowledge from the courses. I will follow up on their progress by encouraging them to come to office hours to make individual contacts. This way, I can follow up on their progress. This principle will be also true to graduate students. I will brainstorm with them every week to see if they and I can prepare for the next projects they are on by combining results from different projects.

Specific Topics of Interest

Among many classes, I like to teach colloids and surface science along with AFM. I like to involve AFM, because it demonstrates the surface forces with visual images. With the advance of nanotechnology and polymer science, the small forces or relatively insignificant properties became important in understanding physical phenomena and chemical engineering processes. I believe chemical engineering students have to see the importance in these features and be able to manipulate the variables to obtain desired results. If this course is not opened, I am happy to prepare the course.

Another course I am very interested in is Unit Operation Laboratory (UO Lab). I loved teaching UO Lab as a graduate student because it gave me more chance of talking to students and finding out not only their knowledge but also their interest in future as engineers. However, the room for improvement I saw among students gave me strong desire to teach this class. I would like to lead this class with evocation of discussion and analysis of the activity and data. One thing I noticed was student tend to focus on the results, and not the meaning. I like to teach them how to produce meaningful conclusion out of data.

Teaching Experiences

Please refer to my curriculum vitae as well. I have actively engaged in the interaction with undergraduate students by 1) helping them for undergraduate research, 2) lecturing for different classes, and 3) teaching at experimental laboratory classes. These experiences prepared me as an effective teacher. My goal in teaching is helping them independent and I saw such progress in them. I have taught junior, senior, and graduate-level classes as well as lab classes. I also supervised three different undergraduate research projects. In all of them, I had a chance to see them grow toward understanding the problem, finding solution, and being autonomous.

Outreach Effort _ I want to direct my outreach effort mostly toward, but not limited to, K6 to K12 students. Youths, especially girls, around this age lose their interest in science, technology, engineering and mathematics (STEM) area, and especially when they do not make connection between STEM knowledge and the real life. (Ref. 1) Therefore it is important to let them realize that we are living in a world utilizing knowledge from STEM area. I like to make such connection using colloid and surface science. For younger students from K6 to K9, I like let them know daily-life products, shampoo, make-ups, cleaning agents, and even diapers, not only work, but also are better with surface science. This demonstration or a set of simple 5 – 10 minute experiments would be presented during a field trip or as one of outreach experiences for school breaks. During this presentation, each student will be given materials and experience on their own hands how surface science is related to their daily life. For older students in K10 to K12, more advanced project is possible. The surface modification with monolayer molecules, atomic force microscope (AFM) with their own samples, or click-chemistry would be a fun and educating experience for them. A simple monolayer-coating experiment can be done by a group of high school students with much care instruction. This way, they will learn safety issue as well. AFM is hard for them. But they will have fun in seeing the nanometer-sized details of the materials they use every day. Safe click-chemistry experiments that can be produced with house chemicals can be a good start for their scientific journey. The focus of my outreach is toward school students with easy and fun experiments. This way, they will relate their daily life to science. Ref. 1 Jeffers, A., Safferman, A., & Safferman, S. (2004). Understanding K-12 engineering outreach programs.

Journal of Professional Issues in Engineering Education and Practice, 130(2), 95-108.

476209 Towards the Understanding of Kinetics and Thermodynamics of Materials

Abstract:

Research Interests:

My research interests lie at the nexus of thermodynamics, kinetics, and material science. Understanding kinetic and thermodynamic properties of materials is fundamentally challenging, yet it is required to solve problems in chemical engineering science. I seek to understand the kinetic and thermodynamic properties that govern the material behavior in solid, solution, and gas phase. My broad research background combined with training in novel experimental tools such as atomic force microscopy, calorimetric instruments, spectroscopic techniques, high speed imaging, and property modeling tools of ASPEN puts me in a unique position to pursue active research in these field. I want to expand my research in the following areas: (i) thermophysical properties and solubility thermodynamics, (ii) material aging and crystal growth kinetics at nanoscale, (iii) mechanisms of kinetic processes and the fragmentation of high energy density materials. Applications of my research include fluid mixtures associated with energy and environment, pharmaceutical compounds, semiconductors, and energetic materials.

Teaching Interests:

My teaching philosophy is based on my motivation of shaping young minds to be intelligent by expanding their imagination and creativity through the fundamental understanding of scientific and engineering laws and principles. My graduate and postdoctoral training have given me a broad background that I am well prepared to teach at undergraduate and graduate levels. I taught a senior level chemical engineering course (course title 'Comprehensive Review of Chemical Engineering') in fall 2013 in Texas Tech University as a full time instructor. This course is designed based on seven basic chemical engineering courses, which are Material and Energy Balance,

Thermodynamics I and II, Reaction Engineering, Mass transfer, Fluid Mechanics, Materials science, and Heat Transfer. My future teaching interest includes all these subjects at undergraduate and graduate level. I also want to develop new courses for undergraduate and graduate level based on the currents needs in chemical industries. In future, I want to develop specialty courses like Applied Thermodynamics in Chemical Engineering, Kinetic Processes in Material Science, Corrosion Engineering, Chemical Process Safety, and Design of Engineering experiments.

Education and Training

Post-Doctoral Fellow: Department of Chemical Engineering, Texas Tech University

Jan 2014-Present

Supervisor: Professor Chau-Chyun Chen

PhD: Department of Chemical Engineering, Texas Tech University

Summer 2013

Supervisor: Professor Brandon Weeks

Instructor: Department of Chemical Engineering, Texas Tech University

Fall 2013

Number of published articles: 26

Citations: 320

Google scholar link for publications and citations: http://scholar.google.com/citations?user=nJvVPvoAAAAJ&hl=en

476101 Nonequilibrium Biophysics and Rheology of the Inner Cell

Abstract:

Research Interests:

The interior of a living cell is a dense, colloidal suspension containing a complex network of interacting protein machines that carry out lifeâ€™s essential functions.Â My research focuses on understanding emergent properties and nonequilibrium processes of the inner cell at the colloidal and molecular scale. I invoke concepts from fluid mechanics, statistical physics, and molecular biophysics to develop mathematical and experimental tools to interrogate the complex hydrodynamic and biochemical processes occurring inside an active cellular cytoplasm.

As a doctoral student, I have developed frameworks to explain the collective motion of living systems, as in bacteria colonies, insect swarms, bird flocks, and fish schools.Â Using analytical theory and computer simulations, my research has revealed that all active matter systems generate a unique mechanical pressure, which is the fundamental physical mechanism that controls collective behavior, self-assembly, and pattern formation. This work was important because the understanding of how collective patterns of self-organization emerge from the joint interactions of active individuals has been a challenging problem in soft matter. Furthermore, the same phenomena of collective motion also occur inside a living cell, so my micromechanical model may be used to understand the rheology of active and living suspensions.

My research has evolved from its origin as a simple theoretical and computational framework to one that describes specific biological systems, like the cell interior where confinement plays a key role in the behavior of active motor proteins. I have developed lab experiments to encapsulate proteins, ions, macromolecules, synthetic active particles, bacteria, and other nano/micron-sized objects inside of a phospholipid vesicle, in an effort to create systems that operate close to the biophysical and physiological constraints of a living cell. I corroborate experimental data with Brownian/Stokesian dynamics simulations to understand the mechanical interactions governing intracellular transport within the cytoplasm.

My future research will involve analytical theory, stochastic simulations, and lab experiments to understand molecular-level transport phenomena in complex biological systems. I will use single-molecule manipulation techniques like optical/magnetic tweezers and atomic force microscopy to study the relationship between the mechanical forces that activate protein motors to the macroscopic behavior of a cell. I intend to pursue microrheology measurements to ascertain how motor proteins may be used as drug delivery targets for treatment of human diseases.

Recent experimental work will be the focus of my oral presentations at this meeting.

Selected publications:

1. **Takatori SC**, De Dier R, Vermant J, Brady JF. *Nature Commun*. 2016; 7:10694.
2. **Takatori SC**, Brady JF. *Curr Opin Colloid Interface Sci*. 2016; 21, 24.
3. **Takatori SC**, Brady JF. *Soft Matter*. 2015; 11, 7920.
4. **Takatori SC**, Brady JF. *Phys Rev E*. 2015; 91, 032117.
5. **Takatori SC**, Brady JF. *Soft Matter*. 2014; 10, 9433.
6. **Takatori SC**, Yan W, Brady JF. *Phys Rev Lett*. 2014; 113, 028103.

Teaching Interests:

I am excited to teach a variety of chemical engineering courses as a faculty member, in particular momentum, heat, and mass transfer at the undergraduate and graduate levels. In addition to transport phenomena, I am interested to teach statistical thermodynamics, soft matter physics, and mathematics and numerical methods for engineers and scientists. As a passionate public speaker, I value scientific communication and always strive to deliver coherently articulated lectures inside the classroom.

476132 Soft Materials Engineering of Biological Interfaces

Abstract:

Research Interests:

The cell membrane is a complex two-dimensional fluid that organizes spatially and temporally to orchestrate processes such as cell division, protein signaling, and vesicle trafficking. Achieving an understanding of these processes has extensive applications in many areas, including drug delivery, cellular communication, mass and energy transport, and signal transduction. My approach to studying biological interfaces is to develop model phospholipid bilayers *in vitro*and building in complexity to mimic processes like membrane protein sorting and pore formation, nanoparticle wrapping and translocation, and lipid diffusion and flip-flop in asymmetric membranes. I have developed advanced microscopy, microfluidic, and electrophysiological tools for fabricating and characterizing large area model biomembranes. A distinguishing feature of this advance is the robust creation of tension-controlled membranes with millimeter-scale surface area in a planar geometry that can be simultaneously imaged and manipulated, for example by applying osmotic, thermal, chemical or hydrostatic pressure gradients. In this presentation, I will highlight my recent efforts in developing this platform and my vision for applying my expertise in several areas of inquiry important to modern chemical and biomolecular engineering- including a comprehensive understanding of lipid bilayer mechanics (elasticity, tension, viscosity), nanoparticle-membrane interactions and assembly, and collective membrane protein assembly and function.

Teaching Interests:

I am confident that my varied teaching experiences and educational background has trained me well to teach any core class in the undergraduate and graduate chemical engineering curriculum. During graduate school at Delaware, my teaching experience included being awarded a Teaching Fellowship to co-teach Chemical Engineering Thermodynamics. At ETH Zürich, I have continued developing my teaching interests and skills by lecturing and assisting developing class projects for the Soft Materials course. In any course, it is my objective to emphasize the fundamentals of chemical engineering while promoting a classroom atmosphere of applied learning through discussions of the literature, practical demonstrations, and guest lecturers, if applicable. I would strive to incorporate group projects and assignments using Matlab (if basic programming is already taught) as this teaches algorithmic thinking and teamwork while applying the fundamentals learned in class in a meaningful way.

Post-doctoral Advisor: Prof. Jan Vermant, ETH Zürich

PhD Advisor: Prof. Eric M. Furst, University of Delaware

Selected Publications:

Beltramo, P. J.,Van Hooghten, R. and Vermant, J. "Millimeter-area, free standing, phospholipid bilayers." Soft Matter, 12, 4324-4331, (2016).

Beltramo, P. J.,Schneider, D., Fytas, G. and Furst, E. M., "Anisotropic hypersonic phonon propagation in films of aligned ellipsoids." Phys. Rev. Lett., 113(20), 205503, (2014).

Beltramo, P. J., Roa, R., Carrique, F. and Furst, E. M. "Dielectric spectroscopy of concentrated colloidal suspensions." J. Colloid Interface Sci., 408(1), 54-58, (2013).

Schneider, D., **Beltramo, P. J.**, Mattarelli, M., Crespy, D., Montagna, M., Pfleiderer, P., Vermant, J., Furst, E. M., and Fytas, G. F. "Elongated Polystyrene Spheres as Building Blocks for Anisotropic Colloidal Crystals: A Particle Vibration Spectroscopy Study." Soft Matter, 9, 9129-9136, (2013).

Beltramo, P. J. and Furst, E. M. "Predicting the disorder-order transition of dielectrophoretic colloidal assembly with dielectric spectroscopy." Electrophoresis, 34(7), 1000-1007, (2013).

476144 Colloid Assembly Engineering

Abstract:

Motivation:

Colloid Assembly plays a crucial role across a wide range of disciplines, and particularly finds applications in pharma, fine chemicals and solar cells (photovoltaics) industries. The assembly of colloids can be understood as the formation of structures arising from the combination of small (molecular or colloidal) entities. The structures being formed can either be the desired product (e.g. quantum dots[1]), or a problematic aggregate formation (e.g. solid formation in reactors that promotes clogging). The complex processes regulating the formation of such structures include nucleation, growth, aggregation and breakage, all of which depend on experimental conditions, e.g. temperature and shear rate[2]. Until significant improvements in the understanding of the aforementioned processes are made in the field of colloid science, we will not be able to i) fully exploit continuous flow technology for the production of pharmaceuticals and fine chemicals and ii) employ Colloid Assembly as a well-defined synthetic route for interesting structures (e.g. quantum dots, biomaterials, insulating materials).

Approach:

I am comparing modeling predictions with experimental results to obtain fundamental, quantitative understanding of Colloid Assembly processes. The modeling techniques I am relying on include Population Balance Equations (PBE), Monte Carlo (MC) and Computational Fluid Dynamics (CFD) approaches. The combination of a modeling perspective with experimental results, obtained through literature and experimental data (e.g. from microscopy, scattering, spectroscopy) allows to discriminate among the possible occurring mechanisms and to identify their kinetic rates. The diversity of the present methodology allows me to appreciate the interplay of the ongoing physical mechanisms from different angles.

Personal background:

After receiving my bachelor and master degree at Politecnico di Milano in Italy, I pursued my doctoral studies at ETH Zurich, Switzerland, and I am currently a postdoc at MIT. During my PhD at ETH Zurich in the group of Prof. Massimo Morbidelli, I worked on several projects in the fields of polymers and colloids, combining experimental and modeling techniques to gain a phenomenological wide-angle view on the processes under investigation. In particular, I developed a significant experience in the field of PBE. This allowed me to approach problems such as the formation of multi-active polymer chains, the interplay between aggregation and coalescence of colloidal particles, as well as the aggregation and breakage of protein fibrils. Being awarded with two post-doctoral fellowships from the Swiss National Science Foundation, allowed me to join the group of Prof. Klavs Jensen at the MIT since April 2015. During my postdoctoral studies, I have been working along two main lines, namely i) developing a temperature dependent model for the formation of quantum dots and ii) developing a mathematical framework to describe solid formation and fouling.

Long term plan:

Capitalizing on the experiences gained during my PhD and Post-doctoral studies, my long-term vision focuses on pursuing several important questions persisting in the field of Colloid Assembly:

i) Fully exploit Continuous Flow Technology for pharmaceutical and fine chemicals production:

- How can Agglomeration and Clogging be prevented?
- How can Crystallization in Flow be modeled?

ii) Improve Control over "Colloidal Materials" production:

- How can the process of nanocrystal formation be optimized (e.g. size distribution)?
- How can high-concentration aggregation be modeled?

iii) Develop a general mathematical framework to describe any assembly process

- Which are the general kinetic rate expressions of such a model?
- Which modeling techniques are best suited to build the model?

Research Interests:

- Colloid Assembly, Colloid Engineering
- Solid Formation, Crystallization, Quantum Dots
- Biodegradable Polymers, Polymer Reaction Engineering, Population Balance Equations

Teaching Interests:

Having followed all the Chemical Engineering Classes at Politecnico di Milano and having some experience in this sense at ETH Zurich too, I would feel comfortable to teach any class in the Chemical Engineering curriculum. My personal inclinations and research interests drive me towards Colloid Stability, Polymer Reaction Engineering or Chemical Kinetics: in these courses, I could establish connections with my research, pointing out the outstanding challenges. Moreover, I would be able to involve students in possible research projects or in starting their master/PhD thesis in my research group.

References

[1] van Embden J., Chesman A.S.R, Jasieniak J.J., Chem. Mater. **2015**, 27, 2246–2285

[2] Henry C., Minier JP, Lefèvre G., Adv. Colloid Interface Sci., **2012**, 185-186, 34-76

476171 Soft Matter Physics of Polymeric Fluids, Biofluids, and Granular Media

Abstract:

Research Interests:

Soft matter such as biological membranes, cells, tissues, body fluids, colloidal dispersions, and gels have properties that fall between those of liquids and solids. They are composed of polymers and/or particulates (liquid or solid) dispersed in a continuous phase.Â Biological processes and technological applications often use flow and external forces to alter the organization of soft matter at the microscale, which can manifest as significant changes in the materialâ€™s mechanical behavior. My goal is to establish a research program in soft matter physics that develops theory and simulations to describe the micro to macroscale behavior of polymeric fluids, granular materials, and biological fluids. I plan to complement these studies with experiments and collaborate closely with other research scholars.

My research training is in the area of fluid mechanics and polymer physics. In my PhD under Prof. Eric Shaqfeh, I developed mathematical models, performed boundary integral and multipole simulations (meso-scale), and conducted microfluidic experiments (vesicle synthesis, particle-tracking, image processing) to describe the motion of droplets, red blood cells, and vesicles under flow in capillaries and microfluidic devices. During my postdoctoral work with Prof. Patrick Doyle, I examined the motion of knots in DNA via Brownian dynamics simulations (nano-scale), collaborating closely with experimentalists doing single-molecule microscopy. This work has possible applications in biology (transcription) and next-generation DNA sequencing.

My future research will focus on polymer dynamics, polymer rheology, biofluid mechanics, and granular materials. One research thrust is to develop microscopic models of polymer entanglements. Can one develop simpler theories to describe collisions between polymer chains, and gain insight into what aspects of molecular structure contribute to friction between chains? On the macroscale, I would like to develop theories and perform experiments on polymeric fluids, describing how the presence of solids alters the dispersionâ€™s material properties and how these fluids partition in a bifurcation. Another research area involves the development of mesoscale models describing the transport of leukocytes through lymph nodes, which is where bodyâ€™s immune response initiates. I would also develop particle-tracking experiments of highly agitated granular media (powders) to understand how collisions alter particle transport. The flow of powders is important in a variety of industries (food, pharmaceutical) and 3D printing.

Teaching Interests: Â

My teaching philosophy rests on two beliefs: a teacher should teach students how to become problem solvers, and a teacher should get students excited about a subject area. My goal is to impart a quantitative mindset to students so that they can view a complex phenomenon, distill its essential physics, and characterize it using mathematical or computational tools. I was a teaching assistant for two Chemical Engineering courses: (1) undergraduate thermodynamics (Caltech), and (2) graduate fluid mechanics (Stanford). I was awarded the Stanford Chemical Engineering Outstanding Teaching Assistant Award and was actively involved in mentoring the subsequent batch of teaching assistants. My laboratory mentorship experience includes the supervision of graduate and undergraduate students.

476222 Colloidal and Interfacial Phenomena: From Fundamental Studies to Emerging Applications

Abstract:

Research Interests:

My primary research interest lies at the fundamental studies of colloidal and interfacial phenomena as well as their applications in energy, environment, and health. My doctoral research focuses on multiscale assembly and interaction of colloidal multiphase systems. During Ph.D. training, my eyes were also opened to oil recovery and water purification, to see tremendous opportunities of using colloidal and interfacial science to solve problems in energy and environment. My postdoc training with Professor John A. Rogers added two new dimensions, nanomedicine and bio-integrated electronics, to my research experience.

Here, I will provide a brief overview of my research from both a fundamental point of view and their emerging applications. In particular, this overview includes: (i) Multiscale assembly of a new class of colloidal materials, capillary foams; (ii) The application opportunities of capillary foams, including the fabrication of loadbearing, lightweight, or intensely colored porous materials, enhanced oil recovery, advanced froth flotation, wastewater treatment, and oil spill remediation; (iii) The surfactant-free tuning and dynamic reconfiguration of bubble-droplet interaction in colloidal multiphase systems; (iv) Soft, implantable, wireless devices for programmable drug delivery.

Based on my previous training and expertise, I would like to start my academic career by using colloidal and interfacial science to solve problems in energy, environment, and health. Topics of interests include colloidal assembly, water purification, nanomedicine, nanomanufacturing, and bio-integrated electronics.

Postdoctoral Advisor: Professor John A. Rogers, Northwestern University

Ph.D. Advisors: Professor Sven H. Behrens and Professor J. Carson Meredith, Georgia Tech

Selected Publications:

1. **Y. Zhang**, A. Shitta, J. C. Meredith, S. H. Behrens, "Bubble meets droplet: particle-assisted reconfiguration of wetting morphologies in colloidal multiphase systems," ***Small***, 2016, doi: 10.1002/smll.201600799.

2. **Y. Zhang**, M. C. Allen, R. Zhao, D. D. Deheyn, S. H. Behrens, J. C. Meredith, "Capillary foams: stabilization and functionalization of porous liquids and solids," ***Langmuir***, 2015, 31(9): 2669-2676.

3. **Y. Zhang**, J. Wu, H. Wang, J. C. Meredith, S. H. Behrens, "Stabilization of liquid foams through the synergistic action of particles and an immiscible liquid," ***Angewandte Chemie***, 2014, 126(49): 13603-13607.

Teaching Interests:

My teaching philosophy is stimulating the curiosity of my students and equipping them with an effective way of learning that they can benefit from for their lifetime. In this rapidly changing world, the knowledge in science and engineering is expanding exponentially. Therefore, instead of solely subject-oriented teaching, it is critical to stimulate students' curiosity to explore different fields and teach them learning skills that they can apply to various learning processes.

At Georgia Tech, I have assisted teaching both undergraduate and graduate courses. During such processes, I had opportunities to practice and improve my teaching skills through leading pre-lab meetings, supervising experiments and data analysis, grading laboratory reports, and holding office hours. In addition, I have advised five undergraduate students for two years and resulted in three publications, five posters, and two awards. Throughout these mentoring processes, I have learned how to motivate students' research interest, manage a project, and communicate with students in a clear, concise and effective manner.

I am prepared to teach both undergraduate- and graduate-level courses in chemical engineering and material science and engineering. I feel confident to teach a comprehensive list of courses, including chemical engineering thermodynamics, transport phenomena, kinetics and reactor design, unit operations laboratory, and polymer science and engineering. I will also develop new courses on cutting-edge research topics, such as nanomedicine and bio-integrated electronics. I believe that through my efforts, more people will become interested in engineering and technologies. I am ready to take the responsibility.

476284 Fundamental Studies at the Interface: Specific Vs Non-Specific Bio-Interactions

Abstract:

Research Interests:

Nanoscale bio-recognition is an essential requirement in many biological processes (e.g. enzyme-subtract interaction, immune response, and signaling pathways). Although our knowledge of the molecular and structural basis of these interactions is limited, scientists have found a way to apply these interactions to several applications like affinity chromatography, bio-sensors, and drug delivery.

My Ph.D. work focused on affinity membranes for bio-separations. It centered around the development (selection of ligands, modification of membranes) and testing (batch and dynamic adsorption experiments, modeling, estimation of binding capacity, dissociation and kinetic constants, effect of ionic strength and flux) of affinity membranes for the purification of different classes of biomolecules: monoclonal antibodies, lectins and MBP-fusion proteins.

During my post-doctoral studies and more recently as a Research Associate, I have continued to work on several projects with the common focus of protein-protein interactions, from separation to aggregation. The poster will give an overview of the research conducted in the following topics: (i) the specificity of nuclear transport; (ii) insights into the lytic mechanism of antimicrobial peptide using supported lipid bilayer to mimic bacterial membranes; (iii) the use of AFM in force mode to measure the binding forces between different split-protein partners; (iv) biophysical characterization of amyloid aggregates, from dimers to fibrils; (v) charged membranes for the separation of proteins with similar molecular weight and charge; (vi) characterization of brush polymers.

I have broadened my background as a chemical engineer, integrating it with a variety of bio-tools (e.g. AFM, QCM-

D, SPR, etc.). I plan to continue my career maintaining this interdisciplinary approach in studying protein-protein recognition both as basic science (e.g. better understanding of biological mechanisms) and applied science (e.g. developing bio-sensors).

Teaching Interests:

Teaching is an essential component of a professor's duties. From a research training perspective, an interdisciplinary approach has enabled me to develop a broad knowledge base and implement different viewpoints and strategies to solve problems. My goal is to equip students with a strong foundation in fundamental concepts and encourage their ability to apply their knowledge to creative solutions.

I have collaborated on Graduate Level (Chemical Engineering) Classes while at RPI: (i) downstream processing in biochemical engineering; (ii) molecular separation and recovery processes; (iii) intro advanced membrane concepts; (iv) advanced fluid mechanics. Responsibilities included lectures preparation and/or presentation. I was also the teaching assistant.

I have been a teaching assistant for Undergraduate Level (Chemical Engineering) classes during my Ph.D.: (i) mass transfer and chemical reactor; (ii) fluid-dynamics and heat transfer. Responsibilities included office hours, test grading, lecture presentation.

I have mentored several students during their research projects: created projects suitable for undergraduate students, demonstrated laboratory techniques, guided them through data analysis, report writing, and decision making.

My education has provided me with the necessary foundation to teach a variety of â€œcoreâ€ chemical engineering courses. In particular, I would enjoy teaching mass and heat transport and developing a course closer to my research, based on biomolecular interactions.

476293 Far-from-Equilibrium Soft Matter: Engineering Networks and Chirality for Energy and Health

Abstract:

Research Interests:

My research interest is in the field of soft materials driven far-from-equilibrium.

During my PhD work with Charles Schroeder at the University of Illinois, I developed and demonstrated a framework that allows for the determination of equilibrium properties of single polymers from the far-from-equilibrium behavior of polymers under flow. Using a combination of non-equilibrium statistical mechanics, Brownian dynamics simulations, and single molecule experiments with DNA, my PhD work demonstrated for the first-time the recovery of purely linear and non-linear elastic properties of polymeric materials from hydrodynamic behavior in strong flows.

Network-forming fluids display striking collective behavior evidenced by the anomalous thermodynamic and structural properties of water and silicon. My postdoctoral work with Pablo Debenedetti at Princeton University has focused on a new class of network forming materials. These materials, inspired by the specificity of atomic substances and the flexibility of polymeric systems, display unusual but tunable collective behavior at equilibrium. The rheology of these materials remains unexplored.

Chirality plays a major role in processing, pharmacology, photonics, and the emergence of life. However, a fundamental understanding of the role of chirality in the equilibrium and far-from-equilibrium behavior of materials remains limited. In collaboration with Frank Stillinger and Pablo Debenedetti, I have developed a molecular model for the analyses of chirality phenomena with the tools of statistical mechanics with a focus on the equilibrium behavior.

Going forward as an Assistant Professor, my research program aims to guide the engineering of chiral and network-forming fluids driven far-from-equilibrium by energy and matter fluxes. My program will focus on applications relevant to energy and health.

Teaching Interests:

My teaching interests include, but are not limited to, undergraduate and graduate level thermodynamics, fluid mechanics, polymer science and engineering, applied mathematics, optimization and control theory.

476172 Advanced Rheological and Neutron Methods for the Rational Design of Soft Materials

Abstract:

Advanced rheological and neutron methods for the rational design of soft materials

Michelle Calabrese, University of Delaware

Chemical & Biomolecular Engineering, 5th year PhD candidate

Research Interests: The design of soft materials with optimal structure and flow properties is critical in applications ranging from polymer processing to drug delivery, where materials undergo nonlinear deformations during processing, transport and use. Unlike standard scattering methods, flow-small angle neutron scattering (flow-SANS) enables researchers to study the flow-induced material structure by integrating rheology with time- and spatially-resolved microstructure measurements. Flow-SANS provides the missing link between the flow behavior and molecular topology, and can be tailored for complex flows including shear, extension, and microfluidics. Recent advances in time-resolved data collection and analysis allow the material structure to be measured continuously in time. As neutrons are not destructive to samples, flow-SANS is unique and of growing interest for applications ranging from proteins and biological materials, to consumer products and oil and energy recovery fluids.

Teaching Interests: My core teaching interests are in the fundamental fields of fluid mechanics and transport; I am also interested in teaching applied courses on rheology and materialsÕ characterization.

Successful Proposals & Grants: NIST Center for Neutron Research (NCNR, USA) BTAC - 4 of 5 proposals accepted: neutron spin echo (2015), SANS (2014, 2015), USANS (2014); Institut Laue-Langevin (ILL, France)- Visiting scientist grant (2016), 3 of 3 proposals accepted: SANS (2015, 2016)

PhD Dissertation: *Structure-property relationships of branched wormlike micelles via nonlinear rheology and small angle neutron scattering (SANS),* Chemical & Biomolecular Engineering, University of Delaware & NCNR, under the supervision of Norman J. Wagner

Visiting scientist project: Precise determination of concentration gradient in shear banding wormlike micelle solutions with Lionel Porcar, ILL

Research Experience: My research thus far has focused around two central tenants: the characterization of soft materials via non-linear rheology and neutron techniques, and the development of new neutron techniques and analysis methods for broader use within the scientific community. The work has been a collaborative effort with chemical engineers, physicists, chemists and materials scientists. Specifically, my experimental projects have focused on advanced rheological techniques and their interpretation (shear startup, large amplitude oscillatory shear, orthogonal superposition, extensional rheology) and neutron scattering methods (static SANS, 1-3 plane rheo-SANS, 1-2 plane flow SANS, neutron spin echo) in various self-assembled surfactant and polymer solutions. While this work is primarily experimental, an emphasis has been placed on experimentally validating modeling predictions in regard to flow instabilities. In the instrumentation-based projects, I have worked closely with beam line scientists at NIST and ILL to improve SANS capabilities and measurement resolution in space and time; these developments have been implemented at NIST and the ILL.

Teaching Experience: I have enjoyed many teaching opportunities throughout my scholastic career. I served as a teaching assistant at both the University of Delaware and the University of Pennsylvania (undergraduate), where I occasionally lectured. I held frequent office hours that were well received, as I received a 4.9/5 average in the teaching evaluations from my students. I was also awarded the Robert L. Pigford Teaching Assistant Award from the CHEG department at UD. I currently help younger members of my group learn neutron techniques and design experiments. At Penn, I was a TA, lab instructor and research mentor for a summer course on biotechnology; I also helped develop material for the chemical engineering section of the freshman introduction to engineering course. For several years, I have worked as a tutor and instructor for various afterschool programs, and continue to do STEM outreach when possible.

Future Direction: As a faculty member, I would continue using a combination of neutron and rheological techniques to characterize and design materials, while advancing instrumentation and measurement capabilities for targeted applications. I hope to extend my current work in surfactants and flow instabilities to biological systems. Flow-SANS is ideal for studying biological soft materials and artificial replacements for such materials, as neutrons, unlike x-rays, are not harmful to the materials. I hope to extend my work surrounding flow instabilities to optimize the properties of these materials for sensitive applications. Further, new flow-SANS cells can be easily developed and tailored to approximate injections or flows in the body, for example. My goal is to reproduce flows seen in real applications as closely as possible in my experiments, thus I foresee this work extending into modeling efforts, or collaborating closely with modeling groups to further our understanding of complex flows.

My other areas of interest include developing and optimizing soft materials for energy applications. I hope to continue a new collaboration with a group at Lawrence Berkeley National Lab, where weÕve applied flow-SANS methods to study Nafion solutions for fuel cell applications. Again, these methodologies can be tailored to approximate the necessary dynamic flow conditions, such as cycling or flow reversals. Further, recent developments have enabled the simultaneous measurement of dielectric properties during flow-SANS, which is an area I hope to explore in these energy-related applications. My overall goal is to apply this multi-faceted neutron and rheological approach to interesting and relevant soft materials, using realistic experimental conditions for application-based projects.

Publications:

1. M.A. Calabrese, S.A. Rogers, L. Porcar & N.J. Wagner. ÒUnderstanding steady and dynamic shear banding in a

branched wormlike micellar solution,Ó *Journal of Rheology (accepted)*, 2016.

2. M.A. Calabrese, N.J. Wagner and S.A. Rogers. ÒAn optimized protocol for the analysis of time-resolved scattering experiments,Ó *Soft Matter*, 2016.

3. M.A. Calabrese, S.A. Rogers, R.P. Murphy and N.J. Wagner. ÒThe rheology and microstructure of branched micelles under shear,Ó *Journal of Rheology*, 2015, 59(5), 1299-1328.

4. A. Kusoglu, M. Calabrese and A. Z. Weber. ÒEffect of mechanical compression on chemical degradation of Nafion membranes,Ó *ECS Electrochemistry Letters*, 2014, 3(5), F33-F36.

Reviews, book chapters & other articles:

1. M.A. Calabrese and N.J. Wagner. ÒNew Insights from Rheo-SANS,Ó chapter in: *Wormlike Micelles: Systems, Characterisation, Applications*, Royal Society of Chemistry (2016).

2. M.A. Calabrese, N.J. Wagner, S.A. Rogers and L. Porcar. ÒEffect of branching on shear banding in worm-like micelles (WLMs) under large amplitude oscillatory shear (LAOS),Ó *Instrument & Technical Upgrades - ILL News*, Dec. 2015.

3. S.A. Rogers, M.A. Calabrese, N.J. Wagner. ÒRheology of branched wormlike micelles,Ó *Current Opinion in Colloid & Interface Science*, 2014, 19(6), 530-535.

4. S.A. Rogers, M.A. Calabrese and N.J. Wagner. ÒAdvances in Time Resolved Neutron Scattering from Flowing Complex Fluids,Ó *NCNR Annual Report*, 2014.

476300 Metabolic Engineering and Synthetic Biology for the Renewable Production of Fuels and Chemicals

Abstract:

Research Interests:

Global warming and diminishing fossil fuel reserves have garnered worldwide attention for the renewable production of fuels and chemicals through the metabolic engineering of microbial cell factories. The cyanobacterium *Synechocystis* sp. strain PCC 6803 was engineered to produce fuels and chemicals from carbon dioxide as carbon substrate and sunlight as energy source. First, metabolic engineering of the cyanobacterium, *Synechocystis* sp. PCC 6803 was performed for synthesizing isobutanol as a renewable fuel alternative. With the expression of the heterologous genes from the Ehrlich Pathway, by incorporating an *in situ* isobutanol harvesting system, and also by employing mixotrophic conditions, the engineered *Synechocystis* 6803 strain accumulated a maximum of ~300 mg/L of isobutanol in a shake flask culture. Second, *Synechocystis* 6803 was engineered with a novel D-lactate dehydrogenase (encoded by *gldA101*) for D-lactic acid synthesis. The final titer of D-lactate was improved to 2.2 g/L by employing the following strategies: (i) cofactor balancing, (ii) codon optimization, and (iii) mixotrophic conditions.

Lignin is the only renewable and abundant polymer on the earth with aromatic units as its building blocks; however it remains as an untapped resource. In the first work aimed at lignin valorization, ^{13}C-fingerprinting, ^{13}C metabolic flux analysis (MFA), and RNA-seq differential expression analysis were utilized to uncover the metabolism of *Sphingobium* sp. SYK-6. SYK-6 is a soil bacterium with a well-studied ligninolytic pathway and an improved understanding of its metabolism will help researchers in the engineering of SYK-6 for lignin valorization. A key finding from this study was the establishment of vanillin catabolic pathway as the major contributor for NAD(P)H synthesis. In the second work, a hybrid promoter engineering approach was employed for the construction of higher strength phenolics inducible promoters. The hybrid promoters were constructed by replacing the spacer region of an endogenous promoter, P_{emrAB} that was naturally inducible by phenolics. In the presence of vanillin, the strains RIF02, RIF03 and RIF04, with the engineered promoters P_{vtac}, P_{vtrc}, and P_{vtic} showed fluorescence increases of 4.6, 3.0, and 1.5 fold respectively in comparison to RIF01 with the native promoter P_{emrAB}. A third study will discuss the research work that was employed for the microbial depolymerization of lignin. We hope that all these different synthetic biology and metabolic engineering efforts will enable the biorefinery to become cost competitive with the petroleum and petrochemical industry.

Teaching Interests: I believe teaching as one of my mentor said, will keep us young in the mind since we will be interacting with younger brains throughout the career. Sharing the knowledge I have gained to others provides joy to my soul and teaching provides a perfect opportunity. As a researcher, it will give me immense pleasure in guiding my students to carry out their individual research and also an opportunity to inspire the younger minds towards research. I will be effective in teaching the core courses of chemical engineering and biotechnology at the undergraduate level. At the graduate level I am interested in developing a course on metabolic engineering and synthetic biology.

476306 Effect of Surfactant-Particle Interactions on the Formation and Stability of Emulsions

Abstract:

The use of binary mixtures of surfactant and colloidal particles is of great academic and commercial interest for the stabilization of emulsions and foams. Depending upon the interactions between the surfactant molecules and the colloidal particles, and on the relative concentrations of surfactant to particles, the formed emulsions can exhibit different properties. Surfactant molecules of charge opposite to that of the colloidal particles modify the wettability of the particle surfaces by exposing their hydrophobic tails to the aqueous phase. On the other hand, surfactant molecules of charge similar to that of the colloidal particles, may not interact with the colloidal particles as a result of the electrostatic repulsions between them. Nonionic surfactants can interact with colloidal particle surfaces through hydrogen bonding or hydrophobic interactions. The resulting surfactant adsorption behavior can greatly affect the colloidal particles dispersion stability and wettability, in turn affecting the formation and stability of the emulsions. In this poster, I will discuss general aspects of the formation and stability of emulsions, as they are affected by the addition of surfactants and colloidal particles. I will show that irrespective of the surfactant-particle interactions, surfactant-stabilized emulsions are always favored at high surfactant-to-particles ratios. On the other hand, I will show that phase separation of oil and water, or emulsions stabilized by both particles and surfactant molecules, are observed at intermediate surfactant-to-particle ratios. These findings will be explained using a combination of visual and microscopic observations, as well as surface tension measurements.

Research Interests:

Colloids and Interfacial science, Surfactants and Emulsions, Sol-gel Synthesis, Micro- and nano-fabrication, Patterning of Soft Surfaces, Rheology, Environmental Remediation

Teaching Interests:

Thermodynamics, Colloids and Interfacial phenomenon

476310 Engineering Soft Functional Materials: From Self-Assembly to Field-Assisted Assembly

Abstract:

Research Interests:

Engineering Soft Functional Materials: From Self-assembly to Field-assisted Assembly

The ubiquity of self-assembly - the process of creating organizational order in systems of components - in nature has inspired technological developments towards synthetic building blocks that assemble into desirable structures with a unique set of properties. Isotropic spherical colloids are a simple example of such building blocks where their spatial arrangement yields photonic crystals that exhibit structural color. The key step towards engineering the assembly process is the ability to tune the interparticle interactions. However, the experimental realization of target structures can be challenging owing to the slow kinetics of the self-assembly process. There is a concerted effort in the field to identify the factors that impact the particle-particle interactions and control the assembly dynamics via external fields. What happens when shape or surface anisotropic particles are used as building blocks for assembly? How is the assembly in bulk different from the 2D assembly in the presence of a fluid interface? What interactions are induced in the presence of an external electric field? To address some of these questions, my research focuses on the assembly of shape and surface anisotropic colloidal particles. I present experiments on the application of fluid interfaces as a template for assembly and discuss the role of particle surface properties in tuning the mechanical stability and flow behavior of the assembled monolayer. External direct current fields are used to control the dynamics of assembly in dense colloidal suspensions and measurements of the electrophoretic mobility demonstrate the significance of the dispersion electrokinetic properties and suspension volume fraction. Given the need for generating shape-memory colloidal structures suitable for applications where rapid, on demand reconfiguration is required, I also present experiments on the actuation of dense suspension of surface anisotropic colloidal fibres using an external electric field. Insights obtained from all these studies contribute to our fundamental understanding of the principles central to the assembly processes and serve as a platform for engineering the bottom-up assembly of functional materials. Using this knowledge, I plan to find new avenues for engineering the assembly of soft functional materials.

Teaching Interests:

Teaching is a great opportunity to contribute to the formal education of the students. I have been fortunate to have the experience of both teaching in a classroom and mentoring students in the lab. I have been a teaching assistant in a number of undergraduate courses including the transport phenomena and nanotechnology. As a part of my classroom teaching responsibilities, I have designed homework assignments, compiled some sources for various topics on the course and given lectures on the molecular dynamics simulations and their importance in the chemical engineering and materials science research. I believe that not only it is crucial to educate the future engineers on the importance of the traditional chemical engineering subjects, but is also essential to integrate computational tools as a part of the course material in order to introduce these techniques early on and discuss their relevance in advancing the field. In addition to the classroom experience, I have mentored both graduate and undergraduate students in the lab. Besides training the students on various research tools, I have always been passionate about encouraging critical thinking, developing the students written and oral presentation skills, and training the students to be independent researchers.

476478 Nucleic Acid Self-Assembly in Alternative Solvents

Abstract:

Research Interests: Control of nucleic acid self-assembly to build complex dynamic systems; Thermodynamics and kinetics of nucleic acid self-assembly

Teaching Interests: Chemical kinetics; Chemical thermodynamics; Biochemical engineering

Abstract

Increased control of molecular self-assembly is important for the further development of nucleic acid-based technologies, with applications ranging from drug design and delivery to development of catalytic nucleic acid sequences. My research is guided by two main goals: understanding the fundamental chemical and physical properties which govern nucleic acid self-assembly, and utilizing nucleic acid self-assembly processes to build dynamic systems that solve specific problems.

My graduate work is focused on nucleic acid self-assembly in alternative solvents. While nucleic acid technology research has been carried out almost exclusively in aqueous buffer, alternative solvents could increase our ability to direct nucleic acid assembly. Deep eutectic solvents (DES), closely related to ionic liquids, are characterized by significantly lower melting points than their separate components. Recently, DES have been shown to preserve nucleic acid base pairing and duplex formation, offering a relatively unexplored medium for directing nucleic acid assembly.

I show that DES provide a highly effective means for tuning nucleic acid thermodynamics and kinetics. The power of alternative solvents to alter a nucleic acid assembly pathway is illustrated by our ability to circumvent the â€œstrand inhibition problemâ€ by thermally cycling nucleic acids in a DES. Strand inhibition describes the propensity of a long nucleic acid polymer to form a duplex with its complementary strand rather than to serve as a template for assembly of oligonucleotides. This longstanding problem in the field of prebiotic chemistry has limited demonstration of a prebiotic nucleic acid replication mechanism. I address the problem of nucleic acid strand inhibition by employing highly viscous solvents to control the thermal stabilities and annealing rates nucleic acid polymers in a size-dependent manner. Based on these thermodynamic and kinetic effects, we show that thermal cycling in a viscous environment can be used to copy a gene-length region (over 300 nucleotides) of a longer mixed sequence template, using both DNA and RNA systems.

478610 Release Mechanism of Fluids Under Confinement: New Findings and Applications for Hydrocarbon Recovery

Abstract:

4th Year Postdoctoral Fellow

Research Interests:

Existing approaches for oil and gas recovery are designed based on macroscopic properties of the produced hydrocarbon fluids. However, recent studies on source rocks revealed that properties of fluids stored in nanopores of organic constituent material kerogen deviate from the bulk behavior. Hence, the traditional equation of state and fluid properties correlations are no longer applicable. This, in turn, leads to added uncertainties in hydrocarbon in-place and recovery calculations for the source rocks that are rich in kerogen such as shales.

First, I seek to address the above question at a fundamental level from the thermodynamics standpoint by simulating isothermal expansion of a quinary hydrocarbon mixture in model nanopore under typical subsurface conditions, and measuring the fluid composition and recovery factor. I found that nanopore confinement limits significantly the release of hydrocarbon molecules from the pores with sizes less than 10nm. For each hydrocarbon component, a strong correlation exists between molar fraction of the component in the produced fluid with that remained inside the pore. This correlation can serve as the basis for establishing alternative methods for the recovery calculations.

Second, molecular dynamics simulations were employed to investigate the possibility of mobilizing the oil phase, modeled as pure n-heptane, in nanopores by utilizing solubilized solvent in the form of microemulsion droplets. In the case of solubilized terpene, the swollen micelles adsorb on the surface as one entity which is different from the conventional behavior of micelles. The delivery of surfactant to the interface is controlled by the solvent concentration inside the microemulsion droplet. The solvent, originally solubilized in a microemulsion droplet, penetrates the thin film of oil. Furthermore, those solvent molecules displace the oil molecules in the adsorbed phase and transform that portion of oil into free oil. The results are important for our understanding of microemulsion behavior under confinement and its application to organic rich shale oil recovery.

My research focuses on utilizing multi-scale modeling techniques together with high performance computing to gain insight into various engineering problems and seek answers from the thermodynamics and physical standpoint. I am particularly interested in heat transfer in nanocomposites, phase behavior and flow in porous media, storage and release mechanism of fluid in nanopores, and surfactant-based enhanced oil recovery. My formulation for delivering successful and high quality research work is collaboration and diversity. I look forward to collaborate with other faculty members in the department and industrial partners.

Teaching Interests:

Iâ€™m interested in building up a new course related to properties of materials and fluids at the microscopic scale for engineering practices. This course will equip our students with the knowledge of heat transfer in nanocomposites, phase behavior and flow in nanopores and how to enhance hydrocarbon recovery from unconventional resources. For example, macroscopic properties such as pore pressure, viscosity, interfacial tension and phase behavior of oil and gas inside the nanopores can be investigated by looking at how fluid molecules interact with the rock particles at the fundamental level of thermodynamics and physics. Students will learn both the theories and the practice of using multi-scale modeling techniques to tackle realistic engineering problems. In fact, the material of this course will be based on my research experience and other advanced work in the field of nanotechnology. This course will be a unique addition to the curriculum of not only petroleum but also other engineering programs. More important, I enjoy teaching and mentoring students in the academic environment during my postdoc and my graduate study.

Other experiences and Awards:

- Senior Research Associate at Department of Petroleum Engineering, Texas A&M University, 2014 to current.
- Developing new research topics and co-writing proposals submitted to Crisman Institute 2013-2016, Flotek Inc. 2013-2016.
- Serving in the graduate student committees of Department of Petroleum Engineering, Texas A&M university.
- Active reviewer of Reviews in Chemical Engineering Journal, Society of Petroleum Engineering (SPE) Journal and Applied Surface Science Journal.
- Explorations in Science through Computation Student Award, Supercomputing Conference 2010, New Orleans, LA.

Selected publications:

Bui, K., Akkutlu, I. Y., Natural Gas Recovery from Kerogen Nanopores, **SPE Journal**, revised manuscript submitted 2016.

2. Bui, K., Akkutlu, I. Y., Zelenev, A. et al. Insights into Mobilization of Shale Oil by Use of Microemulsion, **SPE Journal**, SPE-178630-PA, 2016.

3. Dimitrios V. Papavassiliou, Khoa Bui and Huong Nguyen, **Thermal Boundary Resistance Effects in Carbon Nanotube Composites - Book Chapter**, ISBN: 978-1-119-06893-8, Wiley, 2016.

4. Bui, Khoa; Akkutlu, Yucel, Nanopore wall effect on surface tension of methane, **Molecular Physics**, Vol 113, 22, 2015.

5. Feng Gong, Khoa Bui, Dimitrios V. Papavassiliou, Hai M. Duong, Thermal transport phenomena and limitations in heterogeneous polymer composites containing carbon nanotubes and inorganic nanoparticles. **Carbon** 78, 2014.

6. Bui, Khoa; Grady, Brian P.; Saha, Mrinal C.; Papavassiliou, Dimitrios V. Effect of Carbon Nanotube Persistence Length on Heat Transfer in Nanocomposites: A Simulation Approach. **Applied Physics Letters**, vol. 102, 20, 2013.

7. Bui, Khoa; Grady, Brian P.; Papavassiliou, Dimitrios V., Heat transfer in high volume fraction CNT nanocomposites: Effects of inter-nanotube thermal resistance. **Chemical Physics Letters**, vol. 508, 2011.

475551 Toward Understanding the Atmospheric Chromium Chemistry

Abstract:

Research Interests:

Trivalent chromium (Cr(III)) and hexavalent chromium (Cr(VI)) are the two common oxidation states of chromium in the environment. Inhalation exposure to Cr (VI) may lead to cancer, nasal damage, asthma and bronchitis. In contrast, Cr(III) is a trace element essential for the proper function of living organisms. Therefore, it is important to accurately discriminate between these two species in atmospheric particulate matter (PM) samples. In the United States, Cr(VI) is one of the criteria pollutants which has been monitored through the National Air Toxics Trends Station (NATTS) Network. Thus, measurement of Cr(VI), is one of the focal points for hazardous air pollutants (HAPs) monitoring in the US.

The focus of my PhD research at Clarkson University was centered around understanding the atmospheric chromium chemistry and improving the sampling of atmospheric Cr(VI). Chromium is available as soluble or insoluble in the atmosphere. When soluble chromium is dissolved in water, the transformation of Cr(VI) is expected to take place in the water associated with particles (aqueous chemistry of chromium). Airborne particles that include soluble salts like ammonium sulfate or ammonium nitrate can spontaneously take up water (deliquescence) from the air and form a solution on the surface of the particle. This liquid layer allows chemistry to take place. I have shown that keeping the sampling particulate matter below the point at which it can take up water provides significant improvements in the amount of Cr(VI) retained in the sample. I have found that Cr(III) is converted to Cr(VI) in the presence of gaseous species like ozone and reactive oxygen species (ROS) at dry condition (gas-solids chemistry of chromium) implying that Cr(III) conversion can produce false positive measurements for Cr(VI) sampling. I have demonstrated that conversion decreases at lower temperature values. Thus, based on my work, it is likely that a sampler that keeps the sample dry and cool will improve the recovery of Cr(VI) that is present in the sample. I built a prototype sampler and deployed 4 of them in a field study in NJ, side-by-side with 4 of the samplers currently used by USEPA in their HAPs monitoring network. These tests showed that the improved samplers did provide better preservation of Cr(VI). Presently, I aim to develop a dynamic model for the simulation of aqueous chromium chemistry in the atmosphere (1<pH<5) and during the sampling period (24 hours- pH 9).

Recent studies showed that insoluble Cr(III) and insoluble Cr(VI) dominate atmospheric chromium mass. Regardless of few data in my PhD thesis, no information is available in the literature on the reaction kinetics of insoluble chromium speciation in airborne particles. Therefore my future research will concentrate on insoluble chromium chemistry (gas-solids chemistry of chromium) in the atmosphere (within next five years). My long term research plan (beyond next five years) is to investigate the chemistry/toxicity of soluble and insoluble chromium compounds in human lungs and their impact on lung surfactants. I have proposed three projects on atmospheric chromium. The first project investigates the reaction kinetics of insoluble Cr(III) and insoluble Cr(VI) with reactive oxygen species and volatile organic compounds, respectively (gas-solids reactions). The second project is aimed at designing and field testing a new Cr(VI) sampler to preserve both Cr(VI) and Cr(III) during sampling. The third project focuses on modeling Cr reactions (soluble and insoluble reactions) in the atmosphere and during the sampling. The details of my future research plan and potential funding agencies, current research, previous teaching and undergraduate research mentoring experiences are available during the AIChEÂ â€œMeet the Faculty Candidatesâ€ poster session.

Teaching Interests:

I have the past experiences of teaching the following courses:

Capstone Design Projects in Chemical Engineering, Process Simulation using Aspen Plus, Transport Phenomena, Safety Engineering and Risk Management, Process Calculations (mass and energy balance)

475988 Membranes at the Water-Energy-Food Nexus: Experimental and Modeling Approaches

Abstract:

Research Interests:

This poster will provide an overview of my computational and experimental research about the design and fabrications of different types of membrane (microfiltration, ultrafiltration, reverse osmosis and forward osmosis) at the water-Energy-Food Nexus. My Ph.D. dissertation focused on the application of polymeric nano-composite membranes (Mixed Matrix Membrane) to water purification. During the first part of my research, I worked on understanding the mechanism of natural organic matter (NOM) fouling on an ultrafiltration membrane by relating the operational viewpoint for fouling (macroscopic) to the microscopic viewpoint. Then, in the second part of my research, I used the power of nanoparticles to increase the antifouling properties of the polysulfone membrane. The mixture of titanium dioxide nanoparticles and multiwall carbon nanotube with the optimum mass ratio was embedded with the novel and simple method into the polymeric membrane and interestingly increased both the membrane permeability and rejection two times compared to the pure polysulfide membrane. Also, the nanocomposite membranes showed the higher antifouling behavior compared to the pure membrane at different operational conditions and flaunt physicochemical properties. In my post-doctoral research, I have continued to work on the membrane application for water treatment and energy production. Forward osmosis (FO) (also, pressure retarded osmosis) is a prime membrane technology of interest due to its potential for low energy use in water – energy nexus. The main goal of my post-doctoral project was investigation of effective parameters on transportation of water and solute molecules in forward osmosis membrane to improve FO membrane design and process efficiency. This project combines computational fluid dynamics (CFD) modeling with experimental fabrication of FO membranes to examine the assumptions imbedded in current transport equations and identify sources of inaccuracies in water flux predictions. This effort can lead us to the development of transport equations that can be used by all to support FO research, enabling the deployment of FO for real-world problem solving with economically competitive performance. Based on my knowledge of membranes, instrumentation design and componential fluid dynamic (CFD) modeling, my future research will focus on both experimental and modeling aspects of membrane technology issues for the water – energy –food nexus. In details, I plan to mainly focus my academic career on parallel research area of the computational modeling and experimental synthesis, modification and fabrication of different types of novel mixed matrix membranes for using at different connected interdisciplinary areas at the water – energy – food nexuses.

Some research areas that I plan to pursue in my first five years as an Assistant Professor:

Membrane fouling mitigation** Forward osmosis and pressure retarded osmosis design, surface modification and characterization**Mixed matrix ultrafiltration/microfiltration**Nanomaterials environmental safety: transport and fate of nanomaterials

Teaching Interests:

My first preference would be to teach the chemical engineering core courses, such as Transport Phenomena and Chemical Engineering Thermodynamics. However, since my background and previous experience is diverse, I am flexible to handle other courses within the University Chemical Engineering Curriculum. Additionally I would like (provided the opportunity is given) to develop new courses, as the special topics, including topics of water / wastewater treatment, membrane technology and environmental science.

476060 Removal of Contaminants from Water and Wastewater: (Bio)Sorption, Membrane Filtration, Advanced Oxidation

Abstract:

Overcoming the emerging global water shortages and pollution are grand challenges faced by humanity today. This poster will provide an overview of my research upon the development and optimizations of different biological, chemical and physical methods for removal of toxic and carcinogenic contaminants (both organic and inorganic) from water and wastewater. Summary of my top achievements include: During my Master, I worked on optimization of biosorption process which is an eco-friendly, effective and extremely low-cost technology for removal of synthetic dyes and hexavalent chromium which are toxic and carcinogenic contaminants, from water by using available biosorbents/adsorbents (such as algae, aquatic plants, etc.). Finally, potent organic/inorganic-binding biomass types for removal of pollutants form aqueous solution were discovered. My current PhD research is on Optimization of advanced oxidation process (pulsed corona electrical discharge, UV/photocatalysts) as a highly efficient method, for removal of public health dangers contaminants (synthetic dyes, humic substances) from water/wastewater without using any chemicals (eco-friendly) and optimization of a unique hybrid advanced oxidation technology /nanocomposite ultrafiltration membrane for removal of synthetic and natural organic public health dangers contaminants form water.

Research Interests:

- Using low cost biosorbents/adsorbents for wastewater treatment
- Membrane-based hybrid processes for treating impaired quality waters
- Providing important insights into fundamental mechanisms of membrane fouling under a wide range of operating conditions.
- Development of novel approaches to enhance performance of advanced oxidation processes (electrical discharge, UV photocatalysts, ozone) for water and waste water treatment

Future Direction: My research experiences and objectives give me confidence that I will be able to obtain funding and add significant improvements in areas related to the water and wastewater treatment. I will be able to establish a strong collaboration with industry and academia to make my research objectives achievable.

Teaching Interests: Â As a faculty member, I would like to continue my research in the water and wastewater treatment. I am interested to teach graduate courses in the field of Advanced Topics in Environmental Chemistry, Chemical Principles of Environmental Systems, and the graduate laboratory courses on Environmental Instrumentation. I also would like to teach theÂ undergraduate and core courses on kinetic, transport phenomena, environmental material balances and coordinate in the undergraduate Capstone Course.

Selected publications:

- **Negin Koutahzadeh**, Milad Rabbani Esfahani, Pedro E Arce, Removal of Acid Black 1 from water by the pulsed corona discharge advanced oxidation method, *Journal of Water Process Engineering, Volume 10, April 2016, Pages 1â€"8.*
- **Negin Koutahzadeh**, Milad Rabbani Esfahani, Pedro E Arce, A sequential use of UV/H_2O_2 â€" (PSF/TiO_2/MWCNT) mixed matrix membranes for dye removal in water purification: Membrane permeation, fouling, rejection and decolorization, *Journal of Environmental Engineering Science, 2016* (Accepted, *In press).*
- **Negin Koutahzadeh**, E. Daneshvar, M.Kousha, M.S.Sohrabi, A.Bhatnagar, Biosorption of hexavalent chromium from aqueous solutions by six brown macro algae biomasses: Equilibrium and kinetics studies, *Journal of Desalination and Water Treatment*, Volume 51, Issue 31-33, 2013.

476108 Application of Shrinking Core Model Applied for Gas Hydrate-Based CO_2 Capture in Presence of Porous Hydrogels

Abstract:

Research Interests: CO_2 Capture, Gas Hydrate, Process Modelling, Reservoir Simulation, Flow Assurance, Asphaltene Deposition and Precipitation, Enhanced Oil Recovery

Teaching Interests: Thermodynamics, Transport Phenomena, Chemical Engineering Simulation Software, Heat Transfer, Kinetics and Reactor Design

A salient portion of the global pollutant gases emission is contributed to the carbon dioxide (CO_2) released from industrial sectors. Currently, stiff environmental requirements do not allow industries to proceeds beyond the standard limitations for such a harmful exhaust. Hence, deduction of CO_2 released from industrial processes has been the subject of researches for many years seeking novel, efficient and cost-effective technologies for worldwide applications. Among the various methodologies, the hydrate-based CO_2 capture (HBCC) process is considered as a promising approach to achieve this goal. This process, however, is in its infancy and must be examined through a vast range of experiments and modelling studies prior to being well commercialized. Advancements in both the process intensification and the materials and technology development associated with this method can be found in a few recent reports. While most works focus on the development of new materials to enhance the kinetic of this process, a very small number of studies have been focused on the mechanism of the hydrate-based CO_2 capture process in the presence of porous materials which will lead to design and improvement of new materials for industrial application. In this article, porous hydrogel (HYD) microspheres are investigated for application in CO_2 clathrate hydrates. A new variation of so-called shrinking core model is developed and analyzed for this case study in which the intrinsic hydrate formation rate is combined with transport phenomena to comprehensively capture the role of controlling mechanisms including the diffusivity of the CO_2 gas molecules, the water distribution and conversion rate in various phases of the medium and operational parameters involved. The proposed model allows the prediction and explanation of hydrate formation and CO_2 capture rates in hydrate structure under different pressures and temperatures. The model is tested against experimental data which have been obtained in our research group. A range of sensitivity analysis is carried out to investigate the influence of the extracted parameters.

456510 Size-Dependent Chemical Compositions in Particulate Matters from Major Outdoor Sources in a Megacity and Corresponding Inhalation Exposure Assessment

Abstract:

A variety of existing studies have characterized the chemical compositions in particulate matters (PM) for various megacities. However, these studies rarely accounted for the size-dependence of chemical constituents in PM, hindering the application of the data for inhalation exposure assessment and thus epidemiological studies. In this work, a series of field measurements were conducted to characterize the major emission sources of Singapore, i.e. traffic and industry emissions (coal combustion is rare in Singapore and thus disregarded). The particle size distribution and corresponding size-dependent chemical compositions were identified. Existing respiratory deposition models were employed to analyze the deposition distributions of particles and chemical compositions in the human respiratory system, based on the data of particle size and chemical compositions. Finally, potential protection suggestions were proposed to mitigate the PM exposure for susceptible populations. The data in this work will be also representative to other megacities whose particulate air pollution is mainly derived from traffic and industry emissions.

Research Interests:

Dr. Siming Youâ€™s research interests fall under the umbrella of the interactions between environment and energy systems. He is trying to explore the potential environmental problems underlying existing energy systems, and understand and solve environmental problems from an energy point of view. Graduated from Nanyang Technological University as the top PhD student in Thermal and Fluids Engineering in 2014, he accumulated significant interdisciplinary research experiences by studying human exposure to harmful particulate matters (PM), especially, pathogens, and thus infection transmission from a physicochemical point of view. He then pursued his postdoctoral research at the Department of Civil & Environmental Engineering, Massachusetts Institute of Technology, and gained a deeper understanding of the transport dynamics of harmful PM in environments and inhalation exposure-related health implications. He is now a research fellow at Professor Chi-Hwa Wangâ€™s group at the National University of Singapore, where he is investigating the environmental impacts and challenges of sustainable energy systems and proposing solutions to the potential environmental problems facing sustainable energy systems.

Teaching Interests:

I would like to teach Fluid Mechanics for undergraduate students, and Aerosol Technology and Environmental Health for both undergraduate and graduate students.

Fluid Mechanics is one of the most actively developing fields of physics and one of the basic core courses of engineering. Significant parts of my PhD studies and postdoctoral study at MIT have devoted to understanding how harmful particles transport in environments under the effects of aerodynamic forces. Specifically, I have accumulated a significant amount of knowledge about turbulence, such as the characteristics of turbulent bursts and the methods of quantifying turbulence parameters. These past research experiences and knowledge did lay a solid foundation to enable me to participate in this course on Fluid Mechanics.

Aerosol is ubiquitous throughout the world and the related technology has been widely employed in various fields such as microelectronic industry, pharmaceutical industry, environmental sciences, etc. The course of Aerosol Technology would cover the fundamental properties and behaviors of aerosols, aerosol sampling methods, and the application of aerosol technology in various fields such as industry and environmental health sciences. This course would equip students with the ability to sample, analyze, and apply aerosols for their future requirements in either industry or academics.

Environmental Health would cover the knowledge about how natural and built environments may affect human health. It would present an overview of methods used to understand environmental health hazards, ranging from global issues such as climate change to regional issues such as air, water, and energy to local issues such as food safety, and occupational health, and to conduct risk assessment and risk management for environmental exposure-related health implications. This course would not only equip students with necessary knowledge and skills to address the potential concerns related to environmental health, but also encourage them to make positive changes in their behavior and attitude to protect the environment.

470501 Engineering Intelligently Designed Nano- and Microparticles to Control Interactions with the Immune System

Abstract:

Research Interests:

Nano- and micro-particulate carriers enable the site-specific delivery for controlled biological responses and can harness the intrinsic pathways by which the body responds to natural invaders. These particles are in size range which naturally associates with many innate immune cells, including antigen presenting cells (APCs), and can interface with these cells to deliver a stimulus. Through controlled design properties, engineered nano- and microparticle drug delivery vehicles have the potential to expand the breadth of many therapeutic approaches, impacting immunological outcomes through cell-specific targeted delivery. However, in many applications, such as mucosal vaccines or controlled-release lung depots, optimal particle properties have not yet been identified. Many particle design considerations have been established for systemic intravenous administration to create long-circulating drug delivery vehicles, yet less is known about particle design parameters critical to interfacing with and directing the immune system.

My research to date has focused on filling in these gaps by designing particles to engineer a controlled response with the immune system. Importantly, this research has ranged from particle synthesis, functionalization, and characterization to extensive evaluations in a range of *in vitro* and *in vivo* models. During my PhD, I investigated the role of particle properties to increase local residence times and alter the cellular fate of nano- and microparticles delivered to the lung. In these works, particle association in lung APC populations was characterized as a function of particle parameters, including size, surface charge, shape, and surface chemistry. This work demonstrated the potential use of cationic nanoparticles as pulmonary mucosal vaccines, which were shown to produce superior local and systemic immune responses, as well as PEGylated formulations to extend residence times in the airways. Research

during my postdoc has focused on designing particle therapeutics, again at the interface between particle formulations and interactions with blood cells. This work has evaluated particle properties of size, shape, modulus, and surface chemistry to understand interactions with blood cells and inflamed endothelial cells under physiological blood flow, specifically during inflammatory events. Additional projects funded independently under the UM Presidentâ€™s Postdoctoral Fellowship have evaluated controlled presentation of toll-like-receptor (TLR) adjuvant molecules on the surface of particle carriers to direct Th1 immune stimulation. Combined, these works demonstrate a potential for rationally designed particle therapeutics to interface with the immune system. Future research will continue to evaluate particle design features on controlled immune interactions, especially in non-intravenous routes of administration, towards the development of novel treatments for vaccinations, cancer, inflammation, and allergy.

Teaching Interests:

Transport, Mass and Energy Balances, Heat and Mass Transfer. I have experience guest lecturing for undergraduate level Chemical Process Analysis through the Mentored Teaching Award at North Carolina State University, as well as Heat and Mass Transfer at the University of Michigan.

456396 The Effect of Infusion Position on Convection-Enhanced Delivery of Anticancer Drugs to Remnant Brain Tumour after Surgery

Abstract:

Research Interests: biomedical engineering and bioprocessing, with a special focus on transport processes in biological and physiological systems.

Teaching Interests: mass transfer

Convection enhanced delivery is a promising method to overcome the blood-brain barrier and brain-tumour barrier to improve the drug delivery into tumour interior and enhance the treatment efficacy. In this study, a 3-D geometry of a remnant brain tumour and its normal holding tissue is reconstructed from MR images. A developed mathematical model is applied to investigate the effects of infusion position on the spatial distribution of carmustine and paclitaxel. Results show that infusion position is an important impact factor in CED treatments, and the effect of infusion position varies with the drug. For carmustine, most of the drugs accumulate around the infusion site to form a highly heterogeneous distribution in tumour. The heterogeneity and the spatial average concentration can be reduced by infusing carmustine into the cavity. Paclitaxel is capable to penetrate deep into the brain tumour tissue with the enhanced interstitial fluid flow. However, infusing paclitaxel in the downstream can reduce the homogeneity and spatial averaged concentration in the whole remnant tumour.

462670 Yeast Cell Factories: Construction of Platform Strains and Development of Synthetic Biology Tools

Abstract:

Research Interests:

Biological conversion of renewable feedstock into fuels and chemicals has been intensively investigated due to increasing concerns on sustainability and global climate change. Compared with its counterparts, *Saccharomyces cerevisiae*, the bakerâ€™s yeast, is more industrially relevant thanks to its well-studied genetic and physiological background, the availability of a large collection of genetic tools, the compatibility of high-density and large-scale fermentation, the resistance to phage infection, and the high tolerance against toxic inhibitors and products. Therefore, *S. cerevisiae* is one of the most popular cell factories and has been successfully used in the modern fermentation industry to produce a wide variety of products such as ethanol, organic acids, amino acids, enzymes, and therapeutic proteins.

My Ph.D. work was mainly focused on the construction of yeast cell factories for efficient and cost-effective production of biofuels and chemicals using synthetic biology and metabolic engineering approaches. Notably, a wide variety of products with industrial interests are derived from a few precursor metabolites. For example, lactate, 2,3-butanediol, and *iso*-butanol are derived from pyruvate, while *n*-butanol, polyhydroxybutyrate, isoprenoids, fatty acids, fatty alcohols, alkanes, and fatty acid ethyl esters are derived from acetyl-CoA. Therefore, we aim to develop platform yeast cell factories with enhanced supply of these precursor metabolites for efficient production of biofuels and chemicals. To achieve this, it generally involves host genome engineering such as the disruption of the competing pathways and biosynthetic pathway engineering such as the introduction of heterologous pathways. For example, in order to construct an acetyl-CoA overproducing yeast strain, the ethanol and glycerol formation routes were partially or fully blocked and alternative acetyl-CoA biosynthetic pathways, such as the pyruvate dehydrogenase (PDH) from a bacterium and the ATP-dependent citrate lyase (ACL) from an oleaginous yeast, were introduced. The higher efficiency and lower energy requirement of PDH and ACL, when combined with host genome engineering, increased the acetyl-CoA levels by more than three-fold and the production of acetyl-CoA derived product by more than 12-fold.

During my postdoc research, I switched my focus to the development of novel synthetic biology tools to facilitate the development of cell factories. Although a wide variety of cell factories have been constructed using the traditional strategies mentioned above, we are still facing the challenges of constructing efficient cell factories in a short period of time. Therefore, novel synthetic biology tools are demanded to revolutionize cell factory development. In terms of the biosynthetic pathway engineering, we constructed a series of plasmids with different copy numbers, which enabled the fine-tuning of gene expression levels and optimization of multi-gene biosynthetic pathways. In terms of genome engineering, we took advantage of the recently developed CRISPR/Cas9 system to achieve both gain of function (knock-in and gene activation) and loss of function (knock out and gene repression) in yeast. More importantly, we combined gene deletion, gene repression, and gene activation into a single system, in which a full spectrum of gene expression profiles (zero expression, down-regulation, and up-regulation) was achieved. Now we are performing genome-scale studies to develop a rapid and efficient strategy for the construction of cell factories and unveil the synergistic effects between gene deletion, gene repression, and gene activation in yeast.

Teaching Interests:

As a graduate student in the University of Illinois, I have the opportunity to work with and mentor many talented undergraduate and graduate students. Three semesters of TA in the department of Chemical and Biomolecular engineering have given me a better understanding on how to teach effectively in a big classroom, how to mentor students with various academic background, as well as the skills to communicate with students from a wide range of countries. Besides these formal teaching assignments, I have served as a mentor for undergraduate and junior graduate students in the lab, helping them to understand not only the synthetic biology principles and experimental techniques, but also the experimental design and problem-solving skills.

Keywords: Yeast cell factory; Synthetic biology; Genome engineering; CRISPR/Cas9; Plasmid copy number

464165 Engineering Immune Development By Recapitulating Tissue Microenvironments

Abstract:

Research Interests:

Bone marrow transplantation, also known as hematopoietic stem cell transplantation (HSCT), is a process in which healthy blood stem cells within the bone marrow of a donor are extracted and transferred into a recipient. The process is typically performed in patients with advanced blood-based diseases such as leukemia and multiple myeloma. Prior to a HSCT, the patient undergoes radiation and cytotoxic chemotherapy regimen to eradicate as much of the disease as possible. However, the process severely impairs the immune system as a side effect. The profound long-term post-transplant immunodeficiency often results in severe complications in patients such as infections, the development of graft-versus-host disease (GVHD) and the relapse of the underlying disease. Thus, while effective, HSCT continues to be a risky procedure. One way to alleviate the risk is to speed up the formation of new T- and B-cells of the immune system that are responsible in identifying and preventing such complications. Currently, certain proteins and cell-infusions are used in the clinic that can increase the absolute number of immune cells by inducing division of existing mature cells that already recognize certain pathogens. However, the generation of new immune cells that can identify unknown pathogens that the HSCT recipient would encounter is necessary to completely restore the immune system.

The immune itself system arises from blood stem cells, known as hematopoietic stem cells (HSCs). HSCs reside in 'niches,' which are sites within specific tissues in the body in which cells reside at steady-state. In humans, the HSC niche is contained in the bone marrow. Using signals presented in the bone marrow, the HSCs undergo cell differentiation and ultimately end up as T- or B-cells. This project seeks to replenish the T- and B-cells using a synthetic bone marrow that can capture transplanted HSCs after an HSCT and induce their differentiation into specific types of immune cells that prevent the common complications associated with HSCT. To reconstitute bone containing

bone marrow in a mouse, a porous scaffold substrate and a specific bone-forming protein are sufficient to produce bone marrow in a pocket below the skin and mimic a healthy HSC niche environment. In this work alginate was used, a sugar-based polymer that can form a gel-like scaffold. The interweaving polymer chains created a large surface area to provide sufficient space for HSCs to engraft. By adding specific â€˜address labelsâ€™ in the form of biological signals it was possible to specify the differentiation of these HSCs into immune T- and B-cells. The synthetic marrow was able to rapidly induce the development of new T- and B-cells after a HSCT in a mouse model that rapidly restored the immune cell counts. It was observed that the diversity of the T-cell receptor, which recognizes pathogens, was significantly higher both in terms of gene rearrangement and the number of cells with a particular configuration in treated animals. Ongoing work seeks to induce the development of regulatory T-cells that are known to play a role in alleviating GVHD. To prevent the relapse of leukemia, the project will also investigate the combination of the synthetic marrow with a clinical vaccination regimen to induce the formation of T- and B-cells against specific targets that are known to the associated with leukemia. More broadly, there are a number of natural and induced events that can permanently or transiently weaken the immune system such as aging, autoimmune and infectious diseases. The technology development in this project could offer an option to restore the immune defenses in patients with a compromised immune system and to prevent life-threatening conditions.

Teaching Interests:

I am interested in Chemical Engineering and Bioengineering undergraduate as well as graduate level teaching in the areas of biomaterials, kinetics, thermodynamics.

469437 Deconstructing the Tumor Microenvironment and Its Contribution to Metastasis

Abstract:

Research Interests:

Despite aggressive surgical, radiological, and chemotherapeutic intervention, breast cancer recurrence remains a problem in up to 20% of patients. Most studies focus solely on tumors and tumor cells but often ignore the contribution of the normal tissues that surround and interact with tumor cells to disease progression. I am interested in combining my chemical and biomedical engineering expertise toward examining the mechanisms driving tumor recurrence and metastasis. I aim to study the relationship between tumor cells, normal tissues, and the immune system as a means of deconstructing the tumor microenvironment to understand how each component individually and as a whole contributes to metastasis and relapse after surgical and radiation treatment. The emphasis of my laboratory will be on developing bio-inspired ***in vitro*** models that recapitulate the tumor microenvironment by learning from ***in vivo*** mouse models. These studies will ultimately enable the design of optimized breast cancer therapeutics. ***I will build a program around analyzing the interplay between the physical, chemical, and biological cues that influence cancer metastasis and recurrence.*** Specifically, I will study (1) the collateral effect of radiation and surgery of normal tissues on tumor and immune cell migration patterns, (2) the molecular profiles of normal tissues wounded from therapy, and (3) the changes in the biomechanical properties of the tumor microenvironment and surrounding normal tissues following therapy. The idea that tissue response to therapeutic interventions may contribute to cancer recurrence is in its infancy. My research has the potential to challenge current paradigms about breast cancer treatment. My undergraduate and graduate studies provide me with the foundation to use biomimetic materials to study disease, and I will build on my strengths to design novel tumor microenvironment models in my independent research laboratory. My expertise has grown to encompass basic cancer biology, molecular imaging, biomechanical characterization of tissues, and ***in vivo*** experimental design through my postdoctoral work, which uniquely positions me to carry out research at the interface of engineering and medicine.

Successful Proposals:

K99/R00 Pathway to Independence Award, NIH/NCI; 2016-Present

Katherine McCormick Advanced Postdoctoral Fellowship; 2015-2016

T32 NIH NRSA Postdoctoral Fellowship in the Radiation Sciences; 2012-2013, 2015-2016

American Heart Association Founders Affiliate Predoctoral Fellowship; 2009-2011

Postdoctoral Project: "The Role of the Immune System in Modulating Breast Tumor Cell Migration following Radiation-Induced Changes in the Tissue Microenvironment"

Under the supervision of Edward Graves and Amato Giaccia, Department of Radiation Oncology, Stanford University

Ph.D. Dissertation: "Dual Antibody Functionalized Polyvinyl Alcohol and Alginate Hydrogels for Synergistic Endothelial Cell Adhesion"

Under the supervision of Debra Auguste, School of Engineering and Applied Sciences, Harvard University

Education:

Ph.D., Harvard University, Engineering Sciences, 2012

S.M., Harvard University, Engineering Sciences, 2007

S.B., Massachusetts Institute of Technology, Chemical Engineering, 2006

Selected Peer-Reviewed Publications:

1. **Rafat M**, Ali R, Graves EE. Imaging Radiation Response in Tumor and Normal Tissue. ***American Journal of Nuclear Medicine and Molecular Imaging***. 2015; 5(4): 317-332 (Review).
2. You JO*, **Rafat M***, Almeda D, Maldonado N, Guo P, Nabzdyk CS, Chun M, LoGerfo FW, Hutchinson JW, Pradhan-Nabzdyk LK, Auguste DT. pH-Responsive Scaffolds Generate a Pro-Healing Response. ***Biomaterials***. 2015; 57: 22-32. *Equal contribution. ***Highlighted in Materials Today. May 21, 2015.***
3. Vilalta M, **Rafat M**, Giaccia AJ, Graves EE. Recruitment of Circulating Breast Cancer Cells Is Stimulated by Radiotherapy. ***Cell Reports***. 2014; 8(2): 402-409.
4. **Rafat M**, Rotenstein L, Hu J, Auguste DT. Engineered Endothelial Cell Adhesion via VCAM1 and E-selectin Antibody-Presenting Alginate Hydrogels. ***Acta Biomaterialia***. 2012; 8(7): 2697-2703.
5. **Rafat M**, Rotenstein L, You JO, Auguste DT. Dual Functionalized PVA Hydrogels that Adhere Endothelial Cells Synergistically. ***Biomaterials***. 2012; 33(15): 3880-3886.

Teaching Interests:

Teaching Experience

I held two teaching fellow positions during graduate school, where I earned Certificates of Distinction in Teaching. I was a Teaching Fellow in 2007 for a graduate-level course in drug delivery that focused on engineering solutions for delivering drugs to specific sites in the body. In 2009, I served as the Head Teaching Fellow for an undergraduate-level introductory course in fluid mechanics and transport processes.

During my postdoctoral fellowship in 2014, I participated in the Stanford Pre-Collegiate Studies Program, where I lectured students in the Cancer Biology section of the Topics in Biosciences and Biotechnology module.

Mentoring Experience

I have extensive experience teaching and mentoring students in a research capacity. Through the National Science Foundation Research Experiences for Undergraduates (REU) program, I have mentored three students since 2007. I also advised three undergraduate students on their theses while at Harvard. My mentees contributed to co-authored publications and multiple conference presentations. At Stanford, I have supervised a summer veterinary student and a clinical research assistant who was interested in receiving wet lab training before attending medical school. I am currently supervising an undergraduate student on her own project. I emphasize critical thinking and challenge students to be active in the design, implementation, and analysis of their research. Mentoring students has been incredibly rewarding; I am eager to continue advising and instructing students as a junior faculty member.

Teaching Philosophy

My previous teaching experiences have given me a realistic sense of how to teach classes and mentor students in the future. As a professor, I will give students the necessary tools for developing analytical thinking skills and to solve problems beyond cookie-cutter solutions. I will structure classes using a combination of hands-on learning through laboratory projects as well as reading the current literature for students to gain an understanding of cutting-edge research. I will also expose students to problems with unknown solutions to challenge them to think critically and devise solutions creatively.

472132 Bacterial Biofilms: From the Built Environment to Human Diseases

Abstract:

Research Interests: Biofilms are communities of sessile microorganisms ubiquitously found in both natural and built environments, as well as numerous hosts including human. This mode of growth is preferred by bacteria for surviving in challenging environments because the unique structure of biofilms (e.g., extracellular polymeric substrates (EPS)) provide embedded bacteria with protection from harsh environmental factors such as antimicrobials. The enhanced resistance of biofilms against biocidal agents compared to free swimming bacteria (up to 1000 fold) is one of the leading causes of chronic infections in human and persistent biofouling in industrial settings. Although biofilm formation has become an important research topic in the past two decades, there is still a gap between fundamental knowledge obtained from experimental studies focusing on mono- or dual-species biofilms formed under controllable conditions and environmental studies based on multi-species biofilms in natural environments. This gap has severely hindered the mechanistic understanding of biofilm formation and the role of biofilms in human diseases, as well as the development of surfaces for biofilm prevention and removal. To control bacterial biofilms for public health, there is an urgent need to bridge the gap between experimental and environmental studies on biofilm formation and develop 'smart' antifouling surfaces.

Motivated by the significance of biofilms, I set my long-term career goal to control biofilm formation through interdisciplinary research and teaching. To prepare for my career, I pursued my Ph.D. work with a focus on biofilm control using chemically modified surfaces to direct bacterial adhesion. This well-defined surface system was used to study biofilm formation and the development of biofilm-related high antibiotic resistance. The obtained results together with the effects of surface topography on bacterial adhesion were used to engineer 'smart' antifouling surfaces via topographic surface modification. Engineering surface topography is a promising approach for biofilm control because it does not involve biocidal agents, and thus avoids the problem of antibiotic resistance. With my long-term goal to develop an academic career in the field of bacterial control, I extended my postdoctoral research to study biofilm formation and biofilm-associated high drug tolerance in built environments. People on average spend 90% of their time indoors. Surfaces in indoor environments such as air condition systems are all susceptible to colonization by microorganisms growing in biofilms, thus, it is critical to understand the physiology and virulence of biofilms formed by bacterial pathogens in the built environment, especially, in the hospital environment. Therefore, my postdoctoral research focuses on understanding the effects of environmental factors on bacterial biofilm formation and antibiotic resistance, so that the built environment can be better engineered for antifouling purposes. A well-defined surface system for studying the role of synergistic interaction between multispecies bacteria in biofilm formation and biofilm antibiotic resistance was also developed. The obtained results can help fill the gap between experimental and environmental studies on biofilm formation and design antifouling surfaces with enhanced and prolonged antifouling effects for environmental and medical applications.

In my faculty position, I plan to use my knowledge and skills of microbiology, engineering, material science, and chemistry to understand biofilm formation in built environments and the role of biofilms in environmental acquired chronic infections, especially those related to nosocomial infections. My doctoral and postdoctoral training has prepared me for building a successful independent research program that will make major contributions to the understanding and control of biofilms. Specifically, I will identify the microbial composition in biofilms formed in built environments and humans, as well as the transportation of microbes from environments to humans through human activities by following the 'footprints' of microbes. Meanwhile, the interaction between microbes from different species in the survival of surface attached communities under harsh environments is also a significant and interesting question to study. I also want to understand how biofilm pathogen and host interactions affect biofilm formation and biofilm antibiotic resistance in humans.

The results of environmental, ***in vitro***, or ***in vivo*** studies can allow my research group to have a more comprehensive understanding of the role of biofilms in the transportation of bacterial pathogens from built environments to humans and inspire novel and ecofriendly strategies for biofilm control to effectively interrupt this process. The results of these studies can inform future epidemiological work on cystic fibrosis and microbial exposures in vulnerable populations such as patients with compromised immunity and bedridden elderly, as well as inform local or domestic health departments on the microbial composition in biofilms formed in built environments. Additionally, given the interdisciplinary nature of biofilm research and strong medical relevance, there are exciting opportunities to build collaboration with other fields, such as biostatistics, microbiology, architecture science/engineering, ecology and evolution, and immunology to pursue large multidisciplinary grants. Results from these studies also have the potential to inform industry stakeholders in the development of innovative antifouling products. Beyond research studies, the biofilm related problems are also suitable for integrating into engineering courses and for outreach with local K-12 students and teachers to promote STEM education and talent development.

Teaching Interests: With training in Chemical Engineering, I am comfortable to teach all core courses in this discipline and I am interested in developing new courses related to nano- or micro- techniques for cell manipulation, microbes and human life, and techniques for exploring cell microenvironments.

475397 Structure-Guided Protein Engineering for Targeted Immunotherapy

Abstract:

Structure-Guided Protein Engineering for Targeted Immunotherapy

Jamie Berta Spangler

Postdoctoral Fellow, Stanford University School of Medicine

Research Interests:

The repertoire of naturally occurring proteins is finite and many molecules induce multiple confounding effects, limiting their efficacy as therapeutics. Recently, there has been a growing interest in manipulating existing proteins or engineering new proteins to address the deficiencies of molecules found in nature. Researchers have traditionally taken an unbiased approach to protein engineering, but as our knowledge of protein structure-function relationships advances, we have the exciting opportunity to apply molecular principles to guide engineering. Leveraging cutting-edge tools and unique expertise from structural biology and molecular engineering, I aim to implement a structure-based engineering approach to elucidate the determinants of protein function in order to inform the development of novel protein therapeutics. I am particularly interested in engineering immunological proteins such as antibodies and cytokines to bias immune activity for clinically relevant applications including cancer, infectious diseases, and autoimmune disorders. Harnessing molecular insights and new engineering platforms I have developed in my postdoctoral research, I am poised to pursue a wealth of translational directions that will expand the scope of protein activity to address biomedical challenges.

Successful Proposals:

Leukemia & Lymphoma Society Career Development Postdoctoral Fellowship, Repligen Koch Institute for Integrative Cancer Research Graduate Fellowship, National Defense Science and Engineering Graduate Fellowship

Postdoctoral Project:

"Engineering the interleukin-2 cytokine system to therapeutically modulate immune homeostasis."

Supervised by Prof. K. Christopher Garcia, Departments of Molecular & Cellular Physiology and Structural Biology, Stanford University School of Medicine

Ph.D. Dissertation:

"Characterization and informed design of down-regulating epidermal growth factor receptor antibodies."

Supervised by Prof. K. Dane Wittrup, Department of Biological Engineering, Massachusetts Institute of Technology

Research Experience:

I have cultivated a longstanding interest in understanding and manipulating the molecular determinants of protein function in health and disease and have consequently pursued training in a combination of fundamental biophysical and engineering approaches to biochemistry. As an undergraduate in Biomedical Engineering at Johns Hopkins University, I conducted research in Kalina Hristova's lab, elucidating the mechanistic basis for pathogenicity of mutations in fibroblast growth factor receptor-3 (FGFR3). This work equipped me with essential biochemical and biophysical tools and piqued my interest in studying protein interactions relevant to disease. As a Ph.D. student in Biological Engineering at Massachusetts Institute of Technology, I worked under the supervision of Prof. K. Dane Wittrup. In my thesis research, I used epidermal growth factor receptor (EGFR) as a model antigen to establish antibody-mediated receptor down-regulation as an effective therapeutic strategy and designed multispecific antibody fusions that utilized this strategy to inhibit tumor growth in mice. Through this work I developed a vast knowledge of cell biology and biochemistry and gained experience with antibody engineering and in vivo models of cancer. As a postdoctoral researcher in Professor K. Christopher Garcia's group at Stanford, I have adopted a structure-based approach to biochemistry, integrating new techniques in structural biology, cytokine engineering, and molecular and cellular immunology. My work has focused on interleukin-2 (IL- 2), a multifunctional cytokine that regulates immune homeostasis and that has been used therapeutically to treat diseases such as cancer and autoimmune disorders. The efficacy of IL-2 treatment has been limited by its concurrent promotion of both effector and regulatory immune cells, and I have implemented a multi-pronged approach toward engineering the IL-2 system to modulate the balance of its immunostimulatory versus immunosuppressive effects to achieve targeted therapy. As an independent investigator, I plan to combine my experience in both basic and applied biomedical sciences to build a translational research program at the interface of structural biology and molecular engineering.

Teaching Interests:

Throughout my academic research career, I have been fortunate to gain experience in teaching as well. As a teaching assistant for an introductory Biomedical Engineering course at Johns Hopkins University, I helped lead laboratory sessions and advised students on independent design projects, providing me with the opportunity to interactively train students in applying engineering and design approaches to solve medically relevant problems. As a Ph.D. student, I served as a teaching assistant for a graduate-level course in biomolecular kinetics and cellular dynamics at Massachusetts Institute of Technology. The course focused heavily on computer programming in MATLAB and I instructed several tutorials on this software and also led both lecture- and discussion-based recitation sessions.

In addition to classroom teaching, a formative experience in my pedagogical training has been the opportunity to mentor students. During my graduate career, I mentored six undergraduate students, many of whom have gone on to pursue biomedical careers: three are in top bioengineering Ph.D. programs and two are in medical school. I have also supervised three students and a technician during my postdoctoral tenure, and these opportunities to inspire the next generation of scientists have reinforced my desire to pursue a faculty career.

Future Directions:

The research program I will pursue in my own laboratory represents a culmination of the skills I have amassed through previous work, synthesizing engineering and design experience with structural and biophysical tools to create selective and efficacious molecular therapeutics for a broad range of immunological applications. The most prominent challenge to therapeutic development of cytokines is their pleiotropic nature, as stimulation of signaling on multiple immune cell subsets leads to dose-limiting toxicity and harmful off-target effects. By harnessing structural insights to isolate and selectively engineer the cellular activation properties of cytokines, I will construct a blueprint for biasing protein activity that can be tailored to a variety of disease objectives. In addition to affording unprecedented control over protein behavior, my research will deepen our understanding of the molecular mechanisms that underlie cellular responses and inform the design of candidate drugs for ready clinical translation. While initial work will focus on engineering antibodies to tune cytokine activity, my long term vision is to apply the modular engineering and therapeutic platforms I develop to other systems to allow for the direct manipulation of a diverse array of ligand-receptor interactions including hormone, growth factor, and membrane protein systems.

I recently devised structure-guided evolutionary strategies using yeast surface display methodology to establish two novel antibody-based platforms that modulate cytokine behavior: [1] Cytokine-targeted antibodies that allosterically disrupt cytokine-receptor interactions to bias activity; and [2] Agonistic antibody fragments that use a single binding site to bridge two receptor subunits and activate signaling in the absence of cytokine. Both platforms can be exploited to direct cytokine activity toward specific cell subsets based on surface receptor expression profiles, and their modularity allows for targeting of any protein interaction or cell population of interest to elicit either an immunostimulatory response (for treatment of cancer or chronic infection) or an immunosuppressive response (for engraftment or treatment of autoimmune disorders). In addition to building on platforms I have already established, I plan to develop new engineering platforms to isolate antibodies that act through alternative mechanisms such as modulation of internalization or molecular trafficking properties. The unique mechanisms recruited in these platforms will complement other activities inherent to the antibody scaffold such as competitive inhibition and Fc-mediated effector functions. Moreover, the stable and readily engineerable antibody construction will allow for development of bispecific molecules that synergistically combine antibodies with distinct modes of action, and structural feedback on molecules we develop will further inform design. My experience with mouse models of cancer and autoimmune disease and continued collaborations with leading T cell immunologists established during my postdoctoral tenure will facilitate pre-clinical evaluation and ultimate translation of engineered therapeutics. Through integration of expertise in molecular engineering and structural biology, I am uniquely positioned to make rapid progress in developing new approaches toward targeted therapy for immunomodulation and a host of other biomedical applications.

Selected Publications:

Spangler JB, Tomala J, Luca VC, Jude KM, Dong S, Ring AM, Votavova P, Pepper M, Kovar M, Garcia KC. Antibodies to interleukin-2 elicit selective T cell subset potentiation through distinct conformational mechanisms. ***Immunity.*** 2015;42(5):815-25. PMCID: 4439582.

Mitra S, Ring AM, Amarnath S, **Spangler JB**, Li P, Ju W, Fischer S, Oh J, Spolski R, Weiskopf K, Kohrt H, Foley JE, Rajagopalan S, Long EO, Fowler DH, Waldmann TA, Garcia KC, Leonard WJ. Interleukin-2 activity can be fine tuned with engineered receptor signaling clamps. ***Immunity.*** 2015;42(5):826-38. PMCID: 4560365.

Spangler JB, Moraga I, Mendoza JL, Garcia KC. Insights into cytokine-receptor interactions from cytokine engineering. ***Annu Rev Immunol.*** 2015;33:139-67. PMCID: 4445396.

Moraga I, **Spangler J**, Mendoza JL, Garcia KC. Multifarious determinants of cytokine receptor signaling specificity. ***Adv Immunol.*** 2014;121:1-39. PMCID: 4449261.

Spangler JB, Manzari MT, Rosalia EK, Chen TF, Wittrup KD. Triepitopic antibody fusions inhibit cetuximab-resistant BRAF and KRAS mutant tumors via EGFR signal repression. ***J Mol Biol.*** 2012;422(4):532-44. PMCID: 4041985.

Spangler JB, Neil JR, Abramovitch S, Yarden Y, White FM, Lauffenburger DA, Wittrup KD. Combination antibody treatment down-regulates epidermal growth factor receptor by inhibiting endosomal recycling. ***Proc Natl Acad Sci U S A.*** 2010;107(30):13252-7. PMCID: 2922117.

You M*, **Spangler J***, Li E, Han X, Ghosh P, Hristova K. Effect of pathogenic cysteine mutations on FGFR3 transmembrane domain dimerization in detergents and lipid bilayers. ***Biochemistry.*** 2007;46(39):11039-46.

475630 Enhancer-Mediated Regulation of Transcriptional Bursting

Abstract:

Research Interests:

My research interests lie in quantitative analysis of regulation in development, using both experimental and computational approaches. I completed my Ph.D. studies with Professor Stanislav Shvartsman in the department of Chemical and Biological engineering at Princeton University. Using a combination of genetic tools, high-throughput imaging, and mathematical modeling, I reconstructed developmental dynamics of the highly conserved ERK enzymatic activity from snapshots of fixed early *Drosophila* embryos. Reconstruction showed that the dynamic activation profile of ERK regulates the spatial and temporal extent of the target gene expression. Given that increasing number of studies report the link between changes in ERK kinetics and developmental abnormalities, my project contributed to establish a platform where quantitative regulation of enzyme activity can be studied in vivo.

As a postdoctoral researcher with Professor Michael Levine in the Lewis-Sigler Institute for Integrative Genomics at Princeton University, I applied quantitative image analysis and modeling techniques to explore another critical aspect in development: transcriptional regulation. Transcription is a critical yet a complex process that determines when and where a gene needs to be turned on and off to ensure normal development. While the importance of transcriptional regulation is well acknowledged, complex nature of the process makes it challenging to understand the comprehensive underlying mechanisms of transcriptional regulation.

During the past one year as a postdoctoral researcher, I showed that transcription occurs in a series of sequential stochastic bursts, and these bursts can be readily visualized in living fly embryos. Through quantitative image analysis,

I provided evidence that enhancers regulate transcription by modulating the frequency of these bursts, implying that the regulation of bursting frequency is a key parameter of gene control in development. I also showed that a shared enhancer can co-activate two linked reporter genes, arguing against the conventional enhancer-promoter looping models, which would express sequential activation of two reporter genes. The work has contributed to provide new insights on the enhancer-mediated gene regulation in quantitative manner.

Based on these results and my experience in applying quantitative techniques to biological systems, I will further explore the underlying mechanism of transcriptional bursting and the role of enhancer-promoter communication in transcriptional regulation. I will take advantage of the broad range of molecular cloning, optogenetics, and quantitative tools available in the *Drosophila* system. My initial projects will focus on (1) extracting the properties that comprise minimal unit of transcriptional burst, (2) probing the correlation between the cluster size of RNA polymerase II and the transcriptional burst size, and (3) testing the role of chromosomal loop domains in transcriptional regulation. Given that more and more mutations in the enhancer regions in the genome are found to cause developmental abnormalities, successful implementation of the proposed experiments will shed light on the nature of transcriptional bursts, the relationship between enhancers and bursts, and the effect of enhancer-mediated transcriptional regulation in development.

Teaching Interests:

I have served as a TA once for the undergraduate Chemical Reaction Engineering course and twice for the graduate level Chemical Reactor Engineering course, both of which are core classes for undergraduate and graduate students in Chemical and Biological engineering at Princeton University, respectively. I received the Air Products Assistant in Instruction Award in 2012, from the graduate Chemical Reactor Engineering class. I have mentored three undergraduate students (2 engineering and 1 molecular biology), and two rotation graduate students in molecular biology.

Selected Publications:

1. Fukaya T*, **Lim B***, and Levine M. 2016. Enhancer control of transcriptional bursting. *Cell.* 2016. dx.doi.org/10.1016/j.cell.2016.05.025 [epub ahead of print]

2. **Lim B**, Dsilva CJ, Levario TJ, Lu H, Schupbach T, Kevrekidis IG, Shvartsman SY. Dynamics of inductive ERK signaling in the *Drosophila* embryo. *Curr Biol.* 2015; 25: 1784-1790.

3. Dsilva CJ*, **Lim B***, Lu H, Singer A, Kevrekidis IG, Shvartsman SY. Temporal ordering and registration of images in studies of developmental dynamics. *Development.* 2015; 142: 1717-1724.

4. **Lim B**, Samper N, Lu H, Rushlow C, Jimenez G, Shvartsman SY. Kinetics of gene derepression by ERK signaling. *Proc Natl Acad Sci U S A.* 2013; 110: 10330-10335.

* Equal contribution

475682 Understanding and Controlling Protein Stability from Coarse-Grained Protein Models

Abstract:

Understanding and Controlling Protein Stability from Coarse-Grained Protein Models

Marco A. Blanco

National Institute of Standards and Technology, Gaithersburg, MD 20899

Institute for Bioscience and Biotechnology Research, University of Maryland, Rockville, MD 20850

Research Interests:

Protein-based therapeutics represents one of the fastest growing markets in pharmaceutical industry. The high efficacy and specificity of biotherapeutics make them valuable for targeting different type of illnesses. Nonetheless, most, if not all, proteins present a very wide range of behaviors that limit the development of commercially viable products. Small changes in the protein sequence and formulation conditions may cause conformational and physical instabilities that lead to the self-assembly of proteins (e.g., fluid phase separation, oligomerization, and irreversible aggregation). These stability concerns constitute the primary route for the formation of impurities in biotherapeutics, as well as may induce unfavorable immunologic responses in patients. The ability to accurately and unambiguously predict the behavior of proteins in solution will help control instability problems via an adequate design of proteins and selection of formulation conditions.

My research work has focused on understanding the stability of proteins in solution via simple but realistic coarse-grained computational models, which are combined with biomolecular and biophysical experiments (e.g., mutagenesis, chromatography, calorimetry, scattering) to provide both refinement and validation. By implementing advanced Monte Carlo and Molecular Dynamics techniques, these models allow one to evaluate the thermodynamic pathways by which proteins are destabilized, which might be otherwise challenging to probe due to the generally short-living nature of most of the key species in those processes. Thus, the resulting models provide powerful, physics-based tools to assess how different factors (e.g., pH, excipients, protein structure) affect those phenomena that cause instability in proteins such as phase separation, protein unfolding, and nonnative aggregation; moreover, they offer guidelines for rationally designing/selecting proteins with higher physical stability.

The Blanco laboratory will built on the use of both computational and experimental methods applied to topics within the general theme of protein stability, engineering, and self-assembly. A specific emphasis will be placed on developing new and improved theoretical models and experimental techniques to predict the conformational and solution stability of proteins and peptides, and thus to gain insights regarding how the local structure and surrounding molecular environment influence the underlying mechanisms associated to physical degradation of proteins. Such predicting methods will have direct application in fields such as biomedicine (e.g., elucidating critical stages during amyloidosis), drug discovery (e.g., rational design of high-affinity peptide-based drugs), drug development (e.g., identifying optimal formulation and manufacturing conditions for biopharmaceutics), and biomaterials (e.g., developing "smart" protein-based materials with controllable self-assembly properties). Since research in these fields encompasses various disciplines, there will be strong collaborations from both academy and pharmaceutical industry.

Teaching Interests:

In addition to my research career, I have also gained significant teaching experience. I lectured an undergraduate course at the Industrial University of Santander in Colombia about the implementation of statistical and quantum mechanical computational methods to thermodynamics and transport phenomena problems related to Chemical Engineering. I also TAed thermodynamics courses at both undergraduate and graduate level, as well as mentored undergraduate students and visiting scholars about application of molecular simulations to protein systems.

Given my background in protein science and chemical engineering, my teaching interests extend in undergraduate core chemical engineering courses that include thermodynamics, heat and mass transfer, chemical kinetics, and applied math. Furthermore, I would also like to develop and teach undergraduate- and graduate-level courses in topics such as protein engineering, statistical thermodynamics applied to chemical engineering, and computational and experimental methods for protein characterization.

Research Projects:

<u>PhD Projects:</u>

- "Measuring Protein-Protein interaction via Scattering Experiments from Low to High Protein Concentrations"

- "Evaluating the Effect of Solution Conditions and Protein Sequence on the Cluster Formation of γD-Crystallin"

- "A Novel Algorithm for Predicting Amyloidosis and Protein β-Aggregation"

- "Solution Behavior and Aggregation Propensity of Peptide-Polyacrylic Acid Conjugates"

Supervised by Christopher J. Roberts, Department of Chemical and Biomolecular Engineering, University of Delaware.

<u>Postdoctoral Projects:</u>

- "Effect of Specific Interactions on the Fluid Phase Behavior and Self-Assembly of Proteins and Nanocolloids"

- "Characterization of the Solution Structure and Phase Behavior of Monoclonal Antibodies via Coarse-Grained Models"

- "Effect of Electrostatic Interactions on the Oligomerization and Amyloid Formation of the Cardiomyopathy-Related Protein Transthyretin"

Supervised by Vincent K. Shen and Zvi Kelman, National Institute of Standards and Technology and Institute for Bioscience and Biotechnology Research, University of Maryland.

Selected Publications:

1. Marco A. Blanco, Erinc Sahin, Anne S. Robinson, and Christopher J. Roberts. "Coarse-grained model for colloidal protein interactions, B_{22}, and protein cluster formation". J. Phys. Chem. B, 117, 16013 (2013).

2. Marco A. Blanco, Tatiana Perevozchikova, Vincenzo Martorana, Mauro Manno, and Christopher J. Roberts. "Protein-protein interactions in dilute to concentrated solutions: a-Chymotrypsinogen in acidic conditions". J. Phys. Chem. B, 118, 5817 (2014).

3. Bradford Paik, Marco A. Blanco, Xinqiao Jia, Christopher J. Roberts, and Kristi Kiick. "Aggregation of poly(acrylic acid)-containing elastin-mimetic copolymers". Soft Matter, 11, 1839 (2015).

4. Marco A. Blanco and Vincent K. Shen. "Effect of the surface charge distribution on the fluid phase behavior of charged colloids and proteins". J. Chem. Phys. [submitted].

475765 Microbial Biosynthesis of Bioorthogonal Functionalities and Applications

Abstract:

Research Interests:

Natural products biosynthesized by microbes offer unique functionalities that can be used as chemical handles in bioorthogonal chemistry. Bioorthogonal chemistry have recently emerged as one of the most powerful tools in drug discovery and chemical biology. Despite the importance of the unique chemical functionalities involved in biorthogonal chemistry, the biosynthetic routes to bioorthogonal functionalities are not well understood. In my graduate career, we identified and characterized two biosynthetic pathways in microbes that lead to the generation of a bioorthogonal functionality-terminal alkyne through unprecedented pathways. These pathways were further exploited for tagging various natural products with terminal alkyne functionalities through biochemical pathway engineering. The natural product tagging strategy also enabled the development of an ***in situ*** natural products quantification platform based on bioorthogonal chemistry with the newly installed alkyne tag. Using this quantification platform as a screening method, we applied directed evolution strategy to improve the activity of a membrane-bound bifunctional desaturase/acetylenase. This study not only provides alkyne biosynthetic tools that are useful for chemical biology, but it also has applications in which enzymatic generation of terminal alkynes and ***in situ***bioorthorgonal chemical transformation are required or preferred.

I would like to continue studying microbial biosynthesis of structurally and functionally interesting chemical functionalities and find applications in chemical biology and drug development.

Teaching Interests:

Aside from my research, I also have teaching experience as a teaching assistant during my graduate life. I led discussion sections, prepared teaching materials and held office hours for a core course "Kinetics" for undergraduate majored in chemical engineering. I also spent two semesters organizing and preparing teaching materials for a seminar-based course "Berkeley lectures on energy: energy from biomass". I also mentored two graduate students, and two undergraduate students in my group.

Based on my own academic background, I would love to teach general chemical engineering (e.g. Kinetics, Thermodynamics) for undergraduate students. I would also like to develop a course "Biomolecular Engineering", which is a combination of lectures and lab sessions. The lectures would focus on the use of chemical engineering principles in the design of biologically-based processes (e.g. microbial fermentation, bioreactors, etc), and also cover topics in biotechnology such as enzyme kinetics, recombinant DNA technology, protein expression, protein engineering, synthetic biology, etc. Students would also have chances to practice basic biotechnical skills in lab sessions, such as molecular cloning, protein purification, enzyme kinetic assays, bioreactor-based fermentation, etc.

473004 Modulating Antigen-Specific T Cell Immunity with Biomaterials-Based Vaccine

Abstract:

Research Interests: Over the past several decades, biomaterials have been widely engineered for treating or diagnosing diseases such as cancers. My research experience and interests have been focused on using micro/nanofabrication of biomaterials for cancer therapies and diagnosis. In this poster session, I will highlight the major research directions I have been involved in during my graduate and postdoctoral research. The results of these efforts demonstrate my capabilities in using biomaterials for micro/nanofabrication, drug delivery, bio-sensing, and more recently, for vaccine design. Thus far my efforts have led to 21 papers (6 as first author) Building on this training, I will describe my plans to develop a distinct, independent research program at the interface of biomaterial, surface and immunology. Briefly, I plan to build a research lab studying immune detection and biomaterial vaccine for tissue inflammation and fibrosis.

I completed my PhD in Biomedical Engineering at Florida State University in 2013 under the supervision of Dr. Jingjiao Guan. My research was focused on developing non-conventional micro particles with well-defined shapes, diameters and components.[1, 2] These particles have been employed in biomedical applications such as drug delivery, cell tracking, and radiation therapy for cancer.[3-7] Upon completing my PhD, I worked as a postdoctoral researcher at Worcester Polytechnic Institute under the supervision of Dr. Ming Su, studying enhanced radiation therapy of cancer. Briefly, my research focused on modifying gold nanoparticle surface to enhance their uptake by cancer cells. Upon X radiation, X-ray photons in radiation therapy can generate photoelectrons and Auger electrons, which can cause ionization of water and formation of reactive free radicals. These free radicals diffuse through chain reactions in cells, and damage DNA in mitochondria and nuclei by extracting hydrogen atoms from ribose sugars, kill cancer cells by cleaving of the polynucleotide backbone in cancer cells. During this period of time, I gained valuable training in surface modification of biomaterials for radiation therapy cancer.[3, 5, 8, 9] In August 2014, I joined the University of Maryland College Park under the supervision of Dr. Christopher Jewell. My research in the Jewell lab has focused on exploiting biomaterials to generate specific, tunable immune responses.[10, 11] My research in this area has two directions: 1) investigating and understanding how biomaterials impact the structure of major secondary immune organs such as lymph nodes and spleens, and 2) developing biomaterial vaccines for cancer treatment.

Through training in three different labs, I have gained valuable training and experience in grant writing, scientific management and teaching. These experiences will allows me to build and run my future own labs.

References

[1] Wang ZB, **Zhang P**, Kirkland B, Liu YR, Guan JJ. Microcontact printing of polyelectrolytes on PEG using an unmodified PDMS stamp for micropatterning nanoparticles, DNA, proteins and cells. Soft Matter 2012;8:7630-7.

[2] **Zhang P**, Guan J. Fabrication of multilayered microparticles by integrating layer-by-layer assembly and microcontact printing. Small 2011;7:2998-3004.

[3] Wang CM, Sun A, Qiao Y, **Zhang P**, Ma LY, Su M. Cationic surface modification of gold nanoparticles for enhanced cellular uptake and X-ray radiation therapy. J Mater Chem B 2015;3:7372-6.

[4] **Zhang P**, Xia JF, Wang ZB, Guan JJ. Gold nanoparticle-packed microdisks for multiplex Raman labelling of cells. Nanoscale 2014;6:8762-8.

[5] **Zhang P**, Qiao Y, Xia J, Guan J, Ma L, Su M. Enhanced radiation therapy with multilayer microdisks containing radiosensitizing gold nanoparticles. ACS Appl Mater Interfaces 2015;7:4518-24.

[6] Kirkland B, Wang Z, **Zhang P**, Takebayashi S, Lenhert S, Gilbert DM, et al. Low-cost fabrication of centimetre-scale periodic arrays of single plasmid DNA molecules. Lab Chip 2013;13:3367-72.

[7] Liu Y, Kirkland B, Shirley J, Wang Z, **Zhang P**, Stembridge J, et al. Development of a single-cell array for large-scale DNA fluorescence in situ hybridization. Lab Chip 2013;13:1316-24.

[8] **Zhang P**, Qiao Y, Wang CM, Ma LY, Su M. Enhanced radiation therapy with internalized polyelectrolyte modified nanoparticles. Nanoscale 2014;6:10095-9.

[9] Qiao Y, **Zhang P**, Wang CM, Ma LY, Su M. Reducing X-Ray Induced Oxidative Damages in Fibroblasts with Graphene Oxide. Nanomaterials-Basel 2014;4:522-34.

[10] **Zhang P**, Chiu YC, Tostanoski LH, Jewell CM. Polyelectrolyte Multilayers Assembled Entirely from Immune Signals on Gold Nanoparticle Templates Promote Antigen-Specific T Cell Response. Acs Nano 2015;9:6465-77.

[11] Chakrabarti KR, Andorko JI, Whipple RA, **Zhang P**, Sooklal EL, Martin SS, et al. Lipid tethering of breast tumor cells enables real-time imaging of free-floating cell dynamics and drug response. Oncotarget 2016;7:10486-97. (*cover article*)

***Featured on cover of *Genetic Engineering and Biotechnology* News

Teaching Interests:

STATEMENT OF TEACHING PHILOSOPHY

With all of the roles that professors play, I consider teaching and mentoring be two of the most important ones. Through these two roles, professors deliver multiple levels of information to students, ranging from professional skills to problem-solving techniques, such as math and logic that can be employed from class room to daily life. With my past experiences as both a student and mentor, I have established skills and a philosophy (see below) that would allow me to train students at different levels (e.g. undergraduate and graduate). Based on these experiences and skills, I will expand my teaching and training activities to my research group, the department and the surrounding community. My teaching plan includes using a variety of teaching tools to deliver scientific knowledge to the students, which will help train another generation of outstanding scientists and engineers.

PAST EXPERIENCE

Teach students in accordance with their aptitude. During my Ph.D. study, I served as a teaching assistant (TA) from 2008 to 2011 in courses including *Transport Phenomena I (ECH 3204)*, *Mass and Energy Balances I (ECH 3203)*, *Separation Process (ECH3418)*, *Quantitative Anatomy and Systems Physiology (BME4403C)* and *Lab TA in Quantitative Anatomy and Systems Physiology (BME4403C)*. Besides routine and impersonal teaching jobs such as grading, other works including office hours and teaching in the lab allowed me to interact with students directly. During these interactions, my responsibility involved helping students find out what was wrong with their homework or experiment and guiding them to understand how and why their thought processes led to the incorrect answer. These works required me to understand how the students used the knowledge to finish their work. What I found interesting was that, while different students made similar mistakes in their assignments, the reasons and logic behind these mistakes were often completely different. For example, some students might do their homework or experiment carelessly and get the wrong answer, while others could make a similar mistake due to insufficient understanding of the knowledge. Through these interactions with students, I established my teaching philosophy to "teach students in accordance with their aptitude". This philosophy was further enforced in my other teaching and mentoring experiences.

Establish students with a vision and tangible future plan. I worked as a lecturer for students in a middle school in a rural area of China during summer 2006 and 2007. The goal of this work was to offer children from low income families a chance to improve their studies in Math, Chemistry, and English. What surprised me during these teachings activities was that some kids did not study hard simply because they lacked a vision or a visible future plan that was connected to their studying. For example, one student mentioned to me that he did not need to learn the knowledge since he was born to be labor worker, and that this idea was rooted in his mind by his parents. While this student's studying attitude changed when he learned how the knowledge could change his future, I questioned how long this change could last. The lesson I learned from this experience was that teaching efficiency can be improved if knowledge is delivered to students alongside a concrete vision so that they can link what they are learning with a tangible future plan.

Visible benefits are important, but so is inherent happiness during learning. I mentored students at different levels since 2008. These mentees included undergraduates from Florida State University, the University of Maryland College Park, and other universities such as the California Institute of Technology. The mentoring process included training students on micro/nanofabrication for different bio-applications (e.g. drug delivery, bio-sensing). In Maryland, I mentored a high school student from Wheaton High School in a program called "Enhance Participation in Research (EPR, supported by National Science Foundation)". In this one year program, the mentee selected a biomedical topic that she was interested in; in turn, the mentor provided feedback throughout the process, from collecting literature to finalizing the topic into a poster and report at the end of the year. In addition to mentoring students in experimental design and data analysis, I also directed an undergraduate in the University of Maryland College Park in writing proposals based on her research, which was funded with an ASPIRE fellowship from the university ($3000). In these mentoring processes, I found that the motivations for students in research mainly came from three sources: future career development, happiness in research, or both. For students driven by future career development (e.g. attending medical school), a project that can develop into a scientific paper is a very effective strategy; as for students driven by the inherent happiness of performing research, a topic that intrigues him/her is very important. For example, my Wheaton High School student was very interested in why her dog was slow and reluctant to move when it got old. While there was not enough literature reported, she learned knowledge by comparing the disease in her dog with Alzheimer's disease in humans. Through this learning process, the student developed a motivation to become a biomedical engineer in college. While both mentees learned a great deal of intellectual information on their respective subjects, they informed me that they gained a joy though the research.

TEACHING PHILOSOPHY

Teachings, in my point of view, can deliver two fundamental techniques to the students: 1) professional skills in one specific area, and 2) robust problem-solving techniques. For professional skills in chemical and biomedical engineering, the students can be trained with highly refined techniques such as generating micro-devices through photolithography. By obtaining these techniques, the students can start their careers in international companies such as Intel, Eli Lilly and Company, Novartis, and Merck & Co. In addition, while expertise in one specific area can help students to build their professional life, the problem-solving techniques can arm them with common sense skills that can be applied to almost all areas of life. Some of these problem-solving techniques include logic, math, and the fundamental laws of physics and biology. Based on my teaching and mentoring experiences, I would like to help students obtain these techniques through 3 specific teaching tools, as mentioned above: **1) teach students in accordance with their aptitude; 2) establish students with a vision and a tangible future plan; and 3) help students find happiness in learning.**

TEACHING AND MENTORING PLAN

My plan for training young scientists and engineers in my classes and research group will focus on three aspects: 1) Implement problem solving contents; 2) teach and help students according to their career goals and aptitudes; and 3) improve my own communication and teaching skills during interactions with students.

Teach in current and future courses. My (previous) training is in chemical and biomedical engineering. I was mostly interested in teaching transport phenomena, mass and balance transport, and physiology. However, I am enthusiastic about any courses that will allow me to interact with students. My teaching will focus on two aspects: 1) the fundamental sciences inherent to the courses, and 2) problem-solving techniques associated with these courses. In addition to conventional teaching in class, the student will be asked to design questions based on the courses for peers, structure a project toward specific problems associated with the course, and compare the potential solutions to the given problems. I will also try to develop new courses relevant to my research. For example, one course will be "Micro/Nano-Vaccines". This course will cover the necessary skills and techniques in nanotechnology and immunology and how these two areas interact for novel vaccine development. In this course, students will be introduced to conventional (i.e. emulsion) and non-conventional (i.e. top-down techniques, such as photolithography) methods for generating micro/nanoparticles; students will also learn about immunology and how the micro/nanoparticles modulate the immune system *in vivo*. At the end of this course, the student will give a presentation on one of the existing vaccines and propose how they will improve that vaccine with micro/nanotechnology.

Advise mentees' research and professional development. Advising my own research group is one of the most exciting responsibilities for me. I am eager to work with students who interact and learn in different ways. My advising activities will include at least two parts: 1) a regular portion to discuss each student's research project, and 2) a portion that discuss flexible topics that depends on the student's stage of development. The regular meeting will start with a discussion of research difficulties. Students will not be provided a direct answer; instead, their problems will be answered with other questions. For example, if the student cannot synthesize a polymer particles via emulsion, he/she will be challenged by a series of questions such as: What are the components used in the synthesis? What are the functions of each component? How do these components interact? What is the formulation in other literature? By asking these kinds of questions, the students will be expected to think on their own. The flexible part of the meeting will depend on the student's need. For example, the senior students may want discussions on how to apply for national meetings, fellowships, or even learn to write grants. The student's professional plan is also an important part of my advising, in which he/she will create a mutually agreed upon plan including action steps that are to be updated every 6 months.

Improve my own professional skills and technical expertise. Teaching and interacting with students will offer valuable opportunities to improve my own skills as a teacher and expand my technical expertise. For example, the "Micro/Nano-Vaccine" course will expose me to immunology that I have only partially studied in the past, and I will thus use this teaching experience to enrich my knowledge in this area. The experience of guiding students in lab will also allow me to learn how to mentor them effectively. Therefore, through these process, I will become even more effective and responsive to individuals with different backgrounds, research styles, and goals.

475794 Increasing the Scale and Rate of Metabolic Engineering through Systems Synthetic Biology

Abstract:

Research Interests:

Biological systems are fantastic platforms for solving myriad world issues, as they contain a universe of highly specific catalysts, self-replicate, undergo evolution to improve function, and integrate information about their surroundings to adjust their behavior. As we strive for ever-increasing yields of green fuels and chemicals, and as we attempt to move our designs from the laboratory to the real world (e.g. the microbiome), critical gaps are arising between the theoretical behavior of engineered organisms and the ways in which they behave in reality. To fill these gaps, my graduate and postdoctoral work have benchmarked an approach to metabolic engineering which fuses emerging systems-level understanding of processes occurring at the whole-cell/community level with high-throughput, iterative design. In this poster, I will highlight several of my recent efforts which demonstrate the promise of a systems synthetic biology approach to more quickly merge engineering goals with biological reality.

My graduate research focused on expediting the development cycle of engineered yeast, which is a workhorse organism for the production of diverse chemicals, yet can be time-consuming and unpredictable to engineer. In particular, tuning the rate of enzyme synthesis, improving enzyme catalytic properties, and streamlining the genetic background of the yeast host are each critical to optimizing a bioprocess, yet poor quantitative models and slow genetic tools often made these design cycles “shots in the dark”. To improve enzyme synthesis, I quantified the rate-limiting potential of RNA secondary structure and nucleosome occupancy and in response developed computational techniques which exploit these phenomena to enable metabolic engineers to achieve desired enzyme levels in a single design cycle. In addition, because the design rules governing biocatalyst properties and host performance remain undiscovered, I further developed ***in vivo*** continuous evolution and RNA interference, respectively, for high-throughput optimization of these qualities in yeast, enabling such design rules to be inferred as the optimization proceeds. In each case, my innovative approaches expedited the design cycle by several orders of magnitude over the current state-of-the-art and provide a toolset for my future lab to use when exploring alternate behaviors, use locations, and genetic backgrounds of engineered microorganisms.

As one of my goals while an independent researcher, I aim to broaden the scope of metabolic engineering to include the microbial communities that live in the human gut, as this complex ecosystem plays critical roles in the extraction of nutrients, production of vitamins, and the progression of disease, yet whose engineering is currently regarded as a "grand challenge" in human health. The critical and growing importance of the gut microbiome was recently highlighted by the announcement of the National Microbiome Initiative this year. As this organ is, in its simplest sense, a packed bed reactor, chemical engineers are ideally poised to make significant contributions in this area. In my postdoctoral work, I am laying the foundation for engineering this system through the discovery of factors enabling specified residence time and gene expression levels in engineered probiotics using high-throughput experimental and computational approaches. Looking ahead, the techniques and design principles I have developed will enable my future lab to engineer "smart" probiotics which monitor the gut environment for signatures of disease and subsequently synthesize and deliver therapeutics in close proximity to affected areas. Additionally, integration with approaches developed during my graduate work will enable the construction of highly robust therapeutic strains which adjust their genetic content for improved function in response to changing environmental conditions. Third, these engineered strains will enable the elucidation and control of the three-dimensional structure of gut microbiota, enabling further insight and control over disease progression through modeling the currently unknown spatial reaction networks in the gut. It is clear through these examples that developing solutions for this and related microbial communities, as well as advancing microbial production of fuels and chemicals to the commercial scale, requires the use of high-throughput methods to both develop improved strains and also improve quantitative understanding of increasingly complex engineerable biological systems. My lab will be uniquely positioned at the forefront of the development and implementation of this systems synthetic biology approach to address pressing challenges in health, energy, and the environment.

Teaching Interests:

When I was a freshman in high school, a substitute teacher for geometry once asked the class if a student could introduce the material. To my astonishment, the class unanimously agreed that I should do it. It was an unforgettable experience to explain the surface areas and volumes of pyramids to my classmates at the same time I myself was learning the material – and I was pleasantly surprised to find that I enjoyed it. It was satisfying to see the eyes of my fellow students light up as difficult concepts came within their reach, and were then able to use that concept to solve problems. I reasoned that if it was this much fun to teach material "cold", it would be even more rewarding to teach things I know a lot about - and I was right! During graduate school, I gained an enjoyment for the puzzle of designing effective homework and quiz problems. I especially enjoyed leading office hours where I was able to interact face-to-face with my students and match their individual learning style. I am eager to broaden my impact as a professor by improving the engagement of diverse groups of students with chemical engineering and thereby provide a foundation upon which a variety of careers can be built.

Introductory Courses

My prior experience and interests make me ideally suited to teach mass and energy balances, kinetics, and transport. I utilize several strategies for improving student learning over the "information fire hose" format of the traditional lecture. First, problem-based learning engages students and makes often abstract concepts easier to grasp, as I found when I encouraged students to play with silly putty in preparation for a lab involving non-newtonian fluids. Second, peer-teaching formats such as "Think-Pair-Share" have been shown to be effective at enhancing conceptual understanding over traditional formats. I implement these techniques during lecture or recitation by allowing students solve problems in small groups, and then reconvening to discuss the challenges they encountered. Third, through exit surveys in which students briefly describe what they learned after each lecture, I simultaneously facilitate student reflection (critical to improving future recall) and gauge my teaching effectiveness. Fourth, learning how to study is something that students often get no formal training in, and is especially critical during introductory courses in order successfully transition students from high school to college. I teach metacognition (i.e. learning how to learn) through in-class activities, for example, where students write down the point they are most confused about and then form small groups to address each other's confusion. Finally, just as feedback is critical to the consistent performance of a chemical process, mid-semester course evaluations enable me to quickly adapt to the changing needs of my students by refining my teaching style, class format, and content. Through these efforts I provide a robust framework for my students to tackle any problems they face in their chosen field.

Advanced Courses

My expertise makes me well-suited for the development of a course on biochemical and metabolic engineering, including advanced enzyme kinetics, flux balance analysis, and the kinetics of cell growth. I can also develop a bioinformatics course for metabolic engineers which could be used either as a standalone class or as a module (e.g. in a computational methods or the metabolic engineering course mentioned above). This course would cover the principles and techniques of high-throughput sequencing analysis, RNA and protein structure prediction, and flux balance analysis. In these courses, I will use current literature to inspire discussion of underlying theories and techniques. Then, proposal writing will allow students to creatively combine what they have learned with the state-of-the-art in the field. Evaluation of these proposals will be facilitated through the formation of in-class review sections, allowing students to practice both project conception and critique. Taken together, I envision that these courses will provide guidance in areas which too often are left to be picked up on-the-fly after graduation.

Mentorship

My primary goals as an advisor are to develop in my mentees 1) a critical understanding of their field, 2) a framework to advance their field, and 3) the ability to effectively communicate their findings and goals with others. To do so, I will organize regular journal clubs to develop critical reading skills, teach new concepts, and serve as launch points for brainstorming sessions. I will provide my mentees with a model for effective lab management, including the keeping of electronic records (via a lab wiki) and centralized inventories. A wiki enables the creation of a core knowledge base, promoting efficient lab work and allowing new members to acclimate quickly. I will also strongly encourage mentorship of undergraduate researchers to increase focus on experimental design and develop teaching skills. In addition to authoring papers and presenting at conferences, teaching assistantships train clear communication and in this setting I will encourage my mentees to develop evaluation material and lead several lectures with my guidance. To combine experimental design and communication, my mentees will also be involved in grantwriting, which has the secondary benefit of providing familiarity with the various funding agencies and facilitating efficient acquisition of funding in the future. Throughout this process, roadblocks will naturally occur both in and out of the lab, and I view them as opportunities for discovering something new, scientifically and personally. I will re-evaluate shared goals with each mentee through yearly self-evaluations, and use individual development plans to achieve them. Through these efforts, I aim to guide mentees towards a career which is well-suited to their interests and expertise.

Commitment to Diversity and Inclusion

My research is framed by the principle that the most effective and novel solutions to problems arise by assembling a diverse collection of microbes and assessing each in an unbiased fashion. I embrace this principle in the classroom and lab by valuing diversity in those I teach, facilitating collaborations between individuals of different backgrounds, cultivating a growth mindset in my mentees, and rewarding alternative points of view during group discussions. I further aim to increase participation of underrepresented groups in my research through the creation of a day-long synthetic biology workshop connecting high school teachers in underprivileged areas to the concepts and techniques of engineering living systems, and providing digital materials for teaching these concepts their students. This workshop will be complemented by a program in which promising high school students which would not otherwise have the opportunity to do research are connected with a member of my lab to undertake a project over the summer. I aim to integrate these programs with existing outreach efforts to repair the leaky STEM pipeline and increase the diversity of perspectives available for solving the challenges of the future.

475827 Designing Novel Interfaces to Control Beneficial and Pathogenic Microbes

Abstract:

Research Interests: Bacteria that colonize on medically and environmentally relevant surfaces are subject to a variety of physical forces. For example, in potentially pathological cases, bacterial cells withstand shear stresses along blood vessels and electric discharges near medical implants as they develop infectious, mature multicellular aggregates known as biofilms. And in an environmentally relevant example, oil-degrading bacteria overcome complex interfacial energies to colonize fluid interfaces and establish films of bacteria at (oil-water) interfaces (FBI). My research program exploits fundamental knowledge of the physiological responses of bacteria to physical forces to investigate the development of new interfaces that will both help minimize the deleterious and optimize the beneficial effects of bacterial film formation. The outcomes of this research will inform new biomedical and technical strategies to: (1) eradicate multidrug-resistant (MDR) infections, (2) control microbial dynamics relevant to health and biotechnology, and (3) alleviate the ecological impacts of oil spills.

Research Experience: My research has encompassed interrelated areas ranging from food microbiology, to medical bacteriology and biotechnology. I have investigated how beneficial microbes can contribute to human health, as well as how bacteria are involved in many recalcitrant infections. Over the course of my graduate education, I focused on developing technologies targeting the resistance mechanisms of intrinsically quiescent bacterial cells, called persister cells. I studied how manipulating the electrophysiological properties of ***P. aeruginosa*** persisters can sensitize these cells to antibiotics and metallo-drug complexes previously demonstrated to be ineffective against persister cells. My current postdoctoral training builds upon my prior experience to decipher the interfacial properties of bacterial films.

I have gained expertise in characterizing 2- and 3-D bacterial films from physical and metabolic standpoints. Now, I am developing artificial microniches as functional microbial nanoculture systems using soft nanomaterials to advance the general understanding of microbial dynamics involved in complex biofilms. The microniches, when further advanced, will have applications for personalized, therapeutic medicine or for high-throughput screening of unculturable microbial species relevant to biotechnology and drug discovery. My research experience and relevant products are highlighted below:

Postdoctoral Training–Supervisors: Kathleen Stebe and Daeyeon Lee, Chemical and Biomolecular Engineering, University of Pennsylvania

Project 1: "***Artificial Microniches for High Throughput Study of Microbial Communities***."

1) ***Niepa THR, Hou L, Jiang H, Goulian M, Koo H, Stebe K, and Lee D*** (2016) Microbial Nanoculture as an Artificial Microniche (**Scientific reports, Accepted**).

2) ***Hann SD, Niepa THR, Stebe KJ, Lee D. (2016).*** One-step generation of cell-viable compartments via polyelectrolyte complexation in an aqueous two phase system (Submitted)

3) ***Kim D., Sengupta A., Niepa THR, Lee BH, Weljie A., Murata R., Stebe K.J., Lee D., Koo H. Candida albicans*** stimulate ***Streptococcus mutans*** microcolony development via cross-kindom biofilm-derived metabolites (In preparation).

Project 2: "***Physical and Metabolic Properties of Films of Bacteria at Fluid Interfaces (FBI)***."

4) ***Murphy D, Gemmell B, Vaccari L, Li C, Bacosa H, Evans M, Gemmell C, Harvey T, Jalali M, Niepa THR.*** (2016) An in-depth survey of the oil spill literature since 1968: Long term trends and changes since Deepwater Horizon (Submitted).

5) ***Niepa THR, Vaccari L, Leheny R, Goulian M, Lee D, and Stebe K*** (2016). Films of Bacteria at Interfaces (FBI) formation: A dynamic response to interfacial stress (In preparation)

Doctoral Training–Advisor: Dacheng Ren, Biomedical and Chemical Engineering, Syracuse University

PHD. Thesis: *Electrochemical Control of Bacterial Persister Cells*

6) ***Szkotak R, Niepa THR, Jawrani N, Gilbert JL, Jones MB, Ren D.*** Differential Gene Expression to Investigate the Effects of Low-level Electrochemical Currents on ***Bacillus subtilis.*** **AMB Express 1 (2011) 39.**

7) ***Niepa THR, Gilbert JL, and Ren D.*** Controlling ***Pseudomonas aeruginosa*** persister cells by weak electrochemical currents and synergistic effects with tobramycin. **Biomaterials 33(30) (2012) 7356-7365.**

8) ***Niepa THR, Snepenger LM, Hao Wang, Sivan S, Gilbert JL, Jones MB, and Ren D*** (2016) Sensitizing ***Pseudomonas aeruginosa*** cells to antibiotics by electrochemical disruption of membrane functions. **Biomaterials 74 (2016) 267-279.**

9) ***Niepa THR, Hao Wang, Dabrowiak JC, Gilbert JL, and Ren D*** (2015) Synergy between tobramycin and trivalent chromium ion in electrochemical control of ***Pseudomonas aeruginosa***. **Acta Biomaterialia 36 (2016) 286–295**.

10) ***Ren D., Zhang M., Niepa THR, and Gilbert J. (2014).*** Systems and Method for Controlling Bacterial Persister Cells with Weak Electric Currents. (**Awarded Patent** # 8 569 027).

11) ***Niepa THR, Wang H, Gilbert JL and Ren D.*** (2016) Electrochemical Control of ***Pseudomonas aeruginosa*** Persister and Biofilm Cells using TGON™ 805 in single and dual chamber systems (Submitted).

12) ***Niepa THR, Haveman G, Goyal P, and Ren D*** (2016) Electrochemical treatment of ***Pseudomonas aeruginosa***in a rabbit model of sinusitis (In preparation).

Teaching Interests: My vision is to promote active learning among organized team of students and to develop new methods to help them identify emerging research topics and design new solutions from a multidisciplinary approach. Thus, I am looking forward to teaching the core courses (including Transport, Thermodynamics, Reaction Engineering, Separation Processes) and specialty Bioengineering courses such as Biomaterials, Fields, Forces, and Flows in Biological Systems, Biochemical Engineering or Bioelectrochemistry.

Teaching Experience: Over the course of my academic and professional training, I have participated in and contributed to various teaching and research training experiences. These inspired my pursuit of an academic career so that I can continue to support future generations of scholars, especially those from disadvantaged backgrounds. As a research assistant, my main involvement in teaching was through guest lecturing in the areas of biotechnology, medical devices, and entrepreneurship. Lectures were structured within the course framework as taught by a faculty member, or based on my area of expertise in microbial biofilms. My efforts were well-received by students and also effective as indicated by positive reviews in independent evaluations and students' test performance. Another aspect of teaching entails training graduate and undergraduate students to acquire the research skills essential for their completion of their independent study. I helped four graduate and three undergraduate students to quickly become familiar with lab techniques and so independent researchers to contribute to the topic of bacterial electrophysiology. The success of my efforts led to co-authorship with two of my research mentees in scientific publications. Also as a doctoral student, my mentoring experiences reinforced my desire to become a faculty member. In the summers of 2012 and 2013, I coordinated the Syracuse University-Emerging Frontiers in Research and Innovation–Research Experience and Mentoring (EFRI–REM) program. I have pursued new mentoring opportunities during my fellowship at the University of Pennsylvania through the NSF Louis Stokes Alliance for Minority Participation (LSAMP) or through presentations at seminars and keynote addresses during STEM conferences. My commitment to engagement and teaching resulted in my receipt of the ***Appreciation Award for Dedication, Continuous Inspiration, and outstanding Contribution to Science and Engineering Students***from the Greater Philadelphia LSAMP.

Entrepreneurship: I have professional experience from starting up a small business that has both enhanced my creativity and increased my understanding of the challenges of and opportunities for technology transfer. In 2011, I successfully transferred a basic scientific idea developed in the lab into a business idea. I co-founded Helios Innovative Technology, Inc. (now PurpleSun, Inc.), which develops automated sterilization devices to prevent bacterial cross-contamination and fight hospital-acquired infections.

Founder of Helios Innovative Technology (now PurpleSun); Funding: 2011 New York State Plan Business Competition ($60,000), Angel Investor ($500,000).

13) Romo L., Garritano N., **Niepa THR.**(2013). Door Handle Sterilization System, WO Patent 2,013,025,894. (Patent Application).

Future Directions: As faculty member, I will strengthen and grow my research efforts while developing new, effective teaching programs. The goal of my research efforts is to elucidate the physicochemical mechanisms regulating microbial growth in various environments and, based on this knowledge, design sustainable means to control bacteria. My previous research on the electrophysiology of highly drug-tolerant bacterial cells opens the door for new approaches to cure chronic infections. Building on this foundation, my planned translational-research program would target the design of alternate ways to diagnose and eradicate bacteria using conductive-substrate interfaces and medical devices. Another research area I will pursue is the development of soft nanomaterials for biotechnological applications such as drug release and identification. By developing the means for selective diffusion through functional membranes, I propose to control physiological responses of microbes and direct their metabolic activity towards desired ends e.g., creation of new materials and bioproducts. Thus, the development of the functional membrane would allow the creation of functional nanocultures to encapsulate probiotics or unculturable cells as well as upstream bioprocess technologies to facilitate high-throughput screening and characterization of bioactive molecules.

Finally, I aim to elucidate the effects of interfacial properties of oil-water on the metabolic performance of bacteria in bioprocesses involving oil-water emulsions. I propose to characterize the mechanics and metabolic profiles of films of bacteria at interfaces subjected to strong interfacial energies, and elaborate on this understanding to model sustainability in bioremediation or prevent the biodeterioration of economically relevant products.

As a teacher, my vision is to promote active learning among organized team of students and to develop new methods to help them identify emerging research topics and design new solutions from a multidisciplinary approach. One of my goals as a faculty member is to translate my research ideas to medical and industrial technologies and encourage the development of an entrepreneurial spirit among my students so they are aware of this additional pathway through which research outcomes can benefit society.

475850 *Engineering Micro-Flows:Integrated Experimental-Computational Approach*

Abstract:

Research Interests:

Microfluidic systems have emerged as powerful tools for controllable manipulation of micro-scale flows. Ease of fabrication together with high-speed visualization prompt rapid growth in applications of microfluidic tools. Single-phase microfluidics, which is used to guide biological suspensions inside microchannels without addition of an immiscible suspending fluid, can be utilized for flow cytometry, rare cell diagnostic, growth of biofilm streamers and extensional rheometry. Multi-phase, i.e droplet, microfluidic platforms bring together the precise liquid handling capabilities of microfluidics with unprecedented control over size uniformity and the production rate of droplets. Droplet microfluidics are used for micro-particle synthesis, droplet-based digital polymerase chain reaction (ddPCR), droplet-based DNA sequencing, single-cell analysis and enzyme kinetic studies. Owing to the widespread use of microfluidics in various engineering applications, there is a critical demand for developing new microfluidic tools by relying on understanding the mechanisms of micro-scale flow. However, the fluid dynamics of suspensions and emulsions in complex geometries of microfluidics mean that experimental trials are the main approach to gain insight. The alternative path is utilizing both microfluidic experiments and mesoscale computational methods as interacting partners to address challenges of mixture flow at micro-scale.

A main thrust of my research will focus on developing novel platforms for micro-flow engineering enabled by microfluidic technology. I will employ micro-fabrication, microfluidics and microscopy techniques to accompany mesoscale simulations, in particular the lattice-Boltzmann Method (LBM). The insight provided by my integrated experimental-computational approach permits exploiting the inherent forces of the system, such as hydrodynamic and capillary forces, for developing easily deployable microfluidic devices. The outcome of my research will be applied to biomedical micro-filtration in single-phase microfluidics and on-chip liquid handling and droplet encapsulation systems for genomics, single-cell analysis and tissue engineering. Fundamental fluid dynamic questions which have hampered efficient integration of single-phase and multi-phase microfluidic, such as stochastic entrapment of components in each compartment, will be addressed through my efforts to develop microfluidic self-assembly and label-free sorting systems. On the computational front, I will apply the lattice-Boltzmann framework to micro-scale flow geometries to develop high quality computational models of mixture flow. By adding new modules to LBM using an object-oriented design and implementation framework, I will develop a software for micro-flow applications.

I have been working on my integrated microfluidics-LBM approach through my consistent experimental and computational research. The same line of work will be continued in my career as an assistant professor. I will use the capacity of single-phase microfluidic platforms to develop accurate on-chip micro-filtration tools for cancer diagnostic applications. The advantages of my proposed microfluidic filtration systems over other on-chip technologies, including the technology developed during my postdoctoral work, are smaller sample size and higher separation accuracy. I achieve this goal by utilizing the fluid dynamics of inertial mixture flow around obstacles in microchannels. I will also be focused on developing microchannel self-assembly systems for controlled encapsulation of components in microfluidic droplet generators. The inefficiency of the current on-chip technologies caused by stochastic entrapment will be resolved using my integrated experimental-computational approach. My third direction of research concerns with label-free sorting of droplets by exploiting hydrodynamic, capillary and magnetic forces. In addition to microfluidic experiments, the LBM simulations will assist to gain insight into the physics of micro-scale flow.

Teaching Interests:

I believe that modern engineering education should integrate computational methods throughout varied disciplines. Design tools such as AutoCAD and SolidWorks, simulation and scientific pogramming have a place in almost any course and should be introduced early and often during undergraduate education. With this in mind, I am interested and prepared to provide instruction for courses on the following topics, where I would develop a syllabus that included heavy integration of computational tools:

Undergraduate Level:

Physics and Applications of Microfluidics

Microfluidics Lab

Transport Phenomena

Applied Computational Methods in Science and Engineering

Graduate Level:

Advanced Fluid Dynamics

Rheology of Complex Fluids

Applied Microfluidics

Mesoscale Simulation Methods and Applications

475861 Engineering the Spatial Organization of Proteins for Applications in Synthetic Biology and Beyond

Abstract:

Engineering the Spatial Organization of Proteins for Applications

in Synthetic Biology and Beyond

Christopher Jakobson[1], Marilyn F. Slininger[1], Edward Y. Kim[1], Jeff E. Glasgow[2], Michael A. Asensio[3], Norma M. Morella[4], Emily C. Hartman[2], Alex Chien[5], Yiqun Chen[1], Elias Valdivia[4], Matthew B. Francis[2], Niall M. Mangan[6], and Danielle Tullman-Ercek[7]

[1]Dept. of Chemical Engineering, UC Berkeley

[2]Dept. of Chemistry, UC Berkeley

[3]Dept. of Bioengineering, UC Berkeley

[4]Dept. of Plant and Microbial Biology, UC Berkeley

[5]Biophysics Graduate Group, UC Berkeley

[6]Dept. of Applied Mathematics, U. of Washington

[7]Dept. of Chemical Engineering, Northwestern University

Research Interests:

Protein encapsulation, the storage of biologically active molecules inside protein containers to enhance cargo function, is ubiquitous in nature. My graduate research with Prof. Danielle Tullman-Ercek focused on two examples of this strategy: (i) bacterial microcompartments, subcellular organelles of *Salmonella*, *Yersinia*, and other pathogens, and (ii) virus-like particles, such as those based on bacteriophage MS2. By engineering these systems as platform technologies for synthetic biology, we also discover biological design principles that can guide the development of therapeutics and vaccines.

Bacterial microcompartments are protein-bound containers that house specialized metabolic pathways, allowing invading pathogens to proliferate in the host gut [1]. We have focused on engineering these structures to improve the performance of heterologous biosynthetic pathways encapsulated in the organelles. We have learned how to control organelle formation [2,3], control small molecule transport across the protein membrane [4], and localize heterologous cargo using a variety of cellular zip codes [5]. We have identified surprising plasticity in microcompartment-mediated metabolism and used mathematical models to understand microcompartment function in detail [6]. This toolbox allows us orchestrate the formation of custom subcellular organelles in bacteria and to encapsulate heterologous pathways in a well-controlled manner.

Using the bacteriophage MS2, another protein container, we have explored the possibility of using protein membranes to control the diffusion of small molecules. We have demonstrated that changes to the chemical character of the pores of the MS2 virus-like particle can influence the kinetic behavior of an encapsulated enzyme [7]. We have also devised a selection for MS2 virus-like particle assembly in order to comprehensively characterize the mutational landscape of the MS2 coat protein gene, and we have identified a single point mutant that mediates a stable conversion of the viral capsid structure from a 27 nm to a 17 nm icosahedron [8].

Our studies are but two examples of the growing intersection of synthetic biology and the study of host-pathogen interactions, demonstrating that the knowledge gained from engineering complex biological systems is broadly applicable to understanding their function and evolution. My independent research will focus on engineering probiotics to deplete the small molecules metabolized in microcompartment organelles, characterizing the mutational landscapes of complex systems such as viral capsids to understand their evolution, and designing spatial organization strategies to optimize the performance of heterologous biosynthetic pathways. My laboratory will combine chemical engineering expertise with experimental biology and mathematical modeling to take a unique approach to these questions.

Teaching Interests:

I have teaching experience on the instructional teams for the undergraduate Fluid Mechanics and Heat & Mass Transfer courses at Cornell University, as well as the graduate Protein Engineering course at UC Berkeley. In each of these courses, our focus as teachers was on student-driven learning in the form of guided problem-solving sessions, group projects, and facilitated discussion of primary literature [9]. I also serve as a scientific advisor to the K-12 Alliance professional development program for science teachers in California. I am part of a teaching team that leads a weeklong summer institute for teacher leaders from districts throughout the state, providing instruction on science content and modeling effective science pedagogy [10]. I will combine the lessons I have learned from teaching at the university and secondary levels to continue to improve my instruction of core chemical engineering courses.

As an independent investigator, I also plan to develop a course on Mathematical Methods in Biological Engineering for graduate and upper-level undergraduate students. The course will provide students with the analytical and numerical tools required to address chemical and biological phenomena at the molecular, organism, and population scales using models of chemical kinetics, cellular metabolism, and disease epidemiology. A key student outcome will be proficiency in the use of MATLAB as a numerical toolkit to address nonlinear dynamics in biological systems.

1. Jakobson and Tullman-Ercek, *PLoS Pathogens* (2016).

2. Kim*, **Jakobson***, and Tullman-Ercek, *PLoS ONE*(2014)

3. Jakobson, Chen, Slininger, Valdivia, Kim, and Tullman-Ercek, *in revision*.

4. Slininger, **Jakobson**, and Tullman-Ercek, *submitted.*

5. Jakobson, Kim, Slininger, Chien, and Tullman-Ercek, *J. Biol. Chem.*(2015)

6. Jakobson, Slininger, Tullman-Ercek, and Mangan, *in preparation.*

7. Glasgow, Asensio, **Jakobson**, Francis, and Tullman-Ercek, *ACS Synth. Biol.* (2015).

8. Asensio*, Morella*, **Jakobson**, Hartman, Glasgow, Sankaran, Zwart, and Tullman-Ercek, *submitted.*

*These authors contributed equally.

Manuscripts in preparation and submitted available upon request.

9. Sample syllabi, schedules, and course materials for Protein Engineering available upon request.

10. See: https://www.wested.org/service/k-12-alliance-science-content-institutes/

475964 Predicting the Aggregation Behaviour in Biopharmaceuticals

Abstract:

Profile

Sarah Hedberg

Imperial College London and Fujifilm Diosynth Biotechnologies

Academic advisers: Dr Daryl Williams and Dr Jerry Heng

PhD Thesis:

Predicting and Controlling Aggregation for mAbs During Bioprocessing, 2016

Research Interests:

My principal area of research is to develop an improved fundamental understanding of the role of protein phase behaviour that can be utilised in downstream bioprocess engineering. Potential research activities would include:

i) Studying the impact of protein aggregation in the downstream processing of biopharmaceuticals (primarily chromatography) and the formulation process;

(ii) The development and improvement of experimental techniques that can be used to predict protein aggregation and estimate the osmotic second virial coefficient (e.g. self-interaction chromatography and cross-interaction chromatography);

(iii) The development of robust crystallisation approaches for proteins using the second virial coefficient data;

(iv) To employ modelling tools for predictioning and control of these biopharmaceutical downstream operations.

My research has always been driven by the needs of the biopharmaceutical industry, with a current focus on controlling protein aggregation, which is critical in order to avoid immunogenicity issues in biopharmaceuticals. Problems with protein aggregation and other impurities would have the greatest impact on the downstream and formulation process. Previous research has shown that the second virial coefficient, describing the thermodynamics of the protein-protein interactions in solution, can be a useful tool for predicting the aggregation propensity of proteins. Automatic, high-throughput techniques requiring minimal proteins such as self-interaction chromatography, cross-interaction chromatography and dynamic light scattering have previously demonstrated to be very successful in determining the second virial coefficient or the interaction parameter that can be helpful in predicting protein aggregation.

My research work has always been in collaboration with industry and the industry partners for my current research are Fujifilm Diosynth Biotechnologies and JML Biotechnology Consulting. The research undertaken has later been directly implemented by these companies. I am planning to maintain these industrial links in my research and to work with these companies in my future research career.

My research vision is that we need to look closer at the thermodynamics governing the protein-protein interactions and approaching the problem from a physical chemical viewpoint. This may then help us gain a better understanding as well as being able to predict and control protein phase behaviour, protein stability and aggregation, protein crystallisation as well as performance for formulation and downstream operations.

Teaching Interests:

My current teaching interests lies within the biochemical engineering/ bioprocess engineering field. I am interested in promoting more biochemical engineering and biological engineering within the traditional chemical engineering teaching environment. I am especially hoping to teach and incorporate more of the downstream purification process and formulation process courses on biopharmaceuticals into the chemical engineering curriculum.

I have previously been part of designing, planning, teaching and marking a wide range of subjects in Chemical Engineering for the past 4 years, in my Department at Imperial College, but also at Tsinghua University, as a summer school course leader, and University College London, as a teaching associate. This teaching has ranged from fundamental chemistry, separation unit operations and chromatography to sustainable development (with a chemical engineering outlook). I have obtained a teaching recognition as an Associate Fellow of the Higher Education Academy (AFHEA) from the Higher Education Academy in the UK.

Publications

Hou, Y., Hedberg, S. & Schneider, I. C. *Differences in Adhesion and Protrusion Properties Correlate with Differences in Migration Speed under EGF Stimulation*. BMC Biophysics,2012, 5: 1-12.

Hedberg, S.H.M., Heng, J.Y.Y., Williams, D.R., and Liddell, J.M. *Self-Interaction Chromatography of MAbs: Accurate Measurement of Dead Volumes.* Pharmaceutical Research, 2015, 32(12): p. 3975-3985.

Hedberg, S.H.M., Heng, J.Y.Y., Williams, D.R., and Liddell, J.M. *Micro Scale Self-Interaction Chromatography of Proteins: A MAb Case-Study.* Journal of Chromatography A, 2016, 1434: p. 57-63.

Hedberg, S.H.M., Brown L., Heng, J.Y.Y., Williams, D.R., *Optimised Protein Immobilisation Reactions and*

Immobilisation Kinetics for Chromatography Media, submitted to Colloids and Surfaces B, 2016.

Hedberg, S.H.M., Heng, J.Y.Y., Williams, D.R., *Kinetics and Thermodynamics of Protein Aggregation*, in preparation 2016.

476183 Engineering the Plant Microbiome to Complement Host Phenotype

Abstract:

Engineering the plant microbiome to complement host phenotype

Plant roots are a rich source of carbon in an otherwise nutrient starved soil environment, and as a result, a diverse community of microbes is attracted to the plant. The complex network of beneficial and antagonistic interactions of community members, with each other and with the host, forms a microbiome that is generally a beneficial extension of the host plant. The plant microbiome can increase nutrient acquisition rates, increase resistance to pathogens, and alleviate environmental stresses such as drought. Understanding how the microbiome functions and interfaces with the host has applications in food and bioenergy agriculture, production of secondary metabolites from plants for commodity chemicals and pharmaceuticals, and will improve our general understanding of the host-microbiome relationship. My research goal is to engineer the microbiome to optimize plant growth for sustainable agriculture and biochemical production.

Research Interests:

My current research is focused on the bacterial microbiome of *Populus,* a genera of fast-growing trees cultivated as a second-generation biofuel feedstock. Using metabolic models, comparative genomics, and phenotype screening we showed that the extreme diversity within 19 *Pseudomonas fluorescens* strains isolated from *Populus* was driven by the metabolite profile of the host and correlated with spatial organization of the microbiome. To gain further insight into spatial aspects of cooperation and inhibition, we designed a novel patterning approach to study distance dependent interactions between bacterial strains. We are also studying the molecular mechanisms that lead to phenotypic changes in the host when colonized by individual bacteria or mixtures using phenotype measures, metabolomics, and transcriptomics. The results show that bacteria can behave synergistically to help optimize and modify plant growth phenotypes. To study community behavior, we measured microbiome response to host stresses using 16S sequencing and identified candidate bacterial taxa that may help alleviate host response to drought, light limitation, and toxin stresses.

My research group will build on the results described by studying how constructed communities of bacteria respond to the host and environmental stresses. Specifically, we will design communities of genome-sequenced bacterial isolates using individual phenotypes, phylogenetic relationships, and predicted genomic content and test the ability of the community to mimic the natural microbiome response to stress. Our approach will include quantitative analysis of bacterial strains and their interactions, and characterization of community response to host stress. Models of bacterial kinetics, metabolism, and interactions in culture conditions will be used to predict behavior on the host plant. Integrating quantitative experimental data with predictive models will help us understand how the microbiome assembles in nature, ultimately leading to the ability to engineer constructed communities of bacteria with specific and predictable functions.

Postdoctoral Project: Interactions between *Populus* and cultured representatives of its bacterial microbiome

Mentors: Dale A Pelletier and David J Weston, Oak Ridge National Laboratory, Biosciences Division

PhD Dissertation: Kinetics of vesicular stomatitis virus mRNA and genomes during infection

Mentor: John Yin, University of Wisconsin-Madison, Department of Chemical and Biological Engineering

Research Career

My research career began as an undergraduate where I studied gas hydrate formation at the Colorado School of Mines. I helped develop software for predicting hydrate formation in gas pipelines and tested predictions using a novel reactor in the laboratory. This valuable experience familiarized me with quantitative experiments and laboratory techniques, but more importantly convinced me to go to graduate school to continue my academic career. In graduate school I changed my focus to biological engineering, where I learned how to design experiments and control biological systems to make quantitative measurements of viral replication. I used the data I collected to build kinetic models of viral infection, and showed that viral mRNA production is independent from host environment and only depends on viral genome number. This suggests that transcription is robust and insensitive and helps to identify targets for anti-viral strategies for RNA viruses. While studying interactions between two organisms (viruses and cells), I became interested in complex ecological systems with multi-member interactions. I was fortunate to find a post-doctoral position as a part of the Plant-Microbe Interfaces project at Oak Ridge National Laboratory studying the microbiome of *Populus* trees. I have focused specifically on understanding bacterial genetic and phenotypic diversity in the system to help determine how interactions between microbiome members and the host lead to overall system function. In my next step I will study how multi-member systems can be engineered to perform specific functions.

Teaching Interests:

As I have moved through my academic career I have become more active in my role as a teacher and mentor. During my undergraduate studies I worked closely with many of my peers, and began to appreciate the different learning and teaching styles present in the group. I had the opportunity to informally tutor younger students in heat and mass transfer classes; in both cases their grades improved after we began working together. In graduate school I got my first opportunity to formally teach classes to students. As a teaching assistant for the introductory transport course, I enjoyed spending time developing lectures and example problems for students, and focused on how I could engage the entire class in a single topic. I also worked as a TA for the process control laboratory where I helped students run experiments and taught them to present their results in concise reports. As a post-doc I have worked with undergraduate summer students and new graduate students from chemistry, biochemistry, biological systems engineering and chemical engineering. Their unique backgrounds have helped me generalize my teaching methods. I have helped them learn how to work in the laboratory, analyze data, and present results. The students helped identify metabolic pathways, isolated bacteria that grow on *Populus* metabolites, identified genes that affect plant growth, and developed methods for studying root growth in microfluidic devices. Watching students succeed with my input has been extremely important to my development and has helped me appreciate the commitment of my teachers and mentors throughout my academic career.

476193 Microfluidic Line of Attack to Comprehend Biological Systems

Abstract:

Research Interests:

My research interests lie in developing microfluidic systems that have applications in cellular and other biological systems analysis as well as medical diagnostics and drug discovery. My primary goal is to investigate the cellular heterogeneity of primary cancer / patient-derived cells. Assessing cellular-level heterogeneity can provide a better understanding in a patient-specific manner how chemotherapy drugs interact with cancer cells so that personalized treatments can be designed. This goal will be extended to isolate individual cells of a desired phenotype and expand them for manufacturing of therapeutic cell lines. Single-cell isolation and expansion is a broad platform that can impact areas of tissue engineering, regenerative medicine, and disease therapies. Apart from cellular study, I would also like to utilize microfluidic systems to perform biological and biomedical assays with low sample/reagent volumes and parallelization capabilities. One of my targeted area will be the fundamental study of crystal nucleation and growth for different body-stones (crystals) in the gallbladder, kidney, and pancreas. My diverse research background—which includes experience with cell cultures, cell assays, and engineering microfluidic platforms for biological and chemical analysis—combined with training in novel experimental tools—such as advanced high-speed imaging techniques, fluorescence microscopy, confocal microscopy and scanning electron microscopy, particle image velocimetry, and microfabrication—puts me in a unique position to pursue active research in these areas.

In addition to my research expertise, my current position as Research Assistant Professor provides me hands-on experience with proposal writing for external funding along with interactions with prospective collaborators. These additional experiences put me a step ahead to propel my future research endeavors as a tenure-track faculty in the fields of cancer therapy and diagnostics.

Teaching Interests:

Apart from my research career, I have extensive teaching experiences as an instructor and teaching assistant at two different institutions. The first class that I taught as an instructor was 'Chemical Engineering Transport Laboratory' at Texas Tech University. Along with my teaching, I've also designed and developed two new experimental demonstrations that had been included in 'Unit Operation Laboratory' and 'Chemical Engineering Transport Laboratory' courses. The second-class as instructor was 'Introduction to Chemical Processes' a sophomore level course at Texas Tech. I have also held teaching duties as teaching assistant for undergrad fluid mechanics (Texas Tech), transport phenomena (University of North Dakota). I did guest lectures for two consecutive years for 'Introduction to Microsystems I' (senior and graduate level Electrical Engineering course) at Texas Tech. With respect to future teaching plans, I am capable of teaching both undergraduate and graduate level classes in Chemical Engineering. At the graduate level, I will offer new course 'Biomicrofabrication and Biomicrofluidics' that will emphasize microfabrication techniques, microscale flows and approaches for cellular analysis.

Proposal Contributions:NSF AIR-TT, NSF EAGER, and CPRIT.

Education:

Ph.D., Chemical Engineering, Texas Tech University, Lubbock, TX, 2013

M.S., Chemical Engineering, University of North Dakota, Grand Forks, ND, 2008

B.Sc., Chemical Engineering, Bangladesh University of Engineering & Technology, Dhaka, Bangladesh, 2004

Experience:

Research Assistant Professor, Department of Chemical Engineering, Texas Tech University, 03/2016-Present.

Post-doctoral Research Associate, Department of Chemical Engineering, Texas Tech University, 10/2014-03/2016.

Teaching (Instructor), Department of Chemical Engineering, Texas Tech University, 01/2014-08/2014.

Invention Disclosures:

1. Vanapalli, S. A., **Bithi, S. S.** and Sun, M. Methods and devices to control fluid volumes, reagent and particle concentration in arrays of microfluidic drops, US Patent Application 20140051062, 2014.
2. Vanapalli, S. A., **Bithi, S. S.** Microfluidic devices and methods of bioassays, PCT/US2016/036787, filed June 2016.

Publications:

1. **Bithi, S. S.** and Vanapalli, S. A. Collective Dynamics of Non-Coalescing and Coalescing Droplets in Microfluidic Parking Networks, ***Soft Matter***, 11, 5122-5132, (2015)
2. **Bithi, S. S.**, Wang, W. S., Sun, M., Blawzdziewicz, J. and Vanapalli, S. A. Coalescing drops in microfluidic parking networks: A multifunctional platform for drop-based microfluidics, ***Biomicrofluidics***, 8(3), 034118, (2014)
3. **Bithi, S. S.** and Vanapalli, S. A. Behavior of a train of droplets in a fluidic network with hydrodynamic traps, ***Biomicrofluidics***, 4, 044110, (2010) (**Selected as Research Highlight**)
4. Sun, M., **Bithi, S. S.** and Vanapalli, S. A. Microfluidic static droplet arrays with tuneable gradients in material composition, ***Lab on a Chip***, 11, 3949, (2011) (**Cover article**)
5. Maddala. J, Srinivasan, B., **Bithi, S. S.**, Vanapalli, S. A. and Rengaswamy, R. Design of a model-based feedback controller for droplet sorting and synchronization in a microfluidic loop, ***AIChE J***, 58, 2120, (2011)

476206 Biotechnological and Health Applications of Multiscale ME (Metabolism and protein Expression) Models

Abstract:

Research Interests:

- Novel systems biology algorithms to analyze Biomedical Big Data--massive and diverse datasets capturing complex biological interactions and are difficult to analyze using conventional data analysis methods.
- Multiscale models of metabolism, protein expression, and regulation, and multiscale optimization of bioprocesses.
- Antimicrobial strategies using systems biology models and multiple â€“omics data (genomics, transcriptomics, proteomics, epigenomics.
- Computer-aided microbe design for value-added chemicals, nutrients for animal feed, and therapeutic proteins.

Teaching Interests:

- 6 years of Teaching Assistant experience for Process Dynamics and Control
- Certification for completion of a 15-hour course on teaching university courses (at UCSD)
- I make appropriate use of active learning techniques, where student learning is facilitated using in-class clicker (real-time survey) questions on complex concepts that activate existing knowledge, identify misconceptions, and improve knowledge retention. Additional modern pedagogical methods will be used, including the inverted classroom and a balance of formative (ongoing or real-time) assessment and summative (exams, final

presentations, etc.) assessment.

Peer-Reviewed Journal Articles

1. E Brunk, KJ George, J Alonso-Guttierrez, M Thompson, E Baidoo, G Wang, **L Yang**, D McCloskey, T Baath, C Petzold, J Monk, EJ O'Brien, HG Martin, A Feist, P Adams, JD Keasling, BO Palsson, and TS Lee (2016) Characterizing Strain Variation in Engineered E. coli Using a Multi-omics Based Work ow. Cell Systems 2: 335-346.
2. **L Yang**, J Tan, EJ O'Brien, JM Monk, D Kim, HJ Li, P Charusanti, A Ebrahim, CJ Lloyd, JT Yurkovich, B Du, A Drager, A Thomas, Y Sun, MA Saunders, and BO Palsson (2015) A systems biology denition of the core proteome of metabolism and expression is consistent with high-throughput data. Proc Natl Acad Sci USA 112:10810-10815.
3. **L Yang**, S Srinivasan, R Mahadevan, and WR Cluett (2015) Characterizing metabolic pathway diversication in the context of perturbation size. Metab Eng 28: 14-122.
4. K Zhuang, **L Yang**, WR Cluett, and R Mahadevan (2013) Dynamic strain scanning optimization: an ecient strain design strategy for balanced yield, titer, and productivity. DySScO strategy for strain design. BMC Biotechnol 13:8.
5. **L Yang**, WR Cluett, and R Mahadevan (2011) EMILiO: a fast algorithm for genome-scale strain design. Metab Eng 13:272-281.
6. S Garg, **L Yang**, and R Mahadevan (2010) Thermodynamic analysis of regulation in metabolic networks using constraint-based modeling. BMC Research Notes, 3:125.
7. **L Yang**, R Mahadevan, and WR Cluett (2008) A bilevel optimization algorithm to identify enzymatic capacity constraints in metabolic networks. Comput Chem Eng, 32:2072-2085.

Submitted and pre-print articles

1. **L Yang**, D Ma, A Ebrahim, CJ Lloyd, MA Saunders, and BO Palsson. solveME: fast and reliable solution of nonlinear ME models. Under review.
2. **L Yang**, A Ebrahim, CJ Lloyd, MA Saunders, BO Palsson. dynamicME: Dynamic simulation and renement of integrated models of metabolism and protein expression. Under review.
3. **L Yang**, JT Yurkovich, CJ Lloyd, A Ebrahim, MA Saunders, BO Palsson. Principles of proteome allocation are revealed using proteomic data and genome-scale models. Submitted.
4. D Ma, **L Yang**, RMT Fleming, I Thiele, BO Palsson, MA Saunders (2016) Quadruple-precision solution of genome-scale models of Metabolism and macromolecular Expression. arXiv Preprint arXiv:1606.00054v1.

476212 Engineering Proteins and Metabolic Pathways for Biomedical, Bioenergy, and Biomaterial Applications

Abstract:

Research Interests:

Proteins play a key role in building, activating, and regulating what is essential in living organisms. From assembling pilin subunits on the *Staphylococcus aureus* cell surface, to generating cellulose microfibrils in *Arabidopsis* thaliana and converting carbon dioxide to methane in *Methanosarcina barkeri,* enzymes are the main actors in many biological processes. Therefore, understanding enzyme structure and function, and engineering enzyme stability and activity to generate novel bioengineering tools is crucial for solving many biomolecular engineering issues. Recent advances in molecular biology and protein engineering have created a platform to engineer enzymes as more efficient molecular "staplers" or "scissors" with higher activity and more specificity toward desired substrates. Furthermore, metabolic engineering methods are used to regulate enzymatic pathways for generating biofuel or other useful chemicals. My PhD and post-doctorate research has been focused on protein and metabolic engineering. Most of my research has evolved around controlling enzyme-mediated protein ligation and immobilization on the surface to generate site-specific protein assemblies, protein engineering using directed evolution and yeast surface display to create more efficient enzymes, as well as metabolic engineering to generate biofuel. As for my future plan, I intend to utilize recent genetic and protein engineering methods to create novel bioengineering tools with applications in bioenergy, biomedical, and biomaterial research.

Research Experience

Postdoctoral Researcher, University of Delaware

Projects: Engineering anaerobic methanogenic pathway to generate biofuels, Developing a light-activated enzyme.

Mentor: Dr. Wilfred Chen

PhD dissertation, University of Tennessee at Knoxville

Thesis: Engineering Sortase A for generating site-specific protein 3D assemblies

Advisor: Dr. Eric T. Boder

Teaching Interests:

Based on my training in teaching and research the Chemical and Biomolecular Engineering courses that I am readily able to teach include *Bioengineering Fundamentals, Transport Phenomena, and Mathematical Methods in Engineering.* I am also able to develop new biochemical engineering courses in undergraduate or graduate level. These courses will be focused on molecular and cell biology, protein engineering, as well as metabolic engineering.

476226 Towards Treatment of Neurodegenerative Disorders through Nanoparticle Mediated Enzyme Replacement Therapy

Abstract:

Research Interests:

Approximately 1 in 5,000 to 8,000 children are born annually with a lysosomal storage disease (LSD). Current treatment of LSDs, called enzyme replacement therapy (ERT), involves weekly to monthly intravenous (IV) infusions of missing endogenous enzymes. While ERT has proven effective in LSDs without central nervous system (CNS) involvement, the brain has remained untreatable due to the presence of the blood-brain barrier (BBB), which prevents passage of 98% of small molecule drugs, including enzymes, from the blood into the brain. Patients with neuropathic LSDs, which account for 50-70% of diseases in this category, present with severe CNS degeneration, ataxia, and premature death, with no treatment available on the market. The disease of focus, GM1 gangliosidosis, causes premature death between the ages of two and four with no hope available for parents of patients. Encapsulating enzyme into polymeric vesicles made up of polyethylene glycol and polylactic acid, called polymersomes, facilitates the transport of this missing enzyme into the brain through IV injections, effectively extending ERT therapy to the brain for the first time.

The missing enzyme in GM1 gangliosidosis, β-galactosidase (βgal) was loaded into polymersomes at a high efficiency, with 0.08 ± 0.03 mg βgal/mg polymersome, while maintaining its activity at 452 ± 82 a.u. Release of βgal was doubled over 6 hours when loaded polymersomes are placed into citrate buffer (pH=4.1) when compared to HEPES buffer (pH=7.4), indicating increased enzyme release expected in the lysosome. Attachment of ApoE was confirmed with an increase in polymersome diameter to 194.4 ± 3.5 nm and an increase in protein content from 0.06 ± 0.11 mg/mL to 1.03 ± 0.71 mg/mL. GM1SV3 cells, an immortalized cell line from felines with GM1 gangliosidosis, present low density lipoprotein receptors (LDLR) on the surface, as shown in confocal microscopy, indicating that they are a good model for ApoE mediated delivery through the BBB. ApoE increases the uptake of polymersomes into GM1/SV3 fibroblasts when compared to unlabeled control polymersomes.

My work highlights our ability to create polymersomes capable of protecting βgal in the blood stream, targeting the nanocarrier to the BBB, and after transcytosis, releasing βgal in the lysosome of neural cells. This combination of ERT and nanotechnology, demonstrating the capability of our carrier to transport enzymes to the brain while maintaining their activity, will create a paradigm shift in the treatment of CNS disease, providing treatment for currently untreatable and fatal diseases like GM1 gangliosidosis.

Teaching Interests:

As an instructor, it is my privilege to shape future engineers and motivate them to contribute to the direction of the field. I have demonstrated my ability to teach in both a classroom and laboratory setting through Teaching Assistant (TA) experience in Introduction to Chemical Engineering, Senior Design Lab, and Principles of Chemical Engineering at Auburn University. Though each experience was unique, I learned the universal importance of flexibility when teaching a course. It is necessary to constantly evolve in order to service the needs of your students as they progress and their understanding of the material matures.

I am passionate about inspiring students to excel in the chemical engineering field and excited to take on any chemical engineering course. Instruction of Thermodynamics, Principles of Chemical Engineering, Transport Phenomena, and Biomedical Engineering aligns closely with my strengths. My previous research experience developing novel nanocarriers for drug delivery applications and working in both the Chemical Engineering and Veterinary Medicine departments has prepared me for the integration of biomedical projects and assignments into any course, as well as the design of a Drug Delivery course. It is my goal as an educator to provide every student with the opportunity to engage in their own learning through flexible techniques. I want my students to leave the classes I teach with an understanding of the way in which the knowledge they have gained can contribute to the direction of the chemical engineering field.

476236 Microfluidic Technologies in High-Throughput Chemical Screens to Decipher Genetic Basis of Behavior and Development in *C. Elegans*

Abstract:

Research Interests:

The nematode *C. elegans* is of great interest to unravel fundamental biological processes common to the whole animal kingdom and to model human diseases. However, the large number of animals required per assay, complexity of protocols involved, and sensitivity to environmental perturbations render research on *C. elegans*time-consuming and labor-intensive, which slows down progress and limits the scope of addressable questions.

During my post-doc at Georgia Tech, I have exploited the possibilities of microfluidics to enable traditionally challenging types of screens on first larval stage (L1) and adult nematodes. The study of first larval stage *C. elegans*is particularly difficult because of the small size and high motility of the animals. I overcame this challenge by developing a microfluidic platform that uses nanoliter droplets of a reversible hydrogel to offer advanced manipulation of individual animals. This work paves the way for high-throughput screening of L1 nematodes with high-resolution imaging. This will facilitate the study of numerous post-embryonic developmental cellular events such as cell differentiation and cell migration, as well as important developmental decision-making, which specifically take place during the first larval stage.

I have also developed a microfluidic platform for combinatorial chemical screening of adult nematodes. Due to the exponential increase of combinations with the number of individual components, traditional technologies require a large amount of chemicals rendering the screens impractical. I solved this bottleneck by creating a droplet-on-demand platform that enables the preparation of precise mixtures of reagents while handling nanoliter volumes and the delivery of the stimuli to the animals in a robust way. Such a platform offers the possibility of performing combinatorial screens of rare compounds. It will enable deciphering intraspecies communication by screening combinations of pheromones that are difficult to extract or synthesize. It can also benefit drug screens to identify combinatorial therapeutic strategies.

In the future, I plan on building on my experience with *C. elegans* and microfluidics to pursue chemical screens with droplets. In addition, I intend to expand this approach to other model organisms, which will lead to an increase in throughput of assays by several orders of magnitude and open new exciting possibilities.

Teaching Interests:

During my PhD, I have taught undergrad and grad students, leading lab classes and solving-problem classes in addition to the usual TA duties (grading and helping students). I have interacted with students from a wide variety of backgrounds from which I have learned how to adapt to different learning styles. In addition, during my PhD and post-doc, I have mentored several undergrad and grad students on research projects as well.

I can teach Chemical Engineering core classes such as Numerical Methods Applied to Chemical Engineering, Mechanics of Fluids, Thermodynamics, in undergrad and I am willing to design new classes such as Microfluidics or Micro-Nano-Technologies at the graduate level.

476018 Biomaterials and Stem Cell-Based Therapeutics

Abstract:

Research Interests:
My research focuses on making stem cell and molecular therapies as an effective method to model and treat diseases. I believe that this goal can best be achieved by employing engineering fundamentals using biomaterials in stem cell research. At this interface between engineering and stem cell biology, my research leverages many disciplines including polymer synthesis, microfabrication technique, and gene/protein delivery. My goal is to run a multidisciplinary research program that focuses on three broad areas: 1) to fundamentally understand the mechanisms by which chemical and mechanical signals are sensed by stem cells under physiological and pathological conditions, 2) to design biomaterials that provide spatial and temporal control in regulating stem cell differentiation and assembly, and 3) to merge our understanding of stem cell signaling with our engineered stimuli-responsive biomaterials to a range of applications in regenerative medicine, drug delivery, and disease modelling. I envision that students within this research program will have strong collaborative interactions with other scientists, engineers, and clinicians, which will further enrich this training environment.

PhD Dissertation: "Engineering Vascular Networks Using Synthetic Matrices."
Advisor: Sharon Gerecht, Ph.D., Chemical and Biomolecular Engineering, The Johns Hopkins University.

Postdoctoral Project: "Targeted Anti-Thrombotic Therapeutics for Deep Vein Thrombosis."
Advisor: Elliot L. Chaikof, M.D., Ph.D., Department of Surgery, Harvard Medical School.

Research Experience:
My background as a chemical engineer, along with my training in biomaterials, stem cell biology and

vascular surgery lab, uniquely position myself to contribute substantial progress in the fields. During my doctoral work with Dr. Sharon Gerecht at the Johns Hopkins University, I investigated the mechanism of stem cell assembly into functional vascular networks. We developed a synthetic matrix, which properties can be tuned to provide temporal and spatial control for vascular morphogenesis that are useful for therapeutic angiogenesis and tissue regeneration. My postdoctoral training with Dr. Elliot L. Chaikof at Harvard Medical School has further broadened my research horizon in the area of drug delivery to target inflammation and thrombosis associated with vascular injury and stem cell transplantation. We engineered fusion proteins that selectively deliver anti-coagulant agents to activated platelets right at the site of injury, therefore inhibiting thrombosis without increasing the risk of systemic bleeding. I also led collaborations with engineers and surgical residents to evaluate anti-inflammatory and anti-thrombotic therapeutics in a clinically relevant model of deep vein thrombosis and islet transplantation.

Teaching Interests:
I believe that teaching is an integral part of a great research university, as rigorous education for both undergraduate and graduate students is a foundation for an innovative research. Throughout my academic training, I had the opportunity to TA "Senior Design Class for Chemical Engineering," as well as to design and teach advanced course in "Biotechnology for Regenerative Medicine." Additionally, I have also mentored and trained several undergraduate and master students in our research groups. As a faculty member, I am interested in teaching "Introduction to Chemical Engineering" and "Transport Phenomena" among other possible courses within the department. In addition to those classes, I aim to design graduate curricula that will introduce students to the evolving field of biomaterials, tissue engineering, and drug delivery.

Education:
B.S., 2007, Chemical and Biomolecular Engineering, University of Notre Dame.
Ph.D., 2012, Chemical and Biomolecular Engineering, Johns Hopkins University.

Selected Awards and Honors:
2007, *B.S. Magna Cum Laude*, University of Notre Dame.
2011, *U.S. New Investigator Travel Award*, International Society of Thrombosis and Haemostasis (ISTH).
2012, *Siebel Scholar, Class of 2012*, Siebel Scholar Foundation.
2014-2017, *JDRF Postdoctoral Fellowship,* Juvenile Diabetes Research Foundation.

Selected Publications (14 journal articles in total. Citation: 458. H-index: 10):

1. **Hanjaya-Putra, D.**, Yee, J., Ceci, D., Truitt, R., Yee, D., and Gerecht, S. (2010). Vascular endothelial growth factor and substrate mechanics regulate in vitro tubulogenesis of endothelial progenitor cells. *Journal of Cellular and Molecular Medicine*.
2. **Hanjaya-Putra, D.**, Bose, V., Shen, Y.-I., Yee, J., Khetan, S., Fox-Talbot, K., Steenbergen, C., Burdick, J.A., and Gerecht, S. (2011). Controlled activation of morphogenesis to generate a functional human microvasculature in a synthetic matrix. *Blood*.
3. **Hanjaya-Putra, D.**, Wong, K.T., Hirotsu, K., Khetan, S., Burdick, J.A., and Gerecht, S. (2012). Spatial control of cell-mediated degradation to regulate vasculogenesis and angiogenesis in hyaluronan hydrogels. *Biomaterials*.
4. **Hanjaya-Putra, D.**, Shen, Y.-I., Wilson, A., Fox-Talbot, K., Khetan, S., Burdick, J.A., Steenbergen, C., and Gerecht, S. (2013). Integration and Regression of Implanted Engineered Human Vascular Networks During Deep Wound Healing. *Stem Cells Translational Medicine*.
5. Kusuma, S., Shen, Y.-I., **Hanjaya-Putra, D.**, Mali, P., Cheng, L., and Gerecht, S. (2013). Self-organized vascular networks from human pluripotent stem cells in a synthetic matrix. *Proceedings of the National Academy of Sciences*.
6. Krishnamurthy, V.R., Sardar, M.Y.R., Ying, Y., Song, X., Haller, C., Dai, E., Wang, X., **Hanjaya-Putra, D**., Sun, L., Morikis, V., Simon, S.I., Woods, R.J., Cummings, R.D., Chaikof, E.L., (2015). Glycopeptide analogues of PSGL-1 inhibit P-selectin in vitro and in vivo. *Nature Communications*.
7. Naik, N., **Hanjaya-Putra, D.**, Haller C., Allen, M.G., Chaikof, E.L., (2015). Rapid homogenous endothelialization of high aspect ratio microvascular networks. *Biomedical Microdevices*.
8. **Hanjaya-Putra, D.**, Dai E., Haller C., Hagemeyer, C.E., Karlheinz, P., Chaikof, E.L., (2016). Targeted

Anti-Thrombotic Prophylaxis for Deep Vein Thrombosis. *Circulation (submitted)*.

Patents:

1. **Hanjaya-Putra, D.**, Vo E.T., Wanjare, M., Gerecht, S., "Smooth Muscle-Like Cells (SMLCs) Derived from Human Pluripotent Stem Cells," US 13/581,341. (2012).
2. **Hanjaya-Putra, D.**, Gerecht, S., "Hydrogel-Based Vascular Lineage Cell Growth and Uses," US 14/553,442. (2015).

476038 Engineering the Molecular Interactions for Biomedical Applications

Abstract:

Molecular Engineering for Biomedical Applications

Handan Acar

The Institute for Molecular Engineering, The University of Chicago, Chicago, IL 60637, USA

hacar@uchicago.edu

Research Interests:

Interest in peptides as potential drug candidates has naturally increased with recent progress in enhancing their structural and chemical stability, along with their ability to cross the cell membrane. Peptide-based therapeutics that target not only specific cells but also specific protein-protein interactions in the cytoplasm will herald a new era of "personalized medicine". In this approaching new era, chemical tools to engineer peptides into molecular probes will be invaluable. The combination of carefully selected and localized interactions will produce stable assemblies, which can be reversible, highly tunable, dynamic, and modular as required by the specific application. Effective approaches to the design of molecular therapeutics must be safe, sensitive, efficient, and rapidly adapted to new targets with minimal effort and expense. To enable new translational clinical technologies, my research will focus on molecular design that incorporates these aspects into the new materials.

I have extensive research experience in the design and synthesis of self-assembling peptide-based nanostructures, including the modulation of the fundamental driving forces of self-assembly. As a Ph.D student of Prof. Mustafa Ozgur Guler (Bilkent University, Ankara, Turkey), I designed and synthesized self-assembling peptide sequences to generate targeted structures in both organic solvents and water via the specific interactions of rationally placed amino acid sequences. My innovative work produced a novel bottom-up synthesis method for peptide template-directed self-assembled mineralization, generating one-dimensional inorganic nanostructures. I was awarded a patent for this technology.

I sought to broaden the translational focus of my research during my postdoctoral work at Institute for Molecular Engineering at the University of Chicago with Matthew Tirrell, PhD. Taking advantage of the university's rich mix of clinicians and translational scientists, I work closely with James LaBelle MD, PhD, a pediatric oncologist in the Pritzker School of Medicine whose research involves peptide therapeutic translation to refractory malignancies. This cutting-edge collaboration between myself, Dr. Tirrell and Dr. LaBelle led to the development of a self-assembled nanoparticle platform, comprised of enzymatically cleavable peptide amphiphiles, that carries therapeutic peptides into cells. This platform offers the unique ability to follow intracellular drug trafficking and enzymatic cleavage in real time. The broad-ranging application of this technology to clinical medicine, biochemical research, and peptide-based nanoparticle development led us to file a patent application through the University of Chicago.

Teaching Interests:

I have extensive teaching experience, starting as a Ph.D. student; I was the teaching assistant of three graduate courses in Bilkent University. As an independent teacher at university level, I designed and implemented the "Fundamentals of Thermodynamics for Mechanical Engineering" at Iowa State University (Spring 2014), which greatly assisted in my development as an instructor. At the end of the semester my students attained grades higher than the departmental average, a clear indication of my abilities as a teacher. Finally, I have mentored graduate and undergraduate students in my group as a postdoctoral researcher.

Future Direction:

My research lab will establish chemical methodologies to tune the supramolecular interactions of self-assembling molecules to surmount the key challenges for peptide-based therapeutics, diagnostics and delivery platforms. In self-assembly processes, the fundamental building blocks organize themselves into functional structures as driven by the energetics in the system. These behaviors typically emerge from supramolecular phenomena such as hydrogen bonding, hydrophobic effects, electrostatics, metal-ligand and π-π interactions, and van der Waals forces. Sufficient combinations of these rationally placed interactions can produce the reversible self-assembly of stable structures, with these structures presenting a highly tunable and modular platform for a variety of applications.

As an independent investigator, I will combine my experiences to build a lab with a strong focus on designing materials from a molecular perspective for translational clinical technologies. A specific emphasis will be placed on using an interdisciplinary approach from physical and synthetic chemistry for biomedical applications.

Selected Publications: (out of 14)

Acar, H.; Srivastava, S.; Chung E.J.; Scnorenberg, M.R.; Barett; LaBelle, J.L.; Tirrell, M.; *"Self-Assembling Peptide-Based Building Blocks in Medical Applications",* accepted to Advanced Drug Delivery Reviews (2016).

Acar, H.; Cinar, S.; Thunga, M.; Kessler, M.R.; Hashemi, N.; Montazami, R.; 2014 "*Study of physically transient insulating materials for transient and biocompatible electronics*"; Advanced Functional

Materials; 24(26), 4135–4143 (2014).

Acar, H.; Genc, R.; Urel, M.; Erkal, T.S.; Dana, A.; Guler, M.O.; *Self assembled peptide nanofiber template one-dimensional gold nanostructures exhibiting resistive switching"***;** Langmuir*;* 28, 16347-54 (2012).

Acar, H.; Garifullin, R.; Guler, M.O.; *Self-assembled template-directed synthesis of one-dimensional silica and titania nanostructures;* Langmuir*;* 27, 1079-84 (2011).

For more information, please visit: http://tirrell.ime.uchicago.edu/People/Handan_Acar.html

476042 Interfacing Cells and Materials for Advanced Delivery Systems

Abstract:

Research Interests:

The clinical translation of novel therapies and delivery systems is limited by their poor biological performance. Material-based delivery systems (e.g. nanoparticles) perform complex delivery functions such as on-demand or controlled release, but they are limited in performing basic biological functions (e.g. circulating, targeting tissues, penetrating barriers, and entering diseased cells). Cell-based therapies routinely perform these basic biological functions as they have evolved to interact with living systems, but they are not suitable for drug loading and controlled release. My research has led to the development of a number of hybrid delivery systems capable of performing both biological and synthetic functions [1-4]. These advanced delivery systems incorporate findings from my phenomenological studies [5-6], which have elucidated the key biophysical, biochemical, and material parameters that facilitate enhanced delivery.

Ph.D. Research:

My graduate research with Professor Samir Mitragotri at the University of California, Santa Barbara focused on improving the biological performance of nanoparticle delivery systems. During my Ph.D I developed three distinct blood-cell inspired hybrid systems that enhanced the in vivo delivery abilities of nanoparticles by utilizing and mimicking circulatory blood cells, including: (i) red blood cells [2-3], (ii) white blood cells [1], and (iii) platelets [4]. The first two examples utilize a strategy known as "cellular hitchhiking", which involves the attachment of polymeric particles to the surface of circulatory cells so as to transfer innate circulatory (e.g. long circulation of red blood cells) and targeting abilities (e.g. inflammation targeting of white blood cells) from cell to particle. The second strategy involves the design and application of synthetic platelets which incorporate the essential biophysical (elasticity [5] and shape [6]) and biochemical (surface biology) parameters of natural platelets into a single nanoparticle capable of performing hemostasis and preventing blood loss. During my Ph.D. I was an NSF GRFP Fellow and authored more than 25 peer-reviewed research publications in the field of drug delivery.

Postdoctoral Research:

My postdoctoral research with Professor Robert Langer at the Massachusetts Institute of Technology focuses on the development of oral drug delivery systems. I am currently expanding the broad theme of interfacing cells and materials [7] for the improved delivery of live-probiotics to the microbiome. Using a layer-by-layer encapsulation approach, I have shown that encapsulated probiotics exhibit enhanced: (i) protection against harsh gastrointestinal tract conditions, (ii) mucoadhesion and growth on intestines, and (iii) integration with the hostâ€™s microbiome in vivo as compared to non-encapsulated probiotics [8]. Another focus during my postdoctoral work has been on the clinical translation of controlled release polymeric delivery systems. I am currently working with clinicians in an ongoing clinical trial investigating the bioavailability of micronutrients delivered via a novel particle-based oral delivery system.

Future Research:

As a faculty member I will focus on two research areas:

Targeted nanoparticle delivery:

(a) Biologically-active nanoparticles for enhanced tissue targeting and penetration

(b) Determining organ-level and cell-specific distribution of nanoparticles

Microbiome delivery:

(a) Microbiome modulation via delivery of encapsulated probiotics and microbes

(b) Delivery strategies to maintain a healthy microbiome during antibiotic treatments

Teaching Interests:

My teaching philosophy places importance on integrating quantitative mathematical and rational/logical qualitative approaches. Emphasis will be on teaching and understanding the physical laws and fundamentals of chemical engineering to direct and guide quantitative mathematical approaches. As a graduate student I was a Teaching Assistant (TA) for core chemical engineering courses as well as bioengineering electives. As a TA I designed and presented lectures, held office hours, and presented MATLAB- and Mathematica-based approaches for computer-aided methods to solve partial differential equations. During my undergraduate years I served as a tutor for freshman and sophomore chemistry, physics, and mathematics courses.

References:

1. **Anselmo, A. C.**; Gilbert, J. B.; Kumar, S.; Gupta, V.; Cohen, R. E.; Rubner, M. F.; Mitragotri, S., Monocyte-Mediated Delivery of Polymeric Backpacks to Inflamed Tissues: A Generalized Strategy to Deliver Drugs to Treat Inflammation. Journal of Controlled Release 2015, 199, 29-36.
2. **Anselmo, A. C.**; Gupta, V.; Zern, B. J.; Pan, D.; Zakrewsky, M.; Muzykantov, V.; Mitragotri, S., Delivering Nanoparticles to Lungs While Avoiding Liver and Spleen through Adsorption on Red Blood Cells. ACS Nano 2013, 7, 11129-11137.
3. **Anselmo, A. C.**; Kumar, S.; Gupta, V.; Pearce, A. M.; Ragusa, A.; Muzykantov, V.; Mitragotri, S., Exploiting Shape, Cellular-Hitchhiking and Antibodies to Target Nanoparticles to Lung Endothelium: Synergy between Physical, Chemical and Biological Approaches. Biomaterials 2015, 68, 1-8.
4. **Anselmo, A. C.**; Modery-Pawlowski, C. L.; Menegatti, S.; Kumar, S.; Vogus, D. R.; Tian, L. L.; Chen, M.; Squires, T. M.; Sen Gupta, A.; Mitragotri, S., Platelet-Like Nanoparticles: Mimicking Shape, Flexibility, and Surface Biology of Platelets to Target Vascular Injuries. ACS Nano 2014, 8, 11243-11253.

5. **Anselmo, A. C.**; Zhang, M.; Kumar, S.; Vogus, D. R.; Menegatti, S.; Helgeson, M. E.; Mitragotri, S., Elasticity of Nanoparticles Influences Their Blood Circulation, Phagocytosis, Endocytosis, and Targeting. ACS Nano 2015, 9, 3169-3177.
6. *Kolhar, P.; ***Anselmo, A. C.**; Gupta, V.; Pant, K.; Prabhakarpandian, B.; Ruoslahti, E.; Mitragotri, S., Using Shape Effects to Target Antibody-Coated Nanoparticles to Lung and Brain Endothelium. Proceedings of the National Academy of Sciences 2013, 110, 10753-10758.
7. *Jaklenec, A.; ***Anselmo, A. C.**; Hong, J.; Vegas, A. J.; Kozminsky, M.; Langer, R.; Hammond, P. T.; Anderson, D. G., High Throughput Layer-by-Layer Films for Extracting Film Forming Parameters and Modulating Film Interactions with Cells. ACS Applied Materials Interfaces 2016, 8, 2255-2261.
8. **Anselmo, A. C.**, McHugh, K.J., Webster, J., Langer, R., Jaklenec, A. Layer-by-Layer Encapsulation of Probiotics for Delivery to the Microbiome. In revision since July 2016.

476061 Engineering Proteins for Magnetic Resonance Imaging at Molecular and Atomic Resolutions

Abstract:

Research Interests:

Imaging biological processes in intact tissue and cells with molecular specificity and high resolution presents a fundamental challenge to biomedical research. Magnetic resonance imaging (MRI), being non-invasive and not limited by tissue depth, enables whole-body imaging at sub-100μm resolution; however, the main bottleneck is the lack of compatible smart sensors that can recognize specific molecular events and accordingly output an MRI signal. On a different level of resolution, nuclear magnetic resonance (NMR) spectroscopy, a structural biology method compatible with living cells and tissues, allows the visualization of the movement of atoms inside a protein when it performs a biological task. Similarly, what prevents NMR from a wide application to the study of live cells and tissue is the lack of smart agents, which in this case, would restrict NMR signals only from those selective molecules, making atom-specific spectra a feasible goal. **I propose to engineer protein-based smart agents for the two applications that non-invasively image biomolecules at the molecular and atomic resolutions.** The growing genomic data from the sequencing of animals and microbes, together with the understanding and computational prediction of protein structures and functions, provides unique opportunities nowadays to repurpose and modify proteins for applications distinct from their native roles. Since these engineered proteins can be genetically encoded, they can leverage the large amount of genetic engineering methods developed over the decades to achieve functions that are difficult for synthetic agents, such as the precise localization in subcellular compartment, expression-level coupling with genes and circuits, and specificity in molecular recognition. The development of the new non-invasive molecular imaging methods could lead to various applications, such as the functional imaging of the brain in line with the BRAIN initiative, while the development of atomic-resolution imaging of biomacromolecules in live cells could lead to a new imaging paradigm and information that have not been previously accessed.

Postdoctoral Project: ***Genetically encoded gas nanostructures as acoustomagnetic reporters for noninvasive biological imaging.***

Under supervision of Mikhail G. Shapiro, California Institute of Technology

PhD Dissertation: ***Development of Solid-state NMR Spectroscopy for Membrane Proteins: Application to the Mercury Transporter MerF.***

Under supervision of Stanley J. Opella, University of California, San Diego (UCSD)

Research Experience:

My research has been at the interface between biology and physics. In the graduate school, I joined a laboratory that had work spanning from molecular biology, protein chemistry to quantum mechanics theory of the NMR pulse sequences and computational biology on protein structures. The overall goal of my research was to develop a general method for obtaining atomic-resolution structures of membrane proteins in their native lipid bilayer environment. I devoted a large portion of the efforts to the development of new solid-state NMR and computational methods. Besides, I also applied the new methods to solve the structure of a mercury transporter protein from microbial bioremediation pathways.

While the goal for a structural biologist is to understand a protein's native function, my current postdoctoral research in the realm of protein engineering aims to utilize such understanding to repurpose and generate new functions for the proteins. For example, we currently work on gas vesicles, a group of protein-only nanostructures whose native role is to regulate buoyancy of the microbes, and we repurpose them as a genetically encoded MRI reporter gene. During my postdoctoral training, I have also gained insights into the imaging aspects of magnetic resonance methods and the techniques for live animal imaging.

As faculty I would like to leverage my knowledge in protein engineering and magnetic resonance to develop new imaging methodology that can find utilities in various studies on live cells and animals. Just as how green fluorescent protein (GFP) changes the paradigm of optical imaging, I believe the engineering of protein-based probes can break through key bottlenecks in current non-invasive imaging methods and can potentially lead to new platforms.

Selected Publications:

Lu GJ, Tian Y, Vora N, Marassi FM, Opella SJ (2013) The structure of the mercury transporter MerF in phospholipid bilayers: a large conformational rearrangement results from N-terminal truncation. ***J. Am. Chem. Soc.*** 135(25):9299-9302.

Lu GJ, Opella SJ (2014) Mechanism of dilute-spin-exchange in solid-state NMR. ***J. Chem. Phys.*** 140(12): 124201.

Lu GJ, Opella SJ (2014) Resonance assignments of a membrane protein in phospholipid bilayers by combining multiple strategies of oriented sample solid-state NMR. ***J. Biomol. NMR*** 58(1): 69-81.

Lu GJ, Opella SJ (2013) Motion-adapted pulse sequences for oriented sample (OS) solid-state NMR of biopolymers. ***J. Chem. Phys.*** 139(8):084203.

Lu GJ, Park SH, Opella SJ (2012) Improved 1H amide resonance line narrowing in oriented sample solid-state NMR of membrane proteins in phospholipid bilayers. ***J. Magn. Reson.*** 220:54-61.

Lu GJ, Son WS, Opella SJ (2011) A general assignment method for oriented sample (OS) solid- state NMR of proteins based on the correlation of resonances through heteronuclear dipolar couplings in samples aligned parallel and perpendicular to the magnetic field. ***J. Magn. Reson.*** 209(2):195-206.

Lu GJ, Garen CR, Cherney MM, Cherney LT, Lee C, James MNG (2007) Expression, purification and preliminary X-ray analysis of the C-terminal domain of an arginine repressor protein from Mycobacterium tuberculosis. ***Acta Crystallogr Sect F Struct Biol Cryst Commun*** 63(Pt 11):936-939.

Teaching Interests:

At the graduate school, I TAed in analytical chemistry laboratory course, and presented weekly seminars as the only TA for a protein biochemistry course. At CalTech, I presented a 3-hour guest lecture in the Molecular Imaging course offered from chemical engineering department. Lastly, I mentored several undergraduate students during my Ph.D. and postdoctoral studies. Although my undergraduate background was mainly on biology, the exposure to multidisciplinary research in chemistry and physics during my Ph.D. and postdoctoral studies provided me confidence and enthusiasm to teach in various disciplines.

476128 Molecular Engineering for Cellular Imaging: From Fluorescence to Magnetic Resonance

Abstract:

Research Interests: Light-emitting proteins such as luciferase and GFP (green fluorescent protein) have revolutionized biological research by enabling genetically targeted labeling and optical imaging of cells and biomolecules. However, rapid advances in synthetic biology have imposed increasingly challenging requirements for improved imaging technologies that can enable the study of cellular function within several distinct contexts relevant to disease, healthcare, and industrial bioproduction. Addressing these requirements necessitates the development of smart molecular reporters for multiple imaging modalities, providing an exciting opportunity for biomolecular engineering. My research aims to meet this challenge through two distinct approaches involving **molecular engineering of: (1) reporter genes for magnetic resonance imaging (MRI) and (2) O_2-independent fluorescent reporters for optical imaging in low-O_2 conditions.**

Research Experience:

(1) Reporter genes for noninvasive cellular imaging using MRI (*Advisor: Mikhail G. Shapiro, Caltech***)**

MRI is a widely established modality for noninvasive imaging of intact animals and patients. However, the lack of sensitive molecular reporters for MRI (analogous to GFP and luciferase for optics), represents a critical missing capability for extending MRI to the study of cellular functions in whole organisms. In this regard, we recently introduced diffusion altering reporter genes that render cells visible in diffusion-weighted magnetic resonance images simply by increasing the rate of water exchange across the cell membrane[1]. This was achieved by overexpressing the human water channel, aquaporin, which results in robust MRI contrast without any deleterious effects on cell viability. Notably, aquaporin-based contrast is readily detectable at nanomolar levels of expression and in small (≈10%) subsets of aquaporin-labeled cells in mixed populations, which places aquaporin among the most sensitive reporter genes for MRI. Finally, we demonstrated the potential utility of aquaporin by imaging dynamic gene expression in a mouse brain tumor model.

(2) Engineering O_2-independent fluorescent reporter genes (*Advisor: Charles M. Schroeder, UIUC*)

The significance of low O_2 tensions (hypoxia & anoxia) is illustrated by the critical role of hypoxia in regulating bacterial pathogenesis, antibiotic sensitivity, gut microbiome dynamics, and cancer chemoresistance. Consequently, live cell imaging in low-O_2 environments is of tremendous interest in biomedical science. However, widely prevalent GFP and luciferase-based reporters show O_2-dependent light emission, which precludes robust quantitative cell imaging. On the other hand, alternative O_2-independent reporters based on flavin-binding photosensory proteins (FbFPs) are limited by poor brightness, stability, toxicity, a small library size, and an incomplete understanding of their performance and properties as cellular reporters[2]. To this end, a key outcome of my research has been the characterization[3], expansion, and diversification of the O_2-independent FbFP fluorescent palette using directed evolution[4] and genome mining[5]. Notably, we engineered four new FbFP variants, which display improved brightness, photostability, small size (half the size of GFP), and a broad operational pH and thermal range (pH 3-11, up to 60°C).

Future Research Directions:

The transformative impact that optical reporters have had on biology can be attributed to the rich repertoire of cellular biosensors that have been developed based on foundational members of the GFP family. In contrast, there remains a critical dearth of genetically encoded MRI and optical sensors for cellular imaging in live animals and hypoxia respectively. As an independent investigator, I am interested in leveraging the previously described foundational classes of MRI and optical reporter genes to develop viable sensors for studying cellular function. Initial projects in the Mukherjee laboratory will have the following aims.

Aim 1: O_2-independent fluorescent biosensors. I propose to develop FbFP-based biosensors for imaging bacterial second messengers and cellular energy in physiologically realistic hypoxic models of bacterial infection, antibiotic resistance[6], and host-pathogen interactions. Concurrently, we will explore new approaches for further enhancing FbFP performance through improvements in protein folding, solubility, chromophore binding, and enriching cellular flavin content.

Aim 2: Functional MRI reporters. I propose to expand the library of diffusion altering reporter genes through the development, characterization, and application of pH and phosphorylation-sensitive aquaporins for imaging pH fluctuations and cell signaling events in preclinical animal models of infection, disease, and injury. Furthermore, I am interested in repurposing a recently described eukaryotic Mn^{2+} binding protein to develop a genetically encoded MRI-based sensor for calcium imaging *in vivo*.

In summary, I envision the proposed research to culminate in the development of improved imaging tools as well as platform technologies that will enable new approaches to study cellular function and gene expression in the context of realistic *in cellulo* and *in vivo* models.

Teaching Interests:

My teaching interests encompass topics in biomolecular engineering with a focus on imaging and therapy. Specifically, I am interested in developing and introducing a cross-disciplinary course on *Molecular*

Engineering for Imaging and Therapy that integrates fundamental principles from chemical kinetics, thermodynamics, and imaging along with techniques in molecular biology and protein engineering with an overall emphasis on the development of molecular and cell-based sensors and therapeutics. A central goal of my teaching philosophy is to help students identify exciting new avenues for innovation and apply principles from biomolecular engineering to brainstorm creative solutions. Through this initiative, I seek to emphasize the importance of cross-disciplinary approaches in biomolecular engineering while encouraging students to identify and appreciate the foundational engineering and scientific principles that bioengineering is built upon.

Teaching Experience:

In my PhD, I worked as a teaching assistant for four semesters during which time I extensively lectured courses in *Process Control*, *Biomolecular Engineering*, and *Bioenergy Technology* for which I received the campus nomination for "List of Teachers ranked Excellent". During my postdoc, I was a guest lecturer for topics on MRI contrast agents and relaxation theory as part of a course in *Molecular Imaging*. In addition, I have mentored three undergraduate researchers of whom two are currently pursuing graduate studies in chemical and biomolecular engineering.

Publications:

1. Mukherjee, A., et al. Diffusion altering reporter genes for magnetic resonance imaging. In revision (preprint available on bioarXiv)

2. Mukherjee, A. et al. Flavin-based fluorescent proteins: Emerging Paradigms in Biological Imaging. *Curr. Op. Biotech.*. (2014)

3. Mukherjee, A., et al. Characterization of flavin-based fluorescent proteins: an emerging class of powerful fluorescent probes. *PLOS ONE*. (2013)

4. Mukherjee, A., et al. Directed evolution of bright mutants of a flavin-binding fluorescent protein from *Pseudomonas putida*. *J. Biol. Eng.* (2012).

5. Mukherjee, A., et al. Engineering and characterization of LOV-based fluorescent reporter proteins in *C. reinhardtii* & *V. frigida*. *ACS Syn. Biol.*(2014)

6. Mohan, R.*, Mukherjee, A.*, et al. A Multiplexed Microfluidic Platform for Antibiotic Susceptibility Screening. *Biosens. Bioelectron*. (2013) (* co-first)

476130 Harnessing Diverse Microorganisms for Biochemical Production

Abstract:

Research Interests:

Biochemical production of compounds offers advantages over conventional synthesis due to the high specificity and diversity of biological catalysts, reactions occurring under labile conditions, and the use of renewable substrates. Metabolic engineering of microbes has primarily focused on modifying the model organisms *E. coli* and *S. cerevisiae* due to their fast and robust growth on inexpensive fermentation broths and the wealth of genetic tools available to them. Heterologous pathway addition in these platform hosts has had mixed success, often with natural producers reaching higher titers due to improper recombinant enzyme folding and regulation. Additionally, other traits should be considered before selecting a production host, such as substrate utilization, endogenous pathways, product toxicity, secretion of product or enzymes, growth temperature, susceptibility to contamination, pathogenicity of host, and growth environment. These factors have led to the development of alternative hosts for chemical production, including *C. glutamicum* (amino acid synthesis), *Synechocystis* sp. (CO_2 autotroph), *C. thermocellum* (consolidated biomass processing), and *B. subtilis* (GRAS certified, carotenoids, protein secretion), just to name a few. Further, advances in synthetic biology and â€"omics techniques have enabled the rapid modification and characterization of microbes. Our laboratory will use these methods and develop additional tools to harness the diversity of microorganisms for biochemical production. Specifically, we aim to 1) engineer organisms that are uniquely tolerant to reaction conditions, 2) use protein engineering and synthetic biology tools to enhance enzyme production and regulation in these hosts, and 3) discover novel pathways and organisms through bioprospecting to expand the range of chemicals to be synthesized and traits to be selected from.

Product toxicity and removal are two long-standing challenges to bioprocess engineering. Co-solvent culturing aims to alleviate both problems by stripping off products as they are generated, sequestering them from the organism and eliminating them from further conversion. We plan to use sustainable solvents and organisms that grow under these unique reaction conditions to produce compounds of interest, such as aldehydes and alkanes, that are highly toxic to cells and have high volatility. In addition to conventional metabolic engineering to increase biochemical production, we will develop new tools to enhance enzyme expression and create gene knockouts in non-model organisms. In particular, we will develop protein-folding reporters to assess and enhance heterologous enzyme stability in non-model hosts. Combinatorial libraries and metabolic flux analysis will be used to balance enzyme expression. CRISPR-based tools will be created for gene editing and chromosomal integration of pathways. Our lab will seek to not only modify organisms that have already been cultured, but to discover new hosts from environmental samples with unique characteristics to enhance biochemical production. Through sequencing and annotation of pools of organisms from specific environments, we hope to find novel enzymes for both production and remediation of compounds of interest. We aim to study the mechanisms that give rise to the unique traits in these organisms to further improve them for specific processes or transfer them to other species of interest. Through this work, we hope to create a set of design rules for utilizing microbes for chemical production as well as put together a pipeline of methods that is needed to systematically discover, characterize and engineer an organism of interest.

Teaching Interests:

I hope to use teaching of courses such as kinetics, separations, control theory and bioprocess engineering as well as outreach with high school students and teachers to share and further develop our study of biochemical engineering of microorganisms. Additionally, I plan to develop new courses focused around hand-on design of biological systems and implement team-based projects in classes I teach.

476135 Systems Biology and Systems Pharmacology Approaches to HIV Infection and TB/HIV Co-Infection

Abstract:

Systems biology and systems pharmacology approaches to HIV infection and TB/HIV co-infection

Elsje Pienaar, University of Michigan

Research Investigator

Research Interests:

I aim to establish a research group that uses computational approaches to study within-host dynamics and treatment of infectious diseases, such as tuberculosis (TB) and human immunodeficiency virus (HIV). My work will leverage my established collaborations with experts in animal models of HIV and TB-HIV co-infection. I will build multi-scale multi-compartment computational models of HIV infection to determine 1) the roles of cytokine and CD4+/CD8+ T cell imbalances, in the context of different CD4+ T cell levels, in driving immune-deficiency in HIV+ patients; 2) pharmacokinetics and viral dynamics in different physiological compartments which influence HIV treatment efficacy and the persistence of viral reservoirs, and 3) precision treatment options based on patient-specific data.

Ultimately, I will integrate this model with a host-scale model of tuberculosis (TB) infection and treatment to 1) study the natural progression of this globally devastating co-infection, and 2) evaluate different coordinated treatment strategies for TB and HIV.

Through my experience in infectious diseases (experimental and computational), systems biology, systems pharmacology, and productive collaborative experience with experimentalists, I am well positioned to achieve these goals.

Research Proposal Experience:

I have worked with current and past mentors to prepare successful NIH research proposals. I have submitted proposals for independent funding from the Simons Foundation. I am awaiting a decision on an NIH training grant (K01) and I am preparing an application for a PhRMA foundation research starter grant.

Teaching Interests:

My educational experiences include classroom teaching (guest lectures in chemical engineering and mathematical biology), student mentoring (graduate and undergraduate students in chemical engineering and microbiology) and formal training in pedagogy (Certificate in learning, instructing and knowledge (Linköping University, 2012)). I look forward to teaching any core Chemical Engineering courses, as well as developing specialty graduate courses in systems biology/systems pharmacology.

Academic Appointments

2015 – Present. **Research investigator** in Chemical Engineering, Microbiology and Immunology, University of Michigan, Ann Arbor, MI.

Grants

2011 Knut and Alice Wallenberg Travel Grant

2012 Microbiology and Immunology Postdoctoral Travel Grant

Honors and Awards

2014 BMES Innovation and Career Development Award

2010 Outstanding Doctoral Dissertation Award, UNL College of Engineering

2009 NIH-NIAID Scholarship Keystone Symposia

Education and Training

2012 – 2015 **Research fellow**. Chemical Engineering, Microbiology and Immunology, University of Michigan, Ann Arbor, MI. *Advisors: Dr. Jennifer Linderman and Dr. Denise Kirschner.*

2010 – 2012 **Postdoctoral fellow**. Medical Microbiology, Linköping University, Linköping, Sweden. *Advisor: Dr. Maria Lerm.*

2007 – 2010 **PhD**, Chemical and Biomolecular Engineering, Biological Sciences University of Nebraska-Lincoln, Lincoln, NE. *Advisor: Dr. Hendrik Viljoen.*

2005 – 2007 **MSc**, Chemical and Biomolecular Engineering, Biological Sciences University of Nebraska-Lincoln, Lincoln, NE. *Advisor: Dr. Hendrik Viljoen.*

2002 – 2004 **BSc**, Financial and Actuarial Mathematics, University of Pretoria, Pretoria, South Africa

Selected Publications

Pienaar E, Matern W, Linderman JJ, Bader JS, Kirschner DE. Multi-scale model of *Mycobacterium tuberculosis* infection maps metabolite and gene perturbations to granuloma sterilization predictions.

Infection and Immunity. 2016:84(5):1650.

Pienaar E, Dartois V, Linderman JJ, Kirschner DE. *In silico* evaluation and exploration of antibiotic tuberculosis treatment regimens. BMC Systems Biology. 2015;9:79.

Linderman JJ, Cilfone N, **Pienaar E**, Gong C, and Kirschner DE. A Multi-scale Approach to Designing Therapeutics for Tuberculosis. Integrative biology. 2015; 7:591-609.

Pienaar E, Cilfone NA, Lin PL, Dartois V, Mattila JT, Butler JR, Flynn JL, Kirschner DE, Linderman JJ. A computational tool integrating host immunity with antibiotic dynamics to study tuberculosis treatment. J Theor Biol. 2015;367:166-179.

Raffetseder J*, **Pienaar E***, Blomgran R, Eklund D, Patcha Brodin V, Andersson H, Stendahl O, Welin A, Lerm M. Replication Rates of Mycobacterium tuberculosis in Human Macrophages Do Not Correlate with Mycobacterial Antibiotic Susceptibility. PLoS One. 2014;9(11):e112426. *: co-first authors.

Pienaar E, Lerm M. A Mathematical Model of the Initial Interaction between *Mycobacterium tuberculosis* and Macrophages. J Theor Biol. 2014; 342:23-32.

Viljoen S, **Pienaar E**, Viljoen HJ. A state-time epidemiology model of tuberculosis: Importance of re-infection. Computational Biology and Chemistry. 2012;36(0):15-22.

Pienaar E, Fluitt AM, Whitney SE, Freifeld AG, Viljoen HJ. A model of tuberculosis transmission and intervention strategies in an urban residential area. Comput Biol Chem. 2010;34(2):86-96.

Pienaar E, Whitney SE, Viljoen HJ, van Rensburg NF. A model of the complex response of Staphylococcus aureus to methicillin. J Theor Biol. Apr 2009;257(3):438-445.

Pienaar E, Viljoen HJ. The tri-frame model. J Theor Biol. Apr 2008;251(4):616-627.

Pienaar E, Theron M, Nelson M, Viljoen HJ. A quantitative model of error accumulation during PCR amplification. Comput Biol Chem. Apr 2006;30(2):102-111.

Publications in preparation

Pienaar E, Sarathy J, Prideaux B, Dartois V, Kirschner DE and Linderman JJ. Systems Pharmacology Predicts Antibiotic Spatial Distribution And Efficacy In TB Granulomas.

Pienaar E, Kirschner DE and Linderman JJ. Mechanisms driving drug resistance emergence in TB granulomas.

476348 Engineering and Physical Sciences in Oncology: Tumor Cell Adhesion and Treatment in Blood and Bone Marrow

Abstract:

Research Interests: Â

Metastasis through the bloodstream contributes to poor prognosis in many types of cancer. Mounting evidence implicates physical interactions between tumor cells and the local microenvironment as facilitating this process, in a manner similar to leukocyte trafficking during inflammation. In the first part of my talk, I will discuss our work in developing â€œunnatural killer cellsâ€ as an immunotherapy-inspired approach to kill metastatic tumor cells in the bloodstream. Through the use of lipid-based nanomaterials administered intravenously, we were able to coat the surface of circulating white blood cells *in vivo* with the immune cytokine TRAIL and the adhesion receptor E-selectin. This approach is inspired by the cytotoxic activity of natural killer cells, and is effective at killing tumor cells in human blood samples *in vitro* and in the peripheral circulation of mice *in vivo*. This nanoparticle platform holds promise as a means to neutralize circulating tumor cells in the blood and potentially block the progression of metastasis. In the second part of my talk, I will discuss the development of new strategies to deliver RNA therapeutics to the bone marrow microenvironment from the treatment of multiple myeloma, an incurable blood cancer the colonizes within the marrow. Specifically, we have identified novel polymer-lipid hybrid platforms that, for the first time, enable potent gene delivery to the bone marrow microenvironment *in vivo*. These delivery materials were shown to silence key factors within bone marrow that trigger the homing, proliferation, and survival of multiple myeloma *in vivo*. These platforms are now currently being investigated for the prevention of breast and prostate cancer metastasis to bone, the treatment of multiple myeloma, and to increase the therapeutic efficacy of bone marrow stem cell transplantations.

One of the greatest opportunities as a professor is to serve as a mentor, in many different forms, to the next generation of scientists and engineers. Serving as mentor can occur in a variety of settings, ranging from the classroom, the lab, as

well as the next phase of their own independent career. Throughout my academic, I have been committed and actively sought out opportunities to teach future scientists and engineers in various roles including: teaching assistant, guest lecturer, research mentor, and visiting scientist in outreach activities sponsored by federal agencies including the NSF and NIH. I have also had the privilege of mentoring more
than 10 trainees in the laboratory, which has shaped myself both as a research and educator, and has been one of the most rewarding aspects of my academic career to date.

Teaching Interests:

I am interested and qualified in teaching both Chemical Engineering and Bioengineering courses (undergraduate and graduate level) in the areas of biomaterials, reaction kinetics, fluid dynamics, and thermodynamics. Additionally, I am excited to create a graduate course in the emerging fields of Molecular Engineering, Drug Delivery,
and Nanomedicine.

476598 Novel Bio-Ionic Liquid Functionalized Conductive Hydrogel

Abstract:

Research Interests: Material and Biomaterial synthesis, Intelligent Design, Catalysis, Medicine

Teaching Interests: Transport Phenomena, Thermodynamics, Reaction Engineering, Polymer, and Biomaterial

Conducting polymers (CPs) constitute a new generation of organic materials, which exhibit electrical and optical properties that are similar to those of metals and inorganic semiconductors. They also possess the advantageous properties associated with regular polymers, such as ease of synthesis and processability, when compared to metals [1]. CPs are widely used in microelectronics for the manufacture of batteries, photovoltaic devices, light-emitting diodes and electro-chromic displays, as well as in biomedical applications including neural probes and prostheses, and controlled release systems. CPs have also been shown to modulate cellular processes such as cell growth and migration through electrical stimulation (conductivities from 10-4 to 9 S/cm). These properties make CPs and their derivatives remarkably suitable for tissue engineering applications using excitable cell types, including nerve, muscle, and cardiac tissue. However, there are still several technical challenges associated with the use of CPs for tissue engineering applications. The main drawbacks associated with the existing, CP-based biomaterials are: poor polymer-cell interactions due to the absence of cell binding sites, high hydrophobicity, poor solubility and processability, non-biodegradability, as well as the inability to optimize their mechanical properties. Therefore, the synthesis of biomaterials that are both electroactive and biodegradable is highly desirable. Electroconductive properties have been incorporated into biomaterials through the addition of nanomaterials (e.g. silver nanowires, gold nanoparticles, carbon nanotubes (CNTs), and graphene oxide) or conductive polymers (e.g. polyaniline, polypyrole, polythiophene) to the polymeric matrices. However, these approaches suffer from difficult and prolonged processing procedures, harsh reaction conditions, difficulties in the incorporation of nanomaterials into the polymer networks, limited solubility in aqueous media, and cytotoxicity. Therefore, there is an unmet need for the development of biocompatible, biodegradable, and conductive biomaterials, with tunable conductive and physical properties [1].

Here, I introduce a new class of conductive polymer-based biomaterials through conductive bio-ionic liquid (BIL) functionalization of different biopolymers. BIL are low melting organic salts that exhibit several technical advantages, including low volatility, high ionic conductivity and electrochemical stability, and excellent dissolution capabilities. BILs have been previously used as biocompatible and biodegradable materials for various applications such as cancer therapy, multi-responsive drug delivery systems, sensors, batteries, and biomedical implants [2-5]. Functionalization of biopolymers with BIL can provide conductive properties to the polymer networks, while preserving the biological and physical characteristics of the biopolymers. This property makes them unique candidates for various tissue-engineering applications. I performed chemical conjugation of BIL to various synthetic and natural polymers including gelatin, tropoelastin, elastin like polypeptides (ELP), collagen, poly(ethylene glycol) (PEG), hyaluronic acid (HA), poly(glycerol sebacate) polymer (PGS) and combinations of them, in order to generate conductive biomaterials with high biocompatibility and tunable physical and electrical properties. For example, I engineered a novel 3D gelatin-based hydrogel by using a visible light-activated gelatin prepolymer, functionalized with a BIL for heart tissue regeneration. FTIR and NMR analysis confirmed the successful functionalization of BIL to the biopolymer structure. The mechanical and electrical properties of the engineered hydrogel were then optimized to mimic the properties of the native heart tissue. Our *in vitro*tests confirmed that the engineered conductive scaffolds provide a suitable conductive and mechanical substratum for the culture of cardiac cells. This led to the formation of 3D functional cardiac patches, which can potentially be used for heart tissue regeneration.

References:

[1]- Kaur G , Adhikari R, Cass P , Bown M, Gunatillake P. RSC Adv., 2015, 5, 37553-37567

[2] Brock M, Nickel A-C, Madziar B, Blusztajn JK, & Berse B (2007) Brain Research 1145(0):1-10.

[3] Kumar SSD, Surianarayanan M, Vijayaraghavan R, Mandal AB, & MacFarlane DR (2014) European Journal of Pharmaceutical Sciences 51(0):34-44.

[4] Curto VF, et al. (2014). Physical Chemistry Chemical Physics 16(5):1841-1849.

[5] Dias AMA, et al. (2013) ACS Sustainable Chemistry & Engineering 1(11):1480-1492.

478107 Engineering Surfaces through Sequential Stop-Flow Photopatterning

Abstract:

Poster Abstract

This poster describes the design and application of a novel experimental setup to combine light-mediated and flow chemistry for the fabrication of hierarchical surface-grafted polymer brushes. Using light-mediated, surface initiated controlled radical polymerization and post-functionalization via well-established, and highly efficient chemistries, polymer brush films of previously unimaginable complexity are now shown to be accessible. This methodology allows full flexibility to exchange both lithographic photomasks and chemical environments insitu, readily affording multidimensional thin film architectures, all from uniformly functionalized substrates.

Research Interests:

Many diseases of the human nervous system, e.g., Parkinson's, Alzheimer's, and Huntington's, occur as a result of the progressive loss of structure and function of neurons. Charged proteins – naturally occurring polyelectrolytes and polyampholytes – have been identified as key components to the development and evolution of such neurodegenerative processes. My research interests focus on synthetically mimicking and studying the complex electrostatic interactions inherent to such ampholytic polymer architectures. Beyond these fundamental studies, charged macromolecules and their predicted response to electric fields provide fascinating routes towards advanced materials with tunable and switchable properties.

Teaching Interests:

Education is a continuous and essential component of the academic research environment. I value the opportunity to teach and learn from my students as one of the most important and exciting responsibilities in academia. I want to not only transfer knowledge, but also ignite a passion, a strong internal desire to learn and improve. I wish to emphasize critical thinking & scientific questioning, while conveying the joy of discovery, and helping students develop both professionally & personally.

477586 Biological Applications of Fluctuation Solution Theory

Abstract:

Successful Proposals:

NIH F32 â— NSF GRFP â— NSF GK-12 â— NSF REU â— NSF XSEDE/TeraGrid â— Maryland Advanced Research Computing Center â— Assisted in developing/revising rejected and successful NIH R01 proposals

Research Interests:

Thermodynamics and statistical mechanics â— Solution microstructure â— Molecular Dynamics simulations, including methodology development and force field validation/development â— Biothermodynamics/Biophysics, especially proteins and lipids â— Impacts of microenvironment changes and surfaces on conformational and association equilibria of biomolecules â— Lipid organization in multicomponent, physiologically relevant, lipid bilayers â— Kirkwood-Buff Theory / Fluctuation Solution Theory

Teaching Interests:

Thermodynamics and statistical mechanics â— Computational chemistry (classical) â— Physical chemistry â— General chemistry â— Biochemistry/Biophysics â— Proteins â— Linux â— Fortran/MATLAB/basic scripting

481181 Understanding and Improving Biomanufacturing in Chinese Hamster Ovary Cells through New Gene Expression and Systemic Cell Engineering Platforms

Abstract:

Research Interests:

My research experience has focused on the various subjects of Chinese hamster ovary (CHO) cells, such as -omics studies (genomics, transcriptomics, and proteomics), cytogenetics, genome and metabolic engineering, and bioprocess engineering. CHO cells are the most important mammalian host cells for the biomanufacturing of therapeutic proteins, accounting for more than 50% of global annual biopharmaceutical protein sales. In addition, transformative changes in CHO cell line development are required to meet industrial demands, such as lowering manufacturing cost, producing new medicines, and extending culture duration. I plan to continue this work towards improving CHO biomanufacturing platforms.

While significant productivity improvements have been achieved by extensive efforts over three decades, understanding of CHO biology has been limited. For example, what makes a high productivity cell line, how media components/supplements affect culture longevity and productivity, and how stressful culture conditions (low culture temperature, hyperosmolality, or histone deacetylase inhibitor treatment) can lead to higher recombinant protein production are still largely unknown. In addition, although cell line instability issues, such as unexpected changes in productivity, product quality, and host cell protein (HCP) expression, are increasingly recognized to negatively affect biomanufacturing processes, the causes and underlying mechanisms of cell line instability remain poorly understood. Addressing these important questions and challenges will enable us to develop stable, well characterized CHO cell lines producing proteins and biomolecules with desirable quality attributes, thereby providing more effective, better controlled, and safer therapeutics for human health. Therefore, my research interests include (1) developing new gene expression and amplification methods for the stable and high production of therapeutic proteins and (2) mammalian genome and cell engineering based on genome scale modeling and systems biology.

Postdoctoral Projects:

"Knocking out CHO host cell proteins using CRISPR/Cas9 to improve downstream processes" and "Developing a karyotype-based framework to identify and quantify CHO cells' genomic/chromosomal instability that affects protein productivity" under the supervision of Dr. Kelvin Lee at the University of Delaware.
"Metabolic and bioprocess engineering of CHO cells to produce bioengineered heparin" under the supervision of Dr. Susan Sharfstein at SUNY Polytechnic Institute and Dr. Robert Linhardt at Rensselaer Polytechnic Institute.

Ph.D. Dissertation:

"Engineering of recombinant Chinese hamster ovary cells using a proteomic approach" under the supervision of Dr. Gyun Min Lee at Korea Advanced Institute of Science and Technology (KAIST).

Education:

Ph.D., Biological Sciences, KAIST, 2009
B.S., Biological Sciences, KAIST, 2003

Selected Publications:

1. Baik JY, Lee KH. 2016. Toward improved host cell protein impurity assessment. Biotechnol. J. 11(8):998-999.
2. Baik JY, Dahodwala H, Oduah E, Talman L, Gemmill TR, Gasimli L, Datta P, Yang B, Li G, Zhang F, Li L, Linhardt RJ, Campbell AM, Gorfien SF, Sharfstein ST. 2015. Optimization of bioprocess conditions improves production of a CHO cell-derived, bioengineered heparin. Biotechnol. J. 10(7):1067-1081.
3. Baik JY, Lee KH. 2014. Toward product attribute control: Developments from genome sequencing. Curr. Opin. Biotechnol. 30:40-44.
4. Baik JY, Gasimli L, Yang B, Datta P, Zhang F, Glass CA, Esko JD, Linhardt RJ, Sharfstein ST. 2012. Metabolic engineering of Chinese hamster ovary cells: Towards a bioengineered heparin. Metab. Eng. 14(2):81-90.
5. Baik JY, Ha TK, Kim YH, Lee GM. 2011. Proteomic understanding of intracellular responses of recombinant chinese hamster ovary cells adapted to grow in serum-free suspension culture. Biotechnol. Prog. 27(6):1680-1688.

Teaching Interests:

Learning with understanding: Learning is more than just memorizing and practicing; it is about understanding. Through all my teaching and mentoring experience, I challenged my students' critical thinking skills by asking questions, such as what each procedure meant, why it was necessary, and which steps were more critical than others, and by asking them to explain their thought process. Through classes, I will continually emphasize learning with understanding by asking "how can we understand this concept?".

Applicability: As an engineer, I seek to link knowledge to applications. I will provide a platform for students to translate knowledge gleaned from lectures to problems or applications in various engineering fields.
Communication: As collaboration and interdisciplinary research become increasingly important, good communication skills are essential to successful research. Moreover, strong written and verbal skills are critical to convey the importance and impact of research. To improve communication skills, I will foster discussion-oriented class environments, which will also complement my learning with understanding philosophy. In addition to scientific communication skills, I will also pursue outreach activities to connect research to real life in various ways, such as hands-on activities, demonstrations, or engaging tours of

labs/facilities. In conclusion, I believe that when we are able to understand, apply, and propagate what we learn, knowledge becomes alive and inspires future scientists and engineers.

I would apply my teaching principles to undergraduate and graduate courses, including Molecular Biology, Biochemistry, Genetics, Bioprocess Engineering, and Biotechnology. I am also interested in developing advanced cutting-edge courses such as Genomics, Genome Engineering, and Recent Progress in Biochemical Engineering and Technology.

475622 Uranium-Selective Polymer Materials for Water Quality Monitoring and Isotopic Identification

Abstract:

Research Interests: Actinide separations, membrane separations, nuclear fuel reprocessing, nuclear waste management

The manipulation of special nuclear material (i.e. uranium or plutonium) for the express purpose of assembling a weapon of mass destruction will likely result in the contamination of the environment. The ability to monitor environmental waters for gross-activity and perform isotopic analysis would help determine if there is an elevated concentration of actinides present in the environment or an abnormal isotopic distribution; both of which may indicate illicit nuclear activities nearby. Currently, there is no fieldable technique capable of conducting these analyses *directly*from environmental waters. In the past 4 years, I have been involved with 2 projects funded by the Defense Threat Reduction Agency seeking to fill these needs. The goal of these projects has been to develop fieldable techniques to analyze environmental waters for uranium by focusing on near-neutral pH separations that minimize consumables.

In the first project, we synthesized extractive scintillating resins to monitor the activity of radionuclides in groundwater in a continuous, online system. Extractive scintillating resin serves two purposes: (1) selectively concentrate the desired radionuclide and (2) serve as a radiation transducer by emitting light in the presence of a radioactive decay. Through this research, I gained experience in the following technical areas: (1) synthesis and functionalization of polymer resins; (2) characterization of ion-exchange resins; (3) theory and practice of scintillating counting for both liquid and solid-phase scintillators and (4) proper techniques and safety for handling radioactive materials. Specifically, we studied how the physical properties of the resin (pore structure, ligand type, resin diameter, degree of crosslinking and composition) affect the capacity, selectivity and detection efficiency of the material.

In the second project, we developed a rapid sample preparation method for alpha spectroscopy using ultrafiltration membranes. Alpha spectroscopy is an analytical technique that can distinguish isotopes of alpha-emitting radionuclides by the characteristic energy of the alpha particle. Through this research, I gained experience in the following technical areas: (1) Bulk polymerization and purification; (2) Membrane modification by UV-polymerization and other chemistries; (3) Characterization of membranes for molecular weight cut-off, permeability, surface morphology and (4) theory and practice of alpha spectroscopy. Here, we studied how the membrane properties (pore structure, ligand type, method of grafting, degree of grafting, composition and membrane material) affect the capacity, selectivity, detection efficiency and peak resolution in the alpha spectrum.

Proposed future research will apply my expertise in uranium selective-polymeric materials to the field of nuclear fuel reprocessing. Nuclear fuel reprocessing uses the liquid-liquid extraction process called PUREX to separation uranium and plutonium from other fission products. There are many opportunities for membrane technologies downstream from the initial extraction stage. I plan to focus my research around the implementation of polymer inclusion membranes, supported liquid membranes and membrane contactors in the PUREX process. Specially designed membrane technologies can provide low-energy separations that are capable of selectively removing trace-levels of actinides or fission products. Challenges of this research will be choosing ligands with moderate binding affinities to facilitate the "catch and release" of radioisotopes on either side of the membarne as well as desinging membranes with sufficient radiolytic stability for a given waste stream. The results of this research will have implications in the field of membrane science, nuclear fuel reprocessing as well as nuclear waste processing and storage.

Teaching Interests: Introduction to Chemical Engineering (mass and energy balances), transport phenomena, membrane separations, nuclear engineering

Teaching experience: Co-instructor for *Introduction to Chemical Engineering* (2 semesters), Instructor for *Membrane Separations for Clean Water and Energy* (3 semesters), Undergraduate research mentor (4 years, 3 students)

476107 Particle/Process Engineering for Active Novel Formulation and Sustainable Manufacturing

Abstract:

Research Interests:

As breakthroughs in the chemical arena become fewer and farther between, researchers are focusing increasingly on ways to improve existing products and processes. At the heart of this development are the pharmaceutical industries, the biggest markets for human health care in the U.S. It is also an important sector that has contributed toward a 30 year increase in average life span in the USA during the last century, and remains a main line of defense against new diseases. Product development and manufacturing address poor bioavailability which is a major issue for most new drug molecules discovered. Nanotechnology is expanding very fast to handle the bioavailability issues in pharmaceutical industries. The current global market for nanoparticles in biotechnology and pharmaceuticals is at $20B, and is expected to grow to over $50B in five years.

Particle science and technology deals with the production, functionalization, characterization, handling, and utilization of a wide variety of particles in either micro or nano scale. It has a direct impact on a number of industries such as pharmaceutical, advanced materials, environmental, chemical, mineral, agricultural, and food processing. The emerging paradigm in this interdisciplinary area is the recognition of the intrinsic length/time scales at the molecular level, particle-level, powder-level, equipment-level, and process-level.

My research work involved various aspects of particle/process engineering on the proficient in design, development,

and novel formulation for making nano/micro actives applicable to pharmaceutical, cosmetics, agrochemical, and food processing industries. My PhD and postdoctoral research work covers: (1) developing an innovative miniaturized end-to-end on demand pharmaceutical drug manufacturing module; (2) dissolution enhancements of crystalline activeâ€™s via modification into amorphous using mesoporous carriers; (3) stability improvement of drug nano/micro suspensions using dispersants; (4) developing fast dissolving core-shell type nanocomposite powders using fine carriers (<50 microns); (5) developing surfactant free drug nanosuspensions and nanocomposites using swellable cross-linked polymers; (6) developing high drug loaded biocompatible film.

The best practice to solve a problem scientifically is: understanding through basic research, creating innovative solutions, and bringing concept to product through applied research. I have well versed knowledge to apply all three aspects of science to develop solutions for outstanding problems. My research interests center on the application of basic principles of chemical engineering to various particulate systems/processes (applicable for pharmaceutical, bio-pharma, agrochemicals, food industries etc.) via experimentation and modeling to systematic development of a fundamental understanding of mechanisms that lead to tunable sized and functionalized particle/soft materials formation, enhanced stabilization/dispersion, formulate to specific delivery/application, miniature equipment design, and economical manufacturing.

Teaching Interests: Chemical Engineering, Heat/Mass Transfer, Materials & Energy Balance, Nanotechnology, Particle Technology, Pharmaceutical Engineering, Process Engineering, Thermodynamics, Unit Operations.

475793 Reduced & Optimized Chemical Kinetic Mechanisms for Energy & Combustion Applications

Abstract:

Reduced & Optimized Chemical Kinetic Mechanisms for Energy & Combustion Applications

Gudiyella, S., Massachusetts Institute of Technology

2nd Year Postdoctoral Associate

Research Interests:

Solving the current energy/climate problem requires improvements in the energy infrastructure and design of combustion engines. The changing fuel landscape (e.g. natural gas, biomass and reserves of heavy fuel oil) creates the need to improve the feedstock pre-treatment processes, to generate fuels that can be used as 'drop-in' fuels in existing engines. On the other hand, increasingly stringent emission regulations require design changes in the current engines to meet future regulations. The experimental part of my research will focus on developing new chemical processes for producing fuels from heavy crudes and biomass. The modeling part of my research will focus on developing accurate chemical kinetic models for predicting emissions from combustion engines or yield in chemical process industries. The models are built through a combination of experimental work and new fit-for-purpose modeling approaches, for implementation into Computational Fluid Dynamic (CFD) simulations or process simulation software.

The experiments will focus on process improvement to increase the yield of lower molecular weight fuels from upgrading of heavy crudes & biomass feedstocks. An important part of experiments is fuel characterization using gas chromatography (GCxGC-MS) and Nuclear Magnetic Resonance (NMR) techniques. The fuel characterization data will be used develop chemical kinetic models for practical fuels (diesel, gasoline, jet fuel etc.) using the surrogate fuel approach.

The chemical kinetic models will be constructed based on a methodology that exists in literature. For instance, for simpler fuels like jet fuels, the 'comprehensive' approach is to build a surrogate fuel for the jet fuel and develop detailed chemical kinetic mechanisms for the surrogate fuel. However the size of the mechanism is prohibitive for implementation into CFD simulations or process simulation software without performing extreme reduction. Recent studies have used a 'hybrid' approach wherein the fuel decay to major species will be formulated using global steps and the fuel decay is optimized to match the test data. Based on the end-use of the mechanism, global/elementary steps will be included in the mechanism to explain the decay of the major species.

The 'hybrid' approach has an advantage of developing compact mechanisms, whereas the 'comprehensive' approach has an advantage of accurate chemical kinetic description of the fuel. I propose to utilize the best elements of both the 'comprehensive' and the 'hybrid' approach under one framework by building a software called as Global Mechanism Generator (GMG). The software consists of modules to formulate the surrogate fuel (comprehensive approach), to generate a global mechanism and to optimize the mechanism against test data (hybrid approach).

The output of GMG will be a reduced/optimized chemical kinetic models, which can be implemented in CFD or Aspen Plus, to predict emissions from combustion engines or yield in chemical process industries.

Postdoctoral Project: ***"Supercritical water upgrading of Arabian Heavy Crude Oil"***

Under supervision of Prof. William H. Green, Hyot C. Hottel Professor of Chemical Engineering, Massachusetts Institute of Technology (MIT)

PhD Dissertation: ***"An experimental and modeling study of the combustion of aromatic surrogate jet fuel components"***

Under supervision of Prof. Ken Brezinsky, Associate Dean for Research and Graduate Studies in the UIC College of Engineering, Department of Mechanical and Industrial Engineering, University of Illinois at Chicago (UIC)

Teaching Interests:

My teaching philosophy is based on four aspects. They are exposure to academic and industrial research, use of active learning techniques, highlighting the importance of team-work and encouraging the students to learn beyond the classroom.

Students have to be knowledgeable about how to translate the concepts from a class room setting to an industrial setting. Given this need, my courses will start with the fundamentals and move towards real-world applications. I will expose the students to both academic and industrial research, by discussing relevant research publications and patents. I will also educate the students about the tools which can bridge the gap between academic and industrial research.

My mentoring experience in lab & in professional women's networks has made me realize that the overall development of a student continues beyond the walls of the classroom.

Based on the course being taught, I will also introduce the students to relevant research within the university by organizing an educational trip to the labs or nearby chemical process industries. I will also encourage my students to pursue research-oriented higher degrees, obtain internships and participate in conferences. By participating in conferences, the students will gain a better perspective of the problem-solving approaches adopted by accomplished researchers and also build professional connections.

In summary, I will provide students the required knowledge by designing courses that begin with fundamentals and move towards the 'applied' direction.

Research Experience:

As a researcher I am trained in both high pressure, high temperature experimentation and chemical kinetic modeling. My PhD thesis on combustion of jet fuel, enhanced the understanding of soot formation in civil and military aircraft engines, by identifying the species and pathways responsible for the formation of soot. The combustion of jet fuel components was investigated in a High Pressure Single Pulse Shock Tube (HPST). I built detailed chemical kinetic mechanisms for combustion of jet fuel and validated the mechanisms against shock tube experimental data.

After my PhD, I joined General Electric (GE) as a Research Engineer. At GE Global Research, my major focus was developing reduced chemical kinetic mechanisms for fuels such as heavy fuel oil, natural gas and jet fuel. The reduced mechanisms were implemented into CFD to predict the temperature field in the combustor, and emissions and thereby propose combustor design modifications to lower the emissions.

After working for 3 years at GE, I joined the Green Group at MIT as a Postdoctoral Associate. At MIT, my primary project is crude oil upgrading with supercritical water to reduce sulfur impurities and to increase the yield of lower molecular weight fuels. During my postdoc, I expanded my analytical experience by using gas chromatography (GCxGC) and Nuclear Magnetic Resonance (NMR) Spectroscopy for analyzing complex crude oil mixtures.

Teaching Experience:

I worked as a Teaching Assistant for the Chemical Engineering Laboratory I (CHE 381) and Chemical Engineering Laboratory II (CHE 382) courses at UIC. I taught the undergraduate students how to perform experiments on the distillation column, evaporator and solvent extraction setups etc. I also ensured that the students followed the laboratory health & safety protocols, while handling the chemicals and while working on the experimental setups. During my PhD, I mentored two undergraduate students. During my postdoc, I mentored 3 undergraduate students on the crude oil upgrading project and also work with 3 graduate students on diverse projects. I also completed the Kaufman Teachning Certificate Program in June 2016. The workshop has enabled me to design a course, plan and facilitate a class session.

I also explored external mentoring opportunities. I mentored an undergraduate student during my PhD as a part of UIC Women in Science in Engineering (WISE) Mentoring Program. I was also a mentee in 2016 American Women in Science (AWIS) Mentoring Circles.

Publications:

S. Mosbach, J. Hyeong Hong, G. P. E. Brownbridge, M. Kraft, **S. Gudiyella** and K. Brezinsky, "***Bayesian Error Propagation of a Kinetic Model of n-Propylbenzene Oxidation in a Shock Tube***", International Journal of Chemical Kinetics, 46 (2014) 389-404

T. Malewicki, **S. Gudiyella** and K. Brezinsky, "***Experimental and Modeling Study on the Oxidation of Jet-A and the n-Dodecane/Iso-octane/n-Propylbenzene/1,3,5-Trimethylbenzene Surrogate Fuel***", Combustion and Flame 160 (2013) 17-30

S. Gudiyella and K. Brezinsky, "***High Pressure Study of 1,3,5-Trimethylbenzene Oxidation***", Combustion and Flame, 159 (2012) 3264-3285

S. Gudiyella and K. Brezinsky, "***High Pressure Study of n-Propylbenzene Oxidation***", Combustion and Flame, 159 (2012) 940-958

S. Gudiyella, T. Malewicki, A. Comandini and K. Brezinsky, "***High Pressure Study of m-Xylene Oxidation***", Combustion and Flame, 158 (2011) 687-704

S. Gudiyella and K. Brezinsky, "***High Pressure Study of n-Propylbenzene Pyrolysis***", Proceedings of the Combustion Institute, 34 (2013) 1767-1774

476273 Experimental and Numerical Investigations of Particle Flows

Abstract:

Research Interests:

Operations involving solid particles are commonplace in a wide range of industries, such as granulation, heat and mass transfer, fluidized beds, tableting, etc. However, in general, the behavior of particles in these systems is still not well understood. The lack of fundamental understanding leads to poor system performance, ineffective operation design and, in some cases process failure. For instance, particles have a tendency to segregate by size when in motion which can lead to inconsistent product quality (quality-control issues) when particles discharge from hoppers. My research interests include improving the understanding of particle-particle and particle-fluid interactions with the goal of more accurate predictive capabilities, resulting in more efficient process design. Additionally, many industrial operations would benefit from particles with specific properties, such as particle size, cohesion level and/or shape. Thus, I am also interested in improving current methods as well as developing new techniques to control particle properties.

Research Experience:Â

As a postdoctoral research assistant in the Chemical and Biological Engineering Department at the University of Colorado, I am the lead experimentalist in Prof. Christine Hrenyaâ€™s research group. I have worked on projects related to particle cohesion and heat transfer. The overarching objective of these various projects is to develop an improved, fundamental understanding of bulk particle behavior by investigating micro-scale (individual) particle properties and interactions. To achieve this aim, the effects of individual particle properties on bulk particle behavior are studied through the development of experiments and cost-effective characterization techniques. For example, in one project, the effects of cohesion between particles in a fluidized bed are measured and understood via the effects of surface roughness on the cohesion level. Additionally, I have mentored and supervised undergraduates to carry out accurate measurements that can be used for model validation and in applying for grants to pay for their research stipends, as well as gained experience writing proposals and collaborating with industry.

Prior to my postdoc, I received my PhD from the Chemical and Biological Engineering Department of University of Florida where I was advised by Prof. Jennifer Curtis. During my PhD, physical models for dense phase gas-solid multiphase flows using computational fluid dynamics were developed and tested. Models were verified by simulated crater, or scour hole, formation in a particle bed by a subsonic jet of gas. The experimental results from a NASA internship and NASA lunar gravity flight were also used to further develop and verify the physical model. Experiments performed at the University of Florida to investigate particle property effects on cratering allowed for development of a fundamental understanding of the effects of particle shape, benchmark data for computational fluid dynamics simulations and improved crater scaling relationships.

Teaching Interests:

Not only does my background in fluid-particle flows make me a strong candidate for teaching standard fluids based chemical engineering courses, but also allows me to incorporate particle technology into the undergraduate curriculum. Training graduate and undergraduate students in this industrially applicable field will help to prepare a relevant workforce. Additionally, my experience in large-scale, experimental systems provides me with the necessary skills to teach students in unit-operations laboratory and process-control courses.

476143 Multiphase and Multiscale Chemical Reaction Engineering for Clean and Sustainable Production of Fuels, Materials, and Chemicals

Abstract:

Multiphase and Multiscale Chemical Reaction Engineering for Clean and Sustainable Production of Fuels, Materials, and Chemicals.

Staff Scientist at Ames Lab-US DOE

Research Interests: My research interests reside in the area of chemical reaction engineering, the branch of chemical engineering which deals with the design and optimization of chemical reactors and process, with a keen focus on solving the problems with multiphase and multiscale characters. My current research mainly aims to develop advanced computation models for multiphase flow and multiphase turbulence. I firmly believe that a judicious balance of theory, experiments and computation can yield critical insight into multiphase flow physics that enables the development of physics-based models that can be predictive and robust to be used to simulate the performance of real industrial reactors, which in turn can lead to more efficient chemical reactor design. Looking forward, I am expecting to apply my research experience and talent to advance both fundamental research and applications of chemical reaction engineering using both experimental and computational tools, with the ultimate goal of designing better chemical reactors and achieving cleaner and more sustainable production of fuels, materials, and chemicals, thus generating broad impact to industries and society. **Research Projects:** *1. Polydisperse Gas-Particle Flows Modeling Using Quadrature-Based Moments Method, sponsored by DOE-NETL, PI: Rodney Fox, ISU & Ames Lab 2. Kinetic Theory Modeling of Turbulent Multiphase Flow, sponsored by DOE-NETL, PI: Rodney Fox, ISU & Ames Lab 3.*

Development of a Computational Fluid Dynamic Model for Large-scale Three-phase Stirred-Tank Reactor, sponsored by SABIC, PI: Rodney Fox, ISU 4. A Computational Fluid Dynamic Model for Multiphase Loop Polymerization Reactor, sponsored by Chevron Phillips Chemical, PI: Rodney Fox, ISU 5. A Polydispersed Multiphase CFD Model for Veba-Combi-Cracking (VCC) Process, sponsored by BP America, PI: Rodney Fox, ISU 6. Model Validation for Photosynthetically Active Radiation Transport and Multiphase Flow in Algal Photobioreactors, by sponsored NSF, PI: Dennis Vigil and Micheal Olsen, ISU 7. A Novel Taylor-Couette Photobioreactor for Energy Efficient Micro-algae Cultivation, sponsored by Conoco-Phillips, PI: Dennis Vigil and Jacqueline Shanks, ISU

Successful Proposals: *National Science Foundation, CBET-1236676*

Ph.D. Dissertation: *Experimental and Computational Study of Turbulent Mixing in a Confined Rectangular Reactor, under supervision of Michael G. Olsen and James C. Hill, ISU*

Research Experience: All my research effort in last twelve years has been solely focused on the area of chemical reaction engineering, with a gradual shift from using experimental techniques to computational methods. During my Ph.D. years, I studied turbulent mixing and reacting flows in liquids using advanced laser-based fluid diagnostic techniques, such as PIV, PLIF, LDV. In liquid medias, Schmidt number is ofter very high, which means mixing and reactions take place in much smaller length scale than hydrodynamics. The multiscale nature makes it very hard to predict in simulations. Thus, I implemented various reaction models and performed large-eddy simulations to test and validate them against my experimental data. As a postdoc, I worked to develop more energy efficient photobioreactor for microalgae cultivation. Again, I not only designed and conducted various experimental research, but also developed many mathematical models for improving the design of bioreactors. I studied a broad range of issues involved in bioreactors, such as, the gas-liquid hydrodynamics and mass transfer, bubble population balance, photosynthetic active radiation transfer, algae growth kinetics under different light conditions, and coupling difference processes on a large range of time scale together to predict overall reactor performance. Moreover, in the last four years, I have had the privilege of closely working with Prof. Rodney Fox, a world-renowned pioneer in the area of developing and utilizing the computational tools in chemical reaction engineering. Together, we have developed many advanced models for gas-solid and gas-liquid reactor systems, such as polydispersed multiphase solvers based on quadrature-based moment method (QBMM), and new multiphase turbulence models. I implemented many of those models in open-sourced software packages and performed many large-scale simulations on high-performance clusters to study complicated multiphase problems and validate our models. What is more, I also had the opportunities to work on several industrial projects sponsored by major petrochemical companies, BP, SABIC, and Chevron-Phillips, which helped me accumulate some good experiences in working with industries. In summary, I have a solid background in fundamental research in multiphase flow modeling, code development, high-performance simulation, industrial applications, and experiment design and measurement techniques, which can allow me to work in close collaboration with people with a different background to solve some important and complicated issues related to chemical reaction engineering.

Teaching Interests:

Aside from my research career, I also have extensive teaching experience. I TAed and guest-lectured undergraduate lecture and laboratory courses at the Iowa State University, duties included lecture undergrads, holding office hours and developing course material. Lastly, I mentored many graduate students and postdocs. My non-traditional background enables me to teach a broad range of undergraduate as well as graduate courses. I am particularly interested in following courses, thermodynamics, transport phenomenon, numerical analysis, process engineering, reactor design, turbulent reacting flow, multiphase flow.

Future Direction: As faculty, I would like to continue my pursuit in the area of chemical reaction engineering, using both experimental and computational tools. In particular, there are three areas I am very interested in working on. First, model development. There are many knowledge gap and lacks of good models in the broad area of multiphase systems, such as bubble dynamics and gas-liquid mass transfer, multiphase turbulence, close-packed particle behavior et.al. I can identify the modeling need, design well-thought experiments, come up with better models using experiments or DNS, and eventually test them in my codes. Second, code development. With increasing computational capacity of modern computer technique, such as GPU, it has also been becoming feasible to perform large-scale simulations of industrial scale reactor systems, not only just small size lab/pilot-scale reactors. However, high-quality open-source computer code, which can utilize these new computer techniques, can handle all kinds of physical and chemical processes in a real reactor, and more importantly can integrate all the difference process on multi time and length scale together, is in great need right now. Third, bioreactor and biomedical applications. There are many revolutionary new technologies developed in the area of synthetic biology and biomedical areas. However, there are very few biologists have a solid background in chemical reaction engineering, such as transport phenomenon and the reactor/process design. I can foresee many opportunities for me to work with them to bring their research one step further to commercialization.

Selected Publications:

M. Ramezania, B. Kong, X. Gao, M. G. Olsen and R. D. Vigil, ``Experimental measurement of oxygen mass transfer and bubble size distribution in an air-water multiphase Taylor-Couette vortex bioreactor", Chemical Engineering Journal 270 (2015): 508-518.

X. Gao, B. Kong and R. D. Vigil, ``CFD investigation of bubble effects on Taylor-Couette flow patterns in the weakly turbulent vortex regime", Chemical Engineering Journal, vol. 270 pp. 508-518, 2015.

C. Yuan, B. Kong, A. Passalacqua, and R. O. Fox, ``An extended quadrature-based mass-velocity moment model for polydisperse bubbly flows", The Canadian Journal of Chemical Engineering, vol. 92, no. 12, pp. 2053-2066, 2014.

K. M. Nilsen, B. Kong, R. O. Fox, J. C. Hill, and M. G. Olsen, ``Effect of inlet conditions on the accuracy of large eddy simulations of a turbulent rectangular wake", Chemical Engineering Journal, vol. 250, pp. 175-189, Aug. 2014.

B. Kong and R. D. Vigil, ``Simulation of photosynthetically active radiation distribution in algal photobioreactors using a multidimensional spectral radiation model", Bioresource Technology, vol. 158, pp. 141-148, Apr. 2014.

B. Kong and R. D. Vigil, ``Light-limited continuous culture of Chlorella vulgaris in a Taylor vortex reactor", Environ. Prog. Sustainable Energy, vol. 32, no. 4, pp. 884-890, Dec. 2013.

B. Kong, J. V. Shanks, and R. D. Vigil, ``Enhanced algal growth rate in a Taylor vortex reactor", Biotechnology and Bioengineering. vol. 110, no. 8, pp. 2140–2149, Aug. 2013.

B. Kong, M. G. Olsen, R. O. Fox, and J. C. Hill, ``Predictive capability of Large Eddy Simulation for point-wise and spatial turbulence statistics in a confined rectangular jet", Chemical Engineering Science, vol. 69, no. 1, pp. 240-256, Feb. 2012.

B. Kong, M. G. Olsen, R. O. Fox, and J. C. Hill, ``Population, characteristics and kinematics of vortices in a confined rectangular jet with a co-flow", Exp Fluids, vol. 50, no. 6, pp. 1473-1493, Jun. 2011.

476316 Targeted Improvement of Biochemical Processes Via Process Systems Engineering Strategies

Abstract:

Research Interests:

Much of current research is driven toward finding alternative strategies to producing commercial products traditionally made from fossil fuels in an effort to develop a sustainable future. While recent studies have shown the promise of biochemical platforms for the production of fuels as well as commodity and specialty chemicals, such as plastics and pharmaceuticals, multiple challenges exist in the development of commercial-scale processes utilizing these biochemical production pathways. Upstream bioprocessing exhibits the variable and complex nature of biochemical reactions that make the application of traditional chemical engineering practices for continuous operation and reactor scale-up difficult. In addition, many of these technologies are in their infancy and lack the productivity to compete economically with traditional petroleum based processes. However, the focus in biotechnology has been more on improvement through the development of novel recombinant strains than on the development of novel commercial processing strategies that could provide answers to these challenges. Process systems engineering practices have been employed for commodity chemical processes to achieve their optimal performance with respect to economic and environmental objectives, and the application of a systems approach to bio-systems can provide solutions to their commercial implementation over traditional petroleum-based processing methods.

To this end, my research plan is to focus on the implementation a systems approach to develop novel processes in key areas of biochemical manufacturing, specifically sustainable energy, bio-plastics, and pharmaceuticals. My doctoral work at Texas A&M University, under the guidance of Dr. M. Nazmul Karim, has provided me with extensive experience developing novel biochemical processes for the production of fuels and chemicals. I will look to extend this knowledge to my group in an effort to afford students a multi-disciplinary environment capable of providing opportunities to develop both experimental and computational skills in the areas biochemical engineering and traditional chemical engineering. My research will facilitate the bench to commercial scale development of new bioprocesses to increase their reliability and economic viability. Work in renewable liquid transportation fuels will providing a sustainable platform to decrease harmful greenhouse gas emissions and reduce dependence on non-renewable liquid fuel sources. The development of new methods for the production of novel bio-plastics from lignocellulose and then further commercialization to large scale will be explored to provide renewable sources of plastics to consumers. In addition, my group will investigate novel production schemes for the fed-batch and continuous biological production of pharmaceuticals in an effort to utilize the cost-reducing practices of the commodity chemicals industry while guarding against risks such as contamination and product variability. The interdisciplinary nature of the group will also necessitate strong collaborations that will further enrich this training environment.

Teaching Interests:

I am comfortable teaching most general subjects within the chemical engineering curriculum for both undergraduate and graduate students. However, I would prefer to teach process economics, process design, process control, or process analysis courses based on my experience teaching these subjects as a Graduate Teaching Fellow of Texas A&M University.

Based on my own academic background and my experience in the Texas A&M chemical engineering department, I would like to design two new courses that can be offered to either the graduate students or senior undergraduate students: a process synthesis and optimization course and an biochemical processing course.

The first course would cover the various aspects of process synthesis, such as the fundamentals of nonlinear and mixed-integer optimization, the modeling paradigm consistent with the construction of process superstructures, and the use of process integration and intensification strategies to improve upon superstructures. In recent years, process systems engineering has become a special focus in many universities, therefore this topic will be a main focus of this course.

The second course would cover the basics of biochemical processing, both upstream and downstream, and serves to teach essential topics to developing alternative green processes. This course would cover the fundamentals of biochemical reactions (kinetic modeling and bioreactor design), bio-separations (chromatography, membrane extraction, liquid-liquid extraction, etc.), and their applications and challenges in the design of large scale bio-systems.

453440 Composite Materials: Mechanical and Tribological Property Improvement

Abstract:

Research Interests:

Solar cell panels exposed to dessert sands, harsh weather conditions, and contact cleaning method are subjected to scratching force. This factor can lead to loss of structural integrity, tremendously affect the performance of the panel system and greatly lower the energy conversion efficiency. Development of abrasion resistant coatings on solar cell panels can be achieved in composite systems containing both hard polymer and nano scale fillers. Halloysite nanotubes (HNT) encompass high aspect ratio, stiffness and strength while remain understudied about their coating usage. By appropriate treatment, HNT possess high dispersion quality and enable efficient interfacial adhesion to polymer after mixing. Using HNT in hardcoat polymers is supposed to increase strength, modulus and improve mechanical robustness (i.e., effective HNT modulus was estimated to reach ~300 GPa plus a half scratch depth of polymer without HNT). The reinforcement efficiency of HNT is quantitatively expressed using NanoIndentation, which showed up to 60% increase in modulus and hardness for composites with vertically aligned HNT. Anti-abrasion properties are characterized using abrasion test and followed by comparing morphological changes in a sample without undergoing the abrasive test. Consistently the composite coatings with out-of-plane particle orientations showed highest wear resistance. Scanning electron microscopy and atomic force microscopy have been used to verify the morphological and topographical merits after damage. This study aims at development of environmentally friendly and cost effective coating, which is highly potential to be commercialized.

Teaching Interests:

As a polymer material scientist and teacher, I believe it is my responsibility to challenge and support my students, in pursuit of three learning goals: (1) confront misconceptions about material science ideas; (2) master advanced polymer engineering skills; (3) cultivate an interest in science and understanding of how we do science and develop engineering. I developed my teaching practice through the MIT Kaufman Teaching Certificate Program (KTCP), undergraduate mentoring and junior researcher training service, and as a teaching assistant in three courses (1) Introduction to Material Science and Engineering, (2) Mechanics of Composite Materials, and (3) Structure, Property, and Processing of Polymeric Materials. Therefore I am capable and also enthusiastic in teaching materials science, polymer engineering and composite mechanics related courses.

457983 Sythesis of Biomass-Derived Carbon Materials and Their Application on Energy Storage and Fuel Cell

Abstract:

Research Interests:

Synthetic pathway of carbon material from biomass materials such as lignin, protein and lecithin. Application of biomass derived carbon materials on supercapacitor, Li-ion batteries and Oxygen Reduction Reaction

Teaching Interests: Thermodynamic, Transport

463364 Nano-Engineered Functional Materials for Energy Storage and Biomimetic Applications

Abstract:

Nano-Engineered Functional Materials for Energy Storage and Biomimetic Applications

Samanvaya Srivastava

Institute for Molecular Engineering, The University of Chicago, Chicago, IL 60637, USA

samsri@uchicago.edu

A multitude of novel building blocks, including macromolecules and nanoparticles of myriad shapes and sizes, are now available to researchers owing to recent progress in material synthesis and fabrication techniques. However, as technology moves towards smarter materials and complex multifunctional devices, the need for understanding and controlling the interactions and assembly of these mesoscale units also becomes increasingly important. Driven by this motivation, my proposed research program will aspire to employ fundamental approaches from colloid, macromolecular and soft matter sciences to understand and tailor the physical processes that govern the structure-property relationships in these material building blocks and their assemblies. These studies will provide the groundwork for the overarching theme of my research, which is to design bottom-up approaches for assembly of mesoscopic building blocks into nano-engineered materials with *on-demand* properties.

My graduate research with Prof. Lynden Archer at Cornell University primarily focused on developing strategies for synthesis of inorganic-organic hybrid materials with tunable properties. Specifically, I developed strategies for controlling nanoparticle dispersion in polymer hosts, and consequentially synthesizing nanoparticle-polymer composites with desired mechanical and electrochemical properties. Further, in a collaborative effort with Lipson group at Cornell University, I worked on the development of functional inks for unique direct-write applications, including 3-D printed electronic and energy storage devices.

My current research as a postdoctoral researcher with Prof. Matthew Tirrell at The University of Chicago and Argonne National Laboratory involves designing and synthesizing block copolyelectrolytes that self-assemble via electrostatic interactions into polyelectrolyte complex (PEC) based micelles and hydrogels. At low polymer concentrations, oppositely charged triblock copolyelectrolytes form micellar networks that phase separate from the solution. These intriguing structures are understood to be driven via a combination of electrostatic and entropic interactions. At higher polymer concentrations, our investigations show that PEC based hydrogel morphologies span a wide range of features and are very sensitive to external stimuli, thus allowing for facile tuning of the hydrogel properties. I am also investigating the enhancements in the protein activity upon encapsulation in polyelectrolyte complexes. These studies, in collaboration with the Gardel group at The University of Chicago, are particularly important for intelligent design of PEC based micelles and hydrogels as drug carriers and tissue growth scaffolds.

My goal is to establish an independent research program that aims at designing novel bottom-up approaches for self-assembled material. Building on my previous work and harnessing my expertise in soft matter physics, I plan to pursue three areas of research that are set in a similar fundamental landscape, but will serve entirely different realms of applied research. The research themes that I intend to pursue are: i) developing new material platforms for redox flow batteries, which employ electroactive complex fluids that are flown in a parallel flow setup during the charging/discharging process and stored in reservoir tanks otherwise; ii) synthesizing surface-modified nanoparticle based hydrogels that will have superior tunability, stimuli responsiveness and tissue supporting functionalities, and iii) development of smart inks for additive fabrication (also known as 3-D printing) of responsive materials and integrated devices.

Selected Publications (17 total, 3 in preparation)

1. **S. Srivastava** and M. V. Tirrell, Polyelectrolyte Complexation, *Advances in Chemical Physics* **161** (2016), in print.
2. **S. Srivastava**, P. Agarwal, R. Mangal, S. Narayanan, D. L. Koch and L. A. Archer, Hyperdiffusive Dynamics in Newtonian Nanoparticle Fluids, *ACS Macro Letters* **4**, 1149 (2015).
3. R. Mangal, **S. Srivastava**, and L. A. Archer, Phase stability and dynamics of entangled polymer-nanoparticle composites, *Nature Communications* **6**, 7198 (2015).
4. **S. Srivastava**, J. L. Schaefer, Z. Yang, Z. Tu and L. A. Archer, 25th Anniversary Article: Polymerâ€"Particle Composites: Phase Stability and Applications in Electrochemical Energy Storage, *Advanced Materials* **26**, 201 (2014).
5. **S. Srivastava**, S. Narayanan and L. A. Archer, Structure and Transport Anomalies in Soft Colloids, *Physical Review Letters* **110**, 148302 (2013).
6. **S. Srivastava**, P. Agarwal and L. A. Archer, Tethered Nanoparticleâ€"Polymer Composites: Phase Stability and Curvature, *Langmuir* **28**, 6276 (2012).

Research Interests: Self-assembled hybrid materials, polymer and nanoparticle hybrid materials, X-ray based structure and dynamics characterization, polyelectrolyte physics

Teaching Interests: Polymer physics, fluid mechanics, advanced material characterization techniques

For more information, please visit: http://tirrell.ime.uchicago.edu/samsri

467733 Temporally Controlled Release of Platelet-Rich Plasma from Peg Microgels Having Tunable Biodegradation Rate and Size

Abstract:
Research Interests: Biomaterials, Drug Delivery

Teaching Interests: Â Biomaterials

Biomedical applications such as drug delivery and tissue engineering often require polymeric hydrogels which degrade predictably under physiological conditions. Biodegradable injectable drug delivery systems present attractive avenue for site specific and minimally invasive administration, coupled with precise temporal control over drug release, which can be employed for multiple applications. Specifically, poly (ethylene glycol) (PEG) has been thoroughly investigated for the design of biodegradable hydrogel networks because of its exceptional biocompatibility and inertness. In this study we designed biodegradable PEG microgels with a wide range of degradation kinetics using Michaelâ€™s addition chemistry. We further used these microgels for encapsulating and controlling delivery of proteins derived from platelet rich plasma (PRP). PRP is a concentrated assortment of growth factors and cytokines derived from the patientâ€™s own platelets. PRP has been proven useful in a number of regenerative medicine applications. However, an effective method to achieve sustained delivery of PRP is still required.

For the design of degradable PEG hydrogels we used multiarm PEG-acrylate (PEGAc) of varying molecular weight and number of arms (4-arm of 10 kDa, to 8-arm of 20 kDa) and three categories of dithiol crosslinkers, namely, ester containing, non-ester containing and dithiols with neighbouring functional groups of different electronegativities. The PEG hydrogel physical and mechanical characteristics could be modulated depending upon the reaction parameters and gelation time for multiarm PEGAc and dithiol crosslinker structure and molecular weight. By changing these parameters, hydrogels with controlled degradation ranging from 10 hours to 32 days and having mesh size in the range of 9 to 14 nm were obtained. We identified two sets of conditions which yielded fast gelling-fast degrading hydrogels and slow gelling-slow degrading hydrogels. Uniquely, the hydrogel storage moduli could be controlled by the dithiol crosslinker chemical identity independent of the degradation time or mesh size of the hydrogel.

Further, we used the optimized combinations of PEG hydrogels to make PEG microgels of sizes varying between 50 to 700 µm by manipulating several key electrospraying parameters. We also investigated various techniques and times suitable for storing protein-loaded PEG microgels. The microgels were stored at seven conditions: 25°C, 4°C, -80°C, Lyophilization (Lyo), -80°C/Lyo, -80°C/DMSO, and -80°C/DMSO/Lyo for 1, 4, 7, and 28 days. DMSO was introduced at 10% (v/v) in 1x PBS. Our results indicated that neither storage modulus nor the structure of incorporated model protein was affected post storage at different conditions. However, comparing the percent distribution of microgel size pre- and post-storage, we found an increase in the mean diameter indicating increased swelling post-storage. Furthermore, environmental scanning electron microscopy (ESEM) images showed that large pores were created by some of the conditions, mainly by lyophilization.

Finally, we tested the PEG hydrogels and microgels for encapsulation and delivery of PRP-derived growth factors. PRP at 2-10% w/v was encapsulated during gelation of the hydrogels or microgels. Release of growth factors was quantified by multiplexer analysis. Biological activity of PRP released from the hydrogels or microgels were tested *in vitro* using human dermal fibroblast (HDF) or by circular dichroism, respectively. Direct encapsulation of model proteins in the PEG hydrogels during hydrogel gelation preserved their secondary structure. Release kinetics for several model proteins was found to be dependent on the protein size, as well as mesh size and degradation rate of the hydrogel. The amount of total protein release increased over time, corresponding to increase in hydrogel mesh size. Multiplexer growth factor analysis of the releasates showed continuous release of growth factors (PDGF-BB, EGF, RANTES) from PRP-containing hydrogels until degradation. PRP releasates also demonstrated increased proliferation in HDF cells over a period of 72 hours, comparable to that of a bolus PRP injection. Furthermore, similar results could be translated to PEG microgels obtained by electrospraying.

Thus it can be concluded that we could obtain a range of biodegradable PEG hydrogels by a combination of number of PEG arms and dithiol crosslinkers and reaction pH. Further, these hydrogels could be made into injectable microgels which allowed localized and controlled delivery of PRP growth factors in the active form.

471788 Towards the Next Generation of Magnetic Resonance Spectroscopy: Harnessing Light and Spin

Abstract:

Research Interests:

With unmatched chemical specificity and the ability to probe materials across length- and timescales, nuclear magnetic resonance spectroscopy (NMR) and imaging (MRI) are two of the most powerful analytical techniques available to engineers, scientists, and medical professionals. However, they require costly infrastructure and are plagued by low sensitivity resulting from small nuclear spin magnetization. Typically, nuclear spin magnetization is generated by allowing the sample to equilibrate in a magnetic field. However, even with the largest available superconducting magnets, room-temperature equilibrium magnetization is only about 10^{-5} of the theoretical maximum. A general method to generate non-equilibrium nuclear spin magnetization would allow us to dispense with superconducting magnets, decrease sensitivity limits and experimental time by orders of magnitude, and enable a range of previously inaccessible applications.

With this motivation, my laboratory will be organized around three main thrusts:

1. **Enhancement of sensitivity for NMR.**
2. **Exploration of new magnetic resonance applications at low- and zero-fields.**
3. **Fundamental aspects of spectroscopy, chirality, and condensed-matter physics.**

Our primary goal is to harness the pure spin states of photons and electrons to achieve nuclear spin polarization. We will build upon my previous work with optical pumping in semiconductors [1] and nitrogen vacancy (NV^-) centers in diamond [2], where I have demonstrated a 170,000-fold increase in NMR signal from the ^{13}C nuclei in the diamond by coupling the electron and nuclear spin states with microwave irradiation. Future work involves the transfer of this nuclear magnetization to arbitrary samples of interest. This will require development of high-surface area diamond structures, fundamental studies of spin polarization transfer mechanisms, and modification of the diamond surface chemistry.

We also seek to extend NMR beyond the traditional model requiring a high-field superconducting magnet in a dedicated facility. Low-field NMR allows the use of cheap, portable magnets in benchtop and truly portable systems. Zero-field NMR allows the observation of spin-interactions that are invisible in the presence of a magnetic field, providing additional information about material structure, geometry, and dynamics [3]. Furthermore, zero-field experiments are free from the deleterious effects of inhomogeneous magnetic susceptibility, enabling high-resolution study of heterogeneous materials.

My laboratory will study fundamental aspects of spectroscopy including chiral effects of spin-spin couplings and spin dynamics in condensed matter. We anticipate ability to ability to directly resolve molecular chirality through antisymmetric spin-spin interactions without the addition of chiral derivatizing agents. Experimental and theoretical studies of spin dynamics in many-body solid-state systems are important to the understanding of solid-state NMR spectroscopy and also provide a prototypical test bed for concepts in condensed-matter physics.

1. **King, J. P. *et al*.** "Optically rewritable patterns of nuclear magnetization in gallium arsenide." ***Nature Communications*** 3:918 doi: 10.1038/ncomms1918 (2012).
2. **King, J. P. *et al*.** "Room-temperature ***in situ*** nuclear spin hyperpolarization from optically pumped nitrogen vacancy centres in diamond." ***Nature Communications*** 6:8965 doi: 10.1038/ncomms9965 (2015).
3. J. W. Blanchard, T. F. Sjolander, **J. P. King,** M. P. Ledbetter, E. H. Levine, V. S. Bajaj, D. Budker, and A. Pines. "Measurement of untruncated nuclear spin interactions via zero- to ultralow-field nuclear magnetic resonance." Phys. Rev. B **92**, 220202(R) (2015).

Teaching Interests:

My teaching philosophy can be stated concisely: effective teaching is the facilitation of effective learning. While this statement is simple, in practice it presents these challenges: How do we define effective learning? How do we assess if learning has occurred? What steps do we take based on this assessment? I will present my plan to address these challenges in the chemical engineering curriculum. I am interested in teaching all aspects of the core curriculum, but with my background would be especially interested in teaching thermodynamics, fluid dynamics, and transport phenomena at the undergraduate and graduate levels. I would also develop special classes in the theory and application of spectroscopic techniques.

474914 Towards a Greener and Scalable Synthesis of Sodium Titanate Nanorods and Its Application As Anode in Sodium Ion Batteries

Abstract:

Sodium ion batteries are accepted as LIB (lithium ion batteries) alternative to large scale energy storage, but its wide application is currently limited by both the synthesis and electrode materials. Thus, simple and scalable synthesis of electrode materials is desirable to lower the cost of synthesis. In this study, we report a surfactant-free, energy-efficient solid state synthesis to mass-produce uniform, single crystalline 1D metal oxide nanorods with tunable composition. To demonstrate the methodology, a series of single-crystalline $Na_2Ti_3O_7/Na_2Ti_6O_{13}$ nanorods with tunable composition are synthesized and applied in sodium ion batteries. By control of synthesis parameter, it provides an alternative low-temperature route to transform $Na_2Ti_3O_7$ into $Na_2Ti_6O_{13}$. The high theoretical capacity of $Na_2Ti_3O_7$ and the low volume expansion of $Na_2Ti_6O_{13}$ upon charge/discharge are synergistically exploited to achieve high electrochemical performance and stability. This synthesis method opens up a scalable, energy-saving route to mass-produce uniform 1D nanostructured electrode materials and fine-tune the composition and morphology according to performance demands.

The electrochemical performance of $Na_2Ti_3O_7$ and $Na_2Ti_3O_7/Na_2Ti_6O_{13}$ nanorods are compared to conventionally synthesized $Na_2Ti_3O_7$ irregular particles. $Na_2Ti_3O_7$ nanorods outperform $Na_2Ti_3O_7$irregular particles at all current rates in the range of 0.1 to 5C. Without surface modification, the nanorods can still manifest a reversible discharge capacity of 70 mAh/g at a high current rate of 5C. The enhanced electrochemical performance can be attributed to the high surface area of the nanorods as well as efficient ion transport induced in highly uniform one-dimensional nanostructure.

Research Interests: Nanomaterials; Li-ion, Li-air, Na-ion Batteries; Catalysis; Electrochemistry; In-situ Electrochemical Methods

Teaching Interests: Nanotechnology; Materials Science and Engineering

459307 Molecule Separation and Energy Storage Using Novel Porous Material Platform

Abstract:

Research Interests: Metal-organic framework membrane; gas adsorption and separation; energy storage; biomass conversion

Teaching Interests: Thermodynamics, transport, porous materials and applications

Metal-organic frameworks (MOFs) also known as coordination porous polymer are a series of novel porous materials which are composed of metal anodes and organic linkers through coordination bonds. MOFs usually have 3D porous structures, large surface areas and pore volume, which provide promising future in applications including molecule separation and energy conversion. Application example such as CO_2 capture from flue gas using MOFs will be shown in two scenarios, fixed bed and membrane. Unsaturated metal centers have been found to be the key for low partial pressure CO_2 adsorption. A robust hydrophobic MOF membrane showed high CO_2 permeance even under 80%RH, which is close to real flue gas condition. Proposed future work is focusing on oriented asymmetric MOF membrane for molecule separation and catalyst used in fuel cell and other electrochemical energy storage system. The unlimited pairs of metal and organic linker, chemical versatility, and tunable pore size and functionality open up great opportunity for MOFs in research frontiers of energy and environment related research.

473149 Designing Metal Oxide Materials for Reduction/Oxidation Reactions Based on a Fundamental Understanding of Their Behavior

Abstract:

Research Interests:

The release of anthropogenic pollutants such as greenhouse gasses, agricultural and industrial waste, and incomplete treatment of waste and sewage water are increasing and therefore having larger negative impacts on the environment. Dealing with these emissions, wither by avoiding them where possible, and remediating them when avoidance is not possible, is one of the major challenges of the 21st century. My research addresses both of these pathways through the understanding and design of metal oxide reduction/oxidation (Red/Ox) materials for solar thermochemical fuel generation and photocatalytic pollutant degradation. In solar thermochemical fuel generation concentrated solar heat is used to drive water and CO_2 splitting reactions using a metal oxide O-carrier to couple an O_2-generating half reaction to a H_2 or CO generating half reaction. The development of optimal materials is critical for this process. The metal oxide must reduce readily enough to operate at temperatures achievable with concentrated solar power(<1600°C), while not reducing so readily that it is incapable of driving the water or carbon dioxide splitting reaction. My research into solar fuels focuses on developing a fundamental understanding of what controls reducibility of metal oxides using computational chemistry (DFT) and, based on this understating, engineering of new materials with more favorable Red/Ox behavior. The computational design new materials are then verified experimentally. A seemingly different process, but fundamentally coupled research area is photocatalytic pollutant degradation. The photocatalytic remediation of pollutants is inherently a Red/Ox reaction, where a metal oxide, usually TiO_2 acts as catalysts for these reactions. My research in this area has focused on using quantum chemistry to develop a fundamental understanding the rate limiting step of this process, namely the adsorption of O_2 to the TiO_2 surface and simultaneous transfer of photo-excited electrons from the TiO_2 surface to molecular O_2. Based on this understand doping and other modification strategies for TiO_2 can be developed. In both of these cases the understanding developed can be ported to other similar processes where metal oxides facilitate RedOx processes. This includes very similar processes, such as chemical looping, or non-TiO_2photocatalysis or subjects further afield, such as solid oxide electrolysis and fuel cells, photolysis and development of catalytic converter materials. Through the research, my focus is first on a fundamental understanding of the material and its controlling processes followed by novel materials design and improvement.

Teaching Interests:

My approach to chemical engineering teaching is to convey the basics that every student needs to succeed in their careers, namely problem solving skills and a fundamental understanding of physical and chemical processes involved in chemical systems. As such I have interest in teaching at both the undergraduate and graduate levels and aim to instill a love and intuition of chemical engineering processes in the students. Of the major chemical engineering subjects, I am most interested in teaching chemical reaction engineering and transport processes at the undergraduate or graduate levels. However, I enjoy most of the chemical engineering curriculum and am happy to teach everything from basic chemistry through process design and graduate level thermodynamics. Having had the opportunity to design and teach half a semester of a masters level course on renewable fuels at ETH during my Postdoc, I am also interested in developing and teaching new graduate and undergraduate classes such as computational chemistry or renewable energy systems. I believe that the education and proper preparation of the students is the mission of a university and that this includes interactions outside of the classroom as well as within it. Therefore I plan to have an engaged research program where undergraduate as well as graduate students participate and can see the application of the subjects that they are learning about in class.

475158 Multi-Scale Modeling of Bulk Solutions and Solid/Liquid Interfaces

Abstract:

Research Interests:

Today, the close interplay between computational simulations and experiments has proven to be an invaluable tool in providing a fundamental understanding of key aspects of numerous engineering technologies and complex processes, as well as enabling the rational design of novel materials for diverse technological applications. In my proposed research program, I intend to use a multi-scale simulation approach to predict and understand the physicochemical behavior of bulk solutions and solid/liquid interfaces in order to gain a fundamental understanding of synthetic processes of complex materials, the knowledge of which can then be leveraged for cutting edge and novel applications in energy-storage devices. I plan on doing so a using multi-scale simulation approach which will provide indispensable information at the molecular- and mesoscale-levels to understand the physical/chemical behavior of bulk solutions and solid/liquid interfaces, which is crucial in the design and optimization of energy storage devices and the synthesis of complex materials. The development of such an understanding of solutions and interfacial properties is fraught with complications, but crucial to the design and optimization of energy storage devices and material syntheses methods for which experiments alone are insufficient due to difficulties associated with distinguishing between different phases, limitations for experimental resolution and the high cost associated with experimental work. Thus, the synergetic fusion of experimental input and multi-scale simulation will be a key factor in achieving a fundamental, comprehensive understanding of the delicate interplay between the macroscopic properties and events occurring at much smaller length scales.

Teaching Interests: My aim is to facilitate the development of students so that they can formulate and analyze the problems and use appropriate tools to obtain and interpret results. I believe that my students should be critical and indendent thinkers with skills to work in teams. I have five years of teaching experience as teaching assiatant and I beleive that teaching must not be restrictied to mere transfer of facts and equations.

475595 "Click" Polymerizations: From Recycling Polymer to 3D Printing

Abstract:

Research Interests:

Smarter, greener and atomic efficient polymeric materials

Polymer science has seen an amazing century of development. Polymer materials have been mature and ubiquitous, and have completely changed the way we live today. Surprisingly I see vast research opportunities that could change our life for days to come, for good. My research is centered utilizing atomic efficient reactions in polymers that solve problems in cutting-edge areas from nano/micro structured materials, to recycling materials, to 3D printing products. Centered at implementing â€œclickâ€ reactions and many other dynamic covalent systems, I sought to answer tough questions with reliable and widely applicable mythologies. Along the way to pursue novel polymeric materials, I strive to mentor students in both undergraduate and graduate levels, as well as postdoctoral scholars for their professional standings as future researchers.

Teaching Interests:

Â My teaching experiences in graduate school as a teaching assistant and part-time instructor have gave me passion in pursuing life-long career in an academic position. My teaching philosophy is to embrace active classroom, encourage question/discussion among students, and follow each individual in the class. I would like to utilize online systems for sharing teaching materials, grading and assigning reading quiz. I would like to teach reaction engineering, (bio)kinetics, polymer materials, polymer chemistry for both undergraduate and graduate students.

475623 Designing Polymeric & Soft Material Systems Via Inverse Computational Methodologies

Abstract:

2nd Year Postdoctoral Research Fellow

Research Interests:

My research focuses on the design of polymeric and soft material systems where a new material can be designed for a specific application ***in silico*** with new properties not possible from conventional combinatorial experimentation. Computational resource capabilities have increased dramatically over the past few decades both in terms of processing speed and parallelizability, thus enabling simulation methodologies that combine multi-length scale virtual experiments with optimization algorithms to find new systems of interest faster than ever. There have been many successful examples of such design work in the areas of metallic and ceramic materials, but equivalent methodologies for polymeric and soft materials are still in their infancy. By combining and integrating the current state of the art computational tools in polymeric systems such as self-consistent field theory (SCFT), coarse grained Monte Carlo molecular dynamics, atomistic molecular dynamics, dissipative particle dynamics, and finite element methods, polymers can be designed for a given functional purpose considering effects at all length scales. By starting with the desired application such as advanced energy storage, ultrafast electronic devices, or biomedical interfacing materials, a chemical engineering system can be designed computationally to produce a given device with optimal properties to perform that application.

Postdoctoral Research:

– Awarded National Research Council Research Associateship Program Fellowship.

Proposal Name:

"***Enhancing the Measurement and Morphology Control of Thin Film Block Copolymers using Feedback Between Inverse SCFT Simulations and Resonant Soft X-Ray Scattering***"

Under supervision of R. Joseph Kline & Christopher Soles, Materials Science and Engineering Division, National Institute of Standards and Technology (NIST)

Doctoral Thesis:

"***Modeling and Theoretical Design Methods for Directed Self-Assembly of Thin Film Block Copolymer Systems***."

Under supervision of Caroline A. Ross and collaboration with Alfredo Alexander-Katz, Department of Materials Science and Engineering, Massachusetts Institute of Technology (MIT)

Research Experience:

As an undergraduate at the Georgia Institute of Technology (GT) I performed undergraduate research under a President's Undergraduate Research Award (PURA) examining the miscibility of fullerene based molecules with styrene based copolymers through both ultraviolent-visible spectroscopy and molecular dynamic simulations. In graduate school at MIT, I continued work in the self-assembly of copolymers. In particular, I examined thin film block copolymer systems and their incorporation into current manufacturing processes as a method to extend lithography methods for next generation microelectronic devices beyond the sub-10 nm node. I did this work by advancing SCFT simulations for examining directed self-assembly systems with topographical templates under solvent vapor annealing conditions. In this simulation framework, I developed an inverse design algorithm to find post template configurations necessary to produce given target complex morphologies. Currently at NIST I am investigating novel methods for characterizing nanoscale patterns fabricated by processes such as block copolymer directed self-assembly as a National Research Council postdoctoral research fellow. Using X-ray scattering techniques, I am optimizing the inverse computation methods needed to find the real space structure of underlying nanoscale patterns. I am also using physically integrated simulation methods (e.g. SCFT of block copolymers) that predict what structures are formed for different directed self-assembly boundary conditions.

Teaching Interests:

As an undergraduate at GT, I instructed introductory electromagnetism physics labs as well as served as a one-on-one tutor for the Learning Assistance Program. At MIT I taught organic chemistry labs, mechanical testing labs, traditional chalk lectures via recitations for a combined quantum mechanics and solid state physics course, and also helped develop course content for an in class demonstration driven thermodynamics course. I completed a teaching minor in Teaching Materials Science and Engineering for my teaching assistance accomplishments as well as a teaching certificate through the MIT Teaching Learning Laboratory. A copy of the letter for the teaching certification is available upon request. Based on these experiences as well as my research interest, I am well versed to teach any thermodynamic, processing, quantum mechanics, or organic chemistry related courses.

Future Direction:

There are many areas in designing polymeric and soft materials systems I would be excited to investigate in my future research. Here I highlight three particular areas. The inverse design paradigm I have used in my research is applicable to all of these areas. The areas examined in increasing order of difficulty and expected research timeline are (I) controlling the self-assembly of heterogeneous polymer, copolymer, and nanoparticle blends/composites, by finding the optimal processing conditions and chemical structures needed for a given application computationally, (II) characterizing designed polymer samples for computational model validation, and (III) determining the optimal chain sequence in proteins and other complex copolymers that can have precise control over the repeat unit sequence.

Area (I) is important as copolymer based systems naturally form ordered structures and complex morphologies essential to applications such as photovoltaics, light emitting diodes, memory storage, integrated circuits, and batteries. Area (II) is essential in both supporting area (I) by validating the systems designed for the given applications as well as enhancing the computational models needed to design such systems through direct comparison with the model results and experiment. Area (III) is the ultimate goal of any polymer scientist or engineer in being able to program desired properties via a carefully designed polymer sequence which can only be done with robust design algorithms. Each of these research areas naturally lead into the next and provide the pathway for the design, fabrication, and characterization of new polymeric and soft materials.

Selected Publications:

A.F. Hannon, D.F. Sunday, G. Khaira, J. Ren, P.F. Nealey, J.J. de Pablo, R.J. Kline. "Optimizing Self-Consistent Field Theory Block Copolymer Models with X-Ray Metrology," ***In Preparation***

A.F. Hannon, D.F. Sunday, D. Windover, R.J. Kline. "Advancing X-Ray Scattering Metrology Using Inverse Genetic Algorithms," ***Accepted by SPIE: Journal of Micro/Nanolithography, MEMS, and MOEMS***. (2016).

D.F. Sunday*, **A.F. Hannon***, S. Tein, R.J. Kline. "Thermodynamic and Morphological Behavior of Block Copolymer Blends with Thermal Polymer Additives," ***Macromolecules*** doi:10.1021/acs.macromol.6b00651 (2016). *Contributed equally

A.F. Hannon, W. Bai, A. Alexander-Katz, C.A. Ross. "Simulation Methods for Solvent Vapor Annealing of Block Copolymer Thin Films," ***Soft Matter*** **11(19)**, 3794-3805, (2015).

A.F. Hannon, Y. Ding, W. Bai, C.A. Ross, A. Alexander-Katz. "Optimizing Topographical Templates for Directed Self-Assembly of Block Copolymers via Inverse Design Simulations," ***Nano Letters*** **14(1)**, 318–325, (2014).

A.F. Hannon, K.W. Gotrik, C.A. Ross, A. Alexander-Katz. "Inverse Design of Topographical Templates for Directed Self-Assembly of Block Copolymers," ***ACS Macro Letters*** **2(3)**, 251-255 (2013).

A. Tavakkoli K.G., K.W. Gotrik, **A.F. Hannon**, A. Alexander-Katz, C.A. Ross, K.K. Berggren. "Templating Three-Dimensional Self-Assembled Structures in Bilayer Block Copolymer Films," ***Science*** **336**, 1294-1298 (2012).

475636 Thin Films and Two-Dimensional Materials for Energy Applications

Abstract:

2nd Year Postdoctoral Fellow

Research Interests:

New materials are constantly being designed to meet the growing and changing needs of society, such as energy storage, batteries, and electronic devices. Computers are now powerful enough that screening studies are practical, allowing us to do quick tests on a huge variety of materials. I have used density functional theory (DFT) to calculate important electrical and mechanical properties of these systems. Of special interest to me are two-dimensional materials, and interfaces between materials; interfaces allow fascinating phenomena such as two-dimensional conducting gases, and two-dimensional materials are so thin that their physical properties can be very different than that of bulk materials. In both cases, the properties of the system are vastly different for surfaces/interfaces/two-dimensional materials than their corresponding bulk cases.

Postdoctoral Projects: ***"Modeling of two-dimensional transitional metal carbonitrides (MXenes) for catalysis and energy storage"***

Under supervision of Jens Nørskov and Aleksandra Vojvodic, Department of Chemical Engineering, Stanford University and SUNCAT Center for Interface Science and Catalysis, SLAC National Accelerator Laboratory

PhD Dissertation: ***"First-Principles Studies of Perovskite Thin Films and Heterostructures"***

Under supervision of Alexander Demkov, Department of Physics, the University of Texas at Austin

Research Experience:

My academic career bridges physics, chemistry, materials science, and chemical engineering. I have always worked with experimentalists; I was fortunate in that my PhD group was dual theory/experimental. I have experience modeling mechanical, electric, magnetic and optical properties using DFT. I have had experience with multiple high-performance clusters, from massive nationwide level computer centers (TACC at the University of Texas at Austin, NERSC at Lawrence Berkeley National Laboratory) to midsize (Sherlock cluster at Stanford University) to a small cluster used exclusively by our research group (SUNCAT cluster at SLAC).

Teaching Interests:

I taught for four semesters at UT Austin; I taught a lab for premedical students taking physics, with multiple office hours per week to help students and encourage their learning, specially the application of physics towards medicine. UT Austin also gave opportunities to 'coach' students, i.e. a general area where students from any class, at any level, could come and ask questions about physics, which I also participated in for four semesters. I tutored ***pro bono***at my undergraduate and also taught on the side as a graduate student (from high school level to college physics).

Future Direction:

As faculty, I want to continue my research on both MXenes. These fascinating materials show vastly different mechanical and electrical properties that vary with functionalization. I would like to collaborate with experimentalists to show how their properties can be affected by different functionalizations, and demonstrated in the lab. I also want to understand better how to control the functionalization through experimental means (strain, temperature, pH, etc.) I also want to apply screening studies to interfaces (doping, interstitial atoms, vacancies), in order to model interfaces in practical devices such as field effect transistors, spin filters, and semiconducting heterojunctions. Because MXenes are so thin, they have interesting potential in device applications, and they are perfect for screening studies, because a small change in the composition can have drastic effects on conductivity, tensile strength, etc. The logical next step is to understand how their properties are affected by deposition/growth on real substrates.

Selected Publications:

"Two-Dimensional Transition Metal Carbides (MXene) as Efficient Electrocatalysts for Hydrogen Evolution", **K.D. Fredrickson**, Z.W. Seh, B. Anasori, J. Kibsgaard, A.L. Strickler, M.R. Lukatskaya, Y. Gogotsi, T.F. Jaramillo and A. Vojvjodic, submitted to ACS Energy Letters, **2016**

A.A. Demkov, K.J. Kormondy and **K.D. Fredrickson**, "Two-Dimensional Electron Gas at Oxide Interfaces," in ***Oxide Materials at the Two Dimensional Limit***, ed. F. Netzer and A. Fortunelli (Springer, New York) **2016**doi: 10.1007/978-3-319-28332-6

"Spin-polarized, orbital-selected hole gas at the EuO/Pt interface", **K.D. Fredrickson** and A.A. Demkov, J. Appl. Phys. 119, 095309 (**2016**). doi: 10.1063/1.4942837

"Carrier density modulation in a germanium heterostructure by ferroelectric switching", P. Ponath, **K.D. Fredrickson**, A.B. Posadas, Y. Ren, X. Wu, R.K. Vasudevan, M.B. Okatan, S. Jesse, T. Aoki, M.R. McCartney, D.J. Smith, S.V. Kalinin, K. Lai and A.A. Demkov, Nat. Commun. 6, 6067 (**2015**). doi: 10.1038/ncomms7067

"Atomic and electronic structure of the ferroelectric $BaTiO_3$/Ge(001) interface", **K.D. Fredrickson**, P. Ponath, A.B. Posadas, M.R. McCartney, T. Aoki, D.J. Smith and A.A. Demkov, Appl. Phys. Lett. 104, 242908 (**2014**). doi: 10.1063/1.4883883

475662 Layer-By-Layer Assembly for Water Desalination and Gas Separation

Abstract:

1st Year Postdoctoral Fellow

Research Interests:

Layer-by-layer (LbL) assembly is a bottom-up manufacturing technique that uses complementary interactions between components to deposit materials one layer at a time. Since a single layer is usually 1-100 nm thick, the structure of LbL assembly can be controlled with nanometer precision. This level of precision is unobtainable by traditional processing techniques, such as melt blending and solution mixing. Additional customization of the LbL assembly can be realized by changing its components and altering complementary interactions between the components. Besides being precise and versatile, LbL assembly is also very simple, allowing industrial-scale production through conventional dipping, spraying, or spin-coating procedures. These collective advantages make LbL assembly suitable for many applications, including anti-fogging, anti-microbial, drug delivery, and energy generation and storage. I am particularly interested in making advanced LbL assemblies for high-performance water filtration and gas separation applications.

Teaching Interests:

I will kindle students' interest in polymer science by relating classroom content with popular subjects among young people. Customized teaching methods will also be utilized to effectively reach every student as well as encourage the participation of underrepresented groups in science. Throughout my career, I plan to continually improve my teaching by seeking student feedback. Upon completing my course with a final group project, I expect my students to have a strong understanding of the fundamentals of polymers (or even fall in love with polymers), and have improved their critical thinking ability and communication skills. Based on this highly customizable teaching plan, my training in STEM teaching, and my background in polymer science and materials science, I am well prepared for teaching several polymer science and engineering courses, including polymer physics, membrane and separation technology, surface and interfaces, materials science and engineering, and multifunctional materials.

Completed Proposal:

I was a co-principal investigator for a proposal to the U.S. Department of the Interior, Bureau of Safety and Environmental Enforcement (E16PS00024); this proposal currently is under second-round review.

Postdoctoral Project:

"***Layer-by-Layer Assemblies with Advanced Gas Separation Properties***"

Under supervision of Dr. David Hopkinson, Department of Energy, National Energy Technology Laboratory

PhD Dissertation:

"***Improvements in Processing and Stretchability of Super Gas Barrier Multilayer Thin Films***"

Under supervision of Prof. Jaime Grunlan, Department of Mechanical Engineering, Texas A&M University

M.S. Thesis:

"***Modification of Immiscible HDPE/PA6 Blends using Carbon Nanotubes***"

Under supervision of Prof. Yong Wang, School of Materials Science and Engineering, Southwest Jiaotong University

Research Experience:

My research centers on structure-property relationships of polymeric materials. My first research project involved toughening immiscible polymer blends using carbon nanotubes (CNTs). In that project, I improved the toughness by selectively dispersing CNTs onto the interfaces between polymer domains. However, I was unable to control the dispersion of all of the CNTs, due to their strong Brownian motion in molten polymer blends. The desire to fully control the dispersion of nanofillers in polymer matrices propelled me to search for a new solution. I found the answer to this problem in LbL assembly, which can be used to disperse two-dimensional nanosheets in a polymer matrix with very high degrees of exfoliation and orientation. This unique layered-structure imparts exceptional gas barrier by creating a tortuous path for diffusion gas molecules, and improves the mechanical property of the thin film beyond predictions of the Halpin-Tsai model. By replacing rigid nanoparticles with soft polymers that combine with each other through H-bonding, I produced stretchy gas barrier coatings that remained functional under high-strain conditions. My current research on gas-separation membranes helps me better understand the transportation behaviors of different gases within polymeric thin films, and inspires me to develop new LbL assembly membranes for advancing separation technologies for gases and liquids.

Teaching Experience:

I actively sought opportunities to improve my teaching skills. I was a teaching assistant for an undergraduate-level Materials Science and Engineering course for one year, and enjoyed helping students solve problems. I also learned the 'dos and don'ts' of teaching, based on student feedback. I received training from a Science, Technology, Engineering and Math (STEM) teaching program, and guest-lectured an undergraduate-level class on ceramic materials. I have mentored several undergraduate students in my research groups, and almost all of my mentees continued their studies in graduate schools.

Future Direction:

I anticipate building a research group that uses state-of-the-art nanotechnology to counter pressing global challenges, such as water shortage, clean energy generation, and climate change.

Using LbL assembly for preparing water desalination membranes has attracted significant attention because LbL assembly features a multi-bipolar structure. This structure can be used to separate monovalent and divalent ions based on "Donnan exclusion". However, current LbL membranes are either too thick to permit high water flux, or too thin to provide sufficient salt rejection. Using rational post-treatment procedures, we can design a LbL assembly that features a nanoporous skin layer and a macroporous support layer, which impart high salt rejection and high water flux properties, respectively. This technology would improve our understanding on water desalination and should help reduce the global shortage of fresh water.

Besides separation in liquid media, I am also interested in gas separation. Specifically, I plan to produce multilayer membranes for hydrogen purification and carbon dioxide separation (the latter is also known as carbon capture). The hydrogen purification membrane exploits the difference between the kinetic diameters of different gas molecules. Specifically, graphene oxide sheets with tunable nanopores will be used as embedded molecular sieving nanofillers to generate high-purity hydrogen for fuel cells. For carbon capture membranes, I plan to use LbL assembly in conjunction with supported ionic liquid membranes to combine their advantages in mechanical and separation properties. Properly designed and developed films can be used to capture carbon dioxide to help counter global climate change.

The techniques and strategies summarized above can also be used to prepare other kinds of LbL assemblies with designed nanostructures and properties that are suitable for chemical sensing, self-healing, and self-cleaning purposes.

Select Publications:

F. M. Xiang, D. Parviz, T. M. Givens, P. Tzeng, E. M. Davis, C. M. Stafford, M. J. Green, J. C. Grunlan. Stiff and Transparent Multilayer Thin Films Prepared Through Hydrogen-Bonding Layer-by-Layer Assembly of Graphene and Polymer. ***Advanced Functional Materials***, 2016, 26, 2143.

C. Y. Cho, **F. M. Xiang (co-first author)**, K. L. Wallace, J. C. Grunlan. Combined ionic and hydrogen bonding in polymer multilayer thin film for high gas barrier and stretchiness. ***Macromolecules***, 2015, 48, 5723.

F. M. Xiang, T. M. Givens, S. M. Ward, J. C. Grunlan. Elastomeric polymer multilayer thin film with sustainable gas barrier at high strain. ***ACS Applied Materials & Interfaces***, 2015, 7, 16148.

F. M. Xiang, T. M. Givens, J. C. Grunlan. Fast spray deposition of super gas barrier polyelectrolyte multilayer thin films. ***I&EC Research***, 2015, 54, 5254.

F. M. Xiang, S. M. Ward, T. M. Givens, J. C. Grunlan. Structural tailoring of hydrogen-bonded poly(acrylic acid)/poly(ethylene oxide) multilayer thin films for reduced gas permeability. ***Soft Matter***, 2015, 11, 1001. Featured on Soft Matter Blog as a hot article for January 2015.

F. M. Xiang, S. M. Ward, T. M. Givens, J. C. Grunlan. Super stretchy polymer multilayer thin film with high gas barrier. ***ACS Macro Letters***, 2014, 3, 1055.

F. M. Xiang, P. Tzeng, J. S. Sawyer, O. Regev, J. C. Grunlan. Improving the gas barrier property of clay-polymer multilayer thin films using shorter deposition times. ***ACS Applied Materials & Interfaces***, 2014, 6, 6040.

For more information, please visit my personal website http://fangmingxiang.weebly.com

475702 Programmable Assembly and Deformation of Soft Matter

Abstract:

Research Interests:

While surface tension generally plays a negligible role on macroscopic scales, it is often the dominant force on nanometer to micrometer length-scales. The main goal of my research is to understand the role that surface tension plays on sub-millimeter scale objects, especially on soft material systems, and how to utilize this phenomenon to assemble and deform such objects. My research addresses several phenomena of nano-and micro-sized objects at fluid interfaces. On the nano-scale, amphiphilic block copolymers are used to explore interfacial behaviors due to their enhanced stability, mechanical properties, and tunability compared to other interfacially active materials such as small molecule surfactants or lipids. I have investigated the tailoring of amphiphilic block copolymer assemblies through deformation at the oil/water interface by inducing interfacial instabilities to incorporate inorganic nanoparticles into micelles, or by controlling osmotic stresses to prepare multi-compartment emulsions and capsules. Next, thin temperature responsive hydrogel sheets are used to probe elastic properties, buckling instabilities and capillary attraction at the micro-scale. I exploited the competition between surface energy and elastic bending energy to quantitatively determine key elastic properties of thin films that are otherwise challenging to measure. Additionally, I found significant edge imperfections due to the finite resolution of photolithography by observing interfacial deformations. The edge imperfections are then employed to drive the buckling of narrow photo-crosslinked hydrogel ribbons. Further, I introduced a new concept of capillary assembly of soft hydrogel sheets with programmed three-dimensional geometry. This offers opportunities to study correlations between elastic properties and surface tension, as well as capillary interactions between soft materials that can be extended to more complex multi-polar interface deformations.

I plan to form a research group to pursue,

- New lithographic techniques for programming internal heterogeneity in nanocomposite soft materials and characterization of their mechanical properties and photothermal effects.

- Deformation of liquids by external stimuli for the development of energy conversion devices.

Teaching Interests:

I am interested in teaching courses in Chemical Engineering at the undergraduate and graduate levels, as well as developing courses in polymer science and fluid mechanics.

475399 Biohybrid Materials for Applications in Human Healthcare and Sustainability -- Assistant Professor Candidate

Abstract:

R. Helen Zha – Faculty Candidate

Research Vision:

Synthesizing biohybrid materials for applications in human healthcare and sustainability, with focus on establishing relationships between molecular engineering, hierarchical material structure, and function.

Keywords: soft matter, self-assembly, nanomaterials, bio-inspired materials

Research Experience:

Adviser: Sam Stupp (Northwestern University, Chicago IL)

My Ph.D. research focused on the self-assembly of peptide-based amphiphiles into functional nanomaterials. I investigated the formation of ordered "hairy" membranes by peptide amphiphiles and biopolymers using an aqueous-aqueous interface. Through experimental study as well as collaboration with computational researchers, I identified the key factors in controlling membrane morphology and properties. Furthermore, I incorporated anti-cancer peptide motifs into the membrane self-assembly process to create versatile biomaterials for adjuvant cancer therapy. Additionally, I synthesized protein-mimetic nanostructures as potent angiogenesis inhibitors with potential applications in cancer therapy. (10/2007 – 10/2013)

Adviser: Bert Meijer (Eindhoven University of Technology, The Netherlands)

My postdoctoral research in the group of Bert Meijer investigated self-assembling supramolecular polymers and block molecules into ordered nanomaterials with feature sizes less than 5 nm. These materials comprised of a novel class of well-defined siloxanes oligomers with functional end groups. I studied the synergistic and antagonistic intermolecular forces exhibited by these oligomers in order to understand their fundamental self-assembly behavior as well as control material structure and properties. I further used this knowledge to create photoswitchable materials based on the isomerization of azobenzene. (11/2013 – 7/2016)

Adviser: Phil Messersmith (UC Berkeley, Berkeley CA)

My postdoctoral research in the group of Phil Messersmith will work towards developing materials inspired by spider silk and mussel adhesive proteins. (9/2016 - current)

Research Interests:

Molecular materials with defined nanostructure are vital in developing advanced technologies for a variety of emerging global needs. Self-assembly, the spontaneous organization of molecules into well-defined architectures, is key in controlling soft matter at the nano-scale. My research program will utilize self-assembly to create biohybrid materials for applications in human healthcare as well as energy and sustainability. These materials will incorporate versatile biological building blocks, such as peptides and nucleic acids, in conjunction with functional non-natural components. My research approach will incorporate molecular engineering with materials characterization on nano, meso, and macroscopic length scales. Thus, my research will be positioned at the interface between materials science, chemistry, biology, and engineering.

Teaching Interests:

I have gained significant teaching experience on topics in materials science & engineering, biomedical engineering, and chemistry. I look forward to teaching introductory as well as advanced courses in these fields. Furthermore, I have special interest in developing courses that will prepare students for working in nanotechnology, molecular engineering, and soft matter fields (e.g. Nanomaterials Engineering, Experimental Nano-Analysis).

Selected Awards, Publications, & Patents:

NSF Graduate Research Fellowship ($175,000) 9/2008 – 9/2011

Northwestern University Cabell Fellowship ($60,000) 9/2007 – 9/2008

Zha, R. H.; Sur, S.; Stupp, S. I. Self-assembly of cytotoxic peptide amphiphiles into supramolecular membranes for cancer therapy. ***Adv. Healthc. Mater***. **2013**, ***2***, 126-133. doi: 10.1002/adhm.201200118

Stupp, S. I.; **Zha, R. H.**; Palmer, L. C.; Cui, H.; Bitton, R. Self-assembly of biomolecular soft matter. ***Faraday Discuss***. **2013**, ***166***, 9-30. doi: 10.1039/C3FD00120B

Zha, R. H.; Sur, S.; Boekhoven, J.; Shi, H. Y.; Zhang, M.; Stupp, S. I. Supramolecular assembly of multifunctional maspin-mimetic nanostructures as a potent peptide-based angiogenesis inhibitor. ***Acta Biomater***. **2015**, ***12***, 1-10. doi: 10.1016/j.actbio.2014.11.001

Zha, R. H.; Velichko, Y. S.; Bitton, R.; Stupp, S. I. Molecular design for growth of supramolecular membranes with hierarchical structure. ***Soft Matter***. **2016**, ***12***, 1401-1410. doi: 10.1039/c5sm02381e

Zha, R. H.; de Waal, B. F. M.; Lutz, M.; Meijer, E.W. Endgroups of functionalized siloxane oligomers direct block copolymeric or liquid crystalline self-assembly behavior. ***J. Am. Chem. Soc.*** **2016**, ***138***, 5693–5698. doi: 10.1021/jacs.6b02172 ***(ACS Editor's Choice Award)***

Zha, R. H.; Zhang, M.; Stupp, S. I. Anti-angiogenic molecules, nanostructures and uses thereof. US9,169,294, October 27, **2015.**

Research Interests

Understanding and Tailoring Novel Hybrid Nano Materials & Interfaces

Dr. Letian Dou *(ldou@berkeley.edu)*

Currently, one of the major topics in science and engineering is energy. There is a strong demand for new materials and technologies toward clean energy harvesting and energy-efficient devices (e.g. solar cells and light emitting diodes). Semiconducting materials play an essential role in such devices. Both inorganic and organic semiconductors hold great promise, but are currently limited by relatively more complicated fabrication and lower performance. By combining the best of two worlds, hybrid material provides a new solution. From a fundamental science point of view, the bonding between inorganic and organic materials and the charge transport across the interface at different length scales are important to understand. On the other hand, from a practical application point of view, hybridization may lead to superior properties and provide new challenges in materials design and manufacturing.

My research interests lie at the intersection of chemistry and materials science with the target of creating and understanding new hybrid materials. Particularly, I want to answer the questions of how to effectively interface inorganic and organic semiconductors **at different length scales**; how to manipulate the **charge, mass, and heat transport** for achieving better electronic and optoelectronic devices. The material systems of interests are inorganic halide perovskites and organic conjugated molecules/polymers. Using this model system, I will focus on ***materials chemistry*** to develop new hybrid materials with new functionality, investigate the assembly behavior and transport phenomenon at the ***interfaces***, and integrate them into ***electronic and photonic devices*** to improve the efficiency and reliability. The knowledge we obtain and the methodology we create will be generalizable to other material systems, therefore advancing our understanding in the basic chemistry and physics at the inorganic-organic interfaces. These research projects differ significantly from those of my Ph.D. and postdoctoral research advisors yet combine skills and knowledge that I learned from them. Chemical synthesis, advanced structural, optical, and electrical characterization, and electronic device fabrication techniques will be employed in the Dou Research Group.

Teaching Interests

My research background in chemistry, electronic materials and devices has given me experience in most chemical applications, and a vast background in chemical and physical methods and techniques. I am qualified to teach a broad range of topics in chemical engineering at *undergraduate level*; for example, basics in chemical engineering, thermodynamics, energy and mass transport, separation processes, kinetics, polymer science and engineering, etc. My interests in teaching at the *graduate level* are tied to my research work. Specific graduate-level teaching interests include conducting/semiconducting polymers, nano-materials for energy applications, chemical & physical properties of organic/hybrid semiconductors, and the fabrication and characterization of thin film electronic devices (light-emitting diodes/photovoltaics/field-effect transistors).

Letian Dou

315 Hildebrand Hall, Department of Chemistry
University of California, Berkeley, CA, 94720, USA
+1 310 409 5493 (phone); ldou@berkeley.edu
Website: https://letiandou.wordpress.com/

Educational Background

University of California, Berkeley (UC Berkeley)
Chemist Postdoc Fellow — 2014-present
University of California, Los Angeles (UCLA)
Ph.D. in Materials Science and Engineering — 2009-2014
Peking University, Beijing, P. R. China
B.S. in Chemistry — 2005-2009

Honors and Awards

- **Materials Research Society Graduate Student Awards** (2014);
- UCLA Materials Science Department Award (2014), one PhD student awarded annually;
- **The Link Foundation Energy Fellowship** (2013), three recipients in U.S. awarded annually;
- Chinese Government Award For Outstanding Students Abroad (2013);
- Materials Research Society Travel Awards (2013);
- Student Award by Society of Information Display (SID) – LA Chapter (2011);

Selected Publications (†equal contribution; full publication attached)

- **Letian Dou**†, Andrew B. Wong†, Yi Yu†, Minliang Lai, Nikolay Kornienko, Samuel W. Eaton, Anthony Fu, Connor G. Bischak, Jie Ma, Tina Ding, Naomi S. Ginsberg, Lin-Wang Wang, A. Paul Alivisatos, Peidong Yang*
Atomically Thin Two-dimensional Organic-inorganic Hybrid Perovskites
***Science 2015**, 349, 1518-1521.*

- **Letian Dou**, Yonghao Zheng, Xiaoqin Shen, Guang Wu, Kirk Fields, Wan-Ching Hsu, Huanping Zhou, Yang Yang*, Fred Wudl*
Single-Crystal Linear Polymers Through Visible Light-Triggered Topochemical Quantitative Polymerization
***Science 2014**, 343, 272-277.*

- **Letian Dou**†, Jingbi You†, Jun Yang, Chun-Chao Chen, Youjun He, Seiichiro Murase, Tom Moriarty, Keith Emery, Gang Li and Yang Yang*
Tandem polymer solar cells featuring a spectrally matched low-bandgap polymer
Nature Photonics 2012**, 6, 180-185. **(Over 1000 citations)

- Jingbi You[†], **Letian Dou**[†], Ken Yoshimura, Takehito Kato, Kenichiro Ohya, Tom Moriarty, Keith Emery, Chun-Chao Chen, Jing Gao, Gang Li* and Yang Yang*
  A polymer tandem solar cell with 10.6% power conversion efficiency
  Nature Communications* 2013**, *4, 1446.* ***(Over 1500 citations)

- **Letian Dou**, Yongsheng Liu, Ziruo Hong, Gang Li*, Yang Yang*
  Solution-processed Low-bandgap Near-IR Conjugated Polymers/molecules for Organic Electronics
  Chemical Reviews* 2015**, *115, 12633-12665.* ***(Invited review article)

Research Experience

Chemist Postdoc Fellow, UC Berkeley/ Lawrence Berkeley National Lab/ California Research Alliance by BASF 2014-present

Advisor: Prof. Peidong Yang

- **Inorganic and hybrid perovskites nanomaterials**. Develop synthetic chemistry to achieve 0, 1, 2 dimensional nanostructures; study the fundamental optical and electronic properties; and apply the novel materials into photovoltaics, light emitting diodes, lasers, etc.
- **Ultra thin core-shell metal nanowires for transparent electrodes.** Design novel graphene-metal nanowire core-shell structures to achieve highly stable transparent conductors for displays and solar cells.
- **Grant proposal writing (to DoE, AFOSR, BASF)** and PhD student training.

Graduate Student Researcher, UCLA 2009-2014

Advisor: Prof. Yang Yang

- **High efficiency tandem polymer solar cells and visibly-transparent polymer solar cells.** Design and synthesis of novel low-bandgap polymers (near-IR absorbing) specifically for tandem and visibly-transparent solar cells. **Several research grants from NSF (2013-2015), AFOSR (2011-2014) and industry (EFL Tech., Sumitomo Chemical Inc.) were awarded** based on these works.
- Synthesis of fullerene derivatives as acceptor materials for polymer solar cells.
- Thin film morphology and charge transport characterizations in polymer solar cells.

Visiting Student, Univ. of California, Santa Barbara March-August, 2013

Advisor: Prof. Fred Wudl

- **Design, synthesis and characterization of novel polymer single crystals by photochemistry**. Discovery of the first visible-light-triggered topochemical polymerization and a new type of highly perfect polymer single crystal.
- Design and synthesis of new electron acceptors for organic electronic devices.

Teaching and Mentoring Experience

- **Teaching half of the courses** for MSE 248 (*Physics of Solar Cells*, preparing slides, giving lectures), 2014.

- **Teaching Assistant** for MSE 200 (*Principles of Materials Science*, leading discussions, answering questions, grading), 2013.
- **Teaching Assistant** of General Chemistry Lab (mentoring 20 students in the lab), 2009.
- **Mentoring** and training of many **junior graduate students**:
  i. Dr. Wei-Hsuan Chang [UCLA, Ph.D. 2016]: Selenium-containing low band gap polymers for organic photovoltaics.
  ii. Mr. Roger Zhao [UCLA, Ph.D. candidate]: Ladder type building blocks for high mobility conjugated polymers.
  iii. Ms. Ariella Machness [UCLA, undergraduate]: Synthesis of diketopyrrolopyrrole based low band gap polymers.
  iv. Mr. Minliang Lai [UC Berkeley, Ph.D. candidate]: Halide perovskite nanowires and nanosheets synthesis and optical properties.
  v. Mr. Teng Lei [UC Berkeley, Ph.D. candidate]: Controlled doping and alloying of halide perovskite nanowires.

Invited Talks and Conference Presentations

- *Atomically Thin Two-dimensional Organic-inorganic Hybrid Perovskites*
  Letian Dou, Andrew B. Wong, Yi Yu and Peidong Yang
  2016 ECS Meeting, May 2016, San Diego, CA. ***(Invited oral presentation)***

- *Low-Bandgap Polymers Designed for Tandem Polymer Solar Cells*
  Letian Dou, Jingbi You, Chun-Chao Chen, Jing Gao, Gang Li and Yang Yang
  2014 MRS Spring Meeting & Exhibit, April 2014, San Francisco, CA. ***(oral presentation)***

- *Single-Crystal Linear Polymers Through Visible Light-Triggered Topochemical Quantitative Polymerization*
  Letian Dou
  2014 Department of Materials Science Seminar, UCLA ***(oral presentation)***

- *Benzodithiophene- and Diketopyrrolopyrrole-based Low-bandgap Polymers Designed for Tandem Polymer Solar Cells*
  Letian Dou, Jingbi You, Chun-Chao Chen, Jing Gao, Wei-Hsuan Chang, Gang Li and Yang Yang
  2013 MRS Spring Meeting & Exhibit, April 2013, San Francisco, CA. ***(oral presentation)***

- *Recent progress in polymer solar cells and tandem polymer solar cells*
  Letian Dou, Jingbi You, Chun-Chao Chen, Jing Gao, Gang Li and Yang Yang
  ACS 244th National Meeting, August 2012, Philadelphia, PA. ***(Invited oral presentation)***

Patents

- *Active Materials for Electro-Optic Devices and Electro-Optic Devices*
  Yang Yang, **Letian Dou**, Jingbi You (WO2012135527, US20140020739 A1, licensed by

EFL Tech.)

- *Semi-Transparent, Transparent, Stacked and Top-illuminated Organic Photovoltaic Devices*
  Yang Yang, Rui Zhu, Chun-Chao Chen, **Letian Dou**, Gang Li (WO2014007867 A1, US20140084266 A1, licensed by EFL Tech.)

- *Conjugated Polymers for Electronic Devices*
  Yang Yang, **Letian Dou**, Wei-Husan Chang (US20150028265 A1)

- *Electro-Optic Device Having Nanowires Interconnected into A Network of Nanowires*
  Yang Yang, Rui Zhu, Chun-Chao Chen, **Letian Dou**, Gang Li (WO2014007867A1, US20140084266, licensed by EFL Tech.)

- *Ultra Thin Copper-Reduced Graphene Oxide Core-Shell Nanowires based Transparent Conductors*
  Peidong Yang, **Letian Dou**, Fan Cui (BK-2015-154, provisional filing)

- *Synthesis of Ultra-Thin Metal Nanowires Using Organic Free Radicals*
  Peidong Yang, Fan Cui, **Letian Dou** (UCALP073P/BK-2016- 136, provisional filing)

Service and Membership

- **Referee/reviewer** for many scientific journals including: *Journal of the American Chemical Society*; *Advanced Materials*; *Nano Letters*; *ACS Nano*; *Scientific Reports (by NPG)*; *Chemistry of Materials*; *Journal of Materials Chemistry A*; *Nano Energy*; *J. Phys. Chem. C*; *Macromolecules; Polymer Chemistry; Macromolecule rapid communications*; *Macromolecular Chem. & Phys.*; *Chemistry - A European Journal; Organic Electronics; Applied Surface Science; ACS Energy Letters; ACS Applied Materials & Interfaces.*
- **Committee** of 2016 Link Foundation Energy Fellowship.
- **Member** of American Chemical Society (ACS), Materials Research Society (MRS), Electrochemical Society (ECS), American Institute of Chemical Engineers (AIChE), SPIE, Society of Information Display (SID).
- Assistant of 2nd International Symposium on Materials Enabling Nanodevices (2010).
- Student chair of SID-LA chapter (2013).

Full Publication List (*over 6000 citations, H index = 25*)

Google scholar citation: ***https://scholar.google.com/citations?user=nfkBN7UAAAAJ&hl=en***

2016

33. Yi Yu, Dandan Zhang, Christian Kisielowski, **Letian Dou**, Nikilay Kornienko, Yehonadav Bekenstein, A. Paul Alivisatos, Peidong Yang; "*Atomic Resolution Imaging of Halide Perovskites*", ***Nano Letters 2016,*** *submitted.*

32. Samuel W. Eaton, Minliang Lai, Natalie Gibson, Andrew B. Wong, **Letian Dou**, Jie Ma, Lin-Wang Wang, Steve Leone, Peidong Yang; "*Lasing in Robust Cesium Lead Halide Perovskite Nanowires*", ***PNAS 2016,*** *113, 1993-1998.*

31. **Letian Dou**, Fan Cui, Yi Yu, Samuel W. Eaton, Arky Yang, Joaquin Resasco, Christian Schildknecht, Kerstin Schierle-Arndt, Peidong Yang; "*Solution-processed ultra thin copper-reduced graphene oxide core-shell nanowire transparent conductors*", ***ACS Nano 2016,*** *10, 2600-2606.*

2015

30. **Letian Dou**[†], Andrew B. Wong[†], Yi Yu[†], Minliang Lai, Nikolay Kornienko, Samuel W. Eaton, Anthony Fu, Connor G. Bischak, Jie Ma, Tina Ding, Naomi S. Ginsberg, Lin-Wang Wang, A. Paul Alivisatos, Peidong Yang; "*Atomically Thin Two-dimensional Organic-inorganic Hybrid Perovskites*", ***Science 2015****, 349, 1518-1521.*

29. **Letian Dou**, Yongsheng Liu, Ziruo Hong, Gang Li, Yang Yang; "*Solution-processed Low-bandgap Near-IR Conjugated Polymers/molecules for Organic Electronics*", ***Chemical Reviews 2015****, 115, 12633-12665.* ***(Invited review article)***

28. Fan Cui, Yi Yu, **Letian Dou**, Jianwei Sun, Christian Schildknecht, Kerstin Schierle-Arndt, Peidong Yang; "*Ultra-thin colloidal copper nanowires for transparent conductor with high performance*", ***Nano Letters 2015, 2015****, 15, 7610-7615.*

27. Dandan Zhang, Samuel W. Eaton, Yi Yu, **Letian Dou**, Peidong Yang; "*Solution-phase Synthesis of Cesium Lead Halide Perovskite Nanowires*", ***Journal of the American Chemical Society 2015****, 137, 9230-9233.*

26. Andrew B. Wong, Minliang Lai, Samuel W. Eaton, Yi Yu, Elbert Lin, Anthony Fu, **Letian Dou**, Peidong Yang; "*Growth and Anion Exchange Conversion of $CH_3NH_3PbX_3$ Nanorod Arrays for Light-Emitting Diodes*", ***Nano Letters 2015,*** *15, 5519-5524.*

25. Yang (Michael) Yang, Wei Chen, **Letian Dou**, Wei-Hsuan Chang, Hsin-Sheng Duan, Brion Bob, Gang Li, Yang Yang; "*Molecular Compatibility Enabling High-Performance Multiple Donor Bulk Heterojunction Solar Cells*", ***Nature Photonics 2015****, 9, 190-198.*

24. Wei-Hsuan Chang, Lei Meng, **Letian Dou**, Jingbi You, Chun-Chao Chen, Yang (Michael) Yang, Eric P. Young, Gang Li, Yang Yang; "*A Selenophene Containing Benzodithiophene-alt-thienothiophene Polymer for Additive-Free High Performance Solar Cell*", ***Macromolecules 2015***, *48, 562-568.*

23. Yongsheng Liu, Qi Chen, Hsin-Sheng Duan, Huanping Zhou, Yang Michael Yang, Huajun Chen, Song Luo, Tze-Bin Song, **Letian Dou**, Ziruo Hong, Yang Yang; "A dopant-free organic hole transport material for efficient planar heterojunction perovskite solar cells", ***Journal of Materials Chemistry A 2015***, *22, 11940-11947.*

2014

22. **Letian Dou**, Yonghao Zheng, Xiaoqin Shen, Guang Wu, Kirk Fields, Wan-Ching Hsu, Huanping Zhou, Yang Yang, Fred Wudl; "*Single-Crystal Linear Polymers Through Visible Light-Triggered Topochemical Quantitative Polymerization*", ***Science 2014***, *343, 272-277.* (also see *Perspectives* by Nancy Goroff; "*A Clear Path for Polymer Crystallization*", ***Science 2014***, *343, 258.*)

21. **Letian Dou**[†], Yang (Michael) Yang[†], Jingbi You[†], Ziruo Hong, Wei-Hsuan Chang, Gang Li, Yang Yang; "*Solution-Processed Hybrid Perovskite Photodetectors with High Detectivity*", ***Nature Communications 2014***, *5, 5404.*

20. Jing Gao, **Letian Dou**, Wei Chen, Chun-Chao Chen, Xuanrong Guo, Jingbi You, Brion Bob, Wei-Hsuan Chang, Joseph Strzalka, Cheng Wang, Gang Li, and Yang Yang; "*Improving Structural Order for a High-Performance Diketopyrrolopyrrole-Based Polymer Solar Cell with a Thick Active Layer*", ***Advanced Energy Materials 2014***, *4, 1300739.*

19. Jing Gao, Wei Chen, **Letian Dou**, Chun-Chao Chen, Wei-Hsuan Chang, Yongsheng Liu, Gang Li, and Yang Yang; "*Elucidating double aggregation mechanisms in the morphology optimization of Diketopyrrolopyrrole-based narrow bandgap polymer solar cells*", ***Advanced Materials 2014***, *26, 3142-3147.*

18. Wei-Hsuan Chang, Jing Gao, **Letian Dou**, Chun-Chao Chen, Yongsheng Liu, and Yang; "*Side-Chain Tunability via Triple Component Random Copolymerization for Better Photovoltaic Polymers*", ***Advanced Energy Materials 2014***, *4, 1300864.*

17. Qi Chen, Huanping Zhou, Tze-Bin Song, Song Luo, Ziruo Hong, Hsin-Sheng Duan, **Letian Dou**, Yongsheng Liu, and Yang Yang; "Controllable self-induced passivation of hybrid lead iodide perovskites toward high performance solar cells", ***Nano Letters 2014***, *14, 4158-4163.*

2013

16. **Letian Dou**, Jingbi You, Ziruo Hong, Zheng Xu, Gang Li, Robert A. Street, and Yang Yang; "*25th Anniversary Article: A Decade of Organic/Polymeric Photovoltaic Research*", ***Advanced Materials 2013***, *25, 6642-6671.* ***(Invited review article)***

15. Jingbi You†, **Letian Dou**†, Ken Yoshimura, Takehito Kato, Kenichiro Ohya, Tom Moriarty, Keith Emery, Chun-Chao Chen, Jing Gao, Gang Li and Yang Yang; "*A polymer tandem solar cell with 10.6% power conversion efficiency*", ***Nature Communications 2013***, *4, 1446.* ***(Over 1500 citations)***

14. **Letian Dou**, Wei-Hsuan Chang, Jing Gao, Chun-Chao Chen, Jingbi You and Yang Yang; "*A Selenium-Substituted Low-Bandgap Polymer with Versatile Photovoltaic Applications*", ***Advanced Materials 2013***, *25, 825-831.* ***(Over 300 citations)***

13. **Letian Dou**, Chun-Chao Chen, Ken Yoshimura, Kenichiro Ohya, Wei-Hsuan Chang, Jing Gao, Yongsheng Liu, Eric Richard and Yang Yang; "*Synthesis of 5H-Dithieno[3,2-b:2',3'-d]pyran as an Electron-Rich Building Block for Donor-Acceptor Type Low-Bandgap Polymers*", ***Macromolecules 2013***, *46, 3384-3390.* ***(Selected in SYNFACTS for its important insights in polymer chemistry)***

12. Chun-Chao Chen, **Letian Dou**, Jing Gao, Wei-Hsuan Chang, Gang Li and Yang Yang; "*High-performance semi-transparent polymer solar cells possessing tandem structures*", ***Energy & Environmental Science 2013***, *6, 2714-2720.*

11. Jingbi You, **Letian Dou**, Ziruo Hong, Gang Li, Yang Yang; "*Recent trends in polymer tandem solar cell research*", ***Progress in Polymer Science 2013***, *38, 1909-1928.* ***(Invited review article)***

10. Vladan Jankovic, Yang Yang, Jingbi You, **Letian Dou**, Yongsheng Liu, Puilam Cheung, Jane P. Chang and Yang Yang; "*Active Layer-Incorporated, Spectrally-Tuned Au/SiO2 Core/Shell Nanorod-Based Light Trapping for Organic Photovoltaics*", ***ACS Nano 2013***, *7, 3815–3822.*

9. Yongsheng Liu, Yang Yang, Chun-Chao Chen, Qi Chen, **Letian Dou**, Ziruo Hong, Gang Li and Yang Yang; "*Solution-Processed Small Molecules Using Different Electron Linkers for High-Performance Solar Cells*", ***Advanced Materials 2013***, *25, 4657-4662.*

8. Yongsheng Liu, Chun-Chao Chen, Ziruo Hong, Jing Gao, Yang (Michael) Yang, Huanping Zhou, **Letian Dou**, Gang Li, Yang Yang; "*Solution-processed small-molecule solar cells: breaking the 10% power conversion efficiency*", ***Scientific Reports 2013***, *3, 3356.*

2012

7. **Letian Dou**†, Jingbi You†, Jun Yang, Chun-Chao Chen, Youjun He, Seiichiro Murase, Tom Moriarty, Keith Emery, Gang Li and Yang Yang; "*Tandem polymer solar cells featuring a spectrally matched low-bandgap polymer*", ***Nature Photonics 2012***, *6, 180-185.* ***(Over 1000 citations; Top 5 worldwide scientific breakthrough of 2012 highlighted in USA Today; "HOT PAPER IN PHYSICS" highlighted by SCIENCEWATCH; and numerous media reports)***

6. **Letian Dou**, Jing Gao, Eric Richard, Jingbi You, Chun-Chao Chen, Kitty C. Cha, Youjun He, Gang Li and Yang Yang; "*Systematic Investigation of Benzodithiophene and Diketopyrrolopyrrole Based Low-bandgap Polymers Designed for Single Junction and Tandem Polymer Solar Cells*", ***Journal of the American Chemical Society 2012***, *134, 10071-10079.* ***(Over 300 citations)***

5. Chun-Chao Chen[†], **Letian Dou**[†], Rui Zhu, Choong-Heui Chung, Tze-Bin Song, Yue Bing Zheng, Steve Hawks, Gang Li, Paul S. Weiss, and Yang Yang; "*Visibly Transparent Polymer Solar Cells Produced by Solution Processing*", ***ACS Nano* 2012**, *6, 7185-7190.* ***(Over 200 media coverage including Time magazine, Los Angeles Times; most read article in ACS nano 2012)***

4. Jingbi You, Chun-Chao Chen, **Letian Dou**, Seiichiro Murase, Hsin-Sheng Duan, Steve Hawks, Tao Xu, Hae Jung Son, Luping Yu, Gang Li, Yang Yang; "*Metal Oxide Nanoparticles as an Electron-Transport Layer in High-Performance and Stable Inverted Polymer Solar Cells*", **Advanced *Materials* 2012**, *24, 5267-5272.*

3. Youjun He, Jingbi You, **Letian Dou**, Chunchao Chen, Eric Richard, Kitty C Cha, Yue Wu, Gang Li, Yang Yang; "*High performance low band gap polymer solar cells with non-conventional acceptor*", ***Chemical Communications* 2012**, *48, 7616-7618.*

2. Youjun He, Chun-Chao Chen, Eric Richard, **Letian Dou**, Yue Wu, Gang Li, Yang Yang; "*Novel fullerene acceptors: synthesis and application in low band gap polymer solar cells*", ***Journal of Materials Chemistry* 2012**, *22, 13391-13394.*

2011 and before

1. Yijun Zheng, Jia Zheng, **Letian Dou**, Wenqiang Qiao, Xinhua Wan; "*Synthesis and characterization of a novel kind of near-infrared electrochromic polymers containing an anthraquinone imide group and ionic moieties*", ***Journal of Materials Chemistry* 2009**, *19, 8470-8477.*

475884 Functional Polymers for Energy Generation and Storage: Donor-Acceptor Block Copolymers for Photovoltaics and Functional Polyimides for Dielectric Materials

Abstract:

Research Interests:

My research program will develop conjugated block copolymers for power generation and polymer dielectrics for energy storage. We will leverage expertise in polymer and organic chemistry as well as advanced characterization techniques, such as resonant soft X-ray scattering (RSOXS) and energy filtered TEM (EF-TEM). Our unique set of skills are crucial to develop new classes of materials that can serve as (1) model platforms for fundamental studies of optoelectronic and dielectric properties and (2) for demonstrations of breakthrough performance in device applications, furthermore, investigate fundamental structure-property relationship.

Wearable electronics utilized in health care, military, and communication applications require the development of a portable electrical power source. Flexible and light-weight organic photovoltaics are one of the most promising solutions, but challenges remain in performance and stability. My approach will develop a new class of active layer materials capable of exceeding the performance of state-of-the-art blend organic solar cells with improved stability. Donor-acceptor fully conjugated block copolymers (D-A BCPs) will be synthesized as single active layer materials, and their morphologies and solar cell performances will be investigated. Current synthetic approaches, including step-growth polymerization, leads to a mixture of products such as unreacted homopolymers, diblock and triblock copolymers. A focus of my group will be on developing polymerization strategies based on chain-growth mechanisms to prepare well-defined architectures and low dispersities. These well-defined D-A BCPs are expected to enable finer control of the microstructure. In organic electronics, donor/acceptor interfaces are crucial to overall performance, and block copolymers provide a unique approach to control such interfaces. D/A interfaces will be modified with a single bridge unit between donor and acceptor blocks. Furthermore, new chemical structures will be explored for the development of high performance D-A BCPs. Proposed chemical structures for new D-A BCPs are based on thieno-thiophene and cyclopentadithiophene donor blocks with naphthalene diimide acceptor blocks.

Polymeric dielectric materials suitable for electrical energy storage applications are promising because of the potential for flexible, lightweight, photo-patternable, scalable and robust properties compared to ceramic counterparts. But, polymeric dielectric materials have limited impact in various applications, such as hybrid and electric vehicles and aerospace power electronics, due to limited working temperatures. My approach will develop functional polymers to demonstrate improved dielectric properties for high temperature and high electric field applications. Two new diamine monomers of polyimide will be prepared associated with methyl sulfonyl and semi-fluoro alkyl groups. Methyl sulfonyl group is known as one of the strongest polar groups and is expected to enhance the dielectric constant. Semi-fluoro alkyl groups are expected to increase the break-down strength and provide stability in humid environments. A key strategy will be to prepare random copolymers consisting of diamine monomers with both methyl sulfonyl and semi-fluoro alkyl group and corresponding dianhydride monomers to optimize the performance in capacitors. Furthermore, crosslinking of functional polyimdes will enhance thermal stability.

My group will focus on the development of large-scale coating techniques for organic electronics and capacitors. My approach will fabricate solar cells and capacitors through blade or slot-die coating to capture the essential physics of roll-to-roll printing and we will investigate the microstructure within cast polymer films using in-situ X-ray scattering.

Teaching Interests:

I believe it is my responsibility to challenge and support my students to reveal their potential and enjoy their studies. Based on my experience as a student, I will pursue three learning goals in my teaching:

Conceptual learning: Establishing a strong foundation of basic concepts is most important, even more so than understanding other derivative content of a class, because students only understand and evaluate following concepts if they understand the fundamentals. Otherwise, students just memorize at best. I will devote more time, in particular at the beginning of the course, to ensure that students understand fundamental concepts.

Experiential learning: Learning needs momentum, and experience offers strong momentum for learning. When I was an undergraduate, I had a chance to observe a scattering experiment in a synchrotron laboratory. Before then, I had only superficial knowledge of X-ray scattering, but after the scattering experiment I was self-motivated to learn more. As a result, my experience at a synchrotron offered me a good chance to develop my passion for X-ray scattering. In experimental classes especially, I will provide opportunities for students to experience and see concrete examples as much as I can.

Stimulating learning: Students learn best when they are engaged in active inquiry. The level of motivation determines the final level of learning or understanding. It is not as exciting to carry out experiments by simply following instructions without sufficient information about the motivation or fundamentals of the experiments. When working with undergraduates for training or research programs during the summer, I tried to devote time to explain why we do undertake specific experiments and what we are expected to see as results. In case of organic reactions or polymerizations, I explained what we were going to do with the resultant materials in the future. In the same manner, I will help students accept the what and why of their studies, and encourage students to become engaged in classes.

As a researcher and a teacher in Chemical Engineering and Materials Science, I will continue to pursue the three learning goals as stated above, and hopefully contribute to the training of the next generation.

475966 Microfluidic Design of Multi-Phase Emulsion Drops for Functional Materials Production

Abstract:

Research Interests:

Emulsions are attractive for encapsulating, transporting, or controllably releasing a wide variety of chemical and biological active materials. Multi-phase emulsion drops are of particular interest due to the compartments of these emulsion drops that can be enlisted to impart functionality. By using the precise flow control of multi-phasic fluids in microfluidics, in conjunction with fluid-fluid interface design by proper selection of materials in each phase, I have produced novel functional materials that otherwise would have been inaccessible. For example, I developed a new strategy of combining bulk and microfluidic emulsification to achieve highly efficient encapsulation and enhanced retention of fragrance in polymer microcapsules. Additionally, I developed novel triple emulsion drops with an ultra-thin intermediate layer consisting of either hydrogel or fluorocarbon oil that separates the innermost cargo from the polymeric shell. These emulsion drops have interesting properties and great technological potential for encapsulation and controlled release of a wide variety of active materials. Moreover, I have worked on several collaborative projects through which we have designed ideal polymer network in emulsion templated microcapsules for encapsulation and triggered release of challenging actives, and semi-permeable polymeric shell in microcapsules for micro-sensor, and micro-reactor applications, just to name some.

Research Experience:

My academic career path has been a blend of many fields of science and engineering. My formal training at MIT (with Prof. Robert E. Cohen and Prof. Michael F. Rubner) was in Chemical Engineering, which I have studied the mechanisms that govern the structure and function of polymer thin films for developing a diverse range of functional films including antifogging coatings, switchable surfaces and stimuli-responsive hydrogels. Specifically, I have introduced a framework in which poly(vinyl alcohol) can be incorporated into multilayered thin films by investigating hydrogen-bonding interaction of polymers in aqueous media. By tailoring the interlayer diffusion of hydrogen-bonding polymers during the assembly, I also designed a complex hetero-structured architecture that sequentially dissolves with increase in local pH condition. Additionally, I have developed a new concept in controlling water condensation. By thorough characterization and analysis of a model system, a new physical theory, â€œzwitter-wettablilityâ€, was proposed whereby the film readily absorbs water vapor while simultaneously exhibiting hydrophobic character to liquid water. Moreover, this zwitter-wettable concept was further extended to real-life applications where the optimally designed film resulted in a significantly enhanced antifog and even frost-resistant behavior.

Teaching Interests:

Aside from my research career, I also have extensive teaching experience. I TAed an undergraduate course in Chemical Engineering at MIT in which I lectured undergrads, held weekly office hours, and developed course materials; I mentored undergraduate students as a part of their undergraduate thesis at MIT; I also mentored local community college students as a part of an educational outreach program from MIT CMSE; and, lastly, I am currently mentoring and supervising the research of graduate students on a large number of projects.

Future Direction:

As a new faculty, I would like to continue studying soft materials at interfaces, and engineer new hierarchical structures for industrial applications. In particular I would like to leverage the knowledge that I obtained from the two topics that I worked on during my Ph.D. and postdoc: polymer science/engineering and microfluidics, respectively. Many intriguing phenomena occur at fluid-fluid/solid-fluid interfaces and I believe my expertise will provide new perspective in designing functional materials.

My first thrust as a faculty is to take advantage of molecular rearrangement of polymers at surfaces to design novel polymer film structures. It has been demonstrated that configuration of a liquid drop on a given surface depends on the force balance across liquid-solid-air interfaces, which is governed by the surface topology as well as the surface chemistry. However, by separately tuning the surface and bulk properties of a film, unique surfaces can be prepared in which water drops exhibit lower equilibrium contact angle than typical hydrocarbon oils that have much lower surface tension than water. While these engineered surfaces have great potential for applications such as self-cleaning surfaces, and antifogging coatings, how to modulate the rate of rearrangement by materials design is still unclear. I believe that studying this in depth in conjunction with phase change heat and mass transfer will give rise to new applications such as contaminant separation from water vapor for reducing pollutant in air and water.

My second thrust is to utilize the middle phase that separates the innermost drop from the continuous phase in multi-phase emulsion drops to study polymers at interfaces and engineer new functional materials. It has been demonstrated that the middle phase in double-emulsion drops can serve as a template to form polymer microcapsules for encapsulation of actives. Yet, there has been little progress toward the production of microcapsules with membranes that fulfills the practical demands. Moreover, I believe that many well-known scientific findings observed in bulk such as micro-phase separation of block copolymers, and polymer network formation using various chemistries can be translated into this compartment provided by the multi-phase emulsion drops. This will span a wide range of new interesting scientific and engineering questions which will open the door in the field by creating a new route to design and fabricate advanced particles with unique functionalities.

Besides material design my interests include methods to scale up the fabrication of these emulsion drops to produce practical quantities. While microfluidic devices have been massively parallelized to produce highly monodisperse single emulsion drops in high-throughput, extending this framework to produce double-emulsion drops consisting of fluids that have high viscosities have been difficult to achieve. This problem will require extensive fundamental studies of hydrodynamics and interactions of fluids with surfaces. I will use simulation-guided 3D-microfluidic device geometry design, and develop a simple and robust method to selectively modify the surface wetting characteristics of a microfluidic device to solve this challenge. The results of this study will bridge the gap between lab-bench scale to industry-level scalable generation which is both fundamentally and practically important.

475967 Engineering the Surfaces of Tomorrow

Abstract:

Research Interests:

The interface between materials controls many processes of the physical world. Precise control of these interfaces will help solve many of the environmental and energy-related crises that humanity currently faces. Understanding the specific physical and chemical interactions occurring at interfaces allows for exciting, new properties to be realized. For example, controlling the stiffness of a coating between ice and, say, an aircraft wing, allows the interface to cavitate, opening up a new class of materials that cannot adhere to ice. Highly water- and oil-repellent textiles and surfaces can only be fabricated through precise control over the texture and surface energy of the components. Next generation materials cannot be designed by incremental improvements to the current status quo. Revolutionary paradigm shifts manifest when new fundamental principles are uncovered. Leveraging these fundamental scientific discoveries with robust engineering principles allows for exciting new materials to transition from lab-scale innovations to world-changing realities. Designer surfaces are a fascinating new field that, perhaps ironically, exist right at the interface between fundamental science and engineering design.

Research Experience:

I chose materials science and engineering because, even as the name suggests, the discipline combines science and engineering, discovery and implementation, curiosity and practicality. And because everything is made of something, my academic career path has been inherently broad and interdisciplinary. I am formally trained as a surface scientist, which has included engineering zeolite catalysts (ceramics), understanding high-strength fibers (composites), designing solid- and liquid-repellent surfaces (polymers), creating environmentally friendly textiles (natural) and fabricating drag-reducing coatings (metals and semiconductors). Use of these disparate materials and their corresponding interfaces has led to many interdisciplinary and collaborative projects, in the areas of fluid mechanics (Johns Hopkins, US Naval Academy and ONR), surface chemistry (AFRL), biology (University of Michigan and Yale) and rheology (MIT), as well as the many industrial collaborations in the areas of textiles, coatings and designer surfaces. For me, the most valuable research experiences have occurred when combining my core expertise with the expertise of others, leading to several high-impact discoveries (and publications).

Teaching Interests:

My main passion, whether in the laboratory or classroom, has always been teaching. I voluntarily TAed two laboratory classes at Cornell University during my undergraduate tenure. At the University of Michigan I taught several undergraduate classes as well as personally mentored incoming freshmen interested in pursuing a graduate degree. I was fortunate enough to teach under the tutelage of Steve Yalisove, who has been instrumental in pioneering a new type of formative-based learning. I wish to harmonize this new-age teaching style with the traditional methods as a future faculty member. As an NDSEG Fellowship recipient, I have also sat on several fellowship panels and mentored interested graduate students. Lastly, I have mentored many undergraduate and graduate students in my research group.

Future Direction:

As a faculty I will continue to combine fundamental scientific research with engineering principles in the area of surface science. In particular, I would like to further my thesis work on the design and application of liquid- and solid-repellent coatings. I am interested in taking an environmentally conscious approach to surface science, both on the design side as well as the implementation. Recently the EPA and the EU banned long-chain fluorocarbons, and the shorter-chain variants will soon follow. This is a ripe, untapped area of surface science: achieving extreme solid and liquid repellency without the use of fluorinated compounds. The design principles laid out in my previous publications will prove an excellent starting point for attacking this problem. I would also like to combine my graduate and undergraduate work by studying the interfacial phenomenon important to fiber composite body armor. As the failure of ballistic vests typically arises from poor interfacial stress transfer, taking a surface science approach to fiber dynamics should prove insightful. The reverse will also be fruitful: utilizing high-strength fibers in designer coatings will imbue mechanical durability and repellency concomitantly. As a third avenue, I am interested in applying many of the novel surfaces that I designed during my doctoral work as biologically active coatings. Almost every biological process is a surface phenomenon, making the interfacial dynamics critical but amenable to a surface science approach.

My philosophical approach to research is to attack unsolved puzzles in fun and interesting ways. Leveraging creativity, theory and experimentation, my research group will address any and all unanswered questions. These will expand to my broader research interests, which include polymer mechanics, fiber formation, drug delivery, solubility phenomenon, sustainable chemistry and biomimecry. Using my previous experience as a guide, I will forge as many collaborative ties as possible, taking an interdisciplinary approach to tackling problems in a wide variety of fields.

Selected Publications:

K. Golovin, S. P. R. Kobaku, E. T. DiLoreto, J. M. Mabry, and A. Tuteja, "Designing Durable Icephobic Surfaces". *Science Advances*, 2, 3, e1501496 (**2016**).

K. Golovin, D. H. Lee, J. M. Mabry and A. Tuteja, "Transparent, Flexible, Superomniphobic Surfaces with Ultra-Low Contact Angle Hysteresis". *Angewandte Chemie*, 52: 13007–13011 (**2013**).

K. Golovin and S. L. Phoenix, "Effects of extreme transverse deformation on the strength of UHMWPE single filaments for ballistic applications". *Journal of Materials Science:* 1-12 (**2016**).

K. Golovin, J. W. Gose, M. Perlin, S. L. Ceccio and A. Tuteja, "Bio-Inspired Surfaces for Turbulent Drag Reduction". *Philosophical Transactions of the Royal Society A,* 374: 20160189, (**2016**).

R. J. Stafford, K. Golovin, A. Dickinson, T. R. Watkins, A. Shyam and E. Lara-Curzio, "Comparison of Elastic Moduli of Porous Cordierite by Flexure and Dynamic Test Methods". *Ceramic Engineering and Science Proceedings*, 33, 6 197-203 (**2013**).

476006 Adventures in Liquid Crystals

Abstract:

Research Interests:

Liquid crystals are a unique class of synthetic materials with tunable long-range orientations. Cooperative reorientation of liquid crystals (LCs) in response to subtle molecular-level changes at the interface with external stimuli provides amplified responses, which makes them promising materials for numerous industrial applications.

In the first year of my postdocoral work with Prof. Juan de Pablo at the University of Chicago, my research was focused on designing responsive and active interfaces using liquid crystals. Developing a simple set up, we demonstrated that liquid crystal interface is capable of reporting aggregation of polypeptides at lipid-aqueous interfaces at nanomolar concentrations, and therefore offers the potential for early diagnoses of neurodegenerative diseases. The results of this work was published at *Advanced Functional Materials* and appeared on the journal cover page.

Furthermore, in collaboration with the Argonne National Laboratory, we perform synchrotron X-ray reflectivity measurements to obtain structural information of liquid crystal molecules at the interface. Our experimental measurements are accompanied by an advanced theoretical and computational analysis.

In the second year, I have been exploring the topological defects in nematic liquid crystals induced by colloidal particles. Specifically, we have examined experimentally and theoretically, the effect of edge curvature of colloidal particles with homeotropic surface anchoring on their defect configuration and self-assembly. Our studies indicate that the edges sharpness modifies the defect structure, which can significantly affect the interaction between particles and their eventual self-assembly in the nematic liquid crystal.

Blue phase (BP) Liquid crystals are another class of nematic liquid crystals that I am interested in. In fact, BPs are nematic liquid crystals with high chirality that represent unique ordered defect structures over narrow range of temperature. Specifically, I study polymer stabilized blue phases. Using geometrical confinement and organic reagents, I have been able to obtain the blue phases at room temperature and expand the blue phase temperature range.

Although my research in the future will focus mainly on studying fundamental aspects of liquid crystalline materials and their potential applications in electronics and biomedical devices, studying the physics of the collective migration of epithelial and endothelial cells is another field which I am interested to pursue. After accomplishing my Ph.D. degree in polymer physics at ETH Zurich and before joining the de Pablo lab, I have had an opportunity to work on different projects at Harvard University with Prof. Jeffery Fredberg and use my expertise in Polymer Physics and Particle Image Velocimetry to investigate the basic physical processes underlie collective migration of epithelial and endothelial cells in confined geometries and stagnation points. From these works, I have authored and co-authored several manuscripts in high-impact peer-reviewed journals.

Teaching Interests:

With the background in polymer engineering from the Amirkabir University and Polymer Physics from ETH Zurich, I have strong foundation on materials science in particular polymer science related courses including polymer physics, polymer chemistry, crystallization, colloidal systems, rheology and polymer characterization.

475720 Nanostructured Based Lab-on-Chips for Detection of Single Biomolecules

Abstract:

Research Interests:

Nanostructured materials and nanofabricated lab-on-chip devices can play pivotal roles in biomedical research applications including genetic diagnosis and sensitive pathogen detection. Identifying genetic disorders or disease states on-chip requires dynamic manipulation and efficient trapping of a small number of molecules of the target nucleic acid. Current technologies are limited either by sensor resolution or by trapping a low concentration of molecules at the detection sites. Inspired by the state of the art nanofabrication and nano-synthesis methods, my research is focused on designing and fabricating lab-on-chip devices and nanostructured electrodes to overcome these limitations.

In order to enhance the detection sites for biomolecules in low concentration analytes (e.g. Dopamine) I have developed metallic/metal oxide nanostructures with an extended surface area for electro-chemical and biochemical reactions. I have adopted a similar approach to develop nanostructured surfaces to efficiently immobilize DNA and proteins, as part of assays to indirectly detect a wide range of biomolecules. I use electrochemical and lithography techniques to fabricate nanotubular and nanowire structured electrodes as well as nano/microfabricated diagnostics platforms. The platforms have successfully been used for sensitive detection of biomolecules such as dopamine, glucose, uric acid and ascorbic acid as well as pathogens such as Escherichia coli and Staphylococcus aureus.
To further improve trapping efficiency of the target sample on the detection sites, I have developed a sample-delivery system based on reversible, tunable nanofluidic confinement of biomolecules based on the dielectrophoretic force, applied using Silicon Nitride and Indium Tin Oxide nano-patterned electrodes. The resulting device can concentrate and manipulate molecules at high throughput; and can be used to create open-access and simple designs that can be easily interfaced with microfluidic devices for buffer exchange and sample processing; and integrated with optical detectors or on-chip nanostructured electrodes to detect single biomolecules (such DNA fragments up to 100,000 base pairs in length). The ease of fabrication and uncomplicated instrumentation makes this technology a unique point of care instrument for optical detection of biomolecules.

To immediately initiate my research program as a faculty member, I envision three potential projects: (1) Graphene-TiO_2 electrodes for biological and chemical detection, (2) ITO/SiN_x device with programmable electrodes for Cell/DNA sorting and analysis, and (3) 3D Microelectrodes based lab-on-chip for study biochemical reactions.

Teaching Interests:

I would be interested in applying my teaching principles to a wide range of undergraduate and graduate courses in Thermodynamics, Kinetics Fluid Mechanics, Heat Transfer, Transport phenomena, Mass Balances, Biomedical image analysis, Materials properties and characterization, Mechanics of materials, Biomechanics, Fluid Mechanics, Biomedical Instrumentation, tissue engineering, Engineering design, Bio/nanoelectronics. I would like to pursue the development of more specialized courses in biomedical engineering such as Bio/nanomaterials, biosensors, design of medical devices and nanofabrication methods. The fields of materials science along with the current explosions in biomedical engineering and nanotechnology offer many opportunities for the development of interdisciplinary courses.

475739 Harnessing the Power of the Extracellular Matrix to Control Wound Healing and Tissue Regeneration

Abstract:

Research Interests:

Complications arising from current therapies and the increasing prevalence of disease necessitate the development of novel biomaterials for tissue engineering applications. The next generation of materials entering the clinic for cardiac applications includes biomaterials derived from decellularized cardiac extracellular matrix (ECM). While current *in vivo* investigations performed by our labs and others have shown the potential for these materials *in vivo*, the mechanisms by which these materials lead to improved patient prognosis are widely unknown. Classically, the concept of immune response to biomaterials has focused on the negative aspects of a foreign body response. My current postdoctoral work has led to the evaluation of composite silk-cardiac ECM patches for use in the treatment of heart disease. Specific investigations have focused on the role developmental tissue age plays in decellularized ECM composition, immunomodulation, and induced cell response. Using extensive data collected from RNA-SEQ and proteomics evaluation, the future work of the Stoppel laboratory surrounds the hypothesis that ECM proteins and peptides have the ability to modulate host response to biomaterials and wound healing cascades. The long-term goal of the laboratory will be to develop methods and strategies to better understand the biological mechanisms by which these responses occur. Our work will support the development of biomaterials that actively modulate immune response in a controllable and predictable fashion to rapidly advance the development and subsequent utilization of these naturally-derived materials in patients.

To accomplish these goals, we will aim to elucidate the mechanisms of host response to decellularized ECM-containing biomaterials under both healthy and diseased states, focusing on how the wound healing cascade varies in response to biomaterial formulations and ECM composition. A specific emphasis will be placed on therapeutic applications for muscle-based tissues to further advance the fields of cardiac tissue engineering, treatments for muscular dystrophies, and replacements for defects in soft muscle tissue. Students will have the opportunity for interdisciplinary training, as developing new materials requires bioreactor development and optimization via engineering strategies and computational modeling, material property evaluation, and assessment and optimization of designed platforms via traditional biological assays. The interdisciplinary nature of the group will necessitate strong collaborations, particularly with clinicians and immunologists, that will further enrich this training environment. In addition, the Stoppel laboratory will have a focus on development of a diverse group of engineers with strong communication and interpersonal skills to aid in career advancement and contribution to the scientific community.

My own training has included work with alginate hydrogels and cell encapsulation during my PhD in Prof. Susan Robertsâ€™ lab at the University of Massachusetts Amherst, along with postdoctoral training in the laboratories of Profs. David Kaplan and Lauren Black at Tufts University, developing new composite natural materials using silk fibroin and decellularized ECM for cardiac applications.

Teaching Interests:

In the Spring of 2016, I taught Honors 101: Artificial Organs and Prosthetics at UMass Boston as the professor on record as a part of my NIH IRACDA postdoctoral fellowship through the Training in Education and Critical Research Skills (TEACRS) Program at Tufts University. In addition, I have particiated in many workshops and short courses related to diversity and engagement in the classroom, and I hope to have a positive impact on the education of both graduate and undergraduate students. As a fundamentally trained chemical engineer with a focus on biomaterials and bioreactor design, I am excited to teach courses related to materials and energy balances, transport, reaction kinetics, reactor design, biomedical applications, or polymers/materials science.

475758 Polymer Science As a Tool for Materials Design and Biological Discoveries

Abstract:

Research Interests: I envision a research group focusing on emerging questions at the interface of polymer science and biology, with the vision to push the boundary of classical polymer science as well as to solve challenges in energy, environmental science and healthcare. Empowered by the knowledge of polymer science, I will focus on the development of new functional soft (bio) materials for energy harvesting and healthcare, elucidating the roles of biological gels in health and disease, and exploiting the use of synthetic biology in creating designer biomaterials.

Soft materials with extreme mechanical properties and their mechanics under large deformation. In the course of evolution, biological systems have optimized themselves with extraordinary properties. For instance, cells and tissues are extremely soft with typical Youngâ€™s modulus lower than 100kPa, muscles change their stiffness at will, and skins accommodate large deformation easily and self-heal upon damage. Seamlessly interfacing with them offers approaches to solve challenges in healthcare and improving the quality of life, but often requires materials with properties that are comparable, if not the same, to biological objects. Using relatively well-established polymer science as a tool, I will adopt a bottom-up approach to design materials on the molecular-level, enabling new materials with extraordinary mechanical properties and functions. Initially I will focus on (i) extremely soft, yet dry gels and (2) malleable, tough, self-healing polymers. In parallel, I will develop new, fundamental understanding of the mechanics of such compliant materials under large deformation and explore their applications in stretchable electronics and biomedical scaffolds from 3D printing.

Biological gels in health and disease. Biological gels, such as mucus and extracellular matrix, are the major components of mucosal surfaces that line respiratory, gastrointestinal and reproductive tracts. Small changes in biophysical and biochemical properties of these gels result in their abnormal functions that are associated with chronic obstructive pulmonary disease (COPD), cystic fibrosis, inflammatory bowel disease and reproductive tract infection. Restoring the functions of biological gels is critical to treatment of these diseases; this requires elucidating their complex biophysical and biochemical origin. To this end, I will harness relatively well-established knowledge and tools in physical and chemical science to answer emerging clinically and therapeutically important questions related to these biological gels. Using the concept of polymer brush established in polymer physics enables my recent discovery of a new paradigm for mucus clearance, the primary innate defense system for mammalian lungs. I will adapt this unique interdisciplinary approach to elucidate the biophysics and biochemistry of native rather than reconstituted mucus, to study the interaction between mucus and epithelial cells, and to explore the roles of mucus on functionalities of microbiome.

Synthetic biology as a tool for designer biomaterials. Classical polymer science is largely enabled by precise chemical synthesis the can create polymers with controlled structure and thus properties. Yet biopolymers such polypeptides not only have a prescribed structure, but also consist of motifs that possess different strength of interactions such hydrogen bonds, electrostatic and hydrophobic-hydrophilic interactions. Such interactions are relatively weak, yet enable biopolymers with robust and deterministic functions. Due to the diversity and complexity of these interactions, it is challenging to understand how they determine the properties of biopolymers; yet this offers opportunities to classical polymer science. In parallel, emerging synthetic biological techniques allow precise control over both the structure of biopolymers and the types of weak interactions. Harnessing the power of synthetic biology, I will design and create biopolymers with prescribed structure and weak interactions to explore the fundamental polymer science of such systems. Initial research will focus on amphiphilic biopolymers with ionic crosslinkable domains; the amphiphilic nature will allow the polymer to self-assemble to useful structures such as vesicles, whereas the crosslinkable sites can be used to fix the structures after self-assembly. In parallel, I will explore the application of such structures as synthetic vaccines.

Teaching Interests: With my diverse background, B.S. in Physics, Ph.D. in Materials Science, and Postdoctoral experience in Chemistry, Materials Science and Biology, I am well prepared to teach a broad range of topics in materials science. Examples of specific undergraduate courses include polymer chemistry, thermodynamics, mechanics, solid-state physics, and applied mathematics. Examples of advanced undergraduate or graduate courses include polymer physics and chemistry, advanced thermodynamics, materials characterization, biological physics and soft condensed matter physics.

I am also dedicated to develop interdisciplinary courses to contribute to the diversity of curriculum in classical Materials Science, Chemistry, or Chemical Engineering. As modern materials science and engineering is growing into interdisciplinary areas, the emphasis in materials science education should alter accordingly. In addition to fundamental courses, focuses should also be placed on interdisciplinary areas where materials science is coupled with physics, chemistry, and biology; an example of such area is soft matter and polymer science. Knowing the essential concepts and techniques in soft matter and polymer science greatly broadens the view and expertise of students and benefit to their research; according to my experience, however, there is almost no such course dedicated to provide students with an overview of essential techniques through hands-on experience, and only through research projects in laboratories students will have a chance to learn some of them. As a start, I would like to introduce a new course to cover a wide range of techniques and concepts in soft matter and polymer science in combination with hands-on experience. Specific course subjects will cover chemical synthesis, chromatography, light scattering, osmotic measurements, rheology, microscopy, wetting and de-wetting, cell culture, and computer simulation. Each topic will be covered by a two-week lecture; one week focuses on in-class discussion and the other focuses on laboratory training. Students will be required write a short scientific report about each topic and use the knowledge learned during the course to complete a collaborative final project selected from the frontiers of research.

476318 Towards an Understanding of Catalytic Synthesis and Application of Nanomaterials

Abstract:

Research Interests:

My research leverages the intersection between in-situ metrology (XPS, XRD, ESEM, ETEM), process engineering and material science to enable new developments in bottom up novel material design and synthesis. The key scientific questions in my field relate to understanding mechanisms that can be used to achieve atomic level control during bottom up synthesis of novel materials, while the engineering challenges center around developing scalable production processes and interfacing the new materials into functional systems for applications.

2D materials serve as ideal model systems that can be leveraged to develop a fundamental understanding of the mechanisms associated with atomic length scale control for material synthesis using scalable processes while simultaneously enabling high impact applications; e.g. atomically thin membranes (ATMs) offer transformative opportunities for desalination, dialysis, fuel cells, solvent separation, DNA sequencing and can enable in-situ metrology for solid-liquid systems. Such in-situ observations could offer unprecedented insights to help further fundamental understanding of heterogeneous catalytic processes, crystal nucleation/growth/dissolution, material performance in batteries, super-capacitors, bioprocesses, living cells and their interfaces.

My long term vision is to make significant contribution towards moving the design, synthesis and device integration of nanomaterials from their current state of trial and error based experimentation to real time, direct evidence based, fundamental understanding driven rational engineering approaches for atomic length scale control in large-scale manufacturing processes.

Teaching Interests:

I am comfortable teaching a wide range of courses, both introductory and advanced at the undergraduate and graduate levels in Process/Chemical Engineering, Mechanical Engineering, Materials Science and Nanotechnology but have a keen interest towards courses related to material synthesis, catalysis, reaction engineering, metrology and process design & development.

476072 Photovoltaic Processes

Abstract:

Research Interests: The price of photovoltaic (PV) power only depends on cost per area, power conversion efficiency, and rate of efficiency degradation. The microscopic processes and commercial production processes that give rise to those metrics are diverse, yet intimately connected. Research areas include substrates, transparent conductive contacts, absorbers, buffers, power degradation routes, encapsulation, edge sealant, power storage, and energy storage. My research will ultimately advance each of these 9 aspects through the simultaneous study of processes used to produce the PV devices, processes of their operation, and degradation processes. A deep understanding of each aspect will be necessary for long-term success in PV research, for academics and companies alike. In the near-term, several areas will be attacked: metal micro- and nano-grids for transparent conduction in solar cells, thin film absorber development, and new buffer compounds and deposition processes.

My postdoctoral research has been at the National Renewable Energy Laboratory, where I studied the selenization of Cu-Ga-In precursor films, the most popular industry deposition process for $Cu(In,Ga)Se_2$ (CIGS) thin film absorbers [1-3]. To investigate very recent world record CIGS power conversion efficiencies that used a potassium fluoride post-deposition treatment, I chose to do the first basic material science study of $Cu_{1-x}K_xInSe_2$ (CKIS) alloys [4]. I then used this foundational knowledge to fabricate an officially-measured 15.0%-efficient Cu-K-In-Se solar cell (matching the world record $CuInSe_2$efficiency, despite being grown at lower temperature, and without the 'gold standard' three-stage co-evaporation process).

My graduate research was at the University of Florida with Tim Anderson, where I studied thermodynamics of the Cu-Ga-In system. That work culminated in a paper that was recently nominated for the Cahn prize [5]. I also studied diverse aspects for halide chemical vapor deposition (CVD) of CIGS: complex chemical equilibrium calculations, finite element method fluid flow simulations of atmospheric pressure CVD, experimental Cu, Ga, and In volatilization by HCl reaction, with supporting empirical and analytical fluid flow modeling, and the first experimental CVD growth of Cu-Se, In-Se, and Cu-In-Se [6].

Teaching Interests: Along with research, I will teach transport, thermodynamics, electrical engineering, materials science classes . Where possible, I will add emphases on thin film growth, semiconductors, electron devices, and photovoltaics.

[1] **Muzzillo, C.P.**, Mansfield, L.M., DeHart, C., Bowers, K., Reedy, R.C., To, B., Noufi, R., Ramanathan, K., Anderson, T.J., (2014) â€œThe effect of Ga content on the selenization of co-evaporated CuGa/In films and their photovoltaic performance,â€ *40th IEEE PVSC Proc.*, Denver, CO, 1649-1654.

[2] **Muzzillo, C.P.**, Mansfield, L.M., DeHart, C., Bowers, K., Reedy, R.C., To, B., Ramanathan, K., Anderson, T.J., (2015) â€œDifferences between CuGa/In and Cu/Ga/In films for selenization,â€□ *42nd IEEE PVSC Proc.*, New Orleans, LA, 1-6.

[3] **Muzzillo, C.P.**, Mansfield, L.M., Ramanathan, K., McGoffin, J.T., Sites, J.R., Anderson, T.J., (2016) â€œPhotovoltaic properties of selenized CuGa/In films with varied compositions,â€□ 43rd IEEE PVSC Proc., Portland, OR, in press.

[4] **Muzzillo, C.P.**, Mansfield, L.M., Ramanathan, K., Anderson, T.J., (2016) â€œProperties of $Cu_{1-x}K_xInSe_2$ alloys,â€□ *Journal of Materials Science*, 51(14), 6812-6823.

[5] **Muzzillo, C.P.**, Campbell, C.E., Anderson, T.J., (2016) â€œCu-Ga-In thermodynamics: Experimental study, modeling, and implications for photovoltaics,â€□ *Journal of Materials Science*, 51(7), 3362-3379.

[6] **Muzzillo, C.P.**, (2015) â€œChalcopyrites for solar cells: chemical vapor deposition, selenization, and alloying,â€□ Dissertation, Chemical Engineering, University of Florida, Gainesville, FL.

476078 Polymeric Materials for Biomedicine and Nanotechnology

Abstract:

Polymeric Materials for Biomedicine and Nanotechnology

Stephanie Christau,University of Michigan1 st Year Postdoctoral Fellow

Postdoctoral Project:

Preparation of Polymeric Janus Particles for Stimulation of Immune Cells. Advisor:Joerg Lahann, Department of Chemical Engineering, Biointerfaces Institute, University of Michigan.

Dissertation:

Smart Hybrid Materials with Tailored Functions: Immobilization of Gold Nanoparticles into Stimuli-Responsive Polymer Brushes.

Advisor:Regine von Klitzing, Department of Chemistry, Berlin Institute of Technology (TU Berlin).Co-advisor: Jan Genzer, Department of Chemical Engineering, North Carolina State University.

Research Interests:

Although my degree is in chemistry, I have been exposed to different areas of research throughout my academic career. This includes general chemistry, polymer science and engineering, and applied physics. As my PhD thesis focused on the **development of responsive surface coatings** made of **polymer brush/gold particle hybrids,** I have a strong background in polymer and colloid science. In addition to receiving training in standard characterization techniques (ellipsometry, spectroscopy, electron microscopy), I have gained experience in x-ray and neutron scattering techniques. The latter includes writing proposals to get measurement time at the neutron facility, proper planning of the scattering experiments, and the data analysis which requires sophisticated fitting procedures. In my postdoctoral project, I am applying my skill set while also broadening my knowledge base in the area of biomedical science. Yet, I familiarized myself with the preparation of **polymeric particles and their characterization**, and **cell culture experiments**. I have invaluable experience in many research areas including **chemistry, material science, polymer and colloid science and engineering, biomedical engineering, and applied physics**.

Postdoctoral Research Fellow, January 2016 â€“ present

Department of Chemical Engineering, University of Michigan, Ann Arbor, MI, USA.Â *Advisor:* Joerg Lahann

Development of polymeric nanoparticles for T cell activation

Electrohydrodynamic co-jetting of multicompartmental nano-/microparticles

Postdoctoral Researcher, October 2015 â€“ January 2016

Department of Chemical and Biomolecular Engineering, North Carolina State University, Raleigh, NC, USA.Â *Advisor:* Jan Genzer

Preparation of polymer brush charge gradients on flat substrates

Ph.D. (chemistry). DFG-IRTG Fellow, July 2012 â€“ October 2015

Department of Chemistry, Berlin Institute of Technology (TU Berlin), Germany.Â *Advisor:*Regine von Klitzing

Department of Chemical and Biomolecular Engineering, North Carolina State University, Raleigh, NC, USA *Advisor:* Jan Genzer.

Preparation of stimuli-responsive polymer brush surface coatings and immobilization of gold nanoparticles

Self-assembly and plasmonic properties of polymer brush/nanoparticle composite materials for sensor applications

Neutron reflectometry measurements and data fitting to resolve internal structure of polymer brush/nanoparticle hybrids

Preparation of polymer brush gradients (grafting density, charge)

Diploma (chemistry). *Researcher,* 2011 â€“ 2012

Department of Chemistry, Berlin Institute of Technology (TU Berlin), Germany.Â *Advisor:*Â Martin Schoen (TU Berlin)

Department of Chemistry, Eidgenoessische Technische Hochschule (ETH) Zurich, Switzerland.Â *Co-advisor:Â* Martin Quack (ETH Zurich)

High-resolution FTIR spectroscopy of the carbon dioxide hot bands and their contribution to the greenhouse effect (measurements and simulation)

Undergraduate Researcher, 2010 â€“ 2011

Department of Chemistry, Berlin Institute of Technology (TU Berlin), Germany.Â *Advisor:Â* Regine von Klitzing.

Structure properties of polymers in bulk and under confinement (viscometry and atomic force microscopy measurements)

Teaching Interests:

I mentored several undergraduate students at North Carolina State University, TU Berlin and the University of Michigan. I also mentored two bachelor theses; duties included developing a feasible project, organizing overall timeframe and experiments, and training in the laboratory (both, synthesis and characterization techniques) and in scientific writing. I also TAed and guest-lectured undergraduate courses in â€˜polymer and colloid scienceâ€™ at the TU Berlin. Duties included course material, lecturing, and office hours. Furthermore, I am participating in the course â€˜High Impact Principles and Practices for STEM Educationâ€™ at the University of Michigan. The purpose of this course is being prepared to teach effectively as future faculty member in science and engineering.

Successful proposals:

German Science Foundation (DFG, Deutsche Forschungsgemeinschaft) Postdoctoral Research Grant, 2016.

Future Directions:

As a faculty I would like to continue in the field of polymer and colloid science. The projects I am proposing include the **development of polymeric (bio)materials**for

Stimulation of T cells to elicit a specific immune response â€“ applications include cancer immunotherapy and cancer vaccines.

Controlled and sustained *delivery of cytokines/chemokines* â€“ applications include chronic wound healing and cancer immunotherapy.

Development *of composite materials with tunable plasmonic properties* through immobilization of metallic nanoparticles â€“ applications include the development of smart sensors (via polymeric particles or surface coatings) and photothermal therapy (*biodegradable materials*).

Publications:

Journal articles.

(1) Christau, S.; Möller, T.; Brose, F.; Genzer, J.; Soltwedel, O.; von Klitzing, R. Effect of gold nanoparticle hydrophobicity on thermally induced color change of PNIPAM brush/gold nanoparticle hybrids, *Polymer* **2016**, doi: 10.1016/j.polymer.2016.03.088

(2) Kesal, D.; **Christau, S.**; Krause, P.; Moeller, T.; von Klitzing, R. Uptake of pH-Sensitive Gold Nanoparticles in

Strong Polyelectrolyte Brushes, *Polymers* **2016,**8, 134; doi:10.3390/polym8040134

(3) Christau, S.; Möller, T., Yenice, Z.; Genzer, J.; von Klitzing, R. Brush/gold nanoparticle hybrids: Effect of grafting density on particle uptake and distribution within weak polyelectrolyte brushes. *Langmuir,* **2014,** *30*, 13033-13041.

(4) Christau, S.; Thurandt, S.; Yenice, Z.; von Klitzing, R. Stimuli-Responsive Polyelectrolyte Brushes as a Matrix for the Attachment of Gold Nanoparticles: The Effect of Brush Thickness on Particle Distribution. *Polymers (Basel)*, **2014**, *6*, 1877.

(5) Üzüm, C.; **Christau, S.**; von Klitzing, R. Structuring of Polyelectrolyte (NaPSS) Solutions in Bulk and under Confinement as a Function of Concentration and Molecular Weight. *Macromolecules*, **2011**, *44*(19), 7782-7791.

Reviews.

(1) Christau, S.; Genzer, J.; von Klitzing, R. Polymer brush/metal nanoparticle hybrids for optical sensor applications: From self-assembly to tailored functions and nanoengineering. *Z.Phys.Chem.,* **2015,** *229 (7-8)*, 1089-1117.

476086 Design of Advanced Materials for Application in Clean Energy and Carbon Capture and Utilization

Abstract:

Design of Advanced Materials for Application in

Clean Energy and Carbon Capture and Utilization

Peter C. Psarras

Energy Resources Engineering, Stanford University **Introduction** This past half-century has brought much change to the landscape surrounding the chemical processes that sustain our global socio-economic framework. In the midst of this evolving climate, it has become increasingly apparent that new motivations, such as those for greater energy-efficiency and less carbon-intensive processes, are necessary to avert serious global consequences. Fortunately, this call to attention has re-energized the field of materials research, whereby promising new technologies can bridge the gap between out-of-date processes and those better-suited for consequence mitigation. Concurrent to this scientific revitalization has been a related emergence in its own right Ð the dawn of high performance computing environments; thus, theoretical methods represent a powerful approach to the design of materials aimed at resolving the shortcomings of essential industrial processes through revision and innovation. These methods are often favorable in situations where there exists a need to explore and screen a number of materials at low cost and risk, where conditions prove unsuitable or even unfeasible via conventional methods, or where an understanding is sought at molecular levels or timescales that are incompatible with experimental measurement. This presentation will introduce how different theoretical methods can be used to solve four issues related to climate change: nitrogen-selective membranes for indirect CO_2 capture, nitrogen-functionalized carbon sorbents for direct CO_2 capture, the assessment of carbon neutrality for irreplaceable industrial processes, and the effect of late-transition metal doping on methane selectivity in Fischer-Tršpsch (FT) catalysis. **Research Interests: Nitrogen-Functionalized Porous Carbons for Enhanced CO_2 Capture** Traditional amine-based solvent technology is notoriously energy intensive, owing to high regeneration temperatures. Additionally, this process is water-intensive, a factor that is becoming increasingly important in drought-ridden states like California. Solid sorbents represent an alternative method for CO_2 capture, with liquid solvents replaced by (typically) large surface area, carbon-based porous frameworks. These sorbents are generally low-cost, easily fabricated, and are far less energy intensive in terms of regeneration. Further, their design can be customized through the inclusion of surface functional groups. Unfortunately, current sorbents display CO_2 loadings that are deemed too low to be cost-competitive with solvent-based scrubbing. Nitrogen-functionalization of these porous carbons could result in more efficient CO_2 capture through the introduction of surface charge heterogeneity which can selectively interact with the CO_2 quadrupole moment. Similar studies have examined the effects of oxygenated functional groups on CO_2 uptake and selectivity over N_2. This study employs grand canonical Monte Carlo methods to explore the effect of quaternary, pyridinic, pyrrolic, oxidized-pyridinic, and pyridonic groups on CO_2 uptake in porous carbon sorbents. Doping amounts are varied to ascertain the optimum coverage for CO_2 capture. Additionally, gaseous mixtures of CO_2/N_2 and $CO_2/N_2/H_2O$ are examined to assess CO_2 selectivity. Theoretical performance is validated through comparison with experimentally obtained CO_2 and N_2 isotherms over fabricated hierarchical micro/mesoporous N-doped carbon sorbents. The combination of these methods will inform on sorbent design by illustrating which groups are most important for enhanced CO_2 uptake. Design to include more of a particular functional group can be achieved through, for example, a change in carbonization temperature. **Assessment of the CO_2 Capture Potential from Irreplaceable Industrial Sources**

In 2013, the US industrial sector emitted approximately 1.4 gigatonnes of carbon dioxide (Gt CO_2), or 21% of total US CO_2 emissions Ð the third highest figure for any economic sector behind transportation (27%) and electricity (31%). These emissions can be further categorized as direct (fuel combustion accountable on-site, e.g., stationary combustion), indirect (assigned to electricity purchased for power and off-site steam generation) and process (CO_2 liberated as a reaction by-product). Direct and indirect emissions generally constitute ca. 80% of total emissions, with process emissions making up the balance, though the relative contribution of process emissions to total emissions is known to vary by industry.

While these industrial process emissions constitute a smaller percentage of total US CO_2 emissions, they produce commodities like glass, cement, ammonia and steel Ð items that form the irreplaceable fabric of industrialized nations. Unlike the power sector, for example, where mitigation might be achieved through a transition to renewable forms of energy and adoption of best practice technologies, there are few CO_2-free alternative routes to product for most industrial commodities; thus, these irreplaceable industrial processes (IIPs) represent committed CO_2 emissions that remain largely unabatable. Further, these IIPs command attention for two important reasons: 1) there is a committed nature to these emissions by virtue of process chemistry, and perhaps more importantly 2) they can yield exhaust streams with higher CO_2 content when compared to the high volume flue exhausts of the power sector. As the cost of CO_2 separation scales inversely with initial dilution of a mixed feed stream carbon capture technology retrofits have the potential to efficiently and economically divert emissions from industrial exhaust streams to viable CO_2 utilization opportunities, such as enhanced oil recovery (EOR), food processing, refrigeration, and fertilizer production. To assess the capture potential of these irreplaceable industries, it is necessary to geo-reference these sources alongside all current and potential future CO_2 sinks (utilization or market opportunities), with the goal of making economically sound linkages between source and markets of comparable scale. This entails a cost analysis of on-site capture, compression, and transport costs. Geographic information systems (GIS) mapping can assist in identifying regions of high CO_2 demand, geographically-logical source-sink partnerships, and viable routes of CO_2 transport. The aim is to classify these IIPs based on carbon-capture ÒreadinessÓ, which is ultimately a combination of the industry-specific and site-specific factors listed above.

Vanadium-Ruthenium Alloys for Indirect CO_2 Capture Membranes represent a promising technology for carbon capture, yet several material challenges persist. Ideally, a membrane will need to demonstrate excellent CO_2 selectivity, strong mechanical and thermal stability, and high membrane flux. Here, investigate alloyed V/Ru membranes for selective nitrogen capture. In this technology, rather than separating out CO_$_2$ directly, nitrogen is separated from the flue stream. Nitrogen is far more efficient to separate, owing to its greater flue concentration and corresponding higher pressure differential across the membrane. Consequently, after N_2 separation the remaining exhaust components become more concentrated, CO_2 included, making all downstream capture points more efficient by the same virtue. This study will prepare 14 V/Ru alloys in concentrations spanning those considered in real application. Vanadium is known to bind nitrogen strongly, making it a good candidate for activation of the N-N triple bond. However, this tight binding inherently slows membrane diffusion. Nitrogen permeability can be represented as a product of its solubility and diffusivity. The former is sought by considering the effect of Ru concentration on relative binding energies and vibrational frequencies of interstitially-bound nitrogen, while diffusivity can be examined by calculating all potential TS-pathways, tabulating their rates, and employing kinetic Monte Carlo methods. Together, these results will combine to predict nitrogen permeability as a function of Ru content. These findings can inform on membrane design, whereby a theoretically optimal Ru composition can be validated experimentally. **The Effect of Late Transition Metal Doping on Methane Selectivity in Fischer-Tršpsch Catalysis (Dissertation Work)** Fischer-Tršpsch synthesis has endured a long and industrious career as the premier chemical method for the synthesis of hydrocarbons from non-petroleum-based precursors. Historically, a collection of factors has dictated interest in the FT process, including foreign policy, availability of precursor(s), availability of alternate fuel sources and reserves, and environmental considerations. The effect of late transition metal substitution into Fe(100), Ni(111), and Co(0001) surface analogs is investigated using density functional theory (DFT) methods. The surface is modeled first using a 7-atom cluster, with perimeter atoms saturated with hydrogen atoms to approximate surface coordination and mitigate dangling bond artifacts, and next with plane-wave periodic models. *Ab initio* calculations are performed to predict adsorption energies and transition state geometries and energies for eight surface adsorbate species: *C + *H, *CH + *H, *CH_2 + *H, and *CH_3 + *H to represent the hydrogenating steps, while *C + *CH, *CH + *CH, *C + *CH_3, and *CH_2 + *CH_2 represent species involved in four competitive coupling pathways. A review of the effect of Cu, Ag, Au, and Pd on the reaction energies and barriers associated with these critical steps is discussed. Methane selectivity can be described by a single descriptor, the carbon binding strength; thus, a comparison of these energies over different surface/promoter analogs is an effective method to screen for promising candidates. In this representation, a lowering of carbon binding strength is equivalent to the lowering of methane selectivity. **Teaching Interests:** My background in Chemistry has prepared me in the fundamentals of thermodynamics, quantum mechanics, catalysis, molecular simulations and adsorption. However, my research direction has forced me to become as proficient in the understanding of transport phenomena, fluid mechanics, and separations. I feel comfortable to instruct in any of these areas with confidence. Additionally, I plan to design a course centered around the fundamentals of electronic structure methods, tailored for application to climate change mitigation Ð suitable as a standalone course or special topic offering. Additionally, I have experience teaching several undergraduate lecture courses, including Physical Chemistry I and Survey of Physical Chemistry, as well as several laboratory courses. I have also had the great privilege of mentoring several students, many who have since published in high impact journals such as *Journal of Physical Chemistry C* and *Advanced Energy Materials*. Through my experiences in teaching and mentoring, I have developed the valuable skill of balancing research and grant writing with class time and student development.

476124 Characterization of Polymer Particles in Biological Environments for Drug Delivery Applications

Abstract:

Research Interests:

Vehicles for drug delivery are an ideal treatment solution for many diseases, including cancer, where the dosage of drugs is limited by toxic side effects. Creating a successful drug delivery system is difficult, however, and despite much research effort in this field, there are very few examples of drug delivery systems being used in current therapies. This lack of success stems from a limited understanding of how drug delivery vehicles interact with the biological environment and leads to poor design of the drug delivery systems. I intend to establish a research group to investigate the interaction of drug delivery vehicles with the biological environment (such as cells, blood, and proteins) using physical chemistry techniques (such as light scattering and calorimetry) in novel ways. My background in drug delivery and polymer physics will allow my lab to address issues with drug delivery vehicles in unique ways. This work can lead to successful drug delivery vehicles and a better understanding of how materials interact with the biological environment.

Education & Training

2013 – Present Postdoctoral Researcher, Chemical Eng., Biointerfaces Institute, University of Michigan

Advisor: Joerg Lahann

Project: Electrohydrodynamic co-jetting of polymer particles for targeted drug delivery applications

2007 – 2013 **Ph.D.** in Polymer Science and Engineering, University of Massachusetts Amherst

Thesis Advisor: Thomas P. Russell

Thesis Title: Particle behavior at anisotropically curved liquid interfaces

2008 **M.S.** in Polymer Science and Engineering, UMass Amherst

2003 – 2007 **B.S.** in Chemistry, Massachusetts Institute of Technology

Research Advisor: Paula T. Hammond

Project: Layer-by-layer polymer electrolytes for lithium batteries

Fellowships and Awards

Best TA for 'Polymer Characterization' at UMass Amherst as voted by students, Dec 2010

International Center for Materials Research International Research Fellowship, Nov 2009

Graduate Assistance in Areas of National Need (GAANN) Fellowship, Sept 2007 – Aug 2008

Current Research

My current research focuses on electrohydrodynamic co-jetting of polymer nanoparticles to target breast cancer stem cells. By surface modifying the particles with antibodies and incubating the particles with a breast cancer cell line, I have observed an increase in the uptake of the nanoparticles by the cancer stem cells using flow cytometry. The breast cancer cells are a rare population (typically less than 1% of the total population) and very difficult to target, yet an important target for cancer treatment.

Another thrust of my current research is the development of a method for analyzing particle behavior in complex solutions such as blood plasma. By using nanoparticle tracking analysis I have quantified the aggregation of nanoparticles in blood plasma. This method will allow for more thorough testing of particles *in vitro*and allow for better design of drug delivery vehicles.

Teaching Interests:

I am interested in teaching core classes as well as developing more advanced courses about polymers, such as Advanced Polymer Physics and Characterization. I have guest lectured in some graduate level classes at the University of Michigan, and was a TA several times at UMass Amherst (voted best TA for the class in 2010). I have taken several classes on teaching at the college level and I have mentored 12 students (undergraduate, high school, and masters) on research projects in the lab. I have also been involved in many outreach activities involving teaching. For instance, I taught chemistry to blind high school students at a summer camp where I had to tailor my teaching to address non-visual learning styles. The blind student summer camp also gave me much experience in accommodating lessons for those with disabilities as I had to adapt the experiments for blind students. Some of these adapted lessons were published in the *Journal of Chemical Education* in 2009.

Publications

McEnnis, K.; LaPres, C.; Santos Moreira, A.; Burkland, D.; Lahann, J.; Characterization of Nanoparticle Aggregation in Biologically Relevant Fluids. In preparation.

McEnnis, K.; Dinsmore, A. D.; Russell, T. P. Solid Particles Adsorbed on Capillary-Bridge-Shaped Fluid

Polystyrene Surfaces. *Langmuir.* **2015**, *31,* 5299–5305.

Lutkenhaus, J. L.; **McEnnis, K.**; Serghei, A.; Russell, T.P. Confinement Effects on Crystallization and Curie Transitions of Poly(vinylidene fluoride-co-trifluoroethylene). *Macromolecules*. **2010**, *43*, (8), 3844-3850.

Serghei, A.; Lutkenhaus, J. L.; Miranda, D. F.; **McEnnis, K.**; Kremer, F.; Russell, T. P. Density Fluctuations and Phase Transitions of Ferroelectric Polymer Nanowires. *Small.* **2010**, *6*, (16), 1822-1826.

Supalo, C. A.; Mallouk, T. E.; Amorosi, C.; Lanouette, J.; Wohlers H. D.; **McEnnis, K.** Using Adaptive Tools and Techniques to Teach a Class of Students Who are Blind or Low-Vision. *Journal of Chemical Education*. **2009**, *86*, (5), 587-591.

Lutkenhaus, J. L.; **McEnnis, K.**; Hammond, P.T. Nano- and Microporous Layer-by-Layer Assemblies Containing Linear Poly(ethylenimine) and Poly(acrylic acid). *Macromolecules*. **2008**, *41*, (16), 6047-6054.

Lutkenhaus, J. L.; **McEnnis, K.**; Hammond, P. T. The Glass Transition and Ionic Conductivity of Poly(ethylene oxide)/Poly(acrylic acid) Layer-by-Layer Assemblies. *Macromolecules*. **2007**, *40*, (23), 8367-8373.

Lutkenhaus, J. L.; Hrabak, K. D.; **McEnnis, K.**; Hammond, P. T. Elastomeric Flexible Free-Standing Hydrogen-Bonded Nanoscale Assemblies. *Journal of the American Chemical Society*. **2005**, *127*, (49), 17228-17234.

Reviews:

McEnnis, K.; Lahann, J. Nanoparticles for Targeted Drug Delivery. Invited review article in preparation for *Journal of Materials Chemistry B*.

476167 Highly Energy-Dense Rechargeable Alkaline MnO2-Zn Batteries for Grid-Scale Applications

Abstract:

Research Interests:

Current Research

Manganese dioxide (MnO_2) and zinc (Zn) are key constituents of primary or single-use alkaline batteries such as Duracell and Energizer. Single-use batteries are referred to as such because they deliver their entire energy capacity just once, after which they must be discarded. Despite this limitation, primary alkaline batteries are ubiquitous in society on account of abundance, low cost and relative safety of MnO_2 and Zn, and their low cost of manufacturing. Nevertheless, as environmental regulations tighten, the manufacture of these convenient energy cells necessitates a rethink. Producing a primary MnO_2-Zn battery consumes more energy than what it can deliver, and the use of MnO_2-Zn batteries is now a point of contention owing to their large impact on the environment. To this end, electrochemical engineering of MnO_2-Zn batteries towards making these cells rechargeable would represent a quantum advance. Rechargeability greatly improves the overall energy efficiency of these cells, and preserves its cost advantages. However, this objective has proved to be a herculean challenge for a generation of battery researchers owing to fundamental limitations in the material and electrochemical properties of MnO_2 and Zn. Briefly, MnO_2 can theoretically deliver a capacity of approximately 617 mAh/g. It delivers this capacity through a 2-electron electrochemical reaction; wherein each electron provides roughly 308 mAh/g. MnO_2 has been found to be rechargeable when the capacity has been limited to around 5-10% of the 617mAh/g. Its crystal structure breaks as more of the capacity is accessed, and it inherently forms electrochemically irreversible phases. Similar problems are associated with the zinc electrode, where higher utilization of its capacity causes dendrite formation, morphology changes and formation of inactive zinc oxides that ultimately lead to electrode failure. If the entire 2-electron capacity of MnO_2 can be accessed and problems associated with the zinc electrode can be averted, MnO_2-Zn batteries could, in theory, reach energy densities that are comparable to lithium-ion batteries. My current research at the City College of New York has successfully addressed these requirements. Allied with its low cost of manufacturing and safety, a rechargeable MnO_2-Zn battery could be a disruptive technology for energy storage.

Our breakthrough that permits accessing the 2-electron capacity reversibly was achieved by precisely altering the crystal structure of MnO_2 with dopants using a wholly novel and vastly cheaper synthetic route, and we have cycled the MnO_2 electrode to well over 6000 cycles at rates that are of interest to the battery community. Besides innovative technology development, I also unearthed fundamental insights about the material and its complicated electrochemical mechanism, wherein the dopants participate in those reactions involving $MnO_2$to prevent the formation of the electrochemically irreversible phases. Significant performance improvements were also achieved for the Zn electrode through use of complexing agents that inhibit morphological changes in Zn and localize the active material near the current collector. Novel crystal structures doped with elements that play a dual role in reducing charge transfer characteristics and encapsulating Zn were also discovered. The application of these strategies have resulted in cycling the Zn anode at 15-20% of its theoretical capacity for >900 cycles, where Zn has been known to fail at 5-8% of its of theoretical capacity in less than 300 cycles.

The cumulative improvements in the material and electrochemical properties of MnO_2 and Zn have resulted in fabrication of the very first rechargeable alkaline cell that can deliver a record high energy density of ~160 Wh/L, where 100% of the 2-electron capacity of MnO_2 and 15% of theoretical capacity of Zn is utilized. Theoretical calculations have shown that at ~30% utilization of the Zn theoretical capacity, a highly energy dense cell of ~250Wh/L is possible. My current work seeks to achieve these figures, and my work has laid strong foundations for the market entry of rechargeable MnO_2-Zn batteries that are cheaper but equally efficient as lithium-ion batteries. Overall, my postdoctoral work has resulted in the filing of 8 patents and I have recently submitted two manuscripts to high impact factor journals. It is currently under peer-review.

Previous Research

Doctoral Thesis: Design and assembly of nanostructured complex metal oxide materials for the construction of batteries and thermoelectric devices My graduate research at Purdue University served as an excellent primer for my current work. My doctoral work focused on synthesis of one-dimensional, complex metal oxide nanowires for thermoelectric and lithium-ion battery applications. I bridged two distinct technical domains â€“ energy conversion and energy storage â€“ with nanotechnology. Some highlights of my graduate work include the conception of novel synthetic routes for production of ultrathin perovskite nanowires and formalization of template-based approached for fabrication of complex metal oxides such as calcium cobalt oxide ($Ca_9Co_{12}O_{28}$). Saliently, for the latter, my work was the first demonstration of its kind wherein nanowires were synthesized specifically use in thermoelectric devices. Moreover, the thermoelectric characteristics of the nanowires were a significant improvement over competing alternatives reported elsewhere in literature. In addition, I also extended my approach to the synthesis of nanowires and nanostructures of lithium cobalt oxide ($LiCoO_2$) and investigated the impact of nanostructuring on the electrochemical properties of the material. The theme of combining application-driven research with fundamental investigations that seek to expand our understanding about nanomaterials has been a staple of my work since my graduate tenure. My graduate work resulted in publications and 1 patent.

Research Interests

I am a chemical engineer specializing in materials science and electrochemistry. I have a proven track record of advancing current understanding and develop commercializable technologies for energy generation and energy storage. Moreover, I have unique experience in scaling up my research from the nano- to macro scales, as well as from the bench to market. My doctoral work focused on developing new approaches for synthesis of one-dimensional, complex metal oxide nanowires and investigating their suitability for thermoelectric and battery applications. My postdoctoral research, on the other hand, has targeted the enhancement of the electrochemical properties of battery systems through ingenious use of materials science and scaling up these batteries for use in grid-scale applications. My training, research interests and innovation record will be key building blocks of my research program on multiscale engineering for energy conversion and storage. My program will comprise three pillars: (1) materials nanoscience, (2) materials synthesis, (3) device integration and scale-up. Among other projects, I intend to develop new thermoelectric materials for high-temperature applications, and conduct fundamental investigations towards improvement of aqueous battery chemistries. Beyond energy research, I am also interested in the development of catalysts for water oxidation, mesoporous materials for CO_2capture, and water purification. These pursuits will draw upon similar experimental approaches as my research program on energy.

Teaching Interests:

I have considerable teaching experience and previously served as the lead teaching assistant for undergraduate and graduate courses on chemical reaction engineering and transport phenomena in the School of Chemical Engineering at Purdue University. Some of these courses were subscribed by as many as 200 students. I delivered several lectures and led all tutorial and review sessions for these courses. I was praised by the students for the clarity of my presentation and conveyance of learning objectives, and my empathy for student advancement. I was awarded the Magoon Award for Excellence in Teaching, which is the highest accolade for teaching by a student at Purdue University. I can teach any course in the core chemical engineering curricula and technical electives such as electrochemistry and materials science. I value the old virtue of using the blackboard for instruction but also appreciate the power of multimedia and internet-based teaching in broadening the scope and reach of a course.

476173 Harnessing Interfacial Phenomena to Design New Soft Materials

Abstract:

Research Interests:

The intersection between colloidal and interfacial sciences has led to the development of self-assembled colloidal arrays which have been profoundly impactful for lithography, responsive materials, energy migration, and bio-sensing. The field now faces the challenge to transition from synthesis and assembly of proof of concept samples to engineering scalable methods that integrate functional colloidal assemblies into materials over large surface areas. In my research program, I aim to pioneer the incorporation of colloidal assemblies into soft materials by directing interfacial polymerization via chemical vapor deposition (CVD) onto colloidal assemblies at fluid interfaces. This approach will strategically leverage trapped states of colloidal assemblies with dynamic polymer growth to independently control the organization and orientation of colloidal monolayers from the polymer matrix for customization of material properties. Here I present previous works on (i) controlling the morphology of polymers deposited onto liquid substrates and (ii) developing a modular synthesis of colloidal Janus particles with diverse chemical compositions which together provide the platform for realizing my vision to harness interfacial phenomena to build colloidal assemblies into soft materials.

Research Experience:

CVD methods have been exploited for many years to transform the surface chemistry of structured materials by depositing uniform and conformal polymer coatings that preserve the underlying geometry and surface roughness. The initiated CVD (iCVD) process is unique because it uses free-radical initiators which eliminate precursor fragmentation present in other CVD methods, and produces polymers analogous to solution phase free-radical polymerization. This enables the iCVD process to deposit functional coatings incorporating hydrophobic, hydrophilic, or stimuli-responsive polymers. To extend deposition via iCVD beyond conformal thin films, we explored deposition onto liquids with low vapor pressures (such as silicone oil and ionic liquids) to exploit the dynamic diffusion and wetting at liquid—vapor interfaces. We demonstrated that the deposition can result in the formation of either polymer particles, planar films, or microstructured films at liquid surfaces depending on the spreading coefficient of the polymer on the liquid. In addition, precursors that are soluble in the liquid can absorb into the bulk and polymerize resulting in the formation of polymer—liquid gels, layered polymer composite films, and encapsulating polymer shells. This work elucidates the growth mechanism of polymers on liquid substrates and provides an understanding of how process parameters can be controlled to incorporate chemical functionality into a variety of structures and morphologies for potential applications in optics, sensing, and separations.

Inspired by the diversity of chemical functionality that the iCVD technique provides, we sought to bring the same range of chemical compositions to the fabrication of amphiphilic Janus particles which are considered colloidal analogues to molecular surfactants. The chemistry of Janus particles is the fundamental parameter that controls their behavior in bulk solution and at interfaces. To enable their widespread utilization, scalable methods that allow for the synthesis of Janus particles with diverse chemical compositions and shapes are highly desirable. We developed clickable Janus particles that can be functionalized through thiol-yne click reactions with commercially available thiols which widen the palate of chemical compositions. The extent of functionalization can be used to control the particle morphology, and thus the type of emulsion stabilized, as well as to fabricate composite Janus particles through sequential click reactions. Functionalizing Janus particles through thiol-yne click chemistry provides a fast-reacting, scalable synthesis method for the fabrication of chemically diverse Janus particles. Currently, we are working on applying our ability to tune the chemical composition to study their behaviors at fluid—fluid and solid—solid interfaces.

Teaching Interests:

Chemical engineers provide unique perspectives because they are able to exploit fundamental chemical and physical phenomena within real-world systems to produce and transform materials. This was exemplified in my graduate research which showed that the morphology of polymers deposited onto liquid substrates via iCVD could be controlled by the interplay between intrinsic polymer—liquid interactions and deposition conditions. The iCVD technique is a perfect example of how core chemical engineering principles (transport phenomena, reaction kinetics, thermodynamics, and separations) work together to control the polymer composition, thickness, molecular weight, and morphology. My experience developing the design rules for understanding deposition onto liquids via iCVD has prepared me to teach the core curriculum in chemical engineering, and the utilization of iCVD in my research program will provide a hands-on experience for undergraduate and graduate students to apply concepts learned in the classroom.

As an educator of chemical engineering, I also aim to train students how to merge their backgrounds in the central sciences and transport phenomena to construct efficient chemical processes. I will do this by emphasizing ingenious chemical engineering solutions in our everyday lives (such as decaffeinated coffee, pressure cookers, and time-release pharmaceuticals) throughout lectures and course assignments. These themes will culminate by challenging students to work in teams to redesign a process or material for alternative applications. To further promote the integration between fundamental science and process design, I plan to develop a graduate course on CVD methods that is accessible to students across engineering and the sciences. The purpose of the course will be to demonstrate the capabilities and diverse applications of CVD techniques while promoting discussion to foster collaboration between students with different backgrounds. My goal as a faculty member will be to inspire students to look for engineering solutions at the intersections between fields to optimize process design and product performance.

Selected Publications (15 total, 7 first author, 211 citations):

L. C. Bradley, K. J. Stebe, D. Lee. Clickable Janus Particles. *Submitted to J. Am. Chem. Soc.* June 2016.

L. C. Bradley, M. Gupta. Microstructured Films Formed on Liquid Substrates via Initiated Chemical Vapor Deposition of Cross-Linked Polymers. *Langmuir* 2015, 31, 7999-8005.

L. C. Bradley, M. Gupta. Copolymerization of 1-Ethyl-3-vinylimidazolium bis(trifluoromethyl-sulfonyl)imide via Initiated Chemical Vapor Deposition. *Macromolecules* 2014, 47, 6657-6663.

P. D. Haller, **L. C. Bradley**, M. Gupta. Effect of Surface Tension, Viscosity, and Process Conditions on Polymer Morphology Deposited at the Liquid—Vapor Interface. *Langmuir* 2013, 29, 11640-11645.

R. J. Frank-Finney,† **L. C. Bradley**,† M. Gupta. Formation of Polymer—Ionic Liquid Gels Using Vapor Phase Precursors. *Macromolecules* 2013, 46, 6852–6857. †Authors contributed equally to this work.

L. C. Bradley, M. Gupta. Encapsulation of Ionic Liquids within Polymer Shells via Vapor Phase Deposition. *Langmuir* 2012, 28, 10276-10280.

476185 Structure-Property of Polymer and Its Composites: Multiscale Experimental and Computational Studies

Abstract:

Research Interests:

Polymers play vital roles in nearly every application in our current era, which range from mundane food packaging and personal care to advanced energy devices and protective gears. Such versatility of polymer use makes it an attractive platform to incorporate inorganic fillers with novel functionalities, which either enhance the material property of an existing platform or process as an entirely new polymer composite material. Recent studies have found that a strategic placement of nanofillers within a polymer can dictate the composite material properties, whether electrical, mechanical, or thermal. Thus, predictions of the nanostructure and its structure-property relationship are vital to tailor the nanocomposites to advanced applications such as in batteries, molecular sensors, and protective gear.

As with other classes of materials, investigating soft materials by theory and simulation are crucial to contributing towards expedited discovery of new materials. However, computational studies of polymers present more challenges than other classes of materials. Unlike other crystalline materials with unit cells of 10 nm or less, polymers are often amorphous or semi-crystalline entities that can span over microns to millimeters[2]in length scale. These macromolecules also span over a much larger time scale to respond to an external force, which results in higher computational cost. Although predictive equilibrium assemblies of polymer and its composites, or â €œnanocompositesâ€, are still valuable studies, polymeric materials are almost always processed to become a final product. Thus, investigating the structural dynamics of the nanocomposites with nonequilibrium dynamics such as flow and crystallization are crucial to accurately predict its end-use material properties.

To this end, I have sought answers to the following two questions:

First, ***how does the polymer and its composite structure change under nonequilibrium dynamics***? Second, ***what is the structure-property relationship of the nanocomposites and what are the applicable advanced material applications***?

Using **coarse-grained molecular dynamics**, multiple combinations of polymer(homopolymer or block copolymers) and nanofillers(nanoparticles, nanorods, and nanotubes) composite structures under A) **shear and extensional flow** and B) **cylindrical nanoscale confinement** were investigated. Our study reveals that depending on the control variables of the aforementioned dynamics, we can induce novel functional structures such as highly oriented polymer and nanomaterials, evenly placed nanomaterials within polymer matrix, and interconnected nanorods.

As a comparative experiment, electrospinning was used to fabricate polymer/nanofiller nanofibers which undergoes both extensional deformation and nanoscale cylindrical confinement. The structural agreement between the simulation and the experiment result confirms the validity of the model. As such, some of the novel structures observed in simulation were also observed in experiment, such as well-dispersed and well-aligned nanorods within the nano-scale polymer fibers. The novel functional properties that arise from nanostructures of electrospun nanofiber are then readily utilized in various applications such as **molecular sensors** and **high performance battery electrodes**.

I utilized **structure-property relationship of polymer configuration and its molecular orientation** to significantly alter the **mechanical properties of polymeric fibers**. For instance, by improving the molecular orientation and degree of crystallinity from gel-electrospinning ultra high molecular weight polyethylene, I successfully fabricated ultrathin fibers with high stiffness that rivals the commercial protective vest such as Spectra. By defining a fundamental connection between the predicted nanostructure and its functionality, we can expand much further on the capabilities of electrospun nanofibers to significantly impact the polymer processing and its application as a whole.

Teaching Interests:

From my extensive teaching and research experiences, I believe I have the credentials to make significant contributions to teaching in the Chemical Engineering or Material Science Department. My life-long training as a chemical engineer prepared me to analyze open-ended ill-defined processes in a more well-defined and systematic manner. I believe my research and teaching expertise has prepared me to teach transport phenomena (heat, mass, momentum) as well as thermodynamics. Of course, I can also teach statistical mechanics, which is a basis for many computational studies. I believe effective teaching empowers the studentsâ€™ ability to critically think and create their own knowledge, and this ultimately benefits the research community.

476190 Integrating Catalysis and Separations for Energy-Efficient Conversion of Biomass-Derived Feedstocks

Abstract:

Research Interests:

My research career has spanned surface science and catalysis of complex multifunctional biomass-derived molecules to adsorption and separations with a variety of nanoporous materials. I believe my combination of skills will allow for design of unique materials and development of techniques that enable practical catalytic and separations processes, particularly focused around biomass valorization and utilization.

My Ph.D. work at the University of Colorado Boulder under Professor Will Medlin focused on elucidating the fundamental surface reaction mechanisms of biomass-derived molecules, such as furfural, on transition metal surfaces and utilizing those insights to design catalysts for atom-efficient selective transformations. Using surface science techniques, I proposed different surface intermediates that led to either the desired hydrodeoxygenation or the atom-inefficient decarbonylation reactions. This fundamental insight led to the design of surface-modified and bimetallic oxide-supported catalysts with superior hydrodeoxygenation selectivity.

During my tenure as a postdoc at the Georgia Institute of Technology under Professors Chris Jones and Ryan Lively, I have broadened my skill set in the areas of separations, adsorption, and materials synthesis, working with a variety of nanoporous materials from intrinsically porous polymers to metal-organic frameworks to structured porous oxides. The majority of my work has focused on designing sorbent systems for separation of CO_2 and other acid gases from dilute sources such as flue gas and ambient air and controlling the interactions between liquid adsorbent molecules and solid supports. These projects have required engineering at length scales from the atomic-level design of adsorbents and surfaces to macroscopic contactors such as honeycomb-shaped monoliths and hollow fibers.

Research in my independent career will combine these diverse skill sets to develop advanced materials platforms for valorization of biomass through practical catalysis and separations processes. Much of the challenge in directly using biomass-derived feedstocks lies in the diversity of functional groups, which presents multiple sites for substrate-material interaction, or the difficulty in using traditional thermal separation methods to isolate compounds. My research program will aim at simplification of these feedstocks through catalytic or non-thermal membrane separations means, to enable energy- and atom-efficient conversion to high-value products. Materials design will be guided by an understanding of the fundamental interactions between substrate and catalytic or membrane materials with the goal of producing industrially-relevant materials.

Selected Publications (15 total, 9 first author, 130 total citations (Scopus)):

1. S.H. Pang, C.A. Schoenbaum, D.K. Schwartz, J.W. Medlin "Directing Reaction Pathways by Catalyst Active-Site Selection using Self-Assembled Monolayers," ***Nature Communications*** **2013**, 4:2448.
2. S.H. Pang, J.W. Medlin "Adsorption and Reaction of Furfural and Furfuryl Alcohol on Pd(111): Unique Reaction Pathways for Multifunctional Reagents," ***ACS Catalysis*** **2011**, 1, 1272-1283.
3. S.H. Pang, N.E. Love, J.W. Medlin "Synergistic Effects of Alloying and Thiolate Modification in Furfural Hydrogenation over Cu-Based Catalysts," ***J. Physical Chemistry Letters*** **2014**, 5, 4110-4114.
4. S.H. Pang, M.L. Jue, J. Leisen, C.W. Jones, R.P. Lively "PIM-1 as a "Solution-Processable" Molecular Basket for CO_2 Capture from Dilute Sources," ***ACS Macro Letters***, **2015**, 4, 1415-1419.
5. S.H. Pang, C. Han, D.S. Sholl, C.W. Jones, R.P. Lively "Facet-Specific Stability of ZIF-8 in the Presence of Acid Gases Dissolved in Aqueous Solutions," **2016**, ***submitted***.

Teaching Interests:

My teaching philosophy is centered on two major goals for student learning: 1) the students develop a fundamental understanding of the theories and concepts central to Chemical Engineering, and 2) they become critical thinkers with the problem solving and teamwork skills that will be necessary for them to become successful. The achievement of these two goals will be accomplished via a flipped classroom, where time traditionally spent lecturing will be replaced with activities designed to engage students.

Because of my emphasis on conceptual learning and problem solving, I am primarily interested in teaching chemical engineering courses that form the foundation onto which the other courses build, such as Material and Energy Balances and Thermodynamics, though would be comfortable teaching any core Chemical Engineering course. I am also interested in developing upper-level elective courses in Chemical Kinetics and Catalysis or Nanotechnology. Students will be given the opportunity to work in project teams to investigate a topic of their choosing. Over the course of the semester, they will teach their classmates, with guidance from the teacher, for one or two class periods, using the educational tools mentioned above, and give a presentation on their chosen topic. In addition to enhancing their understanding through peer instruction, they will gain valuable practice giving presentations.

476199 2D Materials Assembly for Stretchable Electronics and Smart Fabrics

Abstract:

Assembly and Patterning of 2D Material Building Blocks for Multifunctional Devices in Stretchable Electronics and Smart Fabrics

Po-Yen Chen, *Hibbitt Independent Postdoctoral Fellow*
Massachusetts Institute of Technology & Brown University

Postdoctoral Project:

"Design of Multigenerational Topographies of 2D Building Blocks via Sequential Deformation and Intercalation Templating"
- Under supervision of *Ian Y. Wong* & *Robert H. Hurt*, School of Engineering, Brown University

Ph.D. Dissertation:

"M13 Virus-Enabled Assembly of 3D Nanostructured Composites: Synthesis and Applications in Solar Energy Conversion and Electrochemical Energy Storage Devices"
- Under supervision of *Angela M. Belcher* & *Paula T. Hammond*, Department of Chemical Engineering, Massachusetts Institute of Technology (MIT)

Fellowships:

2015, 2016 Hibbitt Engineering Independent Postdoctoral Fellowship (Brown University)
2013, 2014 Eni Energy Fellowship (MIT)
2013 Taiwan National Study Abroad Fellowship
2010 Saudi Aramco Fellowship (MIT)

Research Experience:

Complex surface topographies are of fundamental importance to materials science, physics, biology, and chemistry, impacting applications from catalysis to energy storage and environmental mediation. However, sophisticated design on topographical structures across multiple length scales and over diverse material systems remains challenging. This poster will present our recent works about the creation of multigenerational wrinkled/crumpled textures of versatile materials via sequential deformation and intercalation templating. First, we developed a fabrication that enables extreme mechanical deformation of 2D nanomaterials (*e.g.*, graphene) through multiple cycles of substrate contraction. In each cycle, the compression can be biaxial or unidirectional, leading to a family of different hierarchical wrinkled/crumpled textures. This method achieves a degree of enhancement on structural complexity that far exceeds that of simple one-step deformation processes. The multiscale patterns of graphene show superhydrophobicity (>160°) and improve the electrochemical current densities over 20 times (in comparison with planar coatings). Next, these multigenerational graphene were further utilized as sacrificial templates, and the multiscale surface patterns can be transcribed into other material systems, even beyond 2D nanomaterials (*e.g.*, metal oxides). By harnessing the intersheet gallery spacing of these graphene templates, a wide variety of hydrated metal ions can be uniformly intercalated and immobilized with the functionalities of templates. Subsequently, annealing these composites results in metal oxide thin films that transcribe the original graphene topography with high fidelity. These new textured metal oxide films exhibit enhanced electrochemical and photocatalytic performance over planar films and serve as high-activity electrodes for energy storage and photocatalysis. Overall, these unconventional approaches are broadly applicable for large-area patterning of other 2D nanomaterials and providing a bridging technology that can transcribe textures into other material systems that cannot sustain large deformation.

Research Interests:

My future research interests will focus on large-area patterning of hierarchical surface structures by using various 2D building blocks and study their physiochemical and barrier properties under large mechanical deformation. 2D building blocks, ranging from carbon-based 2D materials to clays and transition metal oxides, display extraordinary chemical, physical and mechanical properties but remain challenging to manipulate them into higher dimensional architectures with large deformability. To address these challenges, I will first continue the work on organizing 2D materials structurally as soft materials while retaining exceptional physiochemical functionality as hard materials. I will emphasize interfacial and surface engineering to facilitate the self-assembly of multiscale, higher-dimensional architectures with highly convoluted features. In the meantime, a systematic approach will be conducted for achieving a control on the wrinkled/crumpled topographies, selective areas, wavelength/width, length scale, amplitude/height, and dynamic patterning. Through above fundamental understanding, I will develop multiple strategies for the fabrication of complex surface patterns of 2D building blocks: 1) wrinkling and crumpling by mechanical deformation, 2) “snapping” surface via confined swelling, 3) origami and kirigami assembly, and 4) 3D printing.

By understanding the factors that produce these higher dimensional structures, it become feasible to design synthetic materials with exactly the kinds of surfaces needed for specific applications, such as stretchable electronics and smart fabrics. By using the proposed fabrication methods, I will apply the convoluted structures of 2D building blocks with high deformability into electromechanical devices, and these surface patterns will retain exceptional physiochemical functionality as well as still provide high performance during reversible stretching. In addition, based on our recent finding, the intersheet gallery spaces of 2D nanomaterials remain unchanged under large deformation, enabling the design of stretchable environmental barriers for the advance of smart fabrics. The combination of mechanical principles (harnessing instabilities of thin films) and materials selection (from 2D nanomaterials to transition metal oxides) leads to a general technological route that can empower great potential to fabricate various flexible and stretchable devices capable of large deformations.

Teaching Interests:

Aside from my research career, I also have extensive teaching experience. As a part of a Hibbitt Postdoctoral Fellowship, I am currently an instructor at Brown University and teach "*Chemical and Biochemical Reactor Design*". The duties include providing semester-long lecture series, developing course/experiment materials, and holding office hours. Also, I TAed an undergraduate lecture and laboratory course "*Polymer Science Laboratory*" in Chemical Engineering in MIT (for which I was the highest-ranked TA (6.6/7.0) in the semester). In an effort to further improve my teaching skills, I have obtained the Graduate Student Teaching Certificate from MIT's Teaching and Learning Laboratory. Lastly, I have mentored over 10 grads/undergrads in both MIT and Brown University. I found each of these experiences to be very rewarding and believe they have helped prepare me for future mentorship and teaching roles.

I am comfortable with teaching any undergraduate core classes in Chemical Engineering; at the graduate level, I could teach *Thermodynamics* and *Kinetics of Chemical Reactions*. I would also like to develop a new subject on sustainable energy that focuses on the fundamentals of thermodynamics, chemistry, and transport in a context that is applied to energy systems. This class is ideal for bridging several engineering disciplines and demonstrating to the students the importance and need for multidisciplinary approaches to addressing energy related issues.

Selected Peer-Reviewed Publications:

- 21 peer-reviewed journal articles (11 first-authored papers), 2 patents, 1 book chapter
- *Served as corresponding author

1. **P.-Y. Chen***, M. Liu, I. Y. Wong*, R. H. Hurt*, "Multifunctionality of Pressure-Actuated Multiscale Graphene Surface Patterns with Reversible Stretchability." *Nano Letters*. (In Preparation)
2. **P.-Y. Chen***, M. Liu, T. M. Valentin, Z. Wang, R. S. Steinberg, J. Sodhi, I. Y. Wong*, R. H. Hurt*, " Hierarchical Metal Oxide Topographies Replicated from Highly Textured Graphene by Intercalation Templating." *ACS Nano*. (Under Review)
3. **P.-Y. Chen***, Z. Wang, R. H. Hurt*, I. Y. Wong*, "From Flatland to Spaceland: Higher Dimensional Patterning of 2D Nanomaterials." *Advanced Materials*. (Invited Article)
4. **P.-Y. Chen***, J. Sodhi, Y. Qiu, T. M. Valentin, R. S. Steinberg, Z. Wang, R. H. Hurt*, I. Y. Wong*, "Multiscale Graphene Topographies Programmed by Sequential Mechanical Deformation." *Advanced Materials* 28, 3564 (2016).
    1. *- Featured as Inside Back Cover*
    2. *- Highlighted in Brown News (1) & (2), IEEE spectrum, and ScienceDaily*
5. **P.-Y. Chen**, N.-M. Dorval Courchesne, M. N. Hyder, J. Qi, A. M. Belcher*, P. T. Hammond*, "Carbon Nanotube-Polyaniline Core-Shell Nanostructured Hydrogel for Electrochemical Energy Storage." *RSC Advances* 5, 37970 (2015). (Invited Article)
6. **P.-Y. Chen**, X. Dang, M. T. Klug, N.-M. D. Courchesne, J. Qi, M. N. Hyder, A. M. Belcher*, P. T. Hammond*, "M13 Virus-Enabled Synthesis of Titanium Dioxide Nanowires for Tunable Mesoporous Semiconducting Networks." *Chemistry of Materials* 27, 1531 (2015).
7. **P.-Y. Chen**, J. Qi, M. T. Klug, X. Dang, P. T. Hammond*, A. M. Belcher*, "Environmentally Responsible Fabrication of Efficient Perovskite Solar Cells from Recycled Car Batteries." *Energy & Environmental Science* 7, 3659 (2014).
    1. *- Highlighted in MIT News (1)&(2), IEEE spectrum, and Washington Post*
8. **P.-Y. Chen**, M. N. Hyder, D. Mackanic, N.-M. D. Courchesne, J. Qi, M. T. Klug, A. M. Belcher*, P. T. Hammond*, "Assembly of Viral Hydrogels for Three-Dimensional Conducting Nanocomposites." *Advanced Materials* 26, 5101 (2014).
    1. *- Featured as Front Cover*
9. **P.-Y. Chen**, X. Dang, M. T. Klug, J. Qi, N.-M. Dorval Courchesne, F. J. Burpo, N. Fang, P. T. Hammond*, A. M. Belcher*, "Versatile Three-Dimensional Virus-Based Template for Dye-Sensitized Solar Cells with Improved Electron Transport and Light Harvesting." *ACS Nano* 7, 6563 (2013).
10. **P.-Y. Chen**, R. Ladewski, R. Miller, X. Dang, J. Qi, F. Liau, A. M. Belcher*, P. T. Hammond*, "Layer-by-Layer Assembled Porous Photoanodes for Efficient Electron Collection in Dye-Sensitized Solar Cells." *Journal of Materials Chemistry A* 1, 2217 (2013).

11. **P.-Y. Chen**, C.-P. Lee, R. Vittal, K.-C. Ho*, "A Quasi Solid-State Dye-Sensitized Solar Cell Containing Binary Ionic Liquid and Polyaniline-Loaded Carbon Black." *Journal of Power Sources* 195, 3933 (2010).

476216 Structure and Transport in Polymer Membranes for Energy-Efficient Separations

Abstract:

Structure and Transport in Polymer Membranes for Energy-efficient Separations

Hee Jeung Oh

(1) Department of Chemical and Biomolecular Engineering, College of Chemistry, University of California, Berkeley, California 94720, USA

(2) Mckctta Department of Chemical Engineering, Texas Materials Institute, Center for Energy and Environmental Research, The University of Texas at Austin, Austin, Texas 78758, USA

Research Interests:

With the worldâ€™s population growing rapidly, the need for clean water is greater than ever. Among water purification methods, membrane-based separations draw attention because of their energy efficiency (they do not require an energy-intensive phase-change step as in distillation), small footprint, and ability to eliminate trace contaminants. However, only a few membrane materials and formation methods are used in industry today, and current membranes often do not have adequate efficiency (i.e., selectivity), productivity (i.e., permeability), and chemical stability to be used for all sources of water. To overcome these challenges, new polymer membrane materials and formation methods are needed to precisely tune pore size and pore size distributions and to control chemical stability for better separation performance.

Transport of small molecules and ions through polymer membranes is predominantly controlled by polymer nanostructure. Therefore, morphological control of the membrane structure by tuning chain architecture, synthetic routes, and post-synthesis treatment is a critical aspect of designing membrane systems. My graduate research focused on the water and salt transport properties of ion-containing polymers (sulfonated aromatic hydrocarbons) prepared by a new membrane formation method, i.e., solvent-free melt processing. My postdoctoral research focuses on extending my understanding of a polymerâ€™s physical structure by designing and synthesizing highly-structured polymeric systems and characterizing them (e.g., through X-ray scattering, tomography, scanning transmission electron microscopy). Additionally, these porous, crystalline, and conducting polymers can be used as polymer electrolytes in electrochemical systems and as blood filtering devices for biomedical applications.

My research interests are in understanding the fundamental structure-property relationships of water and salt transport in polymers and composite materials. These interests span disciplines from chemical engineering, chemistry, and materials science. The advancement of the technologies mentioned above is highly dependent on the development of polymer membranes, which selectively permeate only desired components while maintaining chemical stability. My goal is to develop correlations between design, synthesis, structure, and transport properties of polymers for targeting applications in energy-efficient separations, batteries and biomedical devices. Synthetic work will focus on highly-structured polymers such as ion-containing polymers, block copolymers, and random copolymers, and polymer characterization will focus on water purification, electrochemical systems, and biomedical devices.

Teaching Interests:

I believe teaching and mentoring the next generation of scientists is the most satisfying aspect of being in academia. Both in classroom instructions and in a research laboratory, my goal is to help students identify important problems and to arrive at intelligent solutions by teaching critical thinking skills.

My education and experiences have prepared me well to teach both undergraduate and graduate courses in the Chemical Engineering department. Also as a graduate student and a postdoctoral researcher, I have supervised research projects for three undergraduate students and several high school students. Seeing these students succeed has been amazingly rewarding to me, and I am looking forward to continue this mentorship with my future graduate students and post-docs.

476220 Engineering Soft Materials with Tunable Structure and Functionalities

Abstract:

Research Interests:

Soft materials encompass a wide range of materials, such as polymers, colloids, emulsions, gels and biological materials, whose properties arise from an interplay of processes across multiple scales. Conquering the major challenges, especially in food, pharmaceuticals and healthcare in the 21st century requires novel and smarter soft materials that can provide unprecedented properties and extraordinary functions, capable of exquisitely interfacing between engineering, chemistry, medicine and biology. The discovery and development of novel materials not only bring about new applications and improved performances, but also enable innovative approaches of material manufacturing in a sustainable way. My vision for this research direction is strongly motivated by the burgeoning need for better multi-scale understanding the structure - dynamic function relation of soft materials and also to develop novel processes for materials chemistry that enable production of such materials ranging in scale from nanocomposites to hierarchically structured polymer *particles* and *gels.*

As future faculty, my goal is to establish an independent research program that aims at designing and developing novel bottom-up approaches for soft materials with tunable architectures and functionalities. I will give emphasis on better understanding on how to precisely control the molecular transport in such materials and surface molecular interactions, besides the understanding their structure-function relation. I intend to leverage my diverse research expertise along with use of innovative approaches at the interface of chemical engineering, polymer and materials science, and emulsion and interface science. The research themes that I intend to pursue are, but not limited to: (i) Understanding and formation of novel structured emulsions with new and remarkable functionality; (ii) Development of nanostructured polymeric materials/ nanocomposites for the capture/separation of industrially, environmentally, and biologically relevant chemicals; (iii) Design and development of the next generation of intelligent hydrogel materials based- delivery devices for applications in (bio)pharmaceutics and healthcare. My poster will highlight the key aspects of some of my past and current research along with an outline of future projects in the above-mentioned platform.

Teaching Interests:

Aside from my research career, I have also substantial teaching experience. I was a Lecturer in the Department of Chemical Engineering at Bangladesh University of Engineering and Technology (BUET), where I had the opportunity to teach one chemical engineering core module and several laboratory courses. At National University of Singapore (NUS), I worked as a tutor in particle technology course. I mentored many undergraduate and grduate students during my graduate studies and postdoc at NUS. Organizing tutoring and review sessions for the students gave me valuable teaching experience, as it was often then that the students asked the most penetrating and insightful questions. At MIT, I participated in â€˜Kaufman Teaching Certificate Programâ€™ to further enhance my teaching skills. I believe that my academic experience at BUET, NUS and MIT has tremendously enriched my perspective, and I would like my students to benefit from it too.

I am well-equipped to teach the core-curriculum for chemical engineering, including separation science, particle technology, mass and heat transfer, process design, and reaction engineering.

476227 Polymer Based Hybrid Materials: From Molecular Design to Applications

Abstract:
Research Interests:

Organic-inorganic nanocomposites with controlled hierarchical structures offer a particularly attractive platform for development of soft functional materials for various applications ranging from health care to energy storage. Here I am going to present how complexation of organic macromolecules and nanomaterials at molecular length scales and under mild conditions, as used by nature, can form advanced materials with exceptional engineering properties. Working with Prof. Matthew Tirrell and Prof. Juan de Pablo at the University of Chicago, I have synthesized patent-pending stable amorphous hybrid nanoparticles of controllable size and surface charge under mild conditions. These hybrid nanoparticles are composed of a polyelectrolyte, either polycation or polyanion and amorphous calcium phosphate nanospheres which are a few nanometer in size. The ACaP nanospheres are formed during synthesis and are uniformly dispersed inside the polymer matrix. We have also shown that through mixing these hybrid nanoparticles with their oppositely charged counterpart, it is possible to prepare injectable pastes with tunable rheological properties. The presence of stable ACaP nanospheres as an inexpensive, bioactive material along with very good handling properties makes these materials beneficial for tissue regeneration and dental applications. Moreover, I have been working on graphene based electrically conductive aerogels with exceptional thermal stability and compatibility with living cells.

In collaboration with prof. Mezzenga from ETH Zurich, we have been able to design gelatinâ€“graphene conductive biopolymer nanocomposites with a record-low electrical percolation threshold of 3.3 Ã— 10$^{\text{â}^{\text{'}}2}$Â vol%, which arises from the homogeneous dispersion of the graphene nanosheets within the gelatin matrix. The results of this study have been published in *Advanced Materials*.

I also had the opportunity to get involved in a pioneering work on the understanding of glass-like dynamic properties of living cells. I joined Professor Fredbergâ€™s lab at the Harvard University in 2012. As a member of a multidisciplinary team there, I was able to use my knowledge in physics of glass transition from my PhD work to study well-characterized endothelial and epithelial cell monolayer systems. We have shown that the mechanics of the cellular monolayer in the lung, and in other organ systems as well, may be dominated by glass transition and/or jamming as in polymeric and colloidal systems. Our findings which offer a unifying new lens into the biology of the cellular monolayers have been recently published in *Nature Materials* and *Differentiation*.

As a faculty member I would like to focus on understanding and fine-tuning the assembly of polymers and 2D nanomaterials into engineering systems and devices. Specifically, my main focus will be on developing compositionally and synthetically simplified organic-inorganic hybrid systems with lower production costs and competitive functionality in wide variety of applications including regenerative medicine and energy storage.

Teaching Interests:

I have several years of teaching experince of polymer courses including polymer physics, polymer chemistry and polymer characterization at both the undergraduate level and graduate level at the University of Tehran. I have also supervised more than 20 MSc and 2 PhD students in polymer science.

In addition to polymer courses, I would also welcome the opportunity to develop and teach classes related to colloidal science and hybrid materials.

476241 Design of Functional Polymeric Materials: From Ion Transport to Bio-Inspired Assembly

Abstract:

Research Interests:

Engineering of polymer-inorganic hybrid materials with structure control from the nanoscale to the mesoscale has remained elusive, especially in comparison to biological composites such as nacre and bone. However, the ability to synthetically design highly crystalline, hierarchically structured, and well-ordered inorganic and composite materials is desired in fields ranging from biomaterials to energy. My research platform will focus on the development of polymer systems with chemical and sequence control focusing on three areas: (1) understanding the fundamental interactions between polymers and inorganics during self-assembly process via *in-situ* techniques, (2) combining these fundamentals with polymer self-assembly to engineer high performance hierarchical inorganic materials for energy storage and catalysis, and (3) responsive assembly of polymeric nanoparticles for healthcare.

This research program builds upon my Ph.D. work, undertaken at UC Santa Barbara under the guidance of Profs. Glenn Fredrickson, Ed Kramer, and Craig Hawker, which elucidated the interactions between polymers and ions for lithium-ion batteries. I focused on the development of poly(allyl glycidyl ether) (PAGE) as an alternative platform for lithium battery polymer electrolytes. The versatility of allyl glycidyl ether as a building block was demonstrated by the preparation and evaluation of various AGE-EO macromolecular architectures that show superior performance to both PAGE and PEO by utilizing copolymerization and thiolene coupling chemistry. Furthermore, in order to develop design rules for polymer electrolytes, a library of structurally similar polyethers, poly(glycidyl ether)s (PGEs) was synthesized by anionic polymerization and investigated by impedance measurements, dielectric spectroscopy, and pulse field gradient NMR to understand the role of polymer chemistry on ion diffusion.

I have taken this interest in controlling polymer chemistry and its effect on composite components to my current postdoctoral position at Cornell University with Prof. Uli Wiesner and Prof. Lara Estroff, where we seek to combine the long-range periodic structure direction of block copolymers with the increased information content provided by sequence control in proteins to generate highly controlled composite materials. Knowledge of inorganic crystallization pathways has been developed through *in-situ* GIWAXS studies of early formation pathways in hybrid organic-inorganic pathways.

Teaching Interests:

My teaching interests are in polymer physics, polymer chemistry, thermodynamics, and introductory materials science.

476424 Engineering Discrete Functional Building Blocks at Molecular Scale for Human-scale Applications

Abstract:

Research Interests:

The preparation of synthetic polymers and nanocomposites with structures, functions and properties that are precise and accurate remains a grand challenge for polymer and material chemists. Synthetic polymers and other material building blocks prepared using bottom-up approach are generally polydisperse, and their use as a statistical ensemble of molecular weights/sizes can be highly problematic in biological and nanoscale applications due to size-dependent effects, aggregations and phase separations.

My research program will focus in the development of an efficient and scalable synthesis of discrete polymer-based materials for applications on a human scale. In specific, there are three areas of research. (1) Synthesis of discrete oligomers/polymers and the utilization of their pendent groups and/or chain-ends to produce discrete macromolecular architectures. Such structures are expected to exhibit accurate and precise properties leading to specific functions. (2) Engineering of discrete polymer building blocks in conjunction with molecules containing swellable, photoswitchable, or degradable moieties to construct designer stimuli-responsive structures (UV, pH, toxic chemicals). (3) Engineering of discrete polymer building blocks with inorganic complexes and nanoparticles to construct hierarchical nanostructures for applications in separation and sensing. Mid-long term goal includes the development of biocompatible nanocarriers containing discrete materials for drug-delivery applications. As such, my research program attempts to address fundamental and technological challenges at the interface of chemical engineering, chemistry, material science and biology.

Postdoctoral Projects:

Department of Materials, University of California Santa BarbaraSupervisor: Prof. Craig J. Hawker

- *"Synthetic Oligomers and Polymers with Discrete Structures and Hierarchical Architectures"*

(1 completed, several sub-projects ongoing)

- *"Polymer Brush Nanostructures Using Visible-Light-Mediated Metal-Free Atom Transfer Radical Polymerization "* (completed)

PhD Dissertations:

Department of Polymer Science and Engineering, University of Massachusetts AmherstSupervisor: Prof. Todd Emrick

- *"Functional Nanostructures from Nanoparticle Building Blocks"*

Research Experience:

Briefly, my postdoctoral work under the guidance of Prof. Craig J. Hawker at UCSB primarily involves the synthesis and development of novel polymer architectures via controlled polymerization techniques, and the preparation of discrete oligomers/polymers. In my PhD work under the guidance of Prof. Todd Emrick at UMass Polymer Science and Engineering, I focused on the synthesis and characterization of nanomaterials and multifunctional polymer-based ligands for making nanostructures using directed self-assembly and covalent assembly techniques. I also investigated the synthesis of zwitterionic polymer covered nanomaterials and their potential applications as nonfouling nanomaterials and photothermal therapeutic agents for *in vivo*studies. Before joining UMass, I was a Research Scientist at Schlumberger Research (Boston) developing optochemical and electrochemical sensing, membrane separation and microfluidic platforms for downhole technologies. I obtained my master and bachelor degree in Chemical Engineering from the University of Tokyo under the guidance of Prof. Takeo Yamaguchi, Prof. Steven Kraines and Prof. Hiroshi Komiyama.

Teaching Experience:

I served as a teaching assistant in the Polymer Synthesis Laboratory in the Department of Polymer Science and Engineering at Univ. of Massachusetts for 2 years, teaching various polymer synthesis to 1styear graduate students. Over my career I have been fortunate to directly mentor 7 undergraduates, 8 graduate students and am currently leading a team of several post-docs and PhD students.

Teaching Interests:

I am comfortable teaching Polymer Chemistry, Polymer Synthesis and Characterization, Polymer Physics, Transport Phenomena, Materials Science, Nanomaterials, and Computational Chemistry.

Awards:

- Dow Materials Institute Travel Award 2015

- Dow Materials Institute Travel Award 2016

- 6 US Patents

- Monbusho Full Scholarship, Japanese Ministry of Education (1998–2003, 2005–2007)

Publications:

1. Lawrence, J.; Lee, S.-H.; Abdilla, A.; Nothling, M. D.; Ren, J. M.; Knight, A. S.; Fleischmann, C.; Li, Y.;

Abrams, A. S.; Schmidt, B. V. K. J.; et al. A Versatile and Scalable Strategy to Discrete Oligomers. *JACS*, **2016**, *138*, 6306

2. Discekici, E.; Pester, C.; Treat, N.; Lawrence, J.; Mattson, K; Narupai, B.; Toumayan, E.; Luo, Y.; McGrath, A.; Clark, P.; Read d. A. J.; Hawker, C. A Simple Bench Top Approach to Polymer Brush Nanostructures Using Visible Light Mediated Metal-Free Atom Transfer Radical Polymerization. *ACS Macro Letters*, **2016**, *5*, 258.

3. Lawrence, J.; Emrick, T. Pentafluorophenyl Ester-Functionalized Nanoparticles as a Ver- satile Platform for Selective and Covalent Inter-Nanoparticle Coupling. *ACS Applied Materials and Interfaces,* **2016**, *8*, 2393.

4. Liu, Y.; Chen, Y-C.; Hutchens, S.; Lawrence, J.; Emrick, T.; Crosby, A. J. Directly Measuring the Complete Stress-Strain Response of Ultrathin Polymer Films. *Macromolecules*, **2015**, *48*, 6534.

5. Lawrence, J.; Yamashita, K.; Yamaguchi, T. Correlating electronic structure and chemical durability of sulfonated poly (arylene ether sulfone)s. *Journal of Power Sources,* **2015**, *279*, 48.

6. Puodziukynaite, E.; Wang, H.W.; Lawrence, J.; Wise, A. J.; Russel, T.P.; Barnes, M.; Emrick, T. Azulene Methacrylate Polymers: Synthesis, Electronic Properties, and Solar Cell Fabrication, *JACS*, **2014**, *136*, 11043.

7. Maiz, J.; Zhao, W.; Gu, Y.; Lawrence, J.; Arbe, A.; Alegria, A.; Emrick, T.; Mijangos, C.; Colmenero, J.; Russell, T.P. Structural and Dynamic Study Of Polystyrene-block-poly(4-vinylpyridine) Copolymer Confined In Self-Ordered Nanoporous Alumina, *Polymer*, **2014**, *55*, 4057.

8. Pham, J. T.; Lawrence, J.; Grason, G. M.; Emrick, T.; Crosby, A. J. Stretching of Assembled Nanoparticle Helical Springs, *Phys. Chem. Chem. Phys.*, **2014**, *16*, 10261.

9. Lawrence, J.; Pham, J.T.; Lee, D. Y.; Liu, Y.J.; Crosby, A. J.; Emrick, T. Highly Conductive Ribbons from Stick-Slip Assembly of Organosoluble Gold Nanoparticles. *ACS Nano*, **2014**, *8*, 1173.

10. Pham, J. T.; Lawrence, J.; Lee, D. Y.; Grason, G. M.; Emrick, T.; Crosby, A. J. Highly Stretchable Nanoparticle Helices Through Geometric Asymmetry and Surface Forces. *Advanced Materials*, **2013**, *25*, 6703.

11. Lee, D. Y.; Pham, J. T.; Lawrence, J.; Lee, C. H.; Parkos, C.; Emrick, T.; Crosby, A. J. Macroscopic Nanoparticle Ribbons and Fabrics. *Advanced Materials,* **2013**, *25*, 1248.

12. Laprade, E. J.; Long, R.; Pham, J. T.; Lawrence, J.; Emrick, T.; Crosby, A. J.; Hui, C.-Y.; Shull, K. R. Large Deformation and Adhesive Contact Studies of Axisymmetric Membranes. *Langmuir*, **2013**, *29*, 1407.

13. Chen, X.*; Lawrence, J.*; Parelkar, S.; Emrick, T. Novel Zwitterionic Copolymers with Dihydrolipoic Acid: Synthesis and Preparation of Nonfouling Nanorods. *Macromolecules*, **2012**, *46*, 119. (*: equal contribution)

14. Bae, J.; Lawrence, J.; Miesch, C.; Ribbe, A.; Li, W.; Emrick, T.; Zhu, J.; Hayward, R. C. Multifunctional Nanoparticle-Loaded Spherical and Wormlike Micelles Formed by Interfacial Instabilities. *Advanced Materials*, **2012**, *24*, 2735.

15. Lawrence, J.; Yamaguchi, T. The Degradation Mechanism of Sulfonated Poly(Arylene Ether Sulfone)s in an Oxidative Environment. *Journal of Membrane Science*, **2008**, *325*, 633.

16. Lawrence, J.; Robinson, K. L.; Lawrence, N. S. Electrochemical Determination of Sulfide at Various Carbon Substrates: a Comparative Study. *Analytical Sciences*, **2007**, *23*, 673.

476850 Flow and Jamming of Particulate Materials

Abstract:

Research Interests:

The majority of industrial processes involve particulate forms of material. The grain sizes range from nano and mico-scale (e.g. colloids) to macro-scale (e.g. granular materials, emulsions, and foams). Even with a comprehensive knowledge of the particle-particle interactions, the prediction of mesoscale and large scale behavior of a collection of these particles is not trivial. A significant effort has been dedicated to understanding the phenomenology, and more importantly, unifying the laws which govern the collective dynamics of particulate systems. These studies cross between several fields such as Physics, Chemical and Mechanical Engineering, and material science. I am in particular interested in the flow of particulate systems, and I would like to pursue this field with experimental studies on the rheology of granular materials and colloids.

Postdoctoral Research:

(i) Reversibility of a Colloidal System: I am currently studying a colloidal system adsorbed at an oil-water interface, which is subject to cyclic shear. The colloidal particles are sub-micron size, such that Brownian motion has significant contribution to the single particle dynamics. Through this study, we found that thermal noise significantly affects the reversibility of material in a nontrivial manner.

[1] S. Farhadi and P. Arratia, "***Shear-induced Reversibility for Thermal Colloids***", to be submitted (2016).

(ii) Dynamics of Active Systems: In active particulate systems, in contrast to passive systems, single particle dynamics is independently present even without external driving. We have designed two species of active granular systems: spinners, and translators. The intra-structure of particles (which are 3D printed) is designed such that they either spin or translate in the presence of an upward air flow. In this ongoing study, we aim at describing the collective dynamics, and in particular, we search for an effective temperature which governs the dynamics. Subsequently, we will be able to extract an equation of state for these systems.

PhD Research:

My PhD research was an experimental study on the jamming and rheology of granular materials, with a specific focus on elliptical-shaped particles. I performed continuous Couette shear (rheology) and cyclic compression (jamming-unjamming) on both systems of circular and elliptical shaped particles. The identical experiment, on both circles and ellipses, enabled us to understand the role of particle anisotropy on jamming and the bulk rheology.

[2] S. Farhadi, A. Zhu, RP. Behringer, "***Stress and Structure Instability in Cyclic Compression of Ellipses and Disks***", Physical Review Letters, (2015).

[3] S. Farhadi, RP. Behringer, "***Dynamics of sheared ellipses and circular disks: effects of particle shape***", Physical Review Letters 112 , 148301 (2014).

[4] S. Farhadi, RP. Behringer, Alex Zhu, "***Slow dynamics for elliptical particles under continuous shear and cyclic compression***", AIP Conf. Proc., 1542, 879-882 (2013).

Future Directions:

I would like to pursue my future research in two directions:

1- I am interested in developing techniques to directly measure force distribution for a disordered matter, in its grain scale. There have been few successful experimental techniques such as photoelasticity and contact area size to measure grain scale stress. However, the filed of colloids lacks experimental tools for this purpose so far. Knowledge of the grain scale forces is useful in connecting the structural properties to the dynamics, and eventually obtaining insight on problems such as self assembly, where both structure and grain scale forces play role.

2- An emerging filed in physical sciences is 'machine learning with big data'. The idea is to extract information by obtaining a large collection of data from a specific system, and create learning models to predict the dynamical behavior. I would like to apply these techniques to understand the underlying mechanisms of flow and transport in particulate systems.

Teaching Interests:

As a graduate student of Physics, I have been involved with teaching several introductory courses, which included lab instruction. I also have experience of teaching nonlinear dynamics, which is a higher level undergraduate course. I should emphasize that I have an undergraduate major degree in Chemical Engineering, which helps me with teaching courses such as thermodynamics, fluid dynamics, and heat and mass transfer. Also, as an expert in granular materials, I can develop new courses such as granular matter, transfer in porous materials, and more mathematical based courses such as nonlinear dynamics, and chaos.

477008 Controlling the Structure of Systems Ordered via Block Copolymer Phase Separation: Simulations and Experiments

Abstract:

Research Interests:

Block polymers and other ordered soft materials have the potential to allow a vast array of structures to be ordered at lengths scales from a few nanometers to several micrometers. This property allows for a varied range of practical applications in, for example, drug delivery, microelectronic materials, advanced plastics, and catalysis. However, only a small subset of the extensive array of possible systems have been explored. This space can best be explored with a concerted effort using both experiment and simulation. My research will start in three main areas:

1. Ordering of nanoparticles via block polymer ligand directed assembly
2. Phase behavior of diblock blends
3. Development and application of protracted colored noise dynamics for polymer systems.

1. Ordering of nanoparticles via block polymer ligand directed assembly

Nanocomposites are widely used for technological and commercial purposes and they have been ordered via a variety of methods. However, precise control is still challenging and the controlling mechanisms are still not understood. Simulations of these systems would provide a theoretical underpinning by which experimental space would be reduced, resulting in greater understanding of the critical properties more quickly.

2. Phase behavior of diblock blends

Control of ordered block polymer structures is most commonly done via precise synthetic techniques, but blending of polymers is a significantly easier way to control structures if the process can be understood. By fundamentally understanding the effect that blends of diblocks have on morphology, domain size, and other properties, new applications and easier production methods could be discovered. **3. Protracted colored noise dynamics for polymer systems**

Molecular dynamics simulations of polymeric systems are significantly limited in terms of the time and length scales that can be studied. A new technique, protracted colored noise dynamics for polymer systems is designed to efficiently sample over energy barriers and can reduce simulation time to equilibrium by more than four orders of magnitude. This technique could be adapted to work with atomistic simulations and a plethora of polymer architectures allowing access to timescales which are currently inaccessible. Systems of particular interest are semi-crystalline polymers such as polyethylene or large polymer systems such as micelle forming systems or star polymers.

Teaching Interests:

As a faculty member in a chemical engineering department, I would be pleased to teach a range of courses that combine my training with the needs of the faculty and the students. I can bring strong preparation and an enthusiastic attitude to any core chemical engineering class. Leveraging my postdoctoral and graduate research experience in polymer physics and simulations, I am also excited to develop new coursework in areas that would fit into both the undergraduate and graduate curriculum related to soft matter engineering and simulation techniques. My experience as a teaching fellow where I co-taught a numerical methods and my experience as a teaching assistant for 3 semesters which required weekly recitation sections has given me the experience I need to be an effective instructor early on.

The reason I want to be faculty is because I love to learn. I love research because itâ€™s never-ending learning and satisfying of curiosity. In teaching I can give this love of learning to my students. By showing them my passion and enthusiasm and enjoyment of learning and of knowledge, I will promote enthusiasm and enjoyment in my students.

477819 Application of Zwitterionic Materials in Stem Cell Expansion and Immunosuppression

Abstract:

Research Interests:

Ultra-low fouling zwitterionic hydrogels offer unique advantages over conventional biomaterials in biomedical applications. By exclusively using zwitterionic monomer and crosslinker, pure zwitterionic hydrogels can be prepared with highly tunable mechanical properties and functionalities for diverse applications. Uniform zwitterionic hydrogels with controllable mechanical and/or chemical gradients can also be developed. My research focuses on the development and application of zwitterionic hydrogels. Examples include zwitterionic hydrogels allow differentiation-free expansion of stem cells, implanted zwitterionic hydrogels resisting foreign-body capsule formation in mice, zwitterionic hydrogel coatings significantly enhancing the longevity and sensitivity of glucose sensors, functionalized hydrogels encapsulating and preserving cells with high viability and long-circulating zwitterionic nanogels.

Research Experience:

1) Development and Applications of Zwitterionic Hydrogel

l Differentiation-free expansion of stem cells and T cells in zwitterionic hydrogels

l Directed differentiation of stem cells in functionalized zwitterionic hydrogels

l Zwitterionic hydrogels resistant to the immunogenicity and fibrous encapsulation in mice l Time-independent self-healing zwitterionic hydrogels

l Long-life zwitterionic material-coated biosensors in whole blood

2) Development of Long-circulating Zwitterionic Nanomaterials

l Long-circulating and therapeutic protein-encapsulated zwitterionic nanogels

l Immunogenicity-free zwitterionic nanoparticles

3) Development and Applications of High-strength Hydrogels

l Construction of ultra-high strength hydrogels

l High-strength hydrogel-based gene transfection platform

l Develop shape-memory effect for high-strength hydrogel

l Directed differentiation of MSCs on a shape-memory hydrogel

Teaching Interests:

In my five-year graduate student career, I have been served as teaching assistance for many different chemical engineering courses like transport and thermodynamic. I am interested in teaching any chemical engineering course as well as polymer science course.

AWARDS

2015 National Award for Outstanding Chinese Students Study Abroad ($6000)

SELECTED JOURNAL PUBLICATIONS (Total: 28, Citations: 800+)

1. **Tao Bai**, Fang Sun, Lei Zhang, Andrew Sinclair, Sijun Liu, Jean Rene Ella Menye, Ying Zheng, and Shaoyi Jiang, 'Restraint of the Differentiation of Mesenchymal Stem Cells by a Nonfouling Zwitterionic Hydrogel', *Angewandte Chemie International Edition,* 126 (2014), 12943-48.

2. Lei Zhang, Zhiqiang Cao, **Tao Bai**, Louisa Carr, Jean-Rene Ella-Menye, Colleen Irvin, Buddy D Ratner, and Shaoyi Jiang, 'Zwitterionic Hydrogels Implanted in Mice Resist the Foreign-Body Reaction', *Nature Biotechnology,* 31 (2013), 553-56.

3. Tao Bai, Andrew Sinclair, Fang Sun, Priyesh Jain, Hsiang-Chieh Hung, Peng Zhang, Jean-Rene Ella-Menye, Wenguang Liu, and Shaoyi Jiang, 'Harnessing Isomerization-Mediated Manipulation of Nonspecific Cell/Matrix Interactions to Reversibly Trigger and Suspend Stem Cell Differentiation', *Chemical Science,* 7 (2016), 333-38.

4. Tao Bai, Sijun Liu, Fang Sun, Andrew Sinclair, Lei Zhang, Qing Shao, and Shaoyi Jiang, 'Zwitterionic Fusion in Hydrogels and Spontaneous and Time-Independent Self-Healing under Physiological Conditions', *Biomaterials,* 35 (2014), 3926-33.

5. Tao Bai, Yanjiao Han, Peng Zhang, Wei Wang, and Wenguang Liu, 'Zinc Ion-Triggered Two-Way Macro-/Microscopic Shape Changing and Memory Effects in High Strength Hydrogels with Pre-Programmed Unilateral Patterned Surfaces', *Soft Matter,* 8 (2012), 6846-52.

6. Tao Bai, Peng Zhang, Yanjiao Han, Yuan Liu, Wenguang Liu, Xiaoli Zhao, and William Lu, 'Construction of an Ultrahigh Strength Hydrogel with Excellent Fatigue Resistance Based on Strong Dipole–Dipole Interaction', *Soft Matter,* 7 (2011), 2825-31.

7. Yanjiao Han, **Tao Bai (Co-first Author)**, and Wenguang Liu, 'Controlled Heterogeneous Stem Cell Differentiation on a Shape Memory Hydrogel Surface', *Scientific reports,* 4 (2014).

8. Yanjiao Han, **Tao Bai (Co-first Author)**, Yuan Liu, Xinyun Zhai, and Wenguang Liu, 'Zinc Ion Uniquely Induced Triple Shape Memory Effect of Dipole–Dipole Reinforced Ultra High Strength Hydrogels', *Macromolecular rapid communications,* 33 (2012), 225-31.

9. Peng Zhang, Fang Sun, Caroline Tsao, Sijun Liu, Priyesh Jain, Andrew Sinclair, Hsiang-Chieh Hung, **Tao Bai**, Kan Wu, and Shaoyi Jiang, 'Zwitterionic Gel Encapsulation Promotes Protein Stability, Enhances Pharmacokinetics, and Reduces Immunogenicity', *Proceedings of the National Academy of Sciences* (2015), 201512465.

10. Xiyang Dai, Yinyu Zhang, Lina Gao, **Tao Bai**, Wei Wang, Yuanlu Cui, and Wenguang Liu, 'A Mechanically Strong, Highly Stable, Thermoplastic, and Self Healable Supramolecular Polymer Hydrogel', *Advanced Materials* (2015).

478365 From Reactive Nano-Particles to Self-Healing Materials: Chemical Research with a Green Twist

Abstract:

Research Interests:

Current and past research interests

My research is at the crossroads between the chemical, oil&gas and environmental engineering sectors, and encompasses colloid, polymer, emulsion and surface science.

Current research interests at SABIC:

Emulsions. Fouling issues in steam crackers may arise due to a number of reasons, including emulsion formation in the quench water tower. Interfacially active species were extracted and fractionated into solubility classes. The study showed the differences in the mechanisms of emulsion stabilization with the different fractions, and highlighted the correlation between interfacial film composition and morphology, and emulsion stability.

Iron-catalyzed radical polymerization. Undesired polymerization may lead to fouling issues in crackers, and should thus be avoided. The research conducted highlighted the role of iron in catalyzing polymerization of styrene and showed the correlation between solvent type and kinetics of radical polymerization.

Corrosion inhibitors. Film forming amines. Film forming amines (FFAs) are commonly used to inhibit corrosion in steam condensate systems, since they form passivating films which can act as a hydrophobic barrier against corrosive species in water. In steam systems FFAs are typically used in conjunction with neutralizing amines, which increase their solubility in water. The research conducted compared the effectiveness of different combinations of FFA and neutralizing amines in inhibiting corrosion, and investigated the factors controlling their performance.

Previous research interests:

Reactive polymer coated nano-particle optimization for groundwater remediation (PhD research). Nano zero-valent iron (nZVI) remediation technology is used to treat aquifers contaminated by chlorinated compounds and heavy metals, which react with iron forming less harmful compounds. The success of nZVI technology relies on adequate particle delivery to the polluted areas, and this was the focus of my research. Particle transport is largely influenced by the surface properties of the particles. Therefore, I investigated the effect of polymer coatings used to stabilize iron particle suspensions on the transport of particles through aquifers having different mineral compositions and water chemistries. I probed the forces of interaction between the particles and the substrates, because large adhesive forces hinder particle transport. In my study I utilized wet-water chemistry to synthesize and electrodeposited iron nano-particles on AFM probes to measure the adhesion forces between single particles and model mineral substrates under no-flow conditions. Furthermore, I probed sorption phenomena under laminar flow conditions using QCM-D. The particles synthesized were characterized using SEM, EDX, XPS and XRD to assess their morphology and composition before and after prolonged exposure to aqueous media, simulating groundwater conditions. My research showed the importance of hydration forces in the interactions between nZVI and mineral surfaces, and demonstrated that polymer coatings with a cationic counter-ion could inhibit particle attachment onto silicate minerals in a wide range of water chemistries.

Emulsions in the upstream oil&gas sector. Problematic emulsions are common in the upstream sector, and specifically in the oil sands industry. The collaborative research I conducted at the University of Alberta demonstrated the role of a specific asphaltene sub-fraction in stabilizing emulsions, and highlighted the factors controlling emulsion stability when using model compounds of asphaltenes. My research further focused on the use of polymeric demulsifiers to destabilize problematic emulsions. My research correlated demulsification performance to the chemistry of the demulsifiers and to their effect on the interfacial films, namely on their compressional and shear rheological properties, and on their morphology (as probed with BAM at the liquid-liquid interface, and with AFM of Langmuir-Blodgett films). A similar approach was successfully adopted to study emulsions stabilization by pH and temperature responsive NIPAM (poly(N-isopropylacrylamide)) micro-gels.

High performance polymer-based non-Newtonian gels: effect of cross-linker and polymer type on the shear and thermal stability of the gels. I developed polymeric gels for enhanced oil recovery applications at high temperature and high shear regimes. I engineered self-healing gels with synthetic and natural polymers, cross-linked with a variety of cross-linkers, and determined the best combination of polymer, cross-linker and water chemistry yielding a highly viscous and stable gel at elevated temperature and high shear rates. Importantly, I was able to produce an additive capable of generating radicals that could break the gel on demand with controlled kinetics, thus reducing its viscosity to extremely low values after a desired time period.

Future research scopes

My objective is to draw knowledge from the oil&gas and chemical sectors to produce materials that can be used for the treatment and the containment of contaminants, investigating at the nano-scale level the factors controlling their performance. My interests further extend to fouling phenomena, which are a major cause for energy losses in industrial processes. I intend to expand on my current research and engineer antifouling surfaces for diverse applications, ranging from petrochemical processes to water treatment using membranes.

Teaching Interests:

As a doctoral researcher at the University of Toronto I had the opportunity to support teaching for a total period of four years in the following disciplines:

- Engineering Mathematics;
- Engineering Graphics;
- Groundwater Flow and Contamination;
- Hydrology and Hydraulics.

I thoroughly enjoyed the role of teaching assistant, and I would be thrilled to pursue my vocation for teaching in the years to follow, engaging in either advanced or undergraduate courses.

Publications

Kuldeep Wadhwa, Juul Hennissen, Sharankumar Shetty, **Erica Pensini**, Carlo Geijselaers. Influence of the substitution of functional groups on the inhibition efficiency of TEMPO analogues on vinyl polymerization. To be submitted.

Erica Pensini, Fabrice Cuoq, Roy van Lier, Wolfgang Hater, Patrick Kraft, Tobias Halthur. Enhanced corrosion resistance of metal surfaces by film forming amines: a comparative study between cyclohexanamine and 2-(diethylamino)ethanol-based formulations. Submitted to I&EC.

Erica Pensini, Leo Vleugels, Martijn Frissen, Kuldeep Wadhwa, Roy van Lier, Gerard Kwakkenbos. A novel perspective on emulsion stabilization in gasoline crackers. Submitted to Colloids and Surfaces A.

Fan Yang, Plamen Tchoukov, **Erica Pensini**, Tadeusz Dabros, Jan Czarnecki, Jacob Masliyah, and Zhenghe XuAsphaltene Subfractions Responsible for Stabilizing Water-in-Crude Oil Emulsions. Part 1: Interfacial Behaviors. Energy Fuels, 2014, 28 (11), pp 6897–6904.

Zifu Li, David Harbottle, **Erica Pensini**, To Ngai, Walter Richtering, and Zhenghe Xu. Fundamental Study of Emulsions Stabilized by Soft and Rigid Particles. Langmuir, 2015, 31 (23), pp 6282–6288.

Jiebin Bi, Fan Yang, David Harbottle, **Erica Pensini**, Plamen Tchoukov, Sébastien Simon, Johan Sjöblom, Tadek Dabros, Jan Czarnecki, Qingxia Liu, and Zhenghe Xu.Interfacial Layer Properties of a Polyaromatic Compound and its Role in Stabilizing Water-in-Oil Emulsions. Langmuir, 2015, 31 (38), pp 10382–10391.

E. Pensini, D. Harbottle, F. Yang, P. Tchoukov, Z. Li, I. Kailey, J. Behles, J. Masliyah, Z. Xu. Demulsification Mechanism of Asphaltene-Stabilized Water-in-Oil Emulsions by a Polymeric Ethylene Oxide −Propylene Oxide Demulsifier. Energy & Fuels, 2014, 28 (11), pp. 6760-6771.

E. Pensini, B.E. Sleep, C. Yip, D. O'Carrol. Effect of water chemistry and aging on iron - mica interaction forces: Implications for iron particle transport. Langmuir, 2012, 28 (28), pp. 10453–10463.

E. Pensini, B.E. Sleep, C. Yip, D. O'Carrol. Forces of interactions between bare and polymer coated iron and silica: effect of pH, ionic strength and humic acids. Environ. Sci. Technol., 2012, 46 (24), pp. 13401–13408.

E. Pensini, B.E. Sleep, C. Yip, D. O'Carrol. Forces of interaction between fresh iron particles and iron oxide (magnetite): Effect of water chemistry and polymer coatings. Colloids and Surfaces A: Physicochemical and Engineering Aspects, 2013, 433, pp. 104–110.

E. Pensini, B.E. Sleep, C. Yip, D. O'Carrol. Carboxy-methyl cellulose binding to mineral substrates: Characterization by atomic force microscopy based - force spectroscopy and quartz-crystal microbalance with dissipation monitoring. Journal of Colloid and Interface Science, 2013, 402, pp. 58-67.

E. Pensini, B.E. Sleep, C. Yip, D. O'Carrol. Forces of interaction between iron particles and aluminum-silicates: Effect of water chemistry, polymer coatings and aluminum-silicate composition. Journal of Colloid and Interface Science, 2013, 411, 8-15.

480140 Energy Solutions through Electrochemical Processing: Electronic Devices,Energy Storage Devices, and Extractive Metallurgy

Abstract:

Research Interests:

Grid scale electrochemical energy storage devices

Multi-valent batteries for automobile propulsion

Electrochemical fabrication for future electronics and optoelectronics

Metal extraction processes and production processes of semiconductor materials

Corrosion for prolonging the life of devices and processes

Research Experience:

In my graduate work at Waseda University, I developed an electrochemical fabrication process for bit patterned magnetic recording media for future hard disk drives. Bit patterned magnetic recording media consist of an ordered highly crystalline and uniform ferromagnetic nanodot array. To fabricate these arrays, I developed a combined process using nano-patterning (nano-imprinting lithography) and electrochemical depositions – electrodeposition, electroless deposition. The process enabled deposition of magnetic materials onto nano-patterned substrates at the single nano-meter scale. I designed deposition conditions by characterizing the crystal structure, magnetic properties, and morphology (nucleation and growth) of the deposits. To better understand the electrochemical deposition reactions, I developed an in-situ technique for characterizing catalytic reactions on substrates at the molecule scale (1-2 nm) using Surface Enhanced Raman Scattering effects. Through this work, I demonstrated the feasibility of electrochemical deposition processes to fabricate nano-structures with both well-controlled crystallinity and uniformity at the single nano-meter scale. I also applied these processes to fabricate Si nano-pillars from ionic liquids.

Currently at MIT, I am leading the research project of liquid metal batteries, which consist of liquid electrodes and a molten salt electrolyte, to address the need for a grid-scale energy storage device. Large-scale electrochemical energy storage devices are valuable for the future grid, enabling intermittent renewable energy technologies to be integrated into the base load and creating a variety of economic benefits via off-peak shifting, load leveling, and frequency regulation. The liquid metal battery is, in principle, a reversible electrorefining cell. In this work, I have investigated the thermodynamics and kinetics of the electrochemical charge/discharge processes at the interfaces of the molten salt electrolytes and liquid metal electrodes. As a result, the applicability of calcium metal and its molten salts for electrochemical energy storage devices was demonstrated (the first calcium metal battery). Furthermore, I identified why and how corrosion occurs in the batteries and developed a solution to prolong their lifetime. These efforts have been the basis for the practical use.

My work contributes new knowledge of the electrochemical behaviors at the interfaces of electrodes and electrolytes (nucleation and growth, alloying and dealloying) at a wide range of scales (atom ~ meter) and temperatures (room temperature ~ high temperature). Tailoring these processes to specific applications is an enduring challenge in electrochemical processes as well as in the operation of the electrochemical devices. The backgrounds behind these topics are multi-disciplinary fields, such as surface science, electrochemistry, electrometallurgy, and multi-scale fabrication processes.

Postdoctoral Project: "Liquid metal battery for grid scale energy storage"

Adviser: Donald R. Sadoway, Materials Science and Engineering, Massachusetts Institute of Technology

PhD Dissertation: "Electrochemical Fabrication of Ordered Ferromagnetic Nanostructures and Control of Initial Deposition Process"

Adviser: Takayuki Homma, Applied Chemistry, Waseda University

Teaching Interests:

As for teaching, I feel competent to handle most any course at the undergraduate and graduate levels based on my multidisciplinary research experiences and extensive teaching experience. I have taught in the Department of Applied Chemistry at Waseda University and in the Department of Materials Science and Engineering at MIT. At both Waseda and MIT, I applied hands-on and heads-on training. At Waseda, I supported several professors in the department as an instructor of undergraduate courses in physical chemistry, inorganic chemistry, and lab courses. I managed teaching assistants and was also responsible for grading exams and reports. At MIT, I worked as an instructor for an undergraduate lab course (3.014 Materials Laboratory) by designing a syllabus, creating modules, giving short lectures, and preparing exam. Furthermore, I completed teaching certificate program at MIT in 2014. In Professor Homma's research group at Waseda, I mentored several undergraduate and master's students in writing their theses while pursuing my own thesis work. In the research group of Professor Sadoway at MIT, I also mentored six undergraduates for their research projects, two visiting students for their short-term projects, and two graduate students for their PhD research.

Future Research Direction:

As a faculty I will aim to contribute to energy research program and materials research program with the success of developing energy storage devices, electronic devices, materials synthesis processes, coating technology, and metal extraction and recycling processes. To do this, I will use the following features of electrochemical processes: precisely control over the composition, crystal structure, and phases; formation of complicated 3D structures by area-selective deposition and etching; rapid production from atom to meter scale. I have the advantage of being able to set the goal of the project from the broader perspectives like in metal extraction and approach the individual issues from understanding and designing basic chemistries at the atomic and molecular levels.

Selected Publications:

1. **T. Ouchi**, H. Kim, B. L. Spatocco, D. R. Sadoway, "Calcium-based multi-element chemistry for grid-scale electrochemical energy storage," *Nature communication*, 7:10999, doi: 10.1038/ncomms10999 (2016).
2. K. Wang, K. Jiang, B. Chung, **T. Ouchi**, P. J. Burke, D. A. Boysen, D. J. Bradwell, H. Kim, U. Muecke, D. R. Sadoway, "Lithium–antimony–lead liquid metal battery for grid-level energy storage," *Nature*, 514, 348–350 (2014).
3. *3.* **T. Ouchi**, H. Kim, X. Ning, D. R. Sadoway, "Calcium-Antimony Alloys as Electrodes for Liquid Metal Batteries," *J. Electrochem. Soc.,* 161(12) A1898-A1904 (2014).
4. **T. Ouchi**, N. Shimano, T. Homma, "CoNiP Electroless Deposition Process for Fabricating Ferromagnetic Nanodot Arrays," *Electrochimica Acta*, 56, 9575-9580 (2011).
5. **T. Ouchi**, Y. Arikawa, T. Homma, "Fabrication of CoPt Magnetic Nanodot Arrays by Electrodeposition Process," *Journal of Magnetism and Magnetic Materials*, 320, (22), 3104-3107 (2008).

476252 Material Interactions and Synergies in Lithium-Air Batteries and Electrochemical Devices

Abstract:

Research Interests:

Energy storage technologies are essential to the security and sustainability of our powered infrastructure. Due to their exceptionally high energy densities, lithium-based batteries have been the beneficiary of intense research and development efforts that now underpin portable electronics, electric vehicles and (soon) the "smart" grid. Research to improve upon or disrupt state-of-the-art Li-ion battery technologies, however, has been painfully slow, especially when compared to advances in semiconductor manufacturing. The complexity of rechargeable batteries, often containing multiple active materials with different cell designs and operating under various load conditions, explains the difficulty of hewing to a straightforward design methodology. My efforts have sought to inject rationality into the development of next-generation battery technologies (Li-air/Li-oxygen systems) with various form factors (ultrathin, transparent Li-ion) by deconvoluting the influences of material interaction and applying those findings to better material selection. My work on Li-air/Li-oxygen batteries has encompassed many aspects of that technology from catalyst-electrolyte synergy, to electrolyte side reactions and anode stability, to *in situ* and *operando* studies of redox processes, and reactant transport during cell operation. My work on designing ultrathin Li-ion battery architectures has involved the development of automated, tunable self-assembly techniques with microgram precision. The experimental tools I use to understand material interactions in electrochemical systems are currently being paired with computer simulations (molecular dynamics, density functional calculations and finite element analysis) from my team at Sandia National Laboratories to build our predictive capabilities and anticipate full device behavior. This holistic approach to understanding and improving lithium batteries showcases the benefits of combining materials science, electrochemistry and chemical engineering disciplines to address design challenges in myriad electrochemical systems (i.e. capacitors, sensors, fuel cells, electrolysers).

Research Experience:

My research experience has involved a deep dive into electrochemistry and materials science with projects touching on many energy-related applications. My doctoral work with Prof. André Taylor at Yale University explored the cutting-edge chemistry of catalysts in Li-oxygen systems as well as nanomanufacturing of ultrathin batteries by layer-by-layer (LbL) self-assembly. I have collaborated on projects including the development of metallic glass electrocatalysts for polymer exchange membrane (PEM) fuel cells, electrodes for microbial fuel cells, as well as electrochemical sensors and conductive thin films. At Sandia National Laboratory, I serve as a co-principal investigator and lead experimentalist for our Li-air battery development team where I interface with computational team members to develop predictive tools for material selection. My experimental focus at Sandia encompasses electrolyte development, particularly involving ionic liquid design, and device prototyping.

Teaching Interests:

My experiences as a teacher and mentor have had a significant impact on my career goals and research program. For three semesters, I served as a teaching assistant in the Department of Chemical and Environmental Engineering at Yale University in the subjects of environmental engineering and professional ethics. Through these roles, which involved teaching weekly discussion sessions and critiquing student reasoning, I gained insight into methods for introducing and refreshing concepts throughout a term. I delivered guest lectures in graduate courses and an annual Climate and Energy symposium, where I sought to relate new concepts to my audience's non-technical knowledge. Most influential to my development has been the experience of mentoring numerous undergraduate students in a lab setting, four of whom I worked with for greater than one year and three of whom are cited as co-authors on peer-reviewed publications. With these students, I developed techniques for initiating newcomers into the discipline of electrochemistry and formulated a groundwork for skill development and research contributions. After my mentees transitioned to graduate programs, a battery startup company and the finance industry, I have continued my support. Mentoring fellow graduate students in my group and serving as a regular editor and internal reviewer for their manuscripts has further extended my impact with junior colleagues. I see mentorship and one-on-one interaction with students as critical to ensure difficult concepts are understood and retained. My future teaching will retain this passion for student growth and the formation of strong mentor/mentee relationships.

Successful Proposals:

Porous Liquid Electrolytes for Metal-Air Batteries, Laboratory Directed Research and Development (LDRD), Sandia National Laboratories, 2016 – Principal Investigator

Selected Publications (of 18):

- Ryu, W.-H.; **Gittleson, F.S.**; Li, J.; Tong, X.; Taylor, A.D. A New Design Strategy for Observing Lithium Oxide Growth-Evolution Interactions Using Geometric Catalyst Positioning. *Nano Lett.*, **2016**. (*in press*)
- **Gittleson, F.S.**; Ryu; W.-H.; Schwab, M.; Tong, X.; Taylor, A.D. Pt and Pd Catalyzed Oxidation of Li_2O_2 and DMSO during Li-O_2 Battery Charging. *Chem. Comm.*, **2016**, *52*, 6605-6608.
- **Gittleson, F.S.**; Hwang, D.; Ryu, W.-H.; Hashmi, S.M.; Hwang, J.; Goh T.; Taylor, A.D. Ultrathin Nanotube/Nanowire Electrodes by Spin-Spray Layer-by-Layer Assembly: A Concept for Transparent Energy Storage. *ACS Nano.*, **2015**, *9*, 10005-10017.
- Ryu, W.-H.; **Gittleson, F.S.**; Schwab, M.; Goh, T., Taylor, A.D. A Mesoporous Catalytic Membrane Architecture for Lithium-Oxygen Battery Systems. *Nano Lett.*, **2015**, *15*, 434-441.
- **Gittleson, F.S.**; Ryu, W.-H.; Taylor, A.D. *Operando* Observation of the Gold-Electrolyte Interface in Li-O_2 Batteries. *ACS Appl. Mater. Inter.*, **2014**, *6*, 19017-19025.
- **Gittleson, F.S.**; Sekol, R.C.; Doubek, G.; Linardi, M.; Taylor, A.D. Catalyst and Electrolyte Synergy in

Li-O_2 Batteries. *Phys. Chem. Chem. Phys.*, **2014**, *16*, 3230-3237.

- **Gittleson, F.S.**; Hwang, J.; Sekol, R.C.; Taylor, A.D. Polymer Coating of Vanadium Oxide Nanowires to Improve Cathodic Capacity in Lithium Batteries. *J. Mat. Chem. A*, **2013**, *1*, 7979-7984.
- **Gittleson, F.S.**; Kohn, D.; Li, X.; Taylor, A.D. Improving the Assembly Speed, Quality, and Tunability of Thin Conductive Multilayers. *ACS Nano*, **2012**, *6*, 3703-3711.

476276 Life at Interfaces: Understanding the Fluid Dynamics, Transport and Surface Translocation of Bacterial Biofilms

Abstract:

2ndYear Postdoctoral Fellow

Research Interests:

In many natural environments, bacteria collectively aggregate to exist in densely populated and structurally diverse communities called biofilms. Within these biofilms, cells exhibit a remarkable division of labor, with various subpopulations of cells involved in active motility by flagellar propulsion, production of extracellular polymeric substances and other functions that collectively results in the initiation, growth, propagation, adhesion and dispersion of biofilms to specific chemical and environmental signals.

By using a combination of experimental techniques such as time-lapse microscopy, constructing fluorescent transcriptional reporters for distinct cell types, confocal microscopy, flow visualization, surface texturing and chemical patterning, my research systematically investigates the collective response that governs the bacterial growth and surface interaction of biofilms under a number of different environmental conditions (such as temperature, pH, agar gel concentration, etc.) with ***Bacillus Subtilis*** as the model organism.

In addition, I also work on theoretically and computationally modeling the stability of thin viscous sheets under shear with a non-uniform temperature gradient. Such a model has practical applications ranging from the manufacture of ultra-shin glass sheets via the float glass technique, to understanding physical and biological phenomena such as the buckling of bacterial pellicles during colony growth, and geological processes involving wrinkling of lava sheets.

Postdoctoral Projects:

(Under the supervision of ***L. Mahadevan*** & ***S. Rubinstein***, School of Engineering & Applied Sciences, Harvard University)

- Buckling instabilities in thin viscous sheets during extensional flows (theory & simulation)
- Characterizing growth & surface translocation in developing ***Bacillus Subtilis*** biofilms (experimental)

PhD Dissertation: "*Exploring Plastron Stability and Fluid Friction Reduction on Robust Micro-Textured Non-Wetting Surfaces.*"

Under the supervision of Robert E. Cohen and Gareth H. McKinley, Department of Chemical & Mechanical Engineering, Massachusetts Institute of Technology.

Research Experience:

My PhD work involved both experimental and theoretical research at the interface of polymer physics, fluid mechanics, wetting and soft matter. In my thesis, I developed a precise and repeatable spray technique for rapid fabrication of drag- reducing liquid-repellent coatings that can be deployed on a large scale. I used these coatings to study the fundamental problem of how structured liquid-repellent surfaces, which support a composite liquid-air interface, can reduce skin friction in both viscous laminar and turbulent flows in a Taylor-Couette setup, and developed a theoretical model that accounts for the effect of wall slip in near-wall turbulent flows. In addition to this central experimental theme, I have also worked on a mathematical analysis of the gravity driven deformation of drops on non-wetting surfaces, correlating the wetting dynamics on the barbs and barbules of aquatic birds with diving and wing-spreading behavior, and on developing a hierarchical model to design oil-repellent woven fabrics.

Teaching Interests:

During my education at MIT, I was a Teaching Assistant for the Graduate Transport Phenomena Course. I organized weekly help sessions and presented extra lectures for a class of 50 graduate students on subjects including fluid mechanics, heat/mass transfer, partial differential equations and mathematical methods. I have also directly supervised undergraduate & graduate research students on specific projects during my postdoctoral fellowship. I am interested teaching modules in fluids mechanics, transport phenomena, solid mechanics, heat transfer, mass transfer and applied engineering mathematics, amongst other courses such as thermodynamics and statistical mechanics.

Future Direction:

As a research faculty, I would like to continue the philosophy of using experimental and theoretical tools to understand the physicochemical hydrodynamics in complex systems. In this regard, I believe the bacterial biofilm system I am currently studying provides a rich experimental toolbox, with a number of interesting and currently unexplored phenomena to investigate. Specific examples include studying the biofilm rheology and mechanical behavior (***e.g.***, the viscoelastic characterization of the extracellular polymer matrix), fluid dynamics (directional flows of surfactant laden films generated by the bacteria during swarming) to surface interactions (influence of structured coatings on bacterial growth)

The uniqueness of my approach is the ability to perform direct in-situ measurements and quantify the dynamic response within the biofilm (such as spatial localization & switching between various cell types, observing transient response to antibiotics and external stresses, measurement of velocities during swarming, etc.) both at a single cell level and at a colony scale, and directly test and develop biophysical mechanisms governing growth and spreading. The ultimate goal is in designing chemically patterned & textured surfaces that selectively inhibit the growth of living matter on surfaces, and has a number of potential applications such as in preventing biofouling in marine vehicles, or spreading in infectious tissue, etc.

Due to the interdisciplinary nature of my work, I envision a strong and active collaboration with fellow faculty in engineering and the basic sciences and/or work with them to setup specific techniques (such as constructing reporter strains, etc.) in my lab.

Selected Publications:

Srinivasan, S., Wei, Z., Mahadevan, L. (**2016**). Buckling instability in extensional flow of a thin non-uniformly viscous sheet. ***(under review)***

Wang, X., Koehler, S. A., Wilking, J. N., Sinha, N. N., Cabeen, M. T., **Srinivasan, S.**, ... others. (**2016**). Probing phenotypic growth in expanding Bacillus subtilis biofilms. ***Applied Microbiology and Biotechnology***, ***100***(10), 4607–4615.

Srinivasan, S., Kleingartner, J. A., Gilbert, J. B., Cohen, R. E., Milne, A. J. B., & McKinley, G. H. (**2015**). Sustainable drag reduction in turbulent Taylor-Couette flows by depositing sprayable superhydrophobic surfaces. ***Physical Review Letters***, ***114***(1), 14501.

Kleingartner, J. A*., **Srinivasan, S**.*, Truong, Q. T., Sieber, M., Cohen, R. E., & McKinley, G. H. (**2015**). Designing Robust Hierarchically Textured Oleophobic Fabrics. ***Langmuir***, ***31***(48), 13201–13213.

Srinivasan, S., Chhatre, S. S., Guardado, J. O., Park, K.-C., Parker, A. R., Rubner, M. F., Mckinley, G.H, Cohen, R. E. (**2014**). Quantification of feather structure, wettability and resistance to liquid penetration. ***Journal of The Royal Society Interface***, ***11***(96), 20140287.

Srinivasan, S., Choi, W., Park, K.-C., Cohen, R. E., & McKinley, G. H. (**2013**). Drag reduction for viscous laminar flow on spray-coated non-wetting surfaces. ***Soft Matter***, ***9***(24), 5691–5702. (featured on cover)

Srinivasan, S., Chhatre, S. S., Mabry, J. M., Cohen, R. E., & McKinley, G. H. (**2011**). Solution spraying of poly (methyl methacrylate) blends to fabricate microtextured, superoleophobic surfaces. ***Polymer***, ***52***(14), 3209–3218.

Srinivasan, S., McKinley, G. H., & Cohen, R. E. (**2011**). Assessing the accuracy of contact angle measurements for sessile drops on liquid-repellent surfaces. ***Langmuir***, ***27***(22), 13582–13589.

476294 Triggerable Tough Hydrogels for Gastrointestinal Biomedical Applications

Abstract:

Research Interests:

My research interests mainly include polymer, biomaterials, nanomaterials, biomedical devices, drug delivery and nanomedicines. In this presentation I will focus on the gastrointestinal biomedical applications of hydrogels.

The introduction of gastric-resident dosage forms has demonstrated significant promise for enhanced drug therapy in humans. However, the lack of precise control over the gastric-retention time remains an important challenge for clinical adoption of these dosage forms. Here we report the development of novel triggerable tough hydrogels (TTHs) that not only show exceptional toughness to overcome gastric pressures achieving long-term residence in stomach but also exhibit remarkable triggerability to be dissolved rapidly with biocompatible agents when required. We demonstrate a proof of concept extended drug delivery from TTHs in a large animal model. Given the dramatic mechanical properties and capable of on-demand dissolution of TTHs we anticipate their implementation in a broad of gastrointestinal biomedical applications.

Teaching Interests:

I have teaching experiences in general material science and chemistry. I would like to design two new courses: a biomaterials course and a nanomaterials course. Both of them can be offered to graduate students or senior undergraduate students.

454451 Bio-Electronic Devices for Healthcare: From Wearable Biosensors to Nanorobots

Abstract:

Research Interests: Wearable Devices, Flexible Electronics, Nanomaterials, Nanomachines, Nanorobotics, Biosensors, Nanomedicine, Nanomotors.

Wearable biosensors are essential to the realization of personalized medicine through continuously monitoring an individualâ€™s state of health. Human sweat, which is rich in physiological information, is an excellent candidate for non-invasive monitoring. Given the complexity of sweat secretion, simultaneous and multiplexed screening of sweat biomarkers is critical. I have developed a mechanically flexible and fully-integrated sensor array for multiplexedÂ *in-situ*Â perspiration analysis, which simultaneously and selectively measures sweat metabolites and electrolytes, as well as the skin temperature (to compensate other sensorsâ€™ readings) (Gao et al. Nature, 2016, 529, 509-514). The platform is the first ever demonstration of a complete system for accurate sweat analysis which merged skin-conforming plastic-based sensors and conventional integrated-circuit components at an unprecedented level of integration, not only to measure the output of multiplexed and selective sensors, but also to obtain an accurate assessment *via* signal processing of the physiological state of the human subjects. This wearable system has been used to measure the detailed sweat profile of human subjects engaged in prolonged indoor and outdoor physical activities. Currently I have greatly expanded the scope of platform to monitor a wide panel of analytes including heavy metals, peptides and proteins. I envision that wearable sweat biosensors can be used for diverse clinical and physiological applications (*e.g.*monitoring pre-diabetes, dehydration, heavy metal poisoning, and major depressive disorder). While wearable devices play an important role in health monitoring on the skin, nanorobots offer great promise to efficiently deliver drug payloads and perform precision diagnosis/surgery inside human body. The nanoscale propulsion is challenging due to low Reynolds fluids. I have invented the first fully biocompatible and biodegradable nanorobots that can move autonomously in human biofluids based on Zn or Mg micromotors. I have also demonstrated the first *in vivo* study of synthetic nanorobots in living mice. The acid-driven propulsion effectively enhances the delivery of cargo payloads onto the stomach wall. In addition, I pioneered a variety of artificial nanomachines that can achieve nanoscale propulsion. After modification with different functional groups or bio-receptors, these nanorobots are able to selectively transport and isolate specific targets (such as protein, DNA, bacteria and cancer cells). These works are anticipated to significantly advance the emerging field of nanomachines and to open the door to *in vivo*evaluation and clinical applications of the synthetic nanorobots.

Teaching Interests: I am interested in teaching most of the the major courses for Chemical Engineering, especially Heat and Mass Transfer, Fluids Dynamics, and Thermodynamics, Separation Process. I am also interested in other courses including Electrochemistry, Nanotechnology and Nanoengineering, Nanomedicine.

Tailoring Inorganic Materials with High Surface Area for Electronic Applications

Wanmei Sun[1], Richard B. Kaner[2], Micah Green[1]

wanmei@tamu.edu

[1] Artie McFerrin Department of Chemical Engineering, Texas A&M Univeristy
[2]Department of Chemistry & Biochemistry, University of California Los Angeles

Synthesis of 2D Materials

Graphene Oxide

--From modified Hummers' method

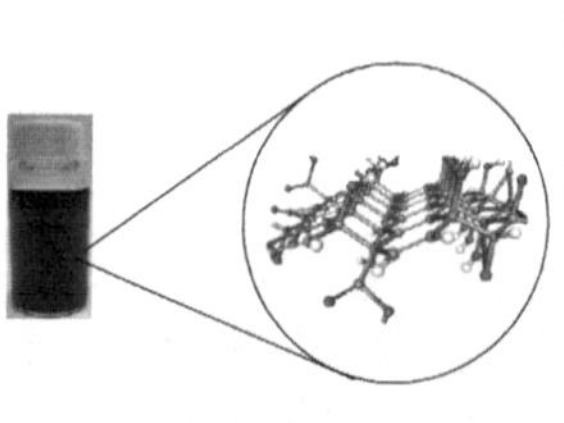

Advantages of Hummers' method

- Easy dispensability in water and other organic solvents
- Sheet size ~ 50 μm
- 1-5 layers

Graphene

--From electrochemical exfoliation

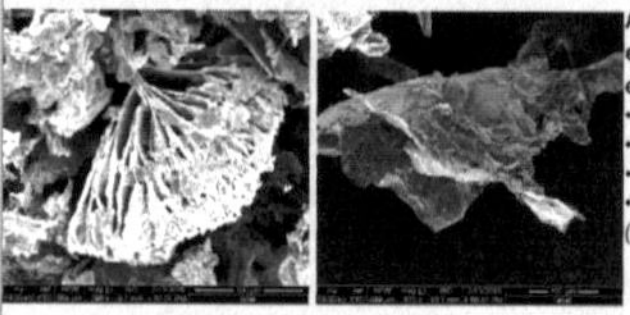

Advantages of electrochemical exfoliation

- Up scalability
- Solution processability
- Eco-friendliness
- Large sheet size (up to 200 μm)

Black Phosphorene (in progress)

--From high energy mechanical milling and solvent exfoliation

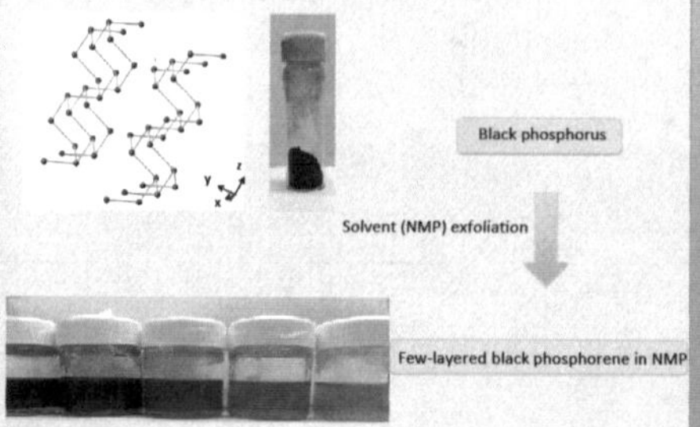

Advantages of black phosphorene

- Tunable direct band gap (0.3 – 2.0 eV for bulk to monolayer)
- High hole mobility (up to 286 $cm^2V^{-1}S^{-1}$)
- Anisotropic electronic, transport, optoelectronic and mechanical properties

Mxene (in progress)

--Selective etching of Al from Ti_2AlC without HF

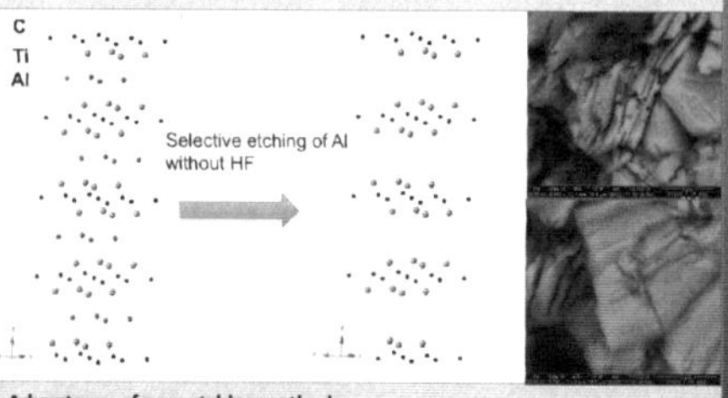

Advantages of new etching method

- No HF involved in the etching (different from all the current synthesis for MXene)
- Tunable functional groups on the surface
- Mxene is hydrophilic and highly conductive

Synthesis of 3D Materials

Hexagonal Tungsten Trioxide

--From hydrothermal reaction

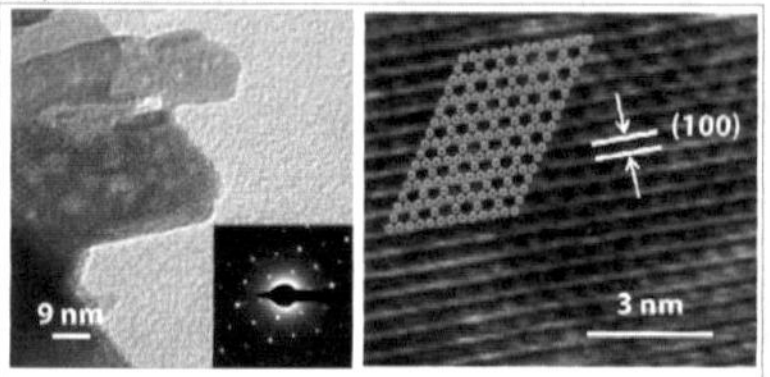

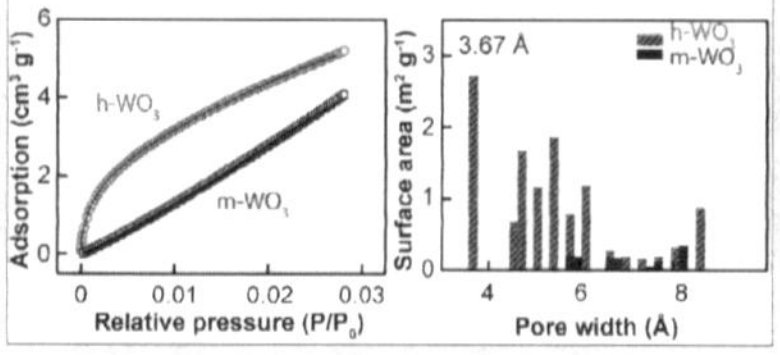

Transmission electron microscopy (TEM) and selected area electron diffraction (SAED) images show high crystallinity. Low pressure CO_2 adsorption isotherms and nonlocal density functional theory fitting shows the existing of intralcrystalline tunnels and the tunnel size is 3.67 Å. Thermal gravimetric analysis(TGA), inductively coupled plasma atomic emission spectroscopy(ICP-OES) show the open and rigid tunnels adsorb H^+ and Li^+, but not Na^+ in aqueous solution.

Synthesis of 0D & 2D Materials

Transition Metal Oxides on Reduced Graphene Oxide

--From *in situ* reduction (in microwave, flash *et al*)

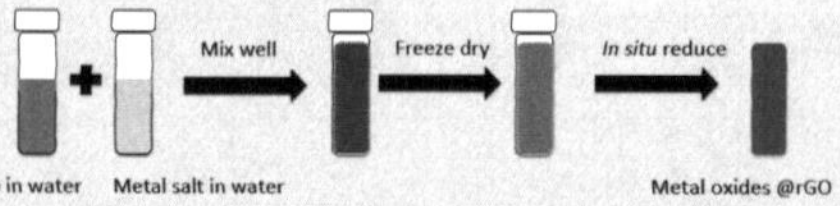

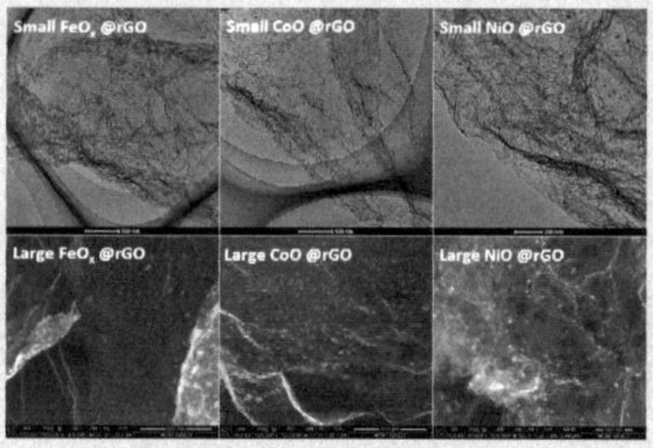

Synthesis of 1D Materials

Tungsten Trioxide, Molybdenum Trioxide *et al*

--From physical vapor deposition

Electrochemical Energy Storage

High Surface Area Improves Electrochemical Prosperities

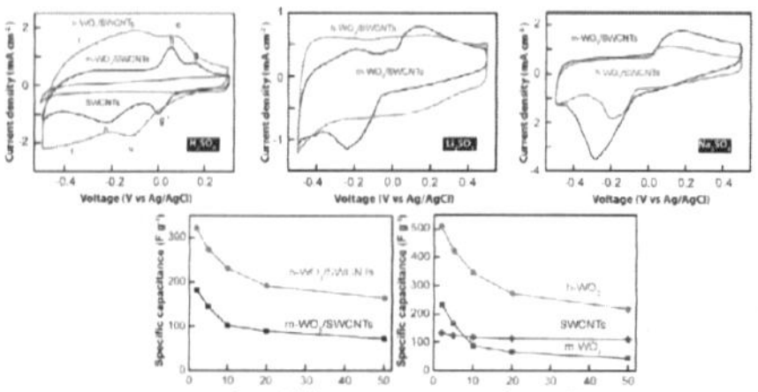

With increased surface area(from intracrystalline tunnels), hexagonal tungsten trioxide processes high specific pesudocapacitance in small ion electrolytes(such as H^+ and Li^+), not for big ion electrolytes(such as Na^+)

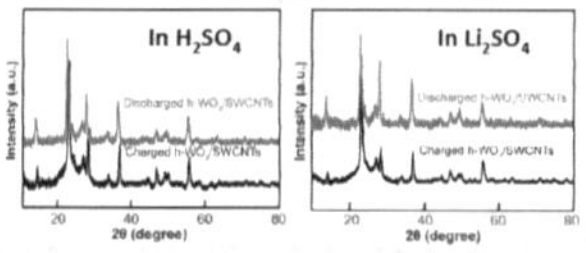

No phase changes for hexagonal tungsten trioxide in small ion electrolytes (such as H^+ and Li^+)

Research Interesting

- Rational design and synthesis of novel 2D materials
- Tailoring the surface of 2D materials for related applications
- 2D materials for electromagnetic interference shielding
- 2D materials and their composites for energy storage
- 0D and 2D materials for biosensor and bio images

Resume

Working experience

- Postdoctoral Research associate — Mar. 2016-Present

Education

- Joint Ph.D in Chemistry — Oct. 2012-Oct.2014
  Department of Chemistry and Biochemistry,
  University of California, Los Angeles
- Ph.D in Physical Electronic — Sept. 2010-Mar. 2016
  Wuhan National Laboratory Optoelectronics,
  School of Optical and Electronic Information,
  Huazhong University of Science and Technology, China
- B.A. in Applied Chemistry — Sept. 2006-Jun. 2010
  School of Chemistry and Chemical Engineering,
  Huazhong University of Science and Technology, China

Publications

- Wanmei Sun, Michael T. Yeung, Andrew T. Lech, Cheng-Wei Lin· Chain Lee, Tianqi Li, Xiangfeng Duan, Jun Zhou, Richard B. Kaner "High Surface Area Tunnels in Hexagonal WO_3" *Nano Letters* 2015, 15 (7), 4834-4838
- Xianghui Zhang, Li Gong, Kang Liu, Yuanzhi Cao, Xu Xiao, Wanmei Sun, Xuejiao Hu, Yihua Gao, Jian Chen, Jun Zhou, Zhong Lin Wang " Tungsten Oxide Nanowires Grown on Carbon Cloth as a Flexible Cold Cathode" *Advanced Materials* 2010, 46 (22), 5292-5296
- Tianqi Li, Majid Beidaghi, Xu Xiao, Liang Huang, Zhimi Hu, Wanmei Sun, Xun Chen, Yury Gogotsi, Jun Zhou "Ethanol Reduced Molybdenum Trioxide for Li-ion Capacitors" *Nano Energy* 2016, 26, 100-107
- Jia Li, Kang Liu, Xiang Gao, Bin Yao, Kaifu Huo, Yongliang Cheng, Xiaofeng Cheng, Dongcheng Chen, Bo Wang, Wanmei Sun, Dong Ding, Meilin Liu, Liang Huang " Oxygen- and Nitrogen-Enriched 3D Porous Carbon for Supercapacitor of High Volumetric Capacity" *ACS Applied Materials & Interfaces* 2015, 7 (44), 24622-24628

459786 Fluorescent Nanosensors for Biomolecular Targets

Abstract:

Research Interests: Single walled carbon nanotubes (SWCNTs) have remarkable optical properties, and they benefit from the ease of surface functionalization and biocompatibility, which render them favorable agents for numerous biomedical applications including imaging, sensing, targeting, delivery, and therapy. Semiconducting SWCNTs have intrinsic bandgap fluorescence of excitonic nature in a wide range of the near infrared (nIR) part of the spectrum. In addition to their high photo-stability, lack of photobleaching, and narrow spectral bandwidth, the SWCNTs nIR fluorescence emission overlaps with the tissue transparency window and benefits from reduced tissue absorption, scattering, and autofluorescence, making them favorable candidates for in vivo imaging and biomedical sensing applications. Upon tailored surface functionalization, SWCNTs can spectrally response to their surroundings, exhibiting a modulation of their fluorescent intensity or a shift of the peak emission wavelength, enabling real-time optical detection. I will present a recent discovery of a SWCNT-heteropolymer complex that recognizes the protein fibrinogen with high selectivity, using the concept of Corona Phase Molecular Recognition (CoPhMoRe). In this approach, a heteropolymer is adsorbed onto and templated by a nanoparticle surface to recognize a specific target analyte. This is the first demonstration of CoPhMoRe for macromolecular targets, opening new avenues for synthetic, non-biological antibody analogues with great promise for medical and clinical applications.

Teaching Interests: I believe that as an educator my role is to catalyze the learning process of the students. I wish for my students to be deep learners and to gain genuine understanding of the material, granting them tools to creatively solve problems they had never encountered before. Since I consider teaching to be an essential aspect of the academic career, I am fully committed to facilitating these goals by setting up a fruitful and inclusive teaching environment, while motivating and engaging the students in class using active learning techniques and educational technologies.

473675 Chemical Engineering Faculty Candidate with Specialization in Nanoscale Science and Engineering

Abstract:

Research Interests:

Research should not always be constrained to theoretical treatise of some interesting phenomenon performed, but should strive toward finding practical solutions for day-to-day problems. My doctoral and postdoctoral research has focused on synthesizing and characterizing organic-inorganic hybrid materials for practical applications. I have developed several multifunctional hybrid-materials and made significant progress toward meeting the necessary requirements for industrial and defense applications. I have broad research interests in materials synthesis and nanocrystal growth, functional materials for energy conversion and antimicrobial activity, high surface area materials for molecular and energy storage and biocompatible materials for controlled drug release. I am especially interested in surface interaction and mechanisms of chemical reaction of interfacial molecules in the broad fields of energy conversion, catalysis and nanotechnology. My research covers wide ranging questions with applications of emerging functional materials from the molecular and nanostructural levels to the applied environment. As a PI and Co-PI on two research grants, I was recently awarded $583k from the Department of Defense to investigate the multifunctionality of quantum-dot incorporated nanostructured functional materials and their optoelectronic sensitivity and photocatalytic activity. I am looking for a tenure-track faculty position in a research oriented university where I can explore my research and teaching potential in chemical engineering.

Teaching Interests:

Proposed Courses
In addition to the basic undergraduate courses required for chemical engineering students, I would like to propose one or more undergraduate/graduate level courses under the following topics: Mass Balances, Fluid Mechanics, Heat Transfer, Kinetics and Reactor Design, and Process Control.

475507 Taking the Temperature of the Interiors of Magnetically Heated Nanoparticles and Optical Biomolecular Chemical Sensing Using Single Wall Carbon Nanotubes

Abstract:

2nd Year Postdoctoral Associate

Research Interests:

My research has been focusing on using nanomaterials for controlled drug delivery and for optical biomolecular sensing. For my Ph. D work, I applied surface modified mesoporous silica nanoparticles for anti-cancer drug delivery, and studied the particle interior magnetic heating effect using temperature-dependent luminescence of nanocrystals. The knowledge on the physical and chemical properties of silica nanoparticles facilitated their biomedical use for in vitro and in vivo studies. During my postdoctoral training, I investigated on using near IR fluorescent carbon nanotubes to detect protein content as well as to examine their binding kinetics. I have developed a miniature microarray platform based on optical signals for label-free multiple protein molecule sensing.

Postdoctoral Project: "***Multiplexed protein detection using single wall carbon nanotube microarray***"

PI: Michael S. Strano, Department of Chemical Engineering, MIT

PhD Dissertation: "Physical Properties of Mesoporous Silica Nanoparticles for Stimuli-Responsive ***Drug***

Delivery"

PI: Jeffrey I. Zink, Department of Chemistry and Biochemistry, UCLA

Research Experience:

- Label-free Single Wall Carbon Nanotube Microarray for Protein Profiling
    - Surface modification of single wall carbon nanotubes with antibody and dsDNA
    - Protein binding kinetic analysis and batch processing
    - Carbon nanotube near IR fluorescence spectroscopy and microscopic imaging
    - Near IR microarray detector construction

- ***In Situ*** Optical Measurement of Nanoparticle Interior Temperature
    - Synthesis of thermally sensitive upconversion lanthanide nanocrystals and dual core nanoparticle
    - ***In situ*** interior temperature study of magnetically heated silica nanoparticles
    - Mechanism investment of nanomachine operation in response to continuous near-IR irradiation
    - Simultaneous spectral detection of nanoparticle interior temperatures and the corresponding cargo release

- Mesoporous Silica Nanoparticles for Controlled Drug Delivery
    - Synthesis of polymer coated silica particles with nanovalves for enhanced biomedical properties
    - Construction of Gd incorporated nanomachines for a multi-functional delivery vehicle
    - Synthesis of functionalized mesoporous silica nanoparticle for ***in vivo*** and ***in vitro*** drug delivery studies

Teaching Experience:

- Teaching certificate, MIT Kaufman Teaching Certificate Program
- Chemistry content expert, MIT BLOSSOMS (outreach program)
- Teaching assistance for general chemistry discussion sessions, UCLA (five quarters)
- Teaching assistance for chemical laboratory sessions, UCLA (two quarters)
- Mentor for two graduate students and one undergraduate student

Teaching Interests:

I believe the goal of a college education is not only to teach the knowledge and skill sets that are needed for work but also to prepare students with the curiosity and methods to continue the learning throughout their lives, for career and personal development. To meet this goal, I will try to improve the learning efficiency by integrating more critical thinking, active learning and connection-building components into our teaching process. Especially for upper level classes, I will introduce advanced projects, in which students are asked to pick research papers that they are interested in, orally present the work to the class and articulate their own opinions about it in the format of a short study report.

Future Direction:

I aim at using carbon nanotube sensing platform to perform early cancer diagnosis by detecting the trace amount of circulating tumor biomarkers. Compared to current immunoassays, nanotube molecular sensors are highly sensitive (down to single molecule detection sensitivity), require no fluorescent labeling and could quantify multiple protein components for correlation studies. Hence, they are especially advantageous in examining the small amount of signature biomolecules for early tumor diagnosis.

In addition, I want to develop a controlled therapeutic delivery platform based on biofunctional molecules. With my experience in using antibody as the recognition site for biomolecular sensing, I am planning on integrating the antibodies with the mesoporous silica nanoparticles. Antibodies offer the flexibility of targeting the receptors of tumor tissues, and hence improve the nanoparticle localization and uptake. In addition, antibody could potentially be the substrate for enzymatic activities and therefore, introduce a new series of nanomachine triggering mechanism. Besides antibody, other biofunctional molecules, such as DNA aptamars and proteins could also be integrated with the silica particles as the next generation of functional delivery vehicles.

Selected Publications:

- **J. Dong** and J. I. Zink. "Simultaneous Spectroscopic Measurements of the Interior Temperature and Induced Cargo Release from Pore-restricted Mesoporous Silica Nanoparticles." ***Nanoscale* 8**, 10558-10563 (2016).
- **J. Dong** and J. I. Zink. "Light or Heat? The Origin of Cargo Release from Nanoimpeller Particles Containing Upconversion Nanocrystals under IR Irradiation." ***Small 11***, 4165-4172 (2015).
- **J. Dong** and J. I. Zink. "Taking the Temperature of the Interiors of Magnetically Heated Nanoparticles." ***ACS Nano 8***, 5199-5207 (2014). **Highlighted** in ***Nature Physics***.
- **J. Dong**, M. Xue and J. I. Zink. "Functioning of Nanovalves on Polymer Coated Mesoporous Silica Nanoparticles." ***Nanoscale 5***, 10300-10306 (2013).

476010 Nanoscale Engineering and Model-Guided Design of Advanced Energy Storage and Conversion Technologies Utilizing Ultrathin Polymer Films

Abstract:

Research Interests:

Due to CO_2-induced climate change, there is currently great interest in using sustainable energy resources such as sunlight. The major challenge is that the collection and storage of solar energy should be economical. A promising approach to address this grand challenge is the application of inexpensive novel polymers in solar power conversion and storage systems. Therefore, my research focuses on scientific exploration of innovative material development in polymers for energy storage and harvesting systems. By developing new chemical vapor deposition (CVD) processes, I have successfully integrated nanoscale polymer thin films (<30nm) into the nanostructured electrodes of dye sensitized solar cells (DSSCs) and supercapacitors, and have improved device performance and stability. Concurrently, much of the current material research in the areas of energy storage and conversion has been based on scientific intuition and trial and error experimentation—a slow and inefficient process with many of the directions taken to address device challenges being without theoretical guidance. In fact, many of the recent advances in the DSSC and supercapacitor fields have relied on the computational design and screening of materials. Therefore, a major focus of my research has been to use first principles modeling to gain insight into the key physical and electrochemical processes occurring in these energy systems, and then apply that knowledge towards enhancing device performance by optimizing material properties and architecture. This holistic synergetic approach between experiment and modeling has allowed me to gain a more complete understanding into the fundamental processes within these devices along with rapidly optimizing material selection, device architecture, and performance.

In the area of energy conversion, DSSCs represent one of the newer generations of solar technologies. Current DSSC technology utilizes an organic liquid electrolyte; however, liquid electrolytes suffer from practical issues of leakage, evaporation, and stability as well as fundamental issues of rapid charge recombination loss at the photoanode-electrolyte interface that limits photocurrent and photovoltage. One promising solution is to replace the liquid with a solid polymer electrolyte material. Polymer electrolytes are an attractive alternative because they do not suffer from the practical disadvantages of liquid electrolytes and could potentially enhance the cell's I-V behavior. However, one of the major challenges with using polymers is the difficulty in integrating them within the tortuous mesoporous TiO_2 photoanode of the DSSC. The photoanode consists of a sintered network of TiO_2 nanoparticles, which results in an electrode layer of at least 10–15 µm thick with 10–25 nm interconnected pores. Conventional liquid processing techniques like spin coating or solvent casting lead to ineffective filling of the pores due to wettability and surface tension issues, resulting in premature pore blockage in this high aspect ratio (>1000) material. My work has addressed these challenges by utilizing a novel initiated chemical vapor deposition (iCVD) technology to directly synthesize and grow polymer electrolytes within the mesoporous TiO_2 photoanode[1]. In this way, there is tight integration of the polymer electrolyte with the photoanode with good contact at the electrode-electrolyte interface. In tandem with these experiments, I've formulated first-principles mathematical models composed of coupled partial differential equations to accurately describe the dynamic internal and interfacial processes during DSSC operation to understand how properties were affected by polymer chemistry and to determine how chemistry affects overall performance. The models accurately predict the J-V behavior of polymer electrolyte DSSCs and has been used for parametric sensitivity studies and parameter estimation to derive cell parameters that can be directly related to polymer properties. The work discovered that the polymer's Lewis acid/base character has a significant impact on the interaction of the polymer with the TiO_2 at the interface, which significantly influences charge conduction and electrode potential. These insights offered for the first time a way to combine modeling and experiments to design polymer electrolytes that would lead to more enhanced DSSCs with much higher power conversion efficiencies, greater than 50% improvement in conversion efficiency when compared to a liquid electrolyte DSSC and this work was published in the ***Journal of Power Sources***[2]. In fact, very recently, I've successfully utilized these insights to carefully select a new polymer electrolyte that enabled an increase in both photocurrent and photovoltage, a breakthrough that was previously not possible with other polymer electrolytes studied that showed only an increase in either the current or voltage but not both. This work has been formally published as an invited paper in a Special Issue on Recent Advances in Energy Conversion and Storage Devices in ***Chemical Engineering Science***[3]. Extending from this work, I've formed successful collaborations that take advantage of my strengths and expertise in DSSC materials, fabrication, and analysis. This has led to another contribution in a Special Issue on Solar Cells in ***AIMS Materials Science***[4] and a submitted paper to the Journal of Physical Chemistry C[5].

In the area of energy storage, there is currently much research interest in developing new, inexpensive, and lightweight energy storage devices that are environmentally friendly. Supercapacitors represent a viable storage solution that provides high power density. However, current supercapacitors suffer from lower energy density compared to other storage technologies like batteries because they only use physical, electrostatic interactions to store charge. A promising approach to enhancing the energy density in supercapacitors is by adding a redox polymer film on the porous carbon electrodes of these devices. These polymers, like polypyrrole, polythiophene and polyaniline, can participate in Faradaic redox reactions that provide further charge storage through more energy-dense chemical reactions, which can double charge storage of these devices. However, integrating polymer coating into these porous electrodes is non-trivial and challenging, but worthwhile because of the possible 2x performance improvement. Current methods for coating polymers onto carbon substrates such as chemical bath deposition, electrodeposition, and casting from suspension are unable to uniformly and conformally coat the microporous (< 2 nm) carbon electrodes throughout the entire electrode thickness, typically 100 µm, without blocking pore accessibility to ions. The high aspect ratios (>10,000) and tortuous pore structure along with liquid surface tension, solution viscosity, poor wettability, and solute steric hindrance make conformal coatings very challenging. Furthermore, poor solubility of conducting polymers in common solvents due to their rigid backbone structure limits processibility, and for electrochemical deposition, the need for a conductive substrate further limits broad utility. Overcoming these challenges is worthwhile because a highly porous structure with an ultrathin uniform conformal polymer coating would maintain the rapid movement of ions through the pores during charge/discharge reactions, preserving the high electrochemical high double layer capacitance (EDLC) while simultaneously dramatically improving the capacitance along with power and energy densities due to the polymer's redox reactions. By developing novel oxidative chemical vapor deposition (oCVD) processes, I was able to synthesize redox (conjugated) polymers and form highly conformal, ultrathin (< 10 nm) coatings around the topologically complex surfaces of porous nanostructured carbons. This initial effort has led to a publication in ***ACS Nano*** in which I was a co-author[6]. Very recently, I have successfully showed, for the first-time, that oCVD can produce polyaniline, which has the highest theoretical capacitance of all redox polymers, and significantly improve the energy capacity and stability of porous carbons derived from carbide materials through a collaborative effort with Prof. Yury Gogotsi and his group, recognized as one of the world's experts in carbon nanomaterials and energy storage. This work is submitted to ***Advanced Materials***[7]. The findings indicate that the performance of the polymer can be tuned based on the reaction conditions and optimal conditions enable oCVD to produce ultrathin (<30 nm) conducting polymer films that conformally coat pores as small as 1.7 nm with the resulting supercapacitors showing >100% increase in specific capacitance compared with bare supercapacitors and exhibiting excellent cyclability, lowering to only 90% of the initially stabilized value (~100 F/g) after 10,000 cycles—which is unprecedented for conducting polymers.

Looking forward, there is a wide range of research areas that could benefit from the integration of ultrathin polymer films using a holistic, synergetic modeling and experimental approach that I have developed and used throughout my research. These areas include functionalizing nanostructured surfaces, energy applications such as solar cells, batteries, fuel cells, and supercapacitors, along with research in separation and membranes. The combined mathematical modeling and iCVD & oCVD expertise is a unique attribute that will facilitate positive research contributions and developing new processing insight particularly with the design of polymeric electronic materials for nanoscale science and engineering.

Teaching Interests:

I received my B.S. in Chemical Engineering at the University of Virginia where I did research for Prof. Teresa B. Culver in sustainable stormwater management. I am currently co-advised by Prof. Soroush and Prof. Lau at Drexel University. Academically, I enjoy learning and have maintaining a perfect 4.0 GPA (grade point average, equivalent to an overall A grade) in all my Ph.D. graduate courses at Drexel, including all the required core graduate chemical engineering courses (Mathematical Methods, Thermodynamics, Transport Phenomena, and Kinetics). I have greatly enjoyed helping teach undergraduate chemical engineering course while I was a TA, some of my favorite course to teach would be Kinetics and Thermodynamics.

I believe that both teaching and research are greatly enhanced when they are integrated together as an educational program. Throughout my Ph.D., I have mentored one high school and nine undergraduate students on research projects related to my research and their educational curriculum. I have trained and guided these students to perform polymer thin film synthesis and processing using CVD approaches as well as materials/device characterization under the broad theme of energy. I have also given back to the Drexel and Philadelphia community by helping to initiate research projects with the Drexel Smart House, a physical building and living environment run by Drexel undergraduate students with the goal of exploring and understanding different sustainable technologies for the future and showcasing those technologies to the larger Philadelphia community. I have also been the President of our department's Graduate Student Association as well as Vice President of the College of Engineering Graduate Association and helped give back to the graduate college.

I am enthusiastic and passionate about research, which is shown tangibly by the number of publications and by the wide range of major professional conferences I have presented at, such as the American Institute of Chemical Engineers (AIChE), American Vacuum Society (AVS), American Chemical Society (ACS), and the Electrochemical Society (ECS)--including one invited talk at the ECS conference in Cancun, Mexico that acknowledged the importance of my DSSC work. I have presented 11 oral presentations and 5 poster sessions. Furthermore, I have formed collaborations with W.L. Gore, Prof. Gogotsi in Drexel's Material Science department, Prof. Ji in Drexel's Chemistry Department, Prof. Li in Penn's Chemical Engineering Department, and Prof. Kalra in the Chemical and Biological Engineering Department at Drexel.

As a testament to my work ethic, I have been the recipient of numerous honors, including the Department of Chemical and Biological Engineering's 2016 Outstanding Ph.D. Student Award, the College of Engineering's 2016 Outstanding Graduate Student Award, and most recently Drexel University's highest honor with the 2016 Research Excellence Award for Outstanding Research given to only four students out of the entire university's graduate population. I have also been the recipient of numerous merit fellowships, including the 2016 Leroy L. Resser Endowed Fellowship, the 2015 George Hill, Jr. Endowed Fellowship, and the 2015 Koerner Family Award. In addition, I have been a GAANN fellow for the Department of Education's National Academy of Engineering Grand Challenges from 2014–2016. I was also one of the six national finalists for the highly competitive W.L. Gore Fellow Award in 2015.

(1) Janakiraman, S., Farrell, S. L., Hsieh, C.-Y., Smolin, Y. Y., Soroush, M. & Lau, K. K. S. Kinetic analysis of the initiated chemical vapor deposition of poly(vinylpyrrolidone) and poly(4-vinylpyridine). ***Thin Solid Films* 595, Part B**, 244-250, (2015). http://www.sciencedirect.com/science/article/pii/S0040609015005088

(2) Smolin, Y. Y., Nejati, S., Bavarian, M., Lee, D., Lau, K. K. S. & Soroush, M. Effects of polymer chemistry on polymer-electrolyte dye sensitized solar cell performance: A theoretical and experimental investigation. ***Journal of Power Sources* 274**, 156-164, (2015). http://www.sciencedirect.com/science/article/pii/S0378775314016371

(3) Kuba, A. G., Smolin, Y. Y., Soroush, M. & Lau, K. K. S. Synthesis and integration of poly(1-vinylimidazole) polymer electrolyte in dye sensitized solar cells by initiated chemical vapor deposition. ***Chemical Engineering Science***. http://www.sciencedirect.com/science/article/pii/S000925091630238X

(4) Johnson, N. M., Smolin, Y. Y., Shindler, C., Hagaman, D., Soroush, M., Lau, K. K. S. & Ji, H.-F. Photochromic Dye-Sensitized Solar Cells. ***AIMS Materials Science* 2**, 503-509, (2015)

(5) Johnson, N. M., Smolin, Y. Y., Hagaman, D., Soroush, M., Lau, K. K. S. & Ji, H.-F. Suitability of N-Propanoic Acid Spiropyrans and Spirooxazines for Use as Sensitizing Dyes in Dye-Sensitized Solar Cells. ***Journal of Physical Chemistry C,* Submitted**, (2016)

(6) Nejati, S., Minford, T. E., Smolin, Y. Y. & Lau, K. K. S. Enhanced Charge Storage of Ultrathin Polythiophene Films within Porous Nanostructures. ***ACS Nano* 8**, 5413-5422, (2014). http://dx.doi.org/10.1021/nn500007c

(7) Smolin, Y. Y., Van Aken, K, Boota, M., Soroush, M., Gogotsi, Y.,Lau, K. K. S. Engineering Ultrathin Polyaniline in Carbide Derived Carbon Supercapacitors using Oxidiative Chemical Vapor Deposition. ***Advanced Materials,* Submitted**, (2016)

476050 Elucidation of Atomic-Scale Structure/Function Relationships: Toward Predictive and Rational Design of Nanoscale Materials

Abstract:

Research Interests: Nanoscale materials, atomic-scale structure/function relationships, synchrotron radiation characterization methods, catalysis, biomimetics

Teaching Interests: With a background in physics and chemistry, and well suited to teach a wide-range of class work. I would be interested in teaching materials related courses, thermodynamics, and/or transport courses.

Elucidation of Atomic-Scale Structure/Function Relationships: Toward Predictive and Rational Design of Nanoscale Materials

Nicholas M. Bedford, National Institute of Standards and Technology, Boulder, CO 80305

Postdoctoral Fellow

Research Interests: The emergence of nanotechnology has produced advances in various application areas of technological importance. Unfortunately, the creation of nanoscale materials is too often achieved serendipitously or via Edisonian approaches, impeding rapid technological development. As the arrangement of atoms within a nanoscale material directly influences the materials properties, it is imperative to have a complete atomic-scale structural understanding to elucidate, predict, and ultimately design/tune materials properties for a given material. As such, my research interests are centered on understanding synthesis-driven structure/function relationships using atomic-scale synchrotron radiation characterization methods. *In-situ* characterization methodologies are also of interest to understand atomic-scale structural changes during relevant chemical processes, *e.g.* charge/discharge cycles of a battery materials or during catalytic turnovers occurring at a catalysts surface.

Successful Proposals: NRC Research Associateship Program Postdoctoral Fellowship (AFRL); DAGSI Graduate Student Research Fellowship Program; 20+ successful synchrotron general user facility proposals

Additional Proposal Experience: Panel reviewer for the Catalysis and Biocatalysis program at CBET-NSF; reviewer for SSRL general user beam time requests; Submission of multiple peer-review proposals to DOE, AFOSR, and ARO.

Postdoctoral Projects: *"Development and Characterization of Peptide-Assembled Nanoparticles"*

Under supervision of Dr. Rajesh Naik, Air Force Research Laboratory

"Elucidation of Structure/Function Relationships for Engineering Nanomaterials"

Under supervision of Dr. Tim Quinn, National Institute of Standards and Technology

Ph.D. Dissertation: *"Electrospun Fibers for Energy, Electrical, and Environmental Applications"*

Under the supervision of Prof. Andrew J. Steckl and Donglu Shi at the University of Cincinnati Department of Material Science and Engineering

Research Experience: My research experience is highly interdisciplinary in nature with a common underlying theme of understanding nano- and atomic-scale phenomena. My formal Ph.D. training was in synthesis and characterization of polymeric and inorganic nanomaterials, while during my postdoctoral appointments I was immersed into the area of synchrotron radiation characterization, including high-energy X-ray diffraction (HE-XRD) coupled to atomic pair distribution function (PDF) analysis, X-ray absorption spectroscopy (XAS), and small angle X-ray scattering (SAXS). Through intensive training at several national laboratories and DOE user facilities, I have become proficient in all aspects of the aforementioned structural characterization methods, which can yield nanomaterial structural models with atomic-scale precision. Using such models, structure/function relationships can be readily examined, providing the needed understanding to create new nanomaterials with improved properties.

Teaching Experience: During my career, I have been involved in various avenues of teaching and mentorship. While at AFRL, I held a dual appointment in the Department of Chemistry at the University of Miami, where I was an instructor for the summer section of Principles of Chemistry I. As the instructor, I was responsible for all aspects of the class. While in graduate school, I was a TA for many engineering courses, including Basic Heat Transfer and Intro to Thermodynamics. During my undergraduate studies, I was laboratory assistant for both organic and general chemistry labs. Additionally, I have mentored several graduate, undergraduate, and high school students during my postdoctoral tenures at both NIST and AFRL.

Future Directions: As faculty I plan to continue applying my expertise in nanomaterials synthesis and synchrotron radiation characterization to establish synthesis-based nanomaterials design rules for a variety of materials systems. My initial focus will involve nanomaterials with emergent catalytic, electrocatalytic, photocatalytic, energy harvesting and/or energy storage properties. Using XAS, HE-XRD coupled to atomic PDF, and SAXS, structural information can be accessed over various length scales, providing the needed data to build full nanomaterial structural models with atom-scale resolution. These models will then be used to assess synthesis-driven structure/function relationships to be implemented in future nanomaterial design. In-situ methodologies will be explored as well, wherein structural transitions occurring during relevant chemical events can provide additional insights in assessing and predicting materials properties.

Synthetically, I plan to explore multiple avenues for the creation of nanomaterials synthesis, where the initial focus will be on biomimetic approaches. Using the specificity of biological molecules, nanomaterials can be synthesized and assembled with unpredicted precision. Additionally, the complexity and modularity of chemical functional groups within biomolecules provide a large basis set of possible biotic/abiotic interactions to manipulate materials properties. In tandem with the aforementioned synchrotron radiation characterization expertise, I fully expect the capability to use biomimetic approaches to achieve nanomaterials with enhanced properties through the development of structure/function relationships.

Selected Publications (*denotes corresponding author), current H-index=11 (Google Scholar):

N. M. Bedford,* Z. Hughes, Z. Tang, B. D. Briggs, Y. Ren, V. G. Petkov, R. R. Naik,* M. R. Knecht,* T. R. Walsh,* "Probing the Sequence-Dependent Structure/Function Relationships of Catalytic Peptide-Enabled Au Nanoparticles", *J. Amer. Chem. Soc.* **2016**, 138, 540.

N. A. Merrill, E. M. McKee, K. C. Merino, L. F. Drummy, S. Lee, B. Reinhart, Y. Ren, A. I. Frenkel, R. R. Naik, **N. M. Bedford**,* M. R. Knecht* "Identifying the Effects of Metal Composition on the Structure and Catalytic Activity of Peptide-Templated Materials", *ACS Nano*, **2015**, 9, 11968.

B. D. Briggs†, **N. M. Bedford†**, S. Seifert, H. Koerner, H. Ramezani-Dakhel, H. Heinz, A. I. Frenkel, R. R. Naik,* M. R. Knecht,* "C-C Coupling of Peptide-Capped Pd Nanoparticle Progresses Through an Atom Leeching Mechanism", *Chem. Sci.* **2015**, 6, 6413. **†Equal author contribution**

N. M. Bedford,* H. Ramezani-Dakhel, J. M. Slocik, B. D. Briggs, Y. Ren, A. I. Frenkel, V. G. Petkov, H. Heinz,* R. R. Naik,* M. R. Knecht,* "Elucidation of Biologically Programmed Atomic-Scale Structure of Nanoparticle Interfaces that Modulates Catalytic Activity", *ACS Nano* **2015**, 9, 5082. **Featured on the front coverN. M. Bedford**, R. Bhandari, J. M. Slocik, S. Seifert, R. R. Naik, M. R. Knecht,* "Peptide-Modified Dendrimers as Templates for the Production of Highly Reactive Catalytic Nanomaterials", *Chem. Mater.* **2014**, 26, 4082.

N. M. Bedford, M. Pelaez, C. Han, D. D. Dionysiou, A. J. Steckl,* "Photocatalytic cellulosic electrospun fibers for the degradation of potent cyanobacteria toxin microcystin-LR", *J. Mater. Chem.* **2012**, 22, 12666. **Published as HOT Article by RSC**

N. M. Bedford, M. B. Dickerson, L. F. Drummy, H. Koerner, K. M. Singh, M. C. Vasudev, M. F. Durstock, R. R. Naik,* A. J. Steckl,* "Nanofiber-Based Bulk-Heterojunction Organic Solar Cells Using Coaxial Electrospinning", *Adv. Ener. Mater.* **2012**, 2, 1136.

476080 Fast Modeling Protein Corona on Nanoparticle Based Biosensors in Complex Solvent Environments/ Cell Membrane By a Coarse Grained Simulation System

Abstract:

Research Interests:

Understand protein behavior in a complex environment is the key in rationalizing the design of many novel techniques such as protein biosensors based on novel nanoparticle (NP) materials. Protein folding and aggregation properties can be largely affected by the environments such as solid abiotic interfaces and complex co-solvents, which would directly affect the expected functionality or performance of the biosensors. Difficulties remain in applying current experimental methods to obtain a detailed picture of protein structures on a solid surface, so our ability is limited by these factors in the design of biosensor surface chemistry a priori for desired protein behavior. To address this point, in this work we present a coarse-grained simulation system that is able to model protein folding, misfolding, and aggregation on various NP surfaces and in the presence of co-solvents based on the Karanicolas and Brooks (KB) Go-like protein model. A flat-surface potential was first included and carefully parameterized based on a large experimental data set. This surface model has been shown in many studies to be able to reproduce and predict protein behavior at different types of surfaces as verified by experimental tests. Further effort in developing the surface potential was applied in describing surface curvature effects on protein behavior as a means of understanding nano-sized biosensor materials. Tanford's transfer free energy model was recently added to this simulation system as a descriptor of co-solvent effects or cell membrane environments. The energy of each part of the underlying folding/surface/solvation potential has been well balanced so that each element works collaboratively as a single complex environment for the protein. Together, this coarse-grained modeling system is expected to accurately predict the protein adsorption, folding/unfolding, aggregation, and protein-protein interactions in a complex environment as it would experience in differing cellular or solvent environments as well as in the context of a NP-based biosensor.

Teaching Interests:

I would like to practice my teaching under the guidance of the student-centered learning philosophy, which emphasizes personal reading, active peer discussing, and collaboration. The eventual goal of my teaching is to help them find their way to acquire knowledge and tools independently but not simply from the class. As a potential new faculty member, I am very interested in keeping optimizing my teaching techniques through my career as an educator.

Specifically, I would like to teach engineering classes that are currently given by the department on undergraduate or graduate levels, such as: Engineering Mathematics, Thermodynamics, Kinetics, and Transport Phenomena. Also, I want to bring classes such as molecular simulation and statistical mechanics for biological engineers, which are directly related to my research experiences and interests. A variety of simulation methods including quantum chemistry calculation, multi-levels molecular dynamics/ Monte Carlo simulation, and quantitative structure-activity relationship (QSAR) modeling will be taught in the classes.

475972 Designing Functional Self-Assembled Structures Via Complex Colloidal Interactions

Abstract:

Research Interests:

Colloidal materials and soft matter in general exist in the forefront, background, and future of our everyday lives. They compose many of the products we interact with and rely on to maintain (or repair) our health, facilitate the harvesting of resources (water, food, energy, etc.), and overall enhance our quality of life. The beauty of soft materials lies in the hierarchical structures that lead to their beneficial properties, which can be tuned by chemically modifying the colloidal building blocks of which they are composed or the medium in which they are suspended. *This chemistry-structure-property nexus has been and will continue to be the crux of my research interests.* In particular, the fundamental relationships between surface chemistry and colloidal interactions and subsequently between solution structure and colloidal dynamics are challenging and intriguing issues I have explored during my PhD. For example, I have identified protein surface charge heterogeneity as the origin of a hierarchical solution structure resulting in a unique gelation hysteresis behavior. Given the implications of the structure-dynamics-properties relationship for designing materials, I am excited to continue exploring similar physical phenomena. Future applications I am interested in include designing soft materials ranging from biocompatible materials for pharmaceutical formulation and delivery, to functional proppants for enhanced oil recovery, to self-assembled networks for separations or electronic applications. Specifically, I intend to exploit the versatility of colloidal chemistry and shape to study the effect of anisotropy and competing interactions on their microscopic behavior and macroscopic properties. While I have previously used proteins as colloids with anisotropic structure and interactions, recent advances in colloid synthesis now provides boundless opportunities to explore their effects on material properties using a range of chemistries that can be tailored for specific applications.

The methodology I have developed to explore my research interests encompasses modelling, simulation, and a range of experimental techniques. During both my PhD and my current post-doctoral position I have developed viscosity and mass transport models, respectively, and utilized Monte Carlo (MC) and Brownian dynamics (BD) simulations, respectively, in conjunction with rheology, small angle scattering, and dynamics measurements. I believe these three techniques are individually important, yet complimentary, and in my experience yield even more powerful insight if implemented in conjunction. In particular, combining simulation and experiment is necessary to understand the dynamics and material properties of anisotropic colloids, for which no models currently exist. I plan to employ this toolbox of research techniques to explore the self-association behavior of anisotropic colloidal building blocks and identify new material properties to engineer more effective soft matter materials relevant to the pharmaceutical and energy industries.

Teaching Interests:

I have enjoyed tutoring and mentoring others during each phase of my education and believe it has enhanced my own understanding of various subjects. My teaching experience is primarily from tutoring students as a TA, first for engineering math courses while at the University of Virginia and then for a pilot scale distillation experiment for the senior lab course in chemical engineering at the University of Delaware, for which I received very good reviews. As a post-doc, I have mentored graduate students in our lab and an undergraduate student on a side project I developed. Also while at MIT, I participated in the Kauffman Teaching Certificate Program, which teaches current best practices in designing courses and implementing active learning techniques in the classroom.

475989 Synthesis and Optimization of Nanomaterials for Sustainable Energy Generation and Catalysis

Abstract:

Research Interests:

The effectiveness of a supramolecular substrate, a solar cell, or a catalyst, heavily depends on the successful manipulation of its constituent components, at the nanoscale. Such manipulation begins at early-stage materials development, whether it is extraction of a protein, synthesis of a new photo-active material, fabrication of a composite energy-generating device, or development of novel catalysts. Understanding the key chemical/physical characteristics of these materials which render them their unique activity is the foundation leading to their successful isolation/fabrication and consequent optimization.

The highly robust, Mycobacterial channel protein â€œMspAâ€, has been isolated in high yield and purity, which served as a building block for synthesis of supramolecular aggregates of protein-organic dye and protein-metal/metal oxide nanoparticle clusters. The specific extraction and purification techniques employed for MspA, had a remarkable impact on its self-organization behavior in aqueous media, which in turn, had a significant impact on its aggregation behavior with external nanomaterials. The optimal aggregate size for various applications (such as drug delivery, hyperthermia and channel activity) of these clusters was studied extensively under various mediums and temperatures. The protein-organic dye clusters in particular, displayed remarkable thermo-stability and showed dynamic surface charge interactions with a buffer medium. These were used successfully in a dye-sensitized TiO_2 based photo-voltaic device, to conduct electricity under simulated sunlight, thus forming the first ever, hybrid, nano-protein solar cell prototype.

Iron/iron(III) oxide core-shell nanoparticles and micro-mesoporous titanosilicates are two types of highly valuable catalysts, with industrial appeal. The core-shell nanoparticles were functionalized with sulfonic acid and used as a highly selective catalyst to hydrolyze cellulose into glucose. A statistical technique was used to systematically optimize the reaction conditions of the above. The statistically-guided optimization led to high selectivity toward glucose, while production of 5-HMF and other monosaccharides were decreased. Reduction of 5-HMF is particularly significant as it is toxic to yeast, which are used industrially, to convert cellulose degradation products into ethanol. The catalyst was found to be highly stable and reusable, due to its magnetic property, thus, leading to a potential sustainable, technology for conversion of plant matter into ethanol. The same statistical approach mentioned above, was utilized to simultaneously probe synthetic parameters of a novel, hierarchically ordered, micro-mesoporous titanosilicate (MMTS), with the aim of achieving enhanced catalytic performance. MMTS was used to catalyze the epoxidation of cyclohexene with *tert*-butylhydroperoxide, and demonstrated unprecedented catalytic activity and high product selectivity. Thus, systematic nanoscale manipulation of reaction conditions and material synthesis, drastically improves the efficiency of catalysis.

The above is a summary of work from my PhD, followed by a short post-doctoral study, conducted at Kansas State University (advisor Prof Stefan Bossmann), and of my current post-doctoral research, conducted at University College London (advisor Prof Marc-Olivier Coppens). My future research plans include development of novel nanomaterials *via* sustainable routes for energy generation and catalysis.

Teaching Interests: I am experienced in conducting lectures and lab classes in areas of physical organic chemistry, synthetic organic chemistry and organometallic chemistry. I am interested in further expanding my scope of teaching into thermodynamics and nanomaterials engineering.

476194 Microfluidic Platform Technologies for Detection of Biochemical Markers

Abstract:

Research Interests: Microfluidics, Bio/chemical sensors, Micro/nanofabrication, Functional sensing materials, Reaction and transport phenomena.

Teaching Interests: Microfluidics, bio/chemical sensors, Interfacial science, Reaction and transport phenomena.

There is a growing interest in microfluidic based technologies for detection of various (bio-)chemical markers in wide variety of applications ranging from healthcare to environment. In microfluidic sensors the microchannel surfaces are functionalized with stack of intermediate molecular layers to the specific probe molecules. The efficiencies of these microfluidic sensors depends on the effective capture of analytes flowing in the carrier fluid by the surface immobilized receptors. The effective diffusion of analytes to the surface bound molecules primarily depends on two important category of parameters i.e. surface engineering and fluid flow characteristics. The surface engineering parameters such as number densities, orientations, surface free energies and lengths of molecular chains/stacks, and interfacial phenomena at fluid-wall interfaces along with the flow characteristics such as nature of liquids, frictional resistance, and flow profiles plays a significant role in defining the performance of the microfluidic sensor. Hence, the design and development of an effective microfluidic sensor requires a rational approach, based on the precise engineering of reactions and transport phenomena which are essential to ensure the optimum performance. Therefore, my research would focus on fundamental understanding of the interfacial phenomena, reaction and transport phenomena along with the biological aspects to engineer the microfluidic platform technologies for detection of various (bio-)chemical markers for applications ranging from healthcare diagnostics to environmental sensors. Being a chemical engineer and having worked in various industries along with semiconductor Fab, I have research interests in microfluidics, bio/chemical sensors, micro/nanofabrication, functional sensing materials, and reaction and transport phenomena. As a future faculty member my goal is to design and develop microfluidic platform technologies by using various smart and soft materials, nanomaterials, electronic materials along with flexible substrates from optical to electrochemical detection methods to meet the demanding needs in the domain of healthcare and environment. Previously, I worked on developing silicon based electrolyte insulator semiconductor based microfluidic immunosensors for early detection of cancer biomarkers and presently my research is focused on developing microfluidic based point of care platform technologies on cellulose materials for detection of various (bio-)chemical markers. These (bio-)chemical markers ranges from proteins, miRNA to DNA for various diseases such as Osteoporosis, Cervical Cancer, Tuberculosis, bacterial and viral infections, and Kidney rejection markers. In my poster, I would be presenting the work on silicon based microfluidic immunosensors, cellulose based point of care devices, and some key highlights from my past and current research projects along with the future projects in the microfluidic platform technologies.

476081 Understanding Structure-Property Relationships for Complex Fluid-Fluid Interfaces

Abstract:

Javen S. Weston

*Postdoctoral Researcher:*Georgetown University Department of Physics and National Institute of Standards and Technology

Research Interests: Material-material interfaces are everywhere around us and understanding how to control and modify them is at the center of many current research problems, from formulating laundry detergents to creating highly efficient photovoltaic solar cells. My research interests primarily revolve around studying how to create, stabilize, and modify these interfaces; with particular interests in surfactant adsorption, emulsion and foam stabilization, nanoparticle dispersions/composites, and nanoparticle assembly at fluid-fluid interfaces. Throughout my career I have both gained experience using a variety of characterization techniques and made valuable contacts at national labs and industrial R&D departments which, when combined, provide me with an excellent foundation to build a research career upon.

Successful Proposals: ADNOC-OSC 2-year Research and Development Grant

Postdoctoral Project: *"Developing Measurements of Structure-Property Relationships for Complex Fluids in Extreme Environments."*

Under supervision of Steve Hudson (NIST, Mat. Sci. and Eng. Division), Daniel Blair (Georgetown, Dept. of Physics), and Katie Weigandt (NIST, Center for Neutron Research)

PhD Dissertation: *"Metal Oxide Nanoparticles: wettability modification for emulsion stabilization and electrostatically induced gelation"*

Under supervision of Jeffrey H. Harwell, Chemical, Biological and Material Engineering, University of Oklahoma

Research Experience: My scientific research journey has introduced me to a wide variety of different topics within the realm of colloid and interface science. During my undergraduate career, I worked as a research assistant in OU's Applied Surfactant Lab (ASL), which balanced industrially-sponsored research projects with more academic grant-funded research. Over my time in the ASL, I worked on a variety of different projects including: formulating a high performance 'green' hard surface cleaner with SC Johnson, investigating the competitive adsorption of surfactants and polyelectrolytes on metal oxide nanoparticles, optimizing nanoparticle propagation in oil reservoir rock with the Advanced Energy Consortium, developing ultra-low interfacial tension surfactant blends for use in a variety of environmental remediation applications, improving performance of erasable inks with Papermate, and incorporating polyelectrolytes into detergent formulations with Clorox. My graduate research was completed as a member of the Center for Interfacial Reaction Engineering with funding from a DoE-EPSCoR Grant. I modified the surfaces of nanoparticles and used them to stabilize Pickering emulsions; investigating how the surface modification alters the emulsion properties. Additional work studying interactions between oppositely charged nanoparticles dispersed in water using microscopy, small-angle scattering, and rheology rounded out my graduate studies. This wide range of experience has left me with a broad understanding of the current state of academic and industrial research in the field of colloids and surfactants.

Teaching Interests : Throughout the course of my graduate and post-graduate work I have had several opportunities to gain hands-on teaching experience. During my first year of graduate school I was given opportunities to act as a guest-lecturer and TA in undergraduate-level laboratory courses. Then, during fall semesters throughout the remainder of my graduate studies, I was asked to act as the Instructor of Record for a section of a First-Year Engineering Experience course in OU's College of Engineering, for which I was awarded an Outstanding Graduate Assistant Teaching Award. As faculty I would be interested in teaching a wide variety of courses and would be excited to teach any course. Specific interests would include materials, fluid mechanics, thermodyanamics, and colloids/surfaces focused courses.

Proposed Future Research: As a faculty member, my research philosophy would rest on the belief that the best way to maintain a well-funded, relevant set of research goals is to strike a balance between basic scientific research that may or may not be of immediate commercial relevance and industry-funded research/consulting projects. Combining these two goals can be difficult, but I believe it is the best way to stay up to date with the cutting edge of academic research while simultaneously anticipating industry needs, which are often an excellent indicator of future research directions in engineering fields. I also believe in using a diverse group of researchers in terms of experience levels; using undergraduates, graduate students, and postdoctoral researchers to maximize the productivity of the research group, as a whole. In particular, I believe bringing in researchers from other STEM majors or creating opportunities for interdepartmental collaboration can help solve problems far more synergistically than can occur when all of the students and faculty involved are approaching the research problem from a similar background.

Scientifically, I plan to focus my research on studying how interfacial properties alter bulk performance in three different applications:

- Nanoparticle stabilized foams - How do interparticle forces, particle morphology, and surface modifications affect foam properties like stability, bubble size, and rheology
- Structure-property relationships in bicontinuous microemulsions - linker molecules are used to modify the interfacial rigidity of microemulsions and alter coalescence time, neutron scattering techniques and rheological measurements can be used to determine how changes in interfacial morphology and dynamics affect kinetics of emulsion coalescence
- Membranes via emulsion polymerization - Developing a technique to produce nanoporous membranes via polymerization of bicontinuous emulsions where the monomer is used as the organic phase could allow for greater control and tunability of pore size distributions and pore densities

Selected Publications:

- **JS Weston**, RE Jentoft, BP Grady, DE Resasco, and JH Harwell. "Silica Nanoparticle Wettability: Characterization and Effects on Emulsion Properties," *Industrial & Engineering Chemistry Research*, DOI: 10.1021/ie504311p, 2015.
- **JS Weston**, D Venkataramani, CP Aichele, BP Grady, JH Harwell, and DE Resasco. “Pseudosolid, Shear-Thinning Gel Formation in Binary Dispersions of Metal Oxide Nanoparticles at Low Volume Fractions,” *Langmuir*, 30 (49), 14982, 2014.
- **JS Weston**, JH Harwell, B-J Shiau, and M Kabir, “Disrupting admicelle formation and preventing surfactant adsorption on metal oxide surfaces using sacrificial polyelectrolytes,” *Langmuir*, 30 (22), 6384-6388, 2014.
- N Briggs, **JS Weston**, B Li, D Venkataramani, CP Aichele, JH Harwell, and S Crossley. “Multiwalled Carbon Nanotubes at the Interface of Pickering Emulsions,” *Langmuir,* 31, 13077, 2015.

476298 Protein Self-Assembly Toward Engineering of Biofunctional Nanomaterials

Abstract:

Research Interests: Making nanomaterials directly from proteins can enable spatial and quantitative controls of protein molecules at the nanoscale. Such controls are expected to extend the utility of proteins through manipulation of physical properties such as stability, molecular transport, and specific binding interactions. Using fundamentals of protein assembly, I plan to run research projects that focus on (1) design of self-assembling protein modules, (2) fundamental study of protein assembly processes, and (3) application of biofunctional protein nanomaterials for nanomedicine, biosensors, and nano-biocatalysis.

Naturally-derived or artificially designed protein motifs present characteristic structural and physical properties, which can be harnessed to design self-assembling modules. For examples, modular combination of protein motifs results in self-assembly of protein suprastructures with various morphologies which include spheres [1], vesicles [2], sheets, porous structures [3], and artificial nano-shapes that are unprecedented in nature. I previously demonstrated protein self-assembly approaches using simple protein motifs such as coiled coils, elastin-like polypeptides, and artificial random coils. I will present protein assembly strategies to retain both self-assembled morphology and biological functionality of protein nanomaterials. To incorporate biological functionality, I will develop functional protein modules. In combination with computational design of protein structures, experimental characterization to engineer self-assembly processes and biological functionality will be discussed. In addition, I will present design strategies that relate nanostructural characteristics and specific applications.

Teaching Interests: At the graduate level, I plan to design a class for protein nanomaterials, which is at the interface between protein engineering and nanotechnology. While I provide knowledge about fundamentals of protein engineering, biochemistry, and nanotechnology, I will focus on teaching students to develop their ability to solve engineering problems in the interdisciplinary area. Also, I am interested in teaching undergraduate chemical engineering classes, using examples of biological systems at the nanoscale. My research background on protein self-assembly covers topics in chemical engineering, such as thermodynamics and kinetics of biomolecules. I would like to introduce examples of engineering problems in bio- and nanotechnology.

[1] W. M. Park and J. A. Champion, *Angew. Chem. Int. Ed.*, **2013**, 52, 8098-8101.
[2] W. M. Park and J. A. Champion, *J. Am. Chem. Soc.*, **2014**, 136, 17906-17909.
[3] W. M. Park, C. M. Yee, J. A. Champion, *J. Mater. Chem. B*, **2016**, 4, 1633-1639.

476317 Graphene and Other Nanosheets: Exfoliation and Processing for Nanocomposites and 3D Macrostructures

Abstract:

Research Background:

The emergence of graphene as a single layer nanosheet with unique properties has paved the way to exploit the optical, electrical, thermal and mechanical properties of other nanosheets such as MoS_2, WS_2, and boron nitride. Many of the existent layered materials have been exfoliated to produce unprecedented nanosheets including Mxenes and black phosphorous. The dependence of the nanosheets properties and potential applications on their structure and morphology, particularly on the number of layers in an individual nanosheet, necessitates a precise control over their production techniques. Current nanosheets exfoliation techniques including the liquid-phase exfoliation lack such a control at atomic level and are challenged by the instability and reaggregation of nanosheets in liquid media. On the other hand, the yield of these techniques is not sufficient for bulk usage of nanosheets in applications such as composites or battery electrodes. Hence, strategies for scalable, high-yield production of nanosheets that offer atomic-level control over their structure and properties are in high demand.

To this point, in Prof. Micah Green's laboratory, I have tried to address the low yield of liquid phase exfoliation by using dispersant molecules for stabilization of colloidal graphene at higher concentrations. I have also tailored the polymer molecules to synthesize polymeric dispersant compatible with pristine graphene, a product with highest quality in graphene family. This allows for the usage of minimal graphene content for mechanical and electrical properties enhancement in polymer nanocomposites. The change of morphology of graphene from 2D nanosheets to 3D crumpled particles is another strategy that I have explored to prevent aggregation during the processing of graphene for other applications. Recently, I have focused on understanding the colloidal interactions in the graphene oxide dispersions in order to control their assembly into 3D macrostructures; ideally, these porous conductive 3D networks could be the optimum electrodes for lithium ion batteries.

Research Interests:

Being aware of the similarities and differences of graphene and other nanosheets, as a faculty member, I will expand my research to new materials and technologies. The research carried in my future laboratory will focus on (i) developing alternative synthesis and/or exfoliation methods for existing and novel 2D nanomaterials, (ii) fundamental study of the processing-structure-property relationships in each nanosheet family, (iii) relate the nanoscale properties of nanosheets to their bulk performance in macrostructures, and (iv) scalable manufacturing strategies to enable usage of these materials in industrial end products in response to real-life needs.

Teaching Interests:

In my laboratory and future classes, I would apply the methodology that I have developed during my research career and my teaching experience as a teaching fellow and teaching assistant (Numerical Analysis for Chemical Engineers & Reaction Design courses) to create a vivid learning environment for my future students. While they will obtain a fundamental understanding of the physics behind their research topic, they will get the opportunity to use their engineering background to design and build products with industrial applications. Obviously, collaboration with both industrial organizations and academic scientists with different computational and experimental skills is essential to accomplish these goals.

478651 Engineered Nanostructured Materials for Efficient Separation and Storage

Abstract:

Postdoctoral Projects:

Georgia Institute of Technology, Chemical & Biomolecular Engineering, 08/2015 – Now

Advisors: Prof. Krista S. Walton, Prof. David Sholl

(1) Synthesis and Characterization of Water-Stable Metal-Organic Frameworks (MOFs)

(2) Structural Stability of MOFs in Hot Liquid Water

(3) Multi-component Gas Adsorption on MOF materials.

(4) MOFs for Adsorption-Based Hydrocarbon Separations and High-Performance Oil/Water Separation

University of South Carolina, Chemical Engineering, 04/2013 –04/2015

Advisor: Prof. Miao Yu

(1) Thin, High-Quality Membranes for Separation. Membrane Materials: Zeolites (NaY, SAPO-34/11, AlPOs, and NaA-Polyamide composite) and Graphene Oxide. Relevant applications: H_2 Separation, N_2/CH_4, CO /CH , Propane/Propylene Separation, Oil/Water, and Desalination.

(2) Nano-valved Adsorbents prepared by Molecular Layer Deposition (MLD) for Efficient Separation and storage (e.g., CH_4 and H_2 Storage, Shape & Size Selective Separation of Organic Molecules (e.g., Ethanol/1-Butanol))

(3) Study the Growth Mechanisms of High-Aspect-Ratio Zeolites (e.g., SAPOs and AlPOs)

University of Oklahoma, Chemical Engineering, 08/2011 – 04/2013

Advisor: Prof. Daniel E. Resasco

(1) Study of The Hydrothermal Stability of FAU Zeolite – A Primary Component in FCC Catalysts

(2) Silylated Hydrophobic Zeolites with Enhanced Tolerance to Hot Liquid Water and Improved Thermal Stability under Reaction Conditions in Water/oil Emulsions

PhD Dissertation: "Zeolite Nanocrystals and Hierarchical Porous Structures: Synthesis, Characterization and Formation Mechanism"

Advisors: *Prof. Huanting Wang* (Monash University, Australia) and *Dr. Anita J. Hill* (Chief of Process Science and Engineering, The Commonwealth Scientific and Industrial Research Organization)

Background and Research Interests:

All across the world, people are facing a wealth of new and challenging problems, particularly the energy and environmental issues. For example, billions of tons of annual CO_2 emissions are the direct result of fossil fuel combustion to generate electricity. To minimize the impact of CO_2 on the environment, design of high-performance separation materials and technologies for efficient carbon capture and sequestration (CCS) is urgent and essential. *My research in this area is creating novel nanostructured (membrane) materials with enhanced transport properties by ordering their nano-architectures via different methods and meanwhile exploring their novel and energy-sustainable scale up.*

On the other hand, enhanced demand for fuels worldwide not only decreased world oil reserves but also increased climate concerns about the use of fossil-based fuel. To address these energy and environmental problems, efforts have been made towards improved utilization of fossil fuel and development of renewable energy production. With the abundant availability and carbon-neutral nature, biomass is recognized as one of the most promising renewable energy resource. A number of transportation fuels can be produced from biomass, helping to alleviate demand for petroleum products and improve the greenhouse gas emissions profile of the transportation sector. Traditional catalysts suffer from many undesirable properties, such as small accessible pore size, low hydrothermal stability, and less controllable active sites. Among these, low hydrothermal stability at upgrading temperatures greatly hinders conversion of lignocellulosic biomass to biofuel. *My proposed research will focus on synthesizing a new class of ultra-stable catalysts with tunable nanostructure and functionalities for efficient bio oil upgrading, with special emphasis on study of their hydrothermal stability.*

Oil pollution is an another serious global issue because of the large amounts of oily wastewater produced by petrochemical and other industries, as well as by frequent off-shore oil-spill accidents. Therefore, it is in great need to develop effective techniques to treat oil-polluted wastewater at such low oil/grease concentrations in order to satisfy the stringent governmental limitations and preserve the environment. Membrane techniques have been widely employed for water purification and are very effective in separating stabilized oil emulsions-especially for removing oil droplets. However, current membranes suffer from membrane fouling both on surface and in internal structures, which significantly limits their service time and degrades separation performance in practical operations. *My proposed research in this field attempts to adopt the concept of biomimetic hierarchical roughness in membrane design for creating superoleophobic membrane surfaces from a vast pool of candidate materials, such as zeolites, metal-organic frameworks (MOFs), and single-layered graphene oxide.* My research also focuses on development of facile, low-cost preparation technique which would open a completely new direction for the membrane society. If the proposed membrane could be successfully fabricated, further investigation on scaling-up will be pursued.

The practice of drug delivery has changed dramatically in the last few decades and even greater changes are anticipated in the near future. It is because controlled delivery systems (CDSs) are one of the promising applications for human healthcare. In pharmaceutical market, the CDS is growing fast with approximate 10% annual increase. However, an important challenge in this area is the efficient delivery of drugs in the body using non-toxic nanocarriers. Most of the existing carrier materials show poor drug loading (usually less than 5 wt% of the transported drug versus the carrier material) and/or rapid release of the proportion of the drug that is simply adsorbed (or anchored) at the external surface of the nanocarrier. Zeolite having a pore size from 0.5 nm to 2 nm and more than 100 different pore structures is a promising candidate. *I aim to produce uniform zeolite nanocrystals and nanostructures for the adsorption of drug molecules with sizes below and above 2 nm, respectively. I will employ an experimental approach to fundamentally study the interaction of drug molecules with microporous zeolite matrices, and realize a controllable releasing of drug molecules to target sites.*

Teaching Interests: I am very comfortable in teaching core Chemical Engineering courses, such as Thermodynamics, Heat Transfer, Transport Phenomena, Reaction Engineering, and Mass Transfer. My teaching interests are also expanded many specialized Chemical Engineering courses, such as Phase Equilibrium and Stage-based Separations, Surfaces and Adsorption, Adsorption and Nanomaterials Characterization, and Process Simulation. I also plan to develop courses in the areas of solid & polymer surface chemistry, hierarchical porous materials design, and membrane separation technologies.

Selected Publications:

[1] K.H. Chu, **Y. Huang**, M. Yu, N. Her, J. Flora, C.M. Park, S. Kim, J. Cho, Evaluation of Humic Acid and Tannic Acid Fouling in Graphene Oxide-coated Ultrafiltration Membranes. *ACS Applied Materials & Interfaces*, 8, 22270–22279 (2016).

[2] **Y. Huang**, L. Wang, Z.N. Song, S.G. Li, M. Yu, Growth of High-Quality, Thickness-Reduced Zeolite Membranes towards N_2/CH_4 Separation Using High-Aspect-Ratio Seeds. *Angewandte Chemie International Edition*, 127, 10993–10997 (2015). *Featured as Back Cover Article.*

[3] J.J. Song, **Y. Huang**, S.W. Nam, M. Yu, J. Heo, N. Her, J.R.V. Flora, Y. Yoon, Ultrathin Graphene Oxide Membranes for the Removal of Humic Acid. *Separation and Purification Technology*, 144, 162–167 (2015).

[4] **Y. Huang**, H. Li, L. Wang, Y.L. Qiao, C.B. Tang, C. Jung, Y.M. Yoon, S.G. Li, M. Yu, Ultrafiltration Membranes with Structure-Optimized Graphene Oxide Coatings for Anti-Fouling Oil/Water Separation. *Advanced Materials Interfaces*, 2, 1400443–1400450 (2015). *Highlighted by the Journal as Top 1 (2014~2015) and Top 4 (2016) Most-Read Full Research Papers.*

[5] Z.N. Song, **Y. Huang**, W.W. Xu, L. Wang, Y. Bao, S.G. Li, M. Yu, Continuously Adjustable, Molecular-Sieving "Gate" on 5A Zeolite for Distinguishing Small Organic Molecules by Size. *Scientific Reports*, 5, Article number: 13981 (2015).

[6] Z.N. Song, **Y. Huang**, W.W. Xu, M. Yu, Composite 5A Zeolite with Ultrathin Porous TiO_2 Coating for Selective Gas Adsorption. *Chemical Communications*, 51, 373–375 (2014).

[7] H. Li, **Y. Huang**, Y.T. Mao, W.W. Xu, H. Ploehn, M. Yu, Tuning Underwater Oleophobicity of Graphene Oxide Coatings via UV Irradiation. *Chemical Communications*, 50, 9849–9851 (2014)

[8] H. Li, Z.N. Song, X.J. Zhang, **Y. Huang**, S.G. Li, Y.T. Mao, H.J. Ploehn, Y. Bao, M. Yu, Ultrathin, Molecular-Sieving Graphene Oxide Membranes for Selective Hydrogen Separation. *Science*, 342, 95–98 (2013).

[9] P.A. Zapata, **Y. Huang**, M.A. Gonzalez-Borja, D.E. Resasco, Sylilated Hydrophobic Zeolites with Enhanced Tolerance to Hot Liquid Water. *Journal of Catalysis*, 308, 82–97 (2013). *Invited Paper for the*

[10] **Y. Huang**, J.F. Yao, X.Y. Zhang, C.H. (Charlie) Kong, H.Y. Chen, D. X. Liu, M. Tsapatsis, M.R. Hill, A.J. Hill, H.T. Wang, Role of Ethanol in Sodalite Crystallization in an Ethanol-Na_2O-Al_2O_3-SiO_2-H_2O System. *CrystEngComm*, 13, 4714–4722 (2011).

[11] **Y. Huang**, D.H. Dong, J.F. Yao, L. He, J. Ho, C.H. Kong, A. J. Hill, H.T. Wang, In Situ Crystallization of Macroporous Monoliths with Hollow NaP Zeolite Structure. *Chemistry of Materials*, 22, 5271–5278 (2010).

[12] **Y. Huang**, K. Wang, D.H. Dong, D. Li, M.R. Hill, A.J. Hill, H.T. Wang, Synthesis of Hierarchical Porous Zeolite NaY Particles with Controllable Particle Sizes. *Microporous and Mesoporous Materials*, 127, 167–175 (2010).

[13] J.F. Yao, **Y. Huang**, H.T. Wang, Controlling Zeolite Structures and Morphologies Using Polymer Networks (Review paper). *Journal of Materials Chemistry*, 20, 9827–9831 (2010).

[14] **Y. Huang**, J. Ho, Z. Wang, P. Nakashima, A.J. Hill, H.T. Wang, Mesoporous Carbon Confined Conversion of Silica Nanoparticles into Zeolite Nanocrystals. *Microporous and Mesoporous Materials*, 117, 490–496 (2009)

[15] D.H. Dong, **Y. Huang**, X.Y. Zhang, L. He, C.Z. Li, H.T. Wang, Shape Forming of Ceramics with Controllable Microstructure by Drying-Free Colloidal Casting. *Journal of Materials Chemistry*, 19, 7070–7074 (2009).

476265 Programmable Peptide-DNA Hybrid Nanomaterials

Abstract:

Research Interests: Natural biomaterials systems evolved over billions of years to solve some of the most challenging engineering problems. These optimized properties result from hierarchical designs exhibiting structure and function on multiple scales. The structure and function of natural materials have inspired scientists in attempts to mimic their designs and construction principles with the goal of mimicking the native qualities using synthetic materials. Moreover, by employing novel chemistries, processing and fabrication tools now available, we might be able to improve on Nature's designs and functions. Although tremendous advance was achieved in mimicking the structural and functional properties of native biomaterials, some key features are still very challenging to reproduce. These include (i) development of matter that is programmable on both temporal and spatial scales (ii) Controlling the Hierarchical Structure (iii) Integrating Sense and Response to the structural design, and more. Specifically, creation of regenerative material and tissue engineering systems with this intricate functionality necessitates new composites that are highly dynamic and can coordinate multiple interacting components. Merging different types of molecules such as peptides and nucleic acids into hybrid materials will generate a new class of materials that are amenable to structural and functional programming. **Postdoctoral Project:** "*Instructing Cells With Programmable Peptide-DNA Hybrid Nanomaterials*."Under supervision of Samuel I. Stupp, Simpson Querrey Institute for BioNanotechnology, Feinberg School of Medicine, Northwestern University, Chicago, IL 60611, United States **PhD Dissertation:** "*Functionalized Semiconductor Quantum Dots for Optical Sensing*" Under supervision of Itamar Willner, The Institute of Chemistry, The Hebrew University of Jerusalem, Jerusalem, Israel. **Research Experience:** I have had highly multidiciplinary training. I obtained my undergraduate degree in chemistry and computer science at Bar-Ilan University in Israel, followed by a Master's degree in chemistry at The Hebrew University of Jerusalem. I then began doctoral studies in chemistry and nanobiotechnology at The Hebrew University of Jerusalem, Israel, in the lab of Prof. Itamar Willner.

My PhD focused on the synthesis of functionalized semiconductor and metallic nanoparticles as well as DNA nanostructures for the multiplexed analysis of biomarkers and probing intracellular processes. During my postdoctoral training at Northwestern University I studied self-assembling nanomaterials for regenerative medicine in the lab of Prof. Samuel Stupp and introduced a new area into the lab, focusing on DNA-peptide hybrids for constructing cell scaffolds with programmable control over structure and function. I also collaborated extensively with medical doctors and biologists to create nanomaterials of biomedical interest in areas like neural stem cell differentiation, regeneration of neurons, and siRNA delivery. As a result of my versatile training and extensive collaborations I have gained experience in many fields such as synthetic chemistry, material science, biology, medicine, optical imaging, and biomedical engineering. **Teaching Interests:** Aside from my research career, I also have extensive teaching experience. At the Hebrew University of Jerusalem I was a teaching assistant of a variety of courses including Organic Chemistry, Biological Chemistry of Materials, Computational Quantum Chemistry and several introductory labs in chemistry and nanotechonolgy. At Northwestern, I served as a co-instructor with my postdoctoral advisor teaching a course on Materials and Nanochemistry intended for undergraduate (senior year) and graduate students. Lastly, I also have had the privilege of mentoring several graduate and undergraduate students. **Fellowships:** 1. European Molecular Biology Organization Long-term Postdoctoral Fellowship. (2013-2015) 2. The Shlomiuk award for outstanding P.hD. research study. (2013) 3. The Charles Clore fellowship for outstanding graduate students in natural sciences. (2010-2012) 4. Converging Technologies fellowship for excellent graduate students. (2008-2009) **Future Direction:** As faculty I would like to continue developing hybrid materials based on nucleic acids and peptides for different applications in nanomedicine. In particular I would like to integrate my PhD experience in developing nanoparticles-based sensing devices, DNA switches and machines with my postdoctoral experience in peptide self-assembly to create biomaterials, or systems of materials, that simultaneously provide structure, sensing, and response - all integrated into function. The oligonucleotide-peptide hybrids will merge the programmability and nanoscale control afforded by DNA nanotechnology with the biological function of peptides and proteins. Applications that will drive and inspire the research include systems that can regulate cell behavior (stem cell; immune cells), injectable materials that "know" to travel to an injury site and trigger tissue regeneration by harnessing the body's own power to heal, and new generations of wearable materials (electronic and others) as “Sense and Treat” systems that could be used as scaffolds to follow certain biomarkers and promote wound healing. **Selected Publications: H-Index: 33; Citations – 4016** 1. **Freeman R.**, Stephanopoulos N., Sur S., Boekhoven J., Lee S.S., Stupp S.I., Instructing cells with programmable DNA peptide hybrids, *Nature*, 2016 (in revisions). 2. **Freeman R.**, Boekhoven J., Dickerson M.B., Naik R.R., Stupp S.I., Biopolymers and supramolecular polymers as biomaterials for biomedical applications, *MRS Bulletin*, 2015, 40, 1089-1101. 3. N. Stephanopoulos, **R. Freeman**, H.N. Scheler, S. Sur, S. Jeong, F. Tantakitti, J.A. Kessler, S.I. Stupp, Bioactive DNA-Peptide Nanotubes Enhance the Differentiation of Neural Stem Cells Into Neurons, *Nano Letters,* 2015, 15, 603-609. 4. **Freeman R.**, Girsh J., Fang-ju Jou A., Annie Ho J., Hug T., Dernedde J, Willner I. Optical Aptasensors for the Analysis of the Vascular Endothelial Growth Factor (VEGF), *Anal. Chem.*, 2012, 84, 6192-6198. 5. **Freeman R.**, Liu X.Q., Willner I. Chemiluminescent and Chemiluminescence Resonance Energy Transfer (CRET) Detection of DNA, Metal Ions and Aptamer-Substrate Complexes Using Hemin / G-Quadruplexes and CdSe/ZnS Quantum Dots, *J. Am. Chem. Soc.*, 2011, 133, 11597-11604. **(cited 287 times)** 6. **Freeman R.**, Finder T., Gill R. Willner I. Probing Protein Kinase (CK2) and Alkaline Phosphatase with CdSe/ZnS Quantum Dots. *Nano Lett.*,2010, 10, 2192-2196. 7. **Freeman R.**, Finder T., Willner I., Multiplexed analysis of Hg2+ and Ag+ ions by nucleic acids-functionalized CdSe/ZnS quantum dots and their use for logic gate operations, *Angew. Chem. Int. Ed.*, 2009, 121, 7958 - 7961. **(cited 344 times)**

476267 Point-of-Care Molecular Detection with Surface Engineering of Nanomaterials for Diagnostic Platforms

Abstract:

Research Interests:

A fundamental axiom of medicine is that faster diagnosis yields earlier intervention, reducing the spread of infection and improving patient outcomes. The development of effective, point-of-care (POC) methods for the detection of disease biomarkers would thus increase the frequency with which many diseases are detected and identified, and decrease the delay between diagnosis and treatment. In order to be efficient, POC diagnostic devices must put clinically actionable information in a doctor's hands during a patient's first visit rather than during a follow-up. Here we proposed a new signaling mechanism for detection of a variety of molecules that displays clinically relevant sensitivity and specificity for variety of targets, and ideally is: 1-rapid relative to the 15 minutes of a typical patient/provider interaction, 2-selective enough to work in unprocessed bodily fluids, 3-amenable to small volumes rather than venous blood draws, 4-quantitative, and 5-able to diagnose multiple analytes simultaneously. This electrochemical DNA-based sensing platform is designed to detect real antibodies in human blood based on steric hindrance effects of the big molecule to influence the DNA-DNA hybridization. We hypothesized that when a large macromolecule binds to a DNA strand, this should, in principle, reduce the number of DNA strands that can hybridize to their complementary strand attached to a surface. We set out to demonstrate this steric hindrance mechanism using electrochemistry by employing a redox-labeled signaling DNA strand and a gold electrode that contains the complementary capturing DNA sequence at high surface density. Upon binding to the capturing strands, the redox-labeled signaling strands generate a large electrochemical signal by bringing the redox labels close to the electrode surface. In the presence of a large target protein that binds a recognition element on the signaling strand, we hypothesize that fewer copies of this strand will be able to reach the surface-bound capturing strand due to steric hindrance, therefore generating lower electrochemical currents. We eventually demonstrate the multiplexing detection of multiple antibodies within less than 10 minutes in whole blood having this detection platform. Additionally, the platform can be broadly adapted to selectively and specifically detect many other proteins having well-designed recognition ligand to that protein. Thus, this detection platform holds great promise to be well positioned for adaptation as, for example, point-of-care diagnostics.

As a faculty candidate, I envision three potential multidisciplinary projects, which are in line with the interdisciplinary and collaborative research environment: (1) Point-Of-Care diagnosis: Single-step methods for rapid and quantitative diagnosis of disease biomarkers. (2) Electrochemical 2D and 3D -sensors for direct and/or indirect molecular detection. Improvement of the catalytic activity of nanostructured Pt (Pd, or Au) by electrochemical alloying, (3) Development of Giant MagnetoResistnace in nanoscale magnetic Multilayer for biosensing application.

Teaching Interests:

I am eager to continue to the undergraduate and graduate education. I would be exited to teach courses across the curriculum of Materials Engineering, Chemical Engineering, Nanotechnology Engineering and Biomedical Engineering, as well as advanced course subjects in the field of nanobiotechnology, surface engineering, electrochemistry and analytical chemistry.

461236 Bioinspired Separation Materials for Environmental and Energy Applications

Abstract:

Research Interests:

My research motivation in membrane science and technology is based on an acute awareness of current environmental issues and the potential of separations to address some of these issues. I have a strong interest in membrane technologies because they have been widely used to supply clean water and there is a growing interest in applications to clean energy and pollution prevention. My Ph.D. program has opened the door of a completely new membrane research field to meâ€"biomimetic membranes and I have helped in bringing the new area of artificial water channels for separations into focus through my PhD work. ***Biomimetic membranes, utilizing biological elements or borrowing concepts from biological systems, can take advantage of processes evolved by nature over billions of years for improving transport efficiency and specificity.*** My Ph.D. study started with aquaporins (AQPs) based biomimetic desalination membranes. Aquaporins are biological water channel proteins that facilitate extremely fast water transport while rejecting all the other ions and solutes. The later part of my Ph.D. work conducted some of the initial work in the exciting area of bioinspired artificial channels. My work showed channels that mimic natural channel proteins self-assemble in lipid and block copolymer bilayers and can approach the performance of biological water and proton channels.

In my future academic career, I am devoted to developing bioinspired energy efficient functional separation materials that can be used in the field of separations important for environmental and energy applications. I want to use artificial channels to fabricate energy-efficient separation materials for liquid and gas separations, and develop environmental sensors for environmental applications. I am also interested in utilizing the bioinspired concepts to mitigate membrane fouling issues in water and wastewater treatment, with a combination of biomimetic topography, controlled surface chemistry and antifouling signals.

Teaching Interests:

During my Ph.D. program, I have volunteered to instruct different classes for 2 years to learn how to teach and develop my own teaching style. Because I have both chemical and environmental engineering backgrounds, I am qualified to teach undergraduate or graduate core courses (such as transport phenomenon, fluid mechanics and reaction engineering), specialized courses (membrane separations, water and wastewater treatment) and experimental courses (environmental or chemical engineering unit operations). ***My teaching will be focused on how to utilize the basic skills acquired in classroom to solve practical problems as this leads to the greatest retention of core concepts. All of the courses I will instruct will be designed to train the students in extending their learning abilities (how to find problems and where to get necessary resources to solve them), communication and collaboration, writing and presenting skills.***

Representative publications:

1. **Shen, Y.**, et al. Highly permeable artificial water channels in block copolymer membranes. (in preparation)
2. Licsandru, E., Kocsis, I., **Shen, Y.**, at al. Salt excluding artificial water channels reveal enhanced dipolar water and proton translocation. *J. Am. Chem. Soc.* 138, 5403-5409, (2016). (**co-first author**)
3. **Shen, Y.**, et al. Highly permeable artificial water channels that can self-assemble into two-dimensional arrays. *Proc. Natl. Acad. Sci. U.S.A.* 112, 9810-9815, (2015).
4. **Shen, Y.**, et al. Biomimetic membranes: A review. *J. Membr. Sci.* 454, 359-381 (2014).
5. Sun, Y., **Shen, Y.**, et al. Linkages between microbial functional potential and wastewater constituents in large-scale membrane bioreactors for municipal wastewater treatment. *Water Res.* 56, 162-171 (2014). (**co-first author**)
6. **Shen, Y.**, et al. Improvement on Modified Lowry Method against Interference by divalent cations for soluble protein measurement in wastewater systems. *Appl. Microbiol. Biotechnol.* 97, 4167-4178 (2013).
7. **Shen, Y.**, et al. Characterization of soluble microbial products in 10 large-scale membrane bioreactors for municipal wastewater treatment in China. *J. Membr. Sci.* 415â€“416, 336-345 (2012).
8. **Shen, Y.**, et al. A systematic insight into fouling propensity of soluble microbial products in membrane bioreactors based on hydrophobic interaction and size exclusion. *J. Membr. Sci.* 346, 187-193 (2010).

Contact Information

Yuexiao Shen, 162 Fenske Laboratory

The Pennsylvania State University

University Park, PA, 16802

Phone: (814)777-8181 Email: yxs174@psu.edu

475849 Demulsification of the Phosphoric Acid-Tributyl Phosphate (W/O) Emulsion By Hydrocyclone

Abstract:

Research Interests: separation by hydrocyclone,demulsification of water-in-oil emulsion, drop coalescence behavior, liquid-liquid dispersion

Teaching Interests: mechanical separation

Emulsions are often generated as unwanted byproducts during the mixing of organic and aqueous phases, but they are required in the emulsification extraction to greatly improve the extraction rate by increasing the contact interfacial area between the two phases. A rapid, convenient, efficient and economic demulsification method is subsequently needed to separate the two phases whether the emulsions are required or not. Liquidâ€"liquid hydrocyclone is efficient for the two-phase separation of emulsion systems with the aid of strong centrifugal force created by the swirling flow. In this study, a demulsification process of phosphoric acidâ€"tributyl phosphate (W/O) emulsion was investigated by hydrocyclone. The drop size distributions and average drop sizes of the emulsions pre and post demulsification were measured. The average drop size differences were used to evaluate the demulsification performance. The force condition and coalescence process of drops in the hydrocyclone were analyzed. The effects of density difference between the two phases, inlet flow rate and structural parameters (large cone part angle, small cone part angle, cylindrical part height, inlet and overflow pipe inner diameter) of the hydrocyclone on the demulsification performance were investigated. Larger density difference between the two phases led to better demulsification performance. The demulsification performance firstly became better and then got worse with the increase of inlet flow rate. The decrease in large and small cone part angle, as well as the increase in cylindrical part height, inlet and overflow pipe inner diameter of the hydrocyclone had positive effects on the demulsification performance. The influence degree of structural parameters for the demulsification performance decreased in the order: the cylindrical part height > the large cone part angle > the inlet inner diameter > the overflow pipe inner diameter > the small cone part angle. The optimum density difference, inlet flow rate and structural parameters were obtained. Under the optimum conditions, the average drop size of emulsion increased from 4.78 to 55.07 Î¼m, and the difference was 50.29 Î¼m. The average drop size differences were associated with these above-mentioned parameters through a dimensionless model. The comparison between the experimental and calculated data showed that this model was reliable to predict the average drop size difference in most demulsification conditions by hydrocyclone.

475710 Membranes for Liquid and Gas Separations

Abstract:

Membranes for liquid and gas separations

Oishi Sanyal

1st year Postdoctoral Researcher

Georgia Tech

Research Interests:

My broad research interest lies in the area of membrane design with an equal focus on both liquid and gas separations. For both liquid and gas feeds, membrane material selection plays a key role in engineering more energy efficient separations based on environmentally benign and energy efficient processes. Conventional thermally driven techniques like distillation have dominated large scale industrial separations; however, membrane-based processes are emerging to replace or debottleneck conventional techniques.

In the water industry, reverse osmosis and nanofiltration are already dominant for desalination and wastewater purification. The major challenge faced by water treatment membranes is their fouling propensity, and I am specifically interested in working on membrane modification techniques to lower fouling tendencies of commercial membranes.In the area of gas separation, I am primarily focused on creating membranes using advanced materials to surpass the current performance "upper bound" for polymeric membranes. Such membranes pose many opportunities for applications such as olefin-paraffin separation, natural gas and flue gas separations.

Teaching Interests:

Having done both my undergraduate and graduate studies in chemical engineering, I am familiar with the usual chemical engineering courses. Specifically, I would be interested in teaching courses like Mass Transfer, Transport Phenomena and Reaction Engineering. I would also like to offer a graduate level course on membrane separations as I personally benefited a lot from a similar course taught during my graduate school. I believe I will also be comfortable in teaching some environmental engineering courses or some electives related to environmental engineering.

Postdoctoral project: Understanding the engineering fundamentals of carbon molecular sieve (CMS) membranes Advisor: Prof. William J. Koros; School of Chemical and Biomolecular Engineering, Georgia Institute of Technology, Atlanta, Georgia, USA.

PhD Dissertation: Design of polyelectrolyte multilayer (PEM) membranes for ion rejection and wastewater effluent treatment

Advisor: Prof. Ilsoon Lee; Department of Chemical Engineering and Materials Science, Michigan State University, East Lansing, MI, USA.

Research experience:

Through my PhD and postdoctoral research, I have gained experience in water treatment and gas separation membranes. For my doctoral research, I worked on the surface modification of membranes using layer-by-layer technology to form polyelectrolyte multilayer (PEM) membranes. In collaboration with another research group in MSU, we extended the usual application of PEM membranes from specific ion rejection to treating a real wastewater effluent. My postdoctoral research focuses on designing asymmetric carbon-molecular sieve (CMS) fibers for advanced gas separation applications. These CMS membranes are produced via controlled high temperature pyrolysis of polyimide hollow fiber polymer precursors. My specific project is about the fundamentals of CMS formation and the effects of different pyrolysis parameters on final separation performance for different gas-pairs.

Teaching experience

During my PhD, I worked as a Teaching Assistant for the course on Transport Phenomena. I was responsible for grading assignments and tests as well as preparing solutions for the same. I hope to further my teaching experience during my stay at Georgia Tech. During my years as a PhD student, I had the opportunity to mentor 5 undergraduate students to perform research in the lab.

Future direction

As a faculty member, I will continue working in the area of membrane-based separation. In the water-treatment area, I will focus on membrane modification but more importantly, I will evaluate the performance of modified membranes under "real" conditions, to extend my doctoral research. Besides traditional water purification, I am interested in other liquid-based separations where the potential of membranes has not been explored in great detail. In the area of gas separation, I will continue my efforts in developing membranes for important separations in the chemical/petrochemical industries like olefin-paraffin mixtures and xylene isomers etc. which are conventionally carried out by highly energy-intensive separation methods.

Selected Publications

1. **Sanyal, O.;** Liu, Z.; Yu, J; Meharg, B. M.; Hong J. S; Liao, W.; Lee, **I**. " Design of fouling-resistant clay-embedded polyelectrolyte multilayer membranes for wastewater effluent treatment," ***Journal of Membrane Science***, **512**, 21-28, (2016).
2. **Sanyal, O.;** Liu, Z.; Meharg, B. M.; Liao, W.; Lee, **I**. "Development of polyelectrolyte multilayer membranes to reduce the COD level of electrocoagulation treated high-strength wastewater," ***Journal of Membrane Science,*** **496**, 259-266, (2015).
3. **Sanyal, O.;** Sommerfeld, A.N.; Lee, **I**. "Design of ultrathin nanostructured polyelectrolyte-based membranes with high perchlorate rejection and high permeability," ***Separation and Purification Technology*** **145**, 113-119, (2015).
4. **Sanyal, O**.; Lee, **I**. "Recent progress in the application of layer-by-layer assembly to the preparation of nanostructured ion-rejecting water purification membranes," ***Journal of Nanoscience and Nanotechnology*** **14**, 2178-2189, (2014). (review)

476028 Elucidating Diffusion, Adsorption, and Catalytic Processes in Oxide Materials

Abstract:

Research Interests:

A conceptual and molecular level understanding of diffusion, adsorption, and catalytic processes forms the basis of most separations and catalytic processes. Herein, I present two case studies in which an improved understanding of physical and chemical rates processes in porous oxide materials enables the design of more effective materials: firstly, amine-modified mesoporous oxide materials for use in large-scale CO_2capture applications, and secondly, aliovalently doped non-reducible bulk oxides for alkane dehydrogenation.

A combination of advanced synthetic techniques, thorough physicochemical characterization, batch and breakthrough adsorption uptake measurements, and mass and heat transfer modeling was used to determine critical factors governing rates of CO_2 adsorption onto supported amine materials. At amine coverages significantly exceeding monolayer coverage, diffusion through the amine-polymer phase was identified as the key step limiting CO_2 adsorption rates. Heteroatom incorporation into the silica framework was used to ameliorate the adverse effect of amine loading on diffusion rates, presumably by resulting in a more open amine-polymer phase. The modified oxide materials thus developed represent *effective*adsorbents exhibiting a combination of both high equilibrium adsorption capacities and rapid diffusion through the amine-polymer network.

In my postdoctoral research work, kinetic and mechanistic details of the non-oxidative conversion of propane to propylene over non-reducible oxides were investigated. High surface area (127m^2/g) bulk yttria doped zirconia materials were demonstrated to be highly active, selective, and regenerable propane dehydrogenation catalysts with

propylene rates comparable to Pt and Cr-based catalysts used commercially. The kinetics and mechanism of alkane dehydrogenation over this new class of catalysts was studied, and methods for assessing the number of catalytic active sites under reaction conditions, developed. Ongoing and future research efforts will focus on extending this concept of highly selective acid-base pair mediated C-H activation on non-reducible oxides to other hydrocarbon conversion reactions.

Building off of my training in separations, catalysis, and industrial R&D, I envisage that my research program will emphasize a molecular level understanding of diffusion, adsorption, and reaction processes, with wide-ranging applications in chemical and energy conversion processes.

Teaching Interests:

Kinetics & Reaction Engineering, Transport Phenomena, Mass Transfer

476129 Advanced Crystallization Technologies for (Bio)Manufacturing

Abstract:

Research Interests: To create novel crystallization strategies and processes for the purification and separation of proteins and other biologically derived molecules.

Teaching Interests: All core courses of chemical engineering. My goal is to help students build a strong background in chemical engineering, sharpen their analytical skills, and apply their knowledge and skills creatively.

Crystallization from solution is a key unit operation for purification, separation, and manufacturing in the environmental, food, pharmaceuticals, chemicals, and materials industries. Challenges that can arise during crystallization include (1) high sensitivities of crystal nucleation and growth kinetics to small variations in the concentrations of contaminating chemicals, (2) high sensitivities of aggregation and agglomeration kinetics to flow fields and particle size, (3) spatial inhomogeneity, (4) sensor limitations, and (5) many orders of magnitude variation in length scales (0.1 nm to 100s of microns) and time scales (20 ps to many hours).

New process designs are described that employ process intensification strategies to greatly enhance the control of the crystal size distribution. These strategies include using noncontact focused ultrasonication, exploiting differences in molecular and thermal diffusion rates at fluid-fluid interfaces to induce nucleation, and selecting crystallizer designs and operations to spontaneously induce multiphase flows that decouple nucleation and growth to enable their individual control. My experiments are supported by mathematical modeling and theoretical analysis, which collectively are also relevant to other chemical process systems.

The growth in demand for new therapeutics and fine chemicals manufactured in bioreactors motivates the invention of technologies to reduce purification and storage costs by at least an order of magnitude compared to existing technologies. A research program is proposed that employs fundamental understanding, mathematical models, and cleverly designed experiments to create novel crystallization strategies and processes for the purification and separation of proteins and other biologically derived molecules.

Keywords: Process intensification, kinetics, separation, biomanufacturing, process design and control, crystallization, purification, protein

479778 Membrane Contactors for Reactive Absorption Process

Abstract:

Research Interests:

I have wide skill set, ranging from membrane science, to reaction engineering, to analytical chemistry. My bachelor's degree is in chemistry, and I work as an antibody chemist and analytical chemist for a number of years prior to graduate school. This experience helped propel my research forward when I decided to go back to school for a doctorate in chemical engineering. My main PhD research focused on developing membrane contract reactors for 3-phase reactions; however, I was also part of an interdisciplinary NSF-IGERT Traineeship, exposing me to a number of collaborations.

Membrane contact reactors: Design and fabricate membranes and membrane reactors. Characterize liquid and gas transport properties of membranes. Characterize structural properties of polymeric and ceramic membranes. Surface modifications of membranes. Reaction engineering: rate laws, reaction order, kinetics, selectivity, and catalyst characterizations.

Liquid-liquid equilibria (LLE): Interdisciplinary project in which phase diagrams of four ternary systems of water-alcohol(s)-methoxycyclopentane (CPME) were determined. Binodal curves were experimentally determined using tie-line measurements and the data was correlated using the universal quasichemical (UNIQUAC) activity coefficient model.

Single cell oil from oleaginous yeast: Interdisciplinary project in which oleaginous yeast was investigated as a renewable alternative to crop-based oil. This project involved investigating scale-up of the fermentation, including cell lysis, separation, purification and economic analysis of the endeavor. Mentoring an undergraduate group involved in the process. And using the above mentioned LLE data to developed a green solvent alternative to the Bligh and Dyer lipid extraction.

Teaching Interests:

I have experience teaching undergraduate classes ranging from freshman to senior levels. In addition to teaching chemical engineering classes, I have also taught Material Science, which includes students from a variety of majors. I am comfortable teaching in both large lecture halls (Material Science), and smaller senior level chemical engineering classes (Chemical Reaction Engineering). My experiences mentoring undergraduates in the labs and coaching the undergraduate rugby team has allowed me to relate very easily to undergraduates and foster fun and welcoming atmospheres in the classroom, laboratories, and office.

Courses taught:

- Current Topics in Chemical Engineering (CHE110)
- Basic Concepts in Materials Science and Engineering (CHE354)
- Fundamentals of Mechanical Properties (CHE355)
    - Fundamentals of Electrical Properties (CHE356)
    - Chemical Engineering Thermodynamics 1 (CHE520)
    - Chemical Reaction Engineering (CHE550)

Professional Preparation:

Ph.D. 2016 Chemical Engineering Kansas State University

B.A. 2011 Chemistry University of California, Santa Cruz

Graduate Certificate- Biobased Products and- Kansas State University

2016 Bioenergy Experience:

2016-Curent Kansas State University, Department of Chemical Engineering

- Postdoctoral Researcher

- Instructor (two classes per semester)

2011-2015: Kansas State University, Department of Chemical Engineering

- NSF IGERT Trainee in Biorefining

- Graduate Research Assistant (PhD Student)

2008-2010 Tainted Love (80's cover band)

- Sound Technician (Roadie)

I was a roadie for a traveling band while I was taking community college classes. I was taking community college classes b/c I was switching from chemistry to engineering.

2004 – 2008: Curtis and Tompkins, LTD. Analytical Laboratories

- GC-MS Chemist

- Total Extractable Hydrocarbon Chemist

- Mercury/Inorganic Chemist

2004 – 2004: Protein Design Labs, Inc

- Antibody Chemist (Internship) PhD Dissertation:

Title: "Membrane Contact Reactors for Three-Phase Catalytic Reactions"

Advisors: Mary E. Rezac, Peter H. Pfromm

Department of Chemical Engineering, Kansas State University

Future Direction:

As faculty I would like to continue applying **Membrane Contactors** for gas-liquid reactions in reactive absorption processes, as well as conducting membrane research in interdisciplinary endeavors. Inorganic nanoparticle (NPs) synthesis is one such process that offers considerable opportunities for future research. These reactions involve a gas being brought into contact with a liquid that contains a dissolved metal salt. The gas must be absorbed into the liquid in order for the gas-metal reaction to take place. In traditional systems, such as stirred tanks or bubble columns, when the gas phase is dispersed into the liquid phase the gas-liquid interfacial area is small and varies with operating conditions, leading to difficulty in control of NP formation and scale-up. However, it is possible to substantially increase the gas-liquid interfacial area and control the gas-liquid mass transfer rate using membrane contactors. By employing a micro porous membrane between the gas and liquid phase, the gas is introduced to the liquid phase through the pores of the membrane. The membrane pores provide a large and stable interfacial surface area, favoring increased absorption rates and faster reaction times. One of the few studies using membrane contactors for NP production was by Jia et al. (J. Membr. Sci., 342 (2009) 1-5). Jia et al. produced $CaCO_3$ NPs by reacting aqueous $Ca(OH)_2$ with CO_2 gas using a membrane contactor described above; additionally, they used a surface renewal model to calculate CO_2absorption per membrane surface area, with the model and empirical data matching well. Proposed future work in this field includes: (1) more examples of noble metal chemistry, (2) systematically varying operating conditions to control size distributions, and (3) experimenting with different NP shapes.

In addition to nanoparticle production and membrane reactors, my interest includes separations and chromatography. With my extensive analytical chemistry background, I am naturally curious about solvent/polymer interactions, as well as physical/chemical interactions in relation to chromatography and separations. Membrane research is sufficiently flexible to adapt to the changes in the funding winds, and I foresee myself using membranes on collaboration research, both intra and interdisciplinary. Selected Awards

- NSF IGERT Trainee 2012-2015

- Carl Storm Underrepresented Minority Fellowships, July 2014

- Accepted to present, and awarded a travel grant to: New York Academy of Science and PepsiCo, Journey through Science Day, December 2014

- Elias Klein Founders' Travel Awards, North American Membrane Society, June 2013, Selected Publications

Wales, M.D.; Huang, C.; Joos, L.B.; Probst, K.V.; Vadlani, P.V.; Anthony, J.; Rezac, M.E.; "Liquid-Liquid Equilibria for Ternary Systems of Water + Methoxycyclopentane (CPME) + Alcohol; Methanol, Ethanol, 1-Propanol, or 2-Propanol"; Journal of Chemical & Engineering Data, 2016, 61 (4), pp 1479–1484, **DOI:** 10.1021/acs.jced.5b00803

Wales, M.D.; Joos, L.B.; Traylor, W.A.; Pfromm P.H.; Rezac, M.E.; "Composite Catalytic Tubular Membranes for Selective Hydrogenation in Three-Phase Systems," Catalysis Today, 2016, 268, pp 12-18, DOI :10.1016/j.cattod.2015.12.011

Zheng, Q.; **Wales, M.D.**; Heidlage, M.G.; Rezac, M.E.; Wang H., Bossmann, S.H.; and Hohn K.L.;

"Conversion of 2,3-Butanediol to Butenes over Bifunctional Catalysts in a Single Reactor," Journal of Catalysis, 2015, 330, pp 222-237, DOI: 10.1016/j.jcat.2015.07.004

Wales, M.D.; Joos, L.B.; Traylor, W.A.; Pfromm P.H.; Rezac, M.E; "Platinum and Palladium Composite Catalytic Membrane for Three-Phase Hydrogenation", Chemical Engineering Journal, (**In Prep**)

Probst, K.V.; **Wales, M.D.**; Rezac, M.E.; Vadlani, P.V.; "Evaluation of Green Solvents: "Oil Extraction from Oleaginous Yeast Lipomyces Starkeyi using Cyclopently Methyl Ether (CPME)", Biotechnology Progress, (**In Prep**)

481045 Development of Green Technologies using Fluid Simulation Knowledge

Abstract:

Research statement

My research interests can be categorized in the broad area of complex fluid simulation and its applications in industrial and environmental aspects. I am predominantly concerned with phase equilibrium and understanding the application of complex systems such as micro-emulsions, aqueous two phase systems, Ionic liquids as green solvents. Theoretical and simulation techniques are utilized to generate important insights into how the inter particle forces influence the behavior of such systems over a broad range of temperature and pressure. Phase equilibrium can play great role in various fields such as polymer synthesis, reservoir engineering, nano materials...

Research Interests:

- **Application of micro-emulsions for synthesis of nanomaterials and nano catalysts**

Microemulsions are colloidal 'nano-dispersions' which stabilized by a surfactant film. These thermodynamically stable dispersions can be considered like a Nano-reactors which can be utilized to carry out chemical reactions to synthesize nanomaterials. In fact using micro emulsions synthesis parameters are easily controlled to produce tailor-made products down to a nanoscale level with new and special properties. Due to their unique properties, namely, ultralow interfacial tension, large interfacial area, thermodynamic stability and the ability to solubilize otherwise immiscible liquids. The uses and applications of microemulsions are numerous in chemical and biological fields.

- **Application of aqueous two phase systems(ATPS) in separation of bio molecules**

Nowadays demands for purified bio-molecules are growing up. For many years chromatography was known as the most common way for purification of both single and double-stranded nucleic acids in industrial scale . But because of its high cost and operational difficulties in scale up, attempts for replacing such systems have been in progress in last decades. Aqueous two-phase systems (ATPSs) as a proper alternative have several preferences in comparison with chromatography technics especially easiness in scale up and lower costs. Furthermore ATPSs have the ability to combine two or more steps of purification process such as mixing, concentration and separation into one which makes the method more applicable.

- **Application of Ionic liquids as a green solvent**

During recent years considerable attention has been paid to the application of IL's as green solvent in various processes such as gas sweetening, CO2 capturing, absorption VOCs and solvent extraction processes. My research in this area can be divided into three main categories: a considerable effort have been focused on the synthesis of IL's for specific applications. In the second category, I am trying to report experimental data on solubility of gases in IL's. finally, modeling of gas solubility in ionic liquids.

- **Phase equilibrium of reservoir fluids in the presence of Surfactant and brine**

Reservoir fluid under extreme conditions of temperature and pressure are of fundamental importance in oil industry, as well as being ubiquitous in nature. Usually surfactants are injected to the reservoir for enhanced oil recovery. Modeling of phase equilibrium in the presence of electrolytes as well as surfactants is really complicated and needs a lot of research in this field. Conventional equations of state need improvement for application in this field. I have focused on theoretical as well as experimental study on phase behavior of reservoir fluids in the presence of surfactants, electrolytes.

- **Thermodynamics of electrolyte solution**

Industrial technologies make use of semi-empirical thermodynamic models extrapolated to extreme conditions (high temperatures, high (azeotropic) concentrations and high pressures). Molten salts are proposed as heat carriers. Pharmaceutical companies are designing processes with organic electrolytes and proteins. Nowadays, the development of accurate correlations and predictive models for thermo-physical properties of electrolyte solutions is highly desirable. Electrolyte thermodynamics concepts can be extended to describe the properties of ionic liquids, radioactive and biological materials. Theories of electrostatics and thermodynamics have to be combined and extended for modeling these novel materials at extreme conditions. Parameter estimation for thermodynamic models can be formulated through the use of advanced optimization algorithms constrained so that they predict accurately the observed number of phase splits, phases and azeotropes.

Teaching Interests:

I have taught several chemical engineering courses during last 10 years at the *Iran University of Science and technology* which gave me teaching ability and a variety of educational skills. At the undergraduate level, I have enough experience in teaching of following courses: *Material and Energy Balances, Thermodynamics I&II, Mass Transfer, Computer aided process engineering (using ASPEN HYSYS simulation environment, Petrochemical processes (3 semesters, Process safety.*

At the graduate level I have taught following courses: *Advanced Thermodynamics, Advanced kinetic & Reaction, PVT simulation, Thermodynamics of reservoir fluids.*

Future Plan

As a long-term plan, I am planning to apply my thermodynamic knowledge in technology development. Knowledge about phase behavior of mixture is an essential tools in process design, material synthesis and so on. In short-term I have planned to commercialize a new oxidative-extractive desulfurization process based on Ionic liquids.

Selected Publications:

1. Partitioning of pyrimidine single stranded oligonucleotide using polyethylene glycol sodium sulfate aqueous two-phase systems: experimental and modeling, **Fluid Phase Equilibria 432 (2017)45-53**
2. Removal of thiophene from model diesel oil with nitrate based ionic liquids at several temperatures, **Journal of Molecular Liquids 221 (2016) 1104-1110**
3. An Efficient Workflow for Production Allocation During Water Flooding, **Journal of Energy Resources Technology 139 (3), 032902 (2016)**
4. A simple accurate model for prediction of Petroleum Fractions physical Properties**, Korean Journal of Chemical Engineering 33: (2016) 2930.**
5. Oxidative desulfurization of Thiophenes using supported Dawson catalyst. **Accepted for publication in Petroleum Research Journal. (2016)**
6. Experimental measurement and thermodynamic modeling of CO2 solubility in aqueous solutions of morpholine, **Journal of Molecular Liquids, 214 (2016) 411-417**
7. Novel liquid–liquid equilibrium data for six ternary systems containing IL, hydrocarbon and thiophene at 25 °C, **Fluid Phase Equilibria, 412 (2016) 21-28**
8. Phase behavior of glycol ether surfactant systems in the presence of brine and hydrocarbon: experiment and modeling, **Fluid Phase Equilibria, 414 (2016) 101-110**
9. Nanostructures in clear and homogeneous mixtures of rapeseed oil and ethanol in the presence of green additives, **Colloid and Polymer Science 293, Issue 11(2015) 3225-3235**
10. Effects of salts and sucrose on the phase behavior of ternary mixtures of water, decane, and mono-ethylene glycol butyl ether, **Colloids and Surfaces A: Physicochemical and Engineering Aspects, 477(2015) 19–25**
11. An investigation of the fish diagrams of water or brine/decane or dodecane/propylene glycol ether (C3P1 or C3P2) systems**, Journal of Molecular Liquids, 206 (2015) 170-175**

12. Modeling adsorption in binary associating solvents using the extended MPTA model, **Adsorption 2014, 1-9**

13. Thermodynamic Modeling of Aqueous Ionic Liquid Solutions Using PC-SAFT Equation of State, **Ind. Eng. Chem. Res., 2012, 51 (30), pp 10274–10282**

14. Application of PHSC Equation of State in Prediction of Gas Hydrate Formation Condition, **Fluid Phase Equilibria, 2012, 333, p.27-37**

15. Modeling CO2 Solubility in Aqueous N-methyldiethanolamine Solution by Electrolyte Modified Peng–Robinson Plus Association Equation of State, **Ind. Eng. Chem. Res., 2012, 51 (29), pp 9875–9885**

1. 16. Prediction of gas hydrate formation condition in the presence of thermodynamic inhibitors with the Elliott-Suresh-Donohue Equation of State , **Journal of Petroleum Science and Engineering , 77 (2011) 93-103**
2. 17. Simplified Local Density Model for Adsorption of Pure Gases on Activated Carbon using Sutherland and Kihara Potentials, **Microporous and Mesoporous Materials 136 (2010) 1–9**

477820 Separation and Reaction Processes in Functional Nanostructured Porous Materials

Abstract:

Research Interests:

The development of new-generation materials that extend the industrial applications of reactionseparation processes is being addressed in some applications like water treatment or fuel cells. The development of functionalized surfaces in highly porous materials can led to stimuli responsiveness such as ionization, electron and ion transfer, swelling, hydrophobicity/hydrophilicity or sorption. These new combination of techniques and materials prove to be effective in selective separations in various environmental conditions, enhancing reactivity, durability and permeation.

The main idea is how to increase the number of functional groups by unit of area on the material used in order to give and augment sensibility, reactivity, selectivity or sorptive characteristics and, at the same time not affecting or even improve the application performance. Polymers and biopolymers may display these different surface functionalities and due to their versatility, they can be applied in diverse research fields and
simultaneously be sustainable technologies.

Current Research

Currently, we are using "green" techniques to create nanocomposites based on polyvinylidene fluoride membranes. This post-manufacture modification relates to the incorporation of various nanomaterials into the structure of the membrane, usually a commercial one. This feature allows better conformation and distribution of the nanomaterial, avoiding aggregation, mixing problems and losses. Membrane pores can be functionalized with hydrophilic materials such as polyvinylpyrrolidone and poly(acrylic acid). Then, by double ion exchange and subsequent reduction, metallic and bimetallic nanoparticles can be synthesized for metal capture or reduction/oxidation of organic contaminants. Through layer-by-layer functionalization, membranes can also be loaded with protein channels (porins) and enzymes. These porins permit the separation of different molecular weights. In the case of enzymes, the objective is the oxidation of aryl halides, avoiding enzyme inactivity.

Dibakar Bhattacharyya, PhD leads this research and the grants associated are the Superfund Research Program of the NIEHS and the KY NSF EPSCoR.

Research Experience

To prepare for the career in separation-reaction process design and engineered materials for different treatment applications, I have worked for more than 10 years in elimination of hard microbial contamination from water, remediation of recalcitrant and harmful pollutants, combination of physicochemical and biological treatments, and plant design for different industries ranging from drinking water to foods to pharmaceutical. In addition, I have good experience in polymeric materials because it is a personal interest, the backbone support of my current research and I had taught this subject during my work as an adjunct professor in Chemical Engineering. My research work (MS, Faculty and PhD) in advanced oxidation processes and membrane functionalization with nanoparticles, respectively, used different platforms in both lab-scale and full-scale for environmentally benign syntheses and manufacture, and the establishment of individual and combined technology strategies to design suitable processes. As an adjunct professor, we succeeded in immobilizing Titania onto activated carbon and polyester for the photocatalytic degradation of wastewater of the printer cartridges reuse process. We advised, assessed equipment, engineered and designed processes for a Drinking Water Treatment and Supply in an Emergency and/or Disaster Scenario in Bogotá D.C., Colombia. In addition, we developed a pilot software Excel app for configuration of water treatment systems in pharmaceutical, food and soft drink applications.

Teaching Interests:

I have an extensive teaching experience. I spent 7 years teaching science and technology to a gifted youngsters in a special education institution, from middle to high school. I was also an adjunct professor in the Chemical Engineering Department at the Universidad de los Andes (Colombia) for 3 years after my master's degree. At los Andes, I taught Separation Processes, Material an Energy Balances, Physicochemical Aspects of
Polymers and Transport Phenomena. At University of Kentucky I was TA and have been guest-lecturer of undergraduate courses in Chemical Engineering such as Separation Processes, Chemical Reactor Design and Chemical Engineering Process Design. To finish, I have mentored graduate and undergraduate students in my group and overseas, including a master's student, REU-NSF and NSF Graduate Fellowship awardees.

Future plans

My research focus on the development of integrated, cost-effective technologies that allow targeted reaction, purification and selective separation of mixtures in different qualities. Likewise, the development of solutions for specific separations and reactions needs intensive research in functional polymers and bioinspired materials at micro and nano scale levels. Specifically I am interested in enabling and designing water treatment systems for specific needs without the need to build conventional, often unsustainable and expensive treatment systems. During the graduate studies and the research as an adjunct professor, I realized that much of the research in water applications focuses on development and study of nano and micro-materials, and how to apply these in current treatment processes. Although this approach has produced excellent developments such as aquaporins, metal nanoparticles or carbon nanotubes inserted into membranes, it has posed also new types of problems like contamination related to nanoparticle activity. In order to address this, it is necessary to develop new technologies based on responsive and immobilized nanostructured materials and to combine reaction and separation processes for each of the applications to increase their selectivity and effectivity. Specifically, start by understanding that the importance of the supporting material and how is functionalized determine the performance of the combined process and the materials used.

Selected Publications

Sebastián Hernández, Anthony Saad, Lindell Ormsbee and Dibakar Bhatacharyya: Nanocomposite and Responsive Membranes for Water Treatment, In Emerging Membrane Technology for Sustainable Water Treatment,
edited by Nicholas Hankins and Rajindar Singh, Elsevier, Boston, 2016, pages 389-431,
ISBN 9780444633125. DOI:10.1016/B978-0-444-63312-5.00016-4

Sebastián Hernández, Shi Lei, Wang Rong, Lindell Ormsbee and Dibakar Bhattacharyya: Functionalization of Flat Sheet and Hollow Fiber Microfiltration Membranes for Water Applications. ACS Sustainable Chemistry & Engineering. 2016; 4 (3): 907-918. DOI: 10.1021/acssuschemeng.5b01005

Sebastián Hernández, Joseph K. Papp and Dibakar Bhattacharyya: Iron-Based Redox Polymerization of Acrylic Acid for Direct Synthesis of Hydrogel/Membranes and Metal Nanoparticles for Water Treatment. Industrial &
Engineering Chemistry Research 2014; 53(3):1130-1142. DOI: 10.1021/ie403353g

Jenny Dussan Garzón, Martha Josefina Vives-Florez, Víctor Manuel Sarria, Oscar Fernando Sánchez Medina, Luis Fernando Delgado, Andrés Fernando González Barrios and Sebastián Hernández Sierra: Aproximaciones Biológicas y Fisicoquímicas en el tratamiento de contaminantes: un resumen del aporte de la Universidad de Los Andes. 2009; DOI: 10.16924/riua.v0i30.233

Sebastián Hernández Sierra, Maryhan Paola Villamil Rodríguez and Lady Liliana Miranda Gómez: Agua Potable. Evaluación de Tecnologías para Dotación de Servicios Vitales en Alojamientos Institucionales, edited by Universidad de Los Andes, Alcaldía Mayor de Bogotá D.C., 07/2010

450979 Chemistry and Engineering of Energy, Environment, and Health

Abstract:

Wicked challenges, such as providing for safe available water, ensuring clean-sustainable energy, and providing for health solutions will require STEM, arts, leadership, and chemistry based solutions! My goal is to provide for progressive education and research via fundamental chemistry and engineering sciences to prepare future generations to solve grand challenges and explore multidisciplinary unknowns with holistic consciousness. To provide for SMART experiential learning, I have developed two teaching tools that can be applied in "choose your own experience" fashion (Jensen 2016, Jensen 2016A). Applied environmental research (Jensen and Müeller 2015, Jensen and Gujarathi 2015), energy research (Jensen *et al*. 2013, Jensen and Lewinski), and Android development (Jensen 2015) contributed to my increasing interest in nanotechnology and nanoinformatics. Synthesis and the evaluation of nanoscale features are state of the art areas of scientific inquiry for the fields of renewable chemicals-fuels, catalysis, medicine and composites to name a few. Nanoinformatics integrates information technology and big data science for multidisciplinary theoretical study and can be used in coordination with laboratory results to provide for layers of validation (*e.g.*, methods, applications, complex computational processes).

References

Jensen CD, et al. (2013) Characterization of Biomass Particles & Geometry During Past Pyrolysis for Systems Models." Proceedings of Conference on Earth & Energy Research. Colorado School of Mines, Golden CO, USA.

Jensen, CD (2015). ChE App Help. Proceedings of the AIChE Annual Meeting, Salt Lake City, UT.

Jensen CD, Gujarathi N. (2015) "Characterization of a Macrophyte Microcosm as a Surface Water Treatment System for Antibiotics." *Environmental Progress and Sustainable Energy*. DOI 10.1002/ep.12158.

In progress, Jensen and Lewinski, Mesoporous Sulphur Doped Titania Photocatalytic Conversion of Oils to Biodiesel.

Jensen CD, Müeller C. (2015) "Benchmark Characteristics of pH based Optoelectronic Enzymatic Biosensors with Non-modified E. coli BL21 pGELAF+." *Applied Biochemistry and Biotechnology*, DOI 10.1007/s12010-015-1623-1.

In press, Jensen, CD. (2016) "Graduate Student Action Research to Support Development of Engineering for Sustainable Development Degree Programs. Part I: A Collaborative Community Action Research Vignette." *Journal of Cleaner Production.*

In submission Jensen, CD (2016A). The Loped: a Multidisciplinary Teaching and Research Tool.

Research Interests: Energy, environment, sustainability, sensors-systems, nano-synthesis and informatics, applications to health, citizen science, and Android apps.

Teaching Interests: Professional development, outreach, innovation, experiential learning, SMART assessment.

460601 A Stabilized, Intrinsically Safe, 10% Efficient, Solar-Driven Water-Splitting Cell Incorporating Earth-Abundant Electrocatalysts with Steady-State pH Gradients and Product Separation Enabled By a Bipolar Membrane

Abstract:

Research Interests:

Materials Innovation and Electrochemistry for Water, Energy and Electronics
- Synthesis, integration, and assembly of 1D/2D nanomaterials with controlled multi-functions, multi-dimensions and multi-compositions for direct production of fuels from sunlight, water and air (N2 and CO2)
- 2D nanomaterial-based electronic, optoelectronic, photonic devices for photodetectors, field-effect transistors, photoelectrochemical biosensors, and electrochromic display
- Materials synthesis and characterization for water quality monitoring, electrochemical water treatment, chemical detection/sensing and seawater desalination driven by renewable energy

Teaching Interests:

- Material Science: Structure and Analysis of Inorganic Solids, Electronic and Optical Properties of Materials, Nanomaterials: Properties and Applications, Semiconductor Physics and Devices
- Electrochemistry and solar energy conversion: Introduction to Analytical Electrochemistry, Electrochemistry for Solar Energy Conversion, Advanced Catalysis for Energy Conversion/Storage, Electrochemical Impedance Spectroscopy, Semiconductor Physics for Photovoltaics and Photoelectrochemistry
- MEMS Materials and Fabrication, Electron Microscopy, Design of Experiments

469261 The Formation of Choline-Lignin Complex By Choline Based Ionic Liquid

Abstract:

Research Interests:

I would dedicate my efforts to lignin valorization to complement the existing strengths of the department in biotechnology and sustainable energy. Three unique characteristics (abundance, energy density and aromatic structure) allow lignin to have great potentials for industrial application.

Teaching Interests:

Teaching Biofuels and Bioenergy is an incredible challenge, because students may have to deal with systematic problems that not only require interdisciplinary knowledge, but also require a long-term effort. My course will address questions such as: Why bioenergy? How will biofuels be produced? What biofuels' challenges need to be resolved? I will also teach students chemistry of biomass and lignin and how to collect and analyze data that will be used in their future careers.

The Formation of Choline-Lignin Complex by Choline Based Ionic Liquid

Jijiao Zeng[1], Michael Kent[2], Jian Sun[1], Feng Xu[1], Tanmoy Dutta[1], Kai Deng[1], Seema Singh[1], Blake Simmons[3] and Kenneth L. Sale[1*]

(1) Deconstruction Division, Joint BioEnergy Institute / Sandia National Laboratories, Emeryville, CA, (2) Sandia National Laboratories, Albuquerque, NM, (3) Lawrence Berkeley National Laboratories, Berkeley, CA

Abstract

Recently ionic liquids have been extensively studied for biomass pretreatment, as they are capable of selectively solubilizing and isolating cell wall components. The deconstruction division within the Joint BioEnergy Institute (JBEI) is developing cheaper, biocompatible and adaptable ionic liquid technologies for use in current biorefineries. Our work has demonstrated that choline-based ionic liquids are highly effective in removing lignin from herbaceous crops such as switch grass. However, the dissociation mechanism of lignin from different biomass by choline-based ionic liquids is still unclear. In this study, modification of the lignin structure by choline glutamate ([Ch][Glu]) and choline α-ketoglutarate ([Ch][aKg]) was investigated. Interestingly, the results showed that choline-based ionic liquids results in the formation a new phenyl-choline ether (4-O-C) bond. Although the yield of isolated lignin from switch grass was significantly higher than for softwood and hardwood, the newly identified signals were present in all isolated lignins. NMR and MALDI-TOF-MS were used to analyze the mechanisms of generating phenyl-choline ether. It is hypothesized that cholination of lignin is a fingerprint for modification of lignin by choline-based ionic liquids.

472159 Can We Use Short Rotation Hybrid Poplar Coppice for Fuels and Chemicals Production Via Biochemical and Thermochemical Conversion?

Abstract:

The economic success of a potential biorefinery is directly related to the use of low cost biomass. Due to the high cost, it is unlikely that a wood-based biorefinery would use only whitewood chips from mature trees in conversion. Using a more heterogeneous raw material is less costly, but efficacy of the conversion using whole short ration coppice (including leaves, bark, branches and white wood) has not been investigated. This study investigated the influence of using whole hybrid poplar coppice on the overall sugar yield via biochemical conversion and the bio-oil production via thermochemical conversion. Results show that converting the short rotation hybrid poplar coppice may be challenging, with up to 50% lower sugar yield and close to 30% lower bio-oil yield than mature poplar wood chips. However, the removal of leaves significantly increased the overall sugar recovery and bio-oil production by 20% and 15%, respectively. These results show that leaf removal from short rotation hybrid poplar coppice prior to harvest may enhance the biorefinery production. With techno-economic analysis and life cycle assessment, we further evaluate the feasibility of converting short rotation hybrid poplar coppice to sugar and bio-oil via biochemical and thermochemical conversions.

Research Interests: Renewable fuels, bioprocess engineering, biorefinery, techno-economic analysis, life cycle assessment, mathematic and mechanistic modeling, fermentation engineering, biochemical engineering

Teaching Interests: Biochemical engineering, bioprocess and biosystem enegineering, natural chemistry, analytical chemistry

475575 Environmental Sustainability through Process Integration, Control and Optimization

Abstract:

Environmental Sustainability through Process Integration, Control and Optimization

Monzure-Khoda Kazi

Post-Doctoral Researcher

Department of Chemical Engineering | Qatar University

Email: kazi.khoda@gmail.com; kazi0001@qu.edu.qa |Tel: (+974) 6640 8300

KEY CONTRIBUTION AREAS

-

§ Process Integration	§ Process Optimization	§ Process Control
§ Process Design	§ Process Intensification	§ Process Simulation
§ Financial Modelling	§ Ad hoc Reporting	§ Market Analysis

VISION AND PERSPECTIVE

The vision of my research is aimed at the improvement of environmental sustainability focusing on energy, water and food security. My research will promote sustainable engineered systems that support human well-being and that lead to economic growth, environmental protection, and social progress for the current generation without compromising the potential of future generations to have an ecosystem that meets their needs. Innovations in industrial ecology, green engineering and large scale engineering research are the prime focuses of my long time research horizons. The ultimate target of my research activities is to conserve natural resources (mass and energy), recycle and reuse materials, prevent pollution, enhance yield, improve quality, advance inherent safety, and increase profitability.

Given the rise in human population and the inevitable consequences of climate change, the challenges of achieving secure and sustainable supplies of energy, water and food are Herculean. Within the next 20 years we will need to produce 50% more food and we will also need 50% more energy, and 30% more fresh water. There is an urgent need to develop long-term strategic programmes of research with integrated and sustainable approaches that will be able to offer potential solutions for managing such challenges. To achieve the desire goals a systematic framework and associated concepts and tools are necessary to guide designer methodically with the global insight of the process, identify root causes of the problems or key areas of opportunities, benchmark the performance of the process, and develop a set of design recommendations that can attain the true potential of the process.

My research background of process integration, process control and specially process optimization assist me for developing new concepts and tools to systematically enhance and reconcile various process objectives, such as cost effectiveness, yield enhancement, energy efficiency, and pollution prevention. I am working last few years to develop such methodical sustainable approaches for various processes starting from small scale production (producing valuable pharmaceutical drags using liquid chromatography) to large scale facility (managing industrial flares during abnormal situations). As my specialization is to work with different kinds of optimization algorithms and control concepts, I would also like to face the challenges from process uncertainty and from the black box problems which can be mitigated using sustainable process integration.

My knowledge of research is not limited to the industrial processes, I have also implemented my acquire optimization knowledge to develop some adaptable financial modelling/algorithms which can be used in the most challenging and unpredictable global financial markets. The unique point of my research activity is that I can use my long experience of capital market to develop long term view for energy and commodity prices; this knowledge can be integrated with my developed framework to make the system/design more robust to face different kind of challenges including unpredictable price fluctuations (price fluctuations of oil, natural gas etc.), unpredictable incidents (industrial flaring) or unknown disturbances (sudden changes in process parameters).

Future Plans

Following are three principal research areas and corresponding projects that I plan to advance in my research group during the first several years:

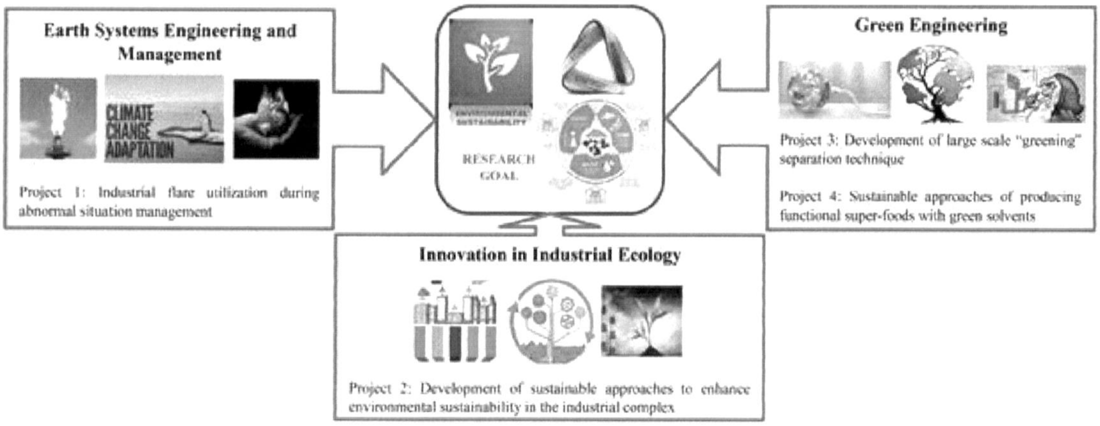

Figure 1. Schematic of proposed research projects and the research goal. Each of these rectangles (representing

research projects) will interact with one another in an iterative manner.

Research Interests:

§ Industrial flare utilization strategies through process integration and energy optimization for abnormal situation management by economic and environmental friendly alternatives

§ Development of energy alternative tools and sustainable process configurations to manage excess resources (i.e., waste energy streams, waste water) in the industrial complex for environmental sustainability

§ Development of fundamentals/technologies/alternatives for large scale "greening" separation technique (i.e., chromatography) to move towards environmental sustainability

§ Sustainable approaches of producing functional super-foods with green solvents for global food security through process integration and intensification

Teaching Interests:

I am comfortable teaching the Chemical Engineering core courses offered within the department at both the undergraduate and graduate level. I would be well suited to teach fundamental course such as: chemical process control, chemical reaction engineering, particle technology, separation technology, chemical & engineering thermodynamics, numerical methods, unit operations, mass transfer technology, process integration etc. I appreciate the need for faculty flexibility and embrace the opportunity to cover a broad spectrum of courses. Based on my research, I also have a strong interest in process integration, control and optimization. I look forward to incorporating fundamental engineering concepts with cutting-edge technologies.

SCHOLARLY ACHIEVEMENTS

SUMMARY OF RECORDS

Peer Reviewed Journal articles: **5** published (**3** first- and 2 co-authored) and **1** (first-authored) in review

Peer Reviewed Conference Proceedings: **3**

Conference Presentations: **14**

Selected Peer Reviewed Publications (*Citations: 16; h-index: 2; i-10 index: 1; total impact factor: 13.90, as in April 2016*)

- **Monzure-Khoda Kazi**, Fadwa Eljack, Nesreen A. Elsayed and Mahmoud M. El-Halwagi. Integration of energy and wastewater treatment alternatives with process facilities to manage industrial flares during normal and abnormal operations – A multi-objective extendible optimization framework. *Journal of Industrial and Engineering Chemistry Research,* ACS publication, 2016, 55, (7), 2020-2034. (doi:10.1021/acs.iecr.5b03938)

- **Monzure-Khoda Kazi**, Fahd Mohammed, Ahmed Mhd Nabil AlNouss and Fadwa Eljack. Multi-objective optimization methodology to size cogeneration systems for managing flares from uncertain sources during abnormal process operation. *Computers and Chemical Engineering,* Elsevier, 2015 vol 76, p.76-86 (doi:10.1016/j.compchemeng.2015.02.012)

- **Monzure-Khoda Kazi**, Bijan Medi, and M. Amanullah. Optimization of an Improved Single-Column Chromatographic Process for the Separation of Enantiomers. *Journal of Chromatography A,* Elsevier, 2012. 1231: p. 22-30. (doi:10.1016/j.chroma.2012.01.057).

GRANT WRITING EXPERIENCE

1. "Water recovery from power plant flue gas: a hybrid technology as a novel source of water for Qatar", PI: Asst. Prof. Mohammad Aman Ullah, 2015 (NPRP no.: 8-2191-2-855)

2. "Experimental evaluation of the optimal operation of a single column continuous chromatography process under overloaded condition: profile-profile interaction", Source: Qatar University (individual grant), Budget: US$ 35,715, PI: Asst. Prof. Mohammad Aman Ullah, 2013

3. "Optimization of a single column continuous chromatographic speration process for the separation of optical isomers", Source: 13th cycle of the QNRF's undergraduate research experience program (UREP), Budget: US$ 58,850, Lead PI: Asst. Prof. Mohammad Aman Ullah, 2013

4. "Design of a simulated moving bed (SMB) process for continuous binary separation of chiral compounds", Source: Qatar University (student grant), Budget: US$ 2,715, PI: Asst. Prof. Mohammad Aman Ullah, 2012

5. "Super food production using supercritical carbon dioxide technology: extraction, micronization, encapsulation and optimization", PI: Assoc. Prof. Timothy Tan, 2012

TEACHING AND MENTORING EXPERIENCE

Lecturer (Green University, Bangladesh)

1. CSE 2102 - Design and Innovation with Computer Simulation
2. TEX 2201 - Fundamental of Textile Testing
3. TEX 2209 – Environmental & Pollution Control

Graduate Teaching Assistant (Nanyang Technological University, Singapore)

1. CH3101 – Chemical & Bioengineering Process Control and Dynamics (AY 2011-2012 Sem 2)
2. CH3001 – Process Control Lab (AY 2011-2012 Sem 2)
3. MATLAB workshop on Numerical Methods (AY 2011-2012 Sem 2)
4. CH1801 - Chemical/Bioengineering Lab (AY 2011-2012 Sem 1)
5. CH1701 - Chemical/Bioengineering Lab (AY 2010-2011 Sem 1)
6. CH3102 – Chemical Reaction Engineering (AY 2009-2010 Sem 2)

Student Advising

1. Fahd Mohammed, MSc Student, Qatar University (2015)
2. Fah Keen Chong, PhD Student, University of Nottingham Malaysia (2015)
3. Viknesh Andiappan, PhD Student, University of Nottingham Malaysia (2015)
4. Calvin Wiratama, Undergraduate Student, Nanyang Technological University Singapore (2012)
5. Xueyuan Tian, Undergraduate Student, Nanyang Technological University Singapore (2010)

REFERENCES

Dr. Mahmoud El-Halwagi

Professor, Artie McFerrin Department of Chemical Engineering, Texas A&M University, College Station, TX 77843-3251

Tel: (979) 845-3484

Email: el-halwagi@tamu.edu

(He is my current research collaborator and mentored me in a workshop)

Dr. Fadwa T. Eljack

Assistant Professor, Department of Chemical Engineering, Qatar University, P.O. Box: 2713 – Doha – Qatar

Tel: +974 4403 4141 Fax: +974 4403 4131 Email: Fadwa.Eljack@qu.edu.qa

(She is my current supervisor)

Dr. Mohammad Aman Ullah

Assistant Professor, Department of Chemical Engineering, Qatar University, P.O. Box: 2713 – Doha – Qatar

Tel: +974 4403 4162 Fax: +974 4403 4131

Email: Aman@qu.edu.qa

(He was my PhD supervisor in Singapore)

Dr. M. M. Faruque Hasan

Assistant Professor, Artie McFerrin Department of Chemical Engineering, Texas A&M University, College Station, TX 77843-3251 Tel: (979) 862-1449

Email: hasan@tamu.edu

(He was my senior colleague in Singapore)

-

476048 Life-Cycle Environmental and Techno-Economic Analysis of Novel Biomass Thermochemical and Electrochemical Energy Systems

Abstract:

Research Interests:

Our focus has shifted from fossil energy to renewable energy sources due to environmental pollution, global warming, and sustainable development issues caused by the consumption of fossil fuels. As a promising renewable alternative, biomass has been playing a critical role in enhancing the nationsâ€™ energy security and mitigating carbon emissions. My research has focused on thermochemical conversion of biomass into fuels, electric power, and chemicals through experimental investigation, process modeling using Aspen Plus, and life-cycle environmental and techno-economic analysis.

I have been involved in several projects working as a post-doc researcher at Iowa State University. We have proposed a novel strategy for reducing carbon emissions from coal-fired power plants based on the utilization of bio-oil and biochar sequestration. This approach provides a sustainable and economical pathway to meeting life-cycle carbon reduction targets in the power sector. We have developed new methods for optimizing revenues from various product portfolios via fast pyrolysis and upgrading in an integrated biorefinery including drop-in transportation biofuels, biobased products, and hydrocarbon chemicals production. The findings from this analysis shed light on product selection and risk mitigation for investors from the bioproduct industry. In more recent work, we explore hybrid thermochemical and electrochemical methods for electricity, fuels, and chemicals generation and focus on three novel system designs: direct biomass fuel cell, biomass electrolyzer, and CO_2electrolyzer. The potential economic advantages and challenges are identified from this analysis.

During my Ph.D. program, I explored transportation biofuel production via thermochemical conversion of agriculture residues (rice husk and corn stover). My work spanned from helping setting up the experimental platform (reactor with a volume of 100mL, 300mL and 10L respectively) to conducting experiments on the catalytic upgrading of fast pyrolysis bio-oil in supercritical ethanol/methanol system. The optimal reaction conditions and catalysts for this bio-oil upgrading method were proposed. I further enhanced my experimental experience by converting model compounds of algae biomass to biofuels through hydrothermal liquefaction. In addition, I focused on employing modeling tools such as GREET and SimaPro to evaluate the environmental performance of different bio-oil upgrading pathways.

My future research interests are applying a combination of experimental and modeling tools to better understand novel biomass conversion technologies and help advance the development of these technologies. Specific research interests include experimental investigation of biofuels, chemicals and power generation from thermochemical processes, process modeling, and evaluation of bioenergy systems from environmental and techno-economic perspectives.

Teaching Interests:

I would like to teach both undergraduate and graduate level courses such as introduction to mechanical engineering, energy and power system, biorenewable technologies, chemical process modeling, life-cycle environmental and techno-economic analysis which are closely related to my research. By integrating my research in my teaching, I can enhance the engagement of students in certain types of research. I hope my students can get some hands-on experience and a better understanding of core engineering concepts by employing state-of-the-art engineering tools to solve real engineering problems after finishing the courses.

476131 Exploring New Boundaries for Sustainable Water and Industrial Wastewater Treatment Technology

Abstract:

Clean and safe water is a basic human need, but it is not available to 1.4 -1.8 billion people around the world. It is essential to direct current research trends toward sustainable water and wastewater treatment technologies that can solve the existing industrial and environmental issues, especially when it comes to solutions that can be successfully commercialized on the global scale. Membrane application technologies are the most effective and sustainable methods of addressing environmental problems in treating water and wastewater to meet or exceed stringent environmental standards. Nevertheless, membrane fouling is one of the primary operational concerns that is currently hindering a widespread application of membrane technology. My major research plans are to optimize synthesis of biomimetic and bioinspired membranes designed with antifouling, self-cleaning, and selective permeation that will pave the way for the production of clean water; to address all of the existing deficiencies of the Forward Osmosis systems; and to control membrane morphology by zeolites incorporation.

Research Interests: Membrane Science and Technology in Water and Wastewater Treatment; Biomimetic and Bioinspired Membranes for Next Generation of Biomaterials and Nanomaterials Engineering; Process Modelling, Simulation, and Optimization of complex systems.

Teaching Interests:

- Received a Formal recognition of completion of the Teaching Development Professional Program (UTDP), accredited by the Staff and Educational Development Association of the United Kingdom (SEDA UK).
- Participated in the curriculum development processes by designing a new effective course for graduate students

  at Ryerson University.
- Strong record of Teaching & Educational Engineering publications

475052 Renewable Transportation Biofuel Production from Wet Biowaste

Abstract:

Research Interests:

- **Summary**

Increasing wet organic waste production has become an unavoidable issue in the US. Although often considered as environmental pollutants, these wet organic feedstocks contain large amounts of unharnessed energy and nutrients. My research aims to recover the energy and nutrients from the wet organic waste with appropriate waste remediation systems. Hydrothermal treatments, including catalytic hydrothermal gasification (CHG) and hydrothermal liquefaction (HTL), is a new energy efficient thermochemical process that can generate high-energy content biogas/biocrude and simultaneously reduce total organic carbon in wastewater. In contrast, anaerobic digestion (AD) is a widely utilized biochemical process that similarly produces renewable biogas. Application of different catalysts and adsorbents in each scenario can increase energy yields and nutrient recovery. Running both systems in parallel on identical wet organic wastes can compare energy returns and pollutant reduction through techno-economic analysis. In addition, effluent waters from each process containing high concentrations of nitrogen and phosphorous, amongst other dilute nutrients, can be further recycled as a desirable substrate for production of crops (e.g. algal biomass or crops grown in hydroponic systems). Secondary algal treatment of wastewaters from CHG/HTL and AD will clean water for environmental discharge meeting EPA standards and also generate renewable resources for further application. Simultaneously, using isotopes with representative model compounds in CHG/HTL process can further elucidate the reaction kinetics, network, and mechanism in this biomass conversion system. Successful completion of these objectives will provide clearer understanding of how advanced thermochemical and biochemical waste remediation technologies can alleviate the ***food-water-energy nexus***.

- **Rationale:**

Wet biowastes have been an increasing interest to researchers for the potentials to produce bioenergy, reuse nutrients and improve the environment. Each year, the United States produces 79 million dry tons of wet biowastes including food-processing waste and livestock manure (DOE, 2011; FAO, 2016; USDA, 2010). These wet biowaste exceeds the â€˜Impactful tonnageâ€™ (>50million ton per year) as a bioenergy feedstock according to DOE, and containing nutrients (N&P) that can substitute about 1.5 million tons of reactive nitrogen as fertilizers. Carbon, nitrogen and phosphorous, along with other minerals contained in wet biowastes are often discharged to environment as pollutants because they cannot be efficiently recovered and utilized. Based on data provided by USDA (USDA, 2011), it is estimated that as much as 45% of nitrogen and 70% of phosphorous could be recycled from such agricultural waste streams. One of the main reasons this energy and nutrients are not effectively harvested is the large water content (20-98%), which greatly reduces the number of bioenergy conversion technologies that can have a net positive energy balance. On the other hand, energy content approximately over 10 quads, equal to all current US renewable energy production, could be recovered for beneficial reuse with appropriate technologies or systems. Moreover, the growth of the global population has placed an ever-increasing stress on food, energy, and water (UN, 2015). Advancing knowledge/technologies that foster more efficient, safe, and secure use of resource within the food-energy-water nexus remains a significant challenge (NSF, 2016).

- **Research Topics**

My research enthusiasm was influenced by my past and present research which encompasses thermochemical conversion processes, biofuel production, biowaste treatment, biomass pretreatment, value-added chemical extraction, techno-economic analysis of the nutrient and bioenergy recovery processes, and reaction network under supercritical fluid. Combining both laboratory and computational methods to link local processes to up-scaled system in our dynamic environment can help address the following questions: (1) How to energetically harness available energy and nutrients from wet organic waste and what are the most relevant factors affecting the biomass conversion process? (2) How to control and fine tune the biomass conversion processes more precisely to obtain more valuable products? (3) How do we combine new technologies with multiple scale (e.g. from laboratory to pilot to demonstration scale) of investigation to demonstrate an affordable system, minimize greenhouse gas emission, maximize the recovered energy, and identified research gaps to strengthen a robust food-energy-water nexus? To answer these questions, my research closely couples innovative thermochemical, biological, environmental, and computational approaches to provide an integrated perspective to establish a solid energy-water-food nexus.

1. ***How to energetically harness available energy and nutrients from wet organic waste?***

Catalytic hydrothermal gasification (CHG) is an emerging new technology alternative for wet biowaste conversions with much faster reaction times (minutes), high conversion efficiencies (>90%), and a cost trajectory that is projected to be competitive with current fossil fuels when the feedstock has a low or negative cost. However, there are no full-scale applications of this technology because commercial-scale equipment has not been developed, and the data available from the current literature highlights several potential practical concerns (e.g. catalysts fouling issues) that need to be addressed before large investments in the technology are justified. Anaerobic digestion (AD) is traditionally a dominant conversion technology used for wet wastes. In order to achieve a higher organic loading rate and energy recovery, a two-stage AD process is proposed in my research program and the results will be compared to those from a conventional AD system. By optimizing the conditions of sludge loading rate, operating temperatures (30-55°C), the dosage of adsorbents, and the types of adsorbents, the yields of hydrogen and methane (hythane) and the energy recovery of the proposed two-stage AD system can be improved. Conventional AD process produces mainly methane, while two-stage AD can produce more hydrogen because it can better control the reaction environment (e.g., pH) and pathways by dividing the fermentation process into hydrolysis, acidogensis, and methanogensis separately (Massanet-Nicolau et al., 2013; Schievano et al., 2014). According to recent studies, CHG and AD can both substantially recover the carbon nutrients in swine manure, but not nitrogen and phosphate (Breinl & Zhang, 2015; Shin & Schideman, 2015; Tommaso et al., 2014). On the other hand, aquatic biomass such as macrophytes and algae has been proven to have removal efficiencies of greater than 80 and 95% for nitrogen and phosphorous, respectively (Möller & Müller,

2012; Polomski et al., 2009; Woertz et al., 2009). In addition, crops (e.g. lettuce and tomatoes) that can grow in the hydroponic system also have the potential to further assimilate the residual nutrients in the CHG and AD effluents. Therefore, a hydroponic farming system is proposed in my research program to simultaneously uptakes the nutrients and remediates the wastewaters coming from AD and CHG. In this case, research efforts on hydroponic cropping development and algal biotechnology for concentrated agricultural wastewater treatment can be advanced to satisfy growing need in the food-water-energy nexus, while the potential of these concentrated agricultural effluents as nutrient sources for aquatic biomass grown in hydroponic systems can be realized. In short, my research program aims to investigate the efficacy of using AD and CHG to harness the chemical energy stored in wet biowastes (e.g. animal manure and citrus processing waste containing value-added chemicals such as flavonoids). The ultimate goal of my research program is to demonstrate the feasibility of a novel concept for integrated bioenergy and biomass production, and wastewater reuse via parallel treatment of CHG and a two-stage AD of wet organic wastes.

2. How to control and fine tune the biomass conversion processes more precisely to obtain more valuable products?
No full-scale applications of the hydrothermal treatment technology have been realized so far, because the data available from the current literature highlights several potential practical concerns that need to be addressed before large investments in the technology are justified.

(i) Catalysts development for effective CHG. Previous literature (Elliott, 2008; Stucki et al., 2009) has extensively focused on using noble or transitional metal catalysts such as Pt, Ru, and Ni for catalytic hydrothermal gasification. To make CHG a more affordable technology for the public, other economic catalysts such as activated carbons, zeolites, calcium carbonates derived from waste shells, and iron recovered from slag will be explored and their catalysts performance (e.g. yield of hydrogen) will be evaluated.

(ii) Catalyst activity under CHG. Fouling of catalyst is another issue under hydrothermal processes and will be investigated in my research program. The previous study (Ong, 2013) using CHG to catalyze waste newspaper shows that unless the catalysts can be regenerated and recycled, CHG can produce more net-energy and present a more efficient energy consumption ratio (ECR, defined as energy input divided by energy output) than anaerobic digestions of waste newspaper. Thereby, in my research program, the catalysts will be characterized before and after the reactions to understand the deactivation mechanisms (e.g. catalysts fouling or catalysts poison) through SEM, TEM, TGA/DSC, XRD, and XRF analyses. Different regeneration methods for catalysts will be explored and the morphology of the regenerated catalysts will be studied. Meanwhile, the life of the catalysts will be understood. Ultimately, methodologies to minimize the catalysts deactivation will be established through my research program.

(iii) ***Reaction Kinetics and Network for bioactive chemicals under CHG.*** Previous studies have demonstrated that CHG can effectively degrade bioactive chemicals, such as environment hormones (e.g. estrogens) and antibiotics (e.g. florfenicol), entrained in swine manure (Shin & Schideman, 2015). However, the reaction kinetics and reaction network of these bioactive chemicals remained unknown. Understanding the reaction kinetics and network for bioactive chemicals may help design the up-scaled CHG reactor in a more practical way and make CHG a more versatile tool for treating different types of bioactive chemicals (e.g. virus and pathogens). In order to achieve this goal, CHG of model bioactive compounds with heavy water (i.e., deuterium oxide) will be conducted at 150-400 °C for 0-2 hr. The CHG products obtained at different reaction time will be analyzed under GC/MS, NMR (1D and 2D), and FTIR essay to further elucidate the reaction pathways of wet biowaste. Sensitivity analysis will also be conducted to explore significant factors governing CHG processes.

3. How do we combine new technologies with multiple scale (e.g. from lab to pilot scale) of investigation to demonstrate an affordable system, minimize greenhouse gas emission, and maximize the recovered energy to strengthen the current food-energy-water nexus?

System analysis of the proposed parallel CHG and AD systems can help understand and identify research gaps to establish a roust food-energy-water nexus. For example, the necessity of reusing residual nutrients in the aqueous products (PHWW) obtained from hydrothermal liquefaction (HTL) processes has been identified through a multi-component mass balance. Conducting the mass balance, energy balance, and economic analyses of the integrated system using the STELLA® and SuperPro software can further elucidate and identify the revenue streams, quantification of cost functions, capital allocations, upstream suppliers, and downstream customers. Simultaneously, life cycle analysis (LCA) through quantitative and computational modeling will be utilized to optimize this system and significantly advance the understanding of the proposed food-energy-water nexus.

- **Indication and Expected Outcomes:**

Successful results of this research program will lead to bridging gaps between biowaste management and energy/nutrient recovery. Modification of two-stage AD processes including fermentation concentration and operating temperatures will help optimize treatment efficiency and yield greater energy content in the form of methane and hydrogen, which is underdeveloped. More recently investigated thermochemical technologies like HTL and CHG show great promise to efficiently convert wet organic feedstocks into high-energy content products such as biocrude oil and biogas, and the combination of different catalysts and reaction parameters can improve overall energy recovery and at the same time lowering energy inputs. Comparison of improvement in two-stage AD and CHG systems will demonstrate the ability to remediate different types of wet biowaste. Once carbonaceous contents are recovered as biogas, the remaining nitrogen and phosphorous will be assimilated by aquatic biomass for more renewable feedstock and/or food production. The reuse of nutrient can amplify the biomass thus the bioenergy production. An overall techno-economic assessment will elucidate the sustainability of the proposed parallel CHG and AD systems.

Teaching Interests:

- **Summary**

My teaching philosophy was molded by my education as an engineer. For me, education is significantly more than imparting knowledge found in textbooks. As an engineering student involved in a multidisciplinary research group, I had the opportunity to gain a great deal of know-hows by participating in different research activities. This experience also motivated and reinforced my teaching philosophy. I truly believe that by involving students with hands-on projects and real-world solid examples, learning new knowledge can be more effective while creative thinking and innovative solutions will be strongly encouraged. As Benjamin Franklin has said, "Tell me and I forget. Teach me and I remember. Involve me and I learn."

- **Teaching Philosophy**

I am deeply committed to effective teaching and mentorship. I have been very fortunate to have had many positive and influential mentors during my education. This experience incentivizes and eagerly drives me to be a supportive and helpful mentor to my students. My objective when teaching and mentoring is to create a learning environment that simultaneously develops critical thinking skills and fosters an open mind to all possibilities by emphasizing the rich conceptual issues that defines sustainable energy science and engineering and how this discipline impacts our lives. By expanding not only the studentsâ€™ scientific understanding of engineering problems but most especially how they apply it to the real world, we can critically review more available sustainable engineering practices and/or stewardship. Advantages and limitations of the established methods would be identified during this critical thinking process, and valid alternative state-of-the-art methods can then be developed. Ultimately, I hope my students can think â€œoutside the boxâ€ and develop their confidence in proposing solution to existing and new problems.

- **Teaching experience**

As I have served as a teaching assistant (TA) in the class of *Transportation Phenomenon*(ABE 341), I have designed weekly worksheets for students, conducted office hours, worked with students on their lab projects, and graded homework and worksheets. Moreover, I realized that effective communication is a key element to help students understand fundamental theories (e.g. fluid dynamics) and accomplish lab projects successfully. A smooth communication can inspire studentsâ€™ potential and establish their confidence. For example, I always encouraged students to come to office hours and work on those difficult problems together. Textbook and homework are not something to beat them but tools to help them learn. Working as a team to solve problem sets usually can help them develop critical thinking skills and address problems more efficiently. For instance, one of my students learnt that units may help her check if the answer makes sense in a fluid dynamics problem. Additionally, I have served as a guest TA for several class projects (ABE 100, ABE 225, CEE 398 and TSM 431, please refer to my CV). As a mentor, I usually ask students some questions and give some solid examples to initiate their interest. Literature review and experimental design are generally required in class projects. Through lab projects, students can gain knowledge on bioenvironmental engineering design, particularly about biomass conversion and sustainable energy engineering, while I always benefit from the mentoring and discussion. Teaching and mentoring are not only a give-and-take process but also an interactive and mutual beneficial discussion with the mentored undergraduates at the same time.

- **Mentoring experience**

By participating a multidisciplinary research programs in University of Illinois, joining the algae club in my home department, and involving with SWE (Society of Women Engineers) on campus, I was very fortunate to have ample opportunities to mentor or work with students from elementary school, high school, to different levels of undergraduate students (please refer to my CV). As providing technical training and helping them develop research topics, I always benefit from the mutual beneficial discussion. Students often inspire me from a different point of view and pay more attention to details than I do. This mutually beneficial discussion has been a cornerstone of my research career. For example, when I worked with my first undergraduate, sparked by her previous experience on centrifugation, we began to develop a more sustainable pretreatment technique, combining centrifugation and ultrasonic processes, for improved bioenergy conversion efficiency from algal biomass. Another instance is that I have further advanced my knowledge on gas-chromatography mass-spectrometry (GC-MS) by instructing several undergraduate and graduate students on using this equipment for analyzing biocrude oil and volatile aqueous productsâ €”mentoring itself is an intensive learning experience.

- **Teaching methods**

When I served as a TA four years ago, I found it was drastically different between knowing the knowledge and transferring to others. For instance, demonstrating a practice example is likely more helpful to the novice than throwing a challenging problem at him/her, which may spoil a studentâ€™s interests and confidence. During my work with undergraduate students from different departments and culture background in University of Illinois, I realized that working on projects can significantly motivate students to learn a specific topic and inspire studentsâ€™ potential as well as establish their confidence. Therefore, I will combine lecture and lab projects in my classes. For fundamental classes (e.g. transport phenomenon and heat and mass transfer), I plan to apply in-class discussion to one of the weekly lectures with a worksheet. I will also separate students into different groups and assign different groups of students to design the worksheet every week. This type of student-centered learning method can further help students think about why the lecturer/textbook asks questions like that way and why those questions matter. For advanced classes (e.g. classes provided for senior undergraduate students or graduate students), I will combine lectures, lab projects, field trips (e.g. visit a zero energy residence building, a bio-ethanol plant, or a pilot scale continuous reactor to generate biofuels/biogas), term-paper writing, and practice proposal writing together. When I took classes and served as a TA in University of Illinois, I found out lab projects usually can make students effectively learn hands-on skills and apply what they learnt in classes to real-world problems while field trips can greatly initiate studentsâ€™ interest on a specific topic. In addition, term paper and practice proposal writing can help students synthesize their thoughts on specific topics and present their ideas in a professional way. For instance, when I wrote my practice proposal for *Green Chemistry* (CHEM 460) and *Polymer Chemistry*(CHEM 480), I read a lot of papers about biorefienries and reviewed different biomass conversion techniques. In the end, I gained deeper insights about renewable energy and value-added chemical production processes, which ultimately contribute to part of my PhD thesis. Furthermore, I will conduct an early feedback, which will happen in the one third or half of the semester, in my classes so that I will have time to fine tune my teaching style or adjust course content during the semester. Overall, by using versatile teaching methods and techniques, I hope to give students a great incentive to learn different levels of courses and understand how they may change and impact the world.

References:

Breinl, J., Zhang, Y. 2015. Hydrothermal Gasification of HTL wastewater.

DOE. 2011. US billion-ton update: Biomass supply for a bioenergy and bioproducts industry. Oak Ridge National Laboratory (ORNL).

Elliott, D.C. 2008. Catalytic hydrothermal gasification of biomass. *Biofuels, Bioproducts and Biorefining*, **2**(3), 254-265.

FAO. 2016. Database of Food and Agriculture Organization of the United Nations

Massanet-Nicolau, J., Dinsdale, R., Guwy, A., Shipley, G. 2013. Use of real time gas production data for more accurate comparison of continuous single-stage and two-stage fermentation. *Bioresource technology*, **129**, 561-567.

Möller, K., Müller, T. 2012. Effects of anaerobic digestion on digestate nutrient availability and crop growth: A review. *Engineering in Life Sciences*, **12**(3), 242-257.

NSF. 2016. Innovations at the Nexus of Food, Energy and Water Systems (INFEWS).

Ong, M. 2013. Evaluation of anaerobic membrane bioreactors and hydrothermal catalytic gasification for enhanced conversion of organic wastes to renewable fuels, University of Illinois at Urbana-Champaign.

Polomski, R.F., Taylor, M.D., Bielenberg, D.G., Bridges, W.C., Klaine, S.J., Whitwell, T. 2009. Nitrogen and phosphorus remediation by three floating aquatic macrophytes in greenhouse-based laboratory-scale subsurface constructed wetlands. *Water, air, and soil pollution*, **197**(1-4), 223-232.

Schievano, A., Tenca, A., Lonati, S., Manzini, E., Adani, F. 2014. Can two-stage instead of one-stage anaerobic digestion really increase energy recovery from biomass? *Applied Energy*, **124**, 335-342.

Shin, Y.H., Schideman, L. 2015. Characterizing the fate and transport of Chemicals of Emerging Concerns (CECs) from integrated bioenergy and manure management system. *2015 ASABE Annual International Meeting*. American Society of Agricultural and Biological Engineers. pp. 1.

Stucki, S., Vogel, F., Ludwig, C., Haiduc, A.G., Brandenberger, M. 2009. Catalytic gasification of algae in supercritical water for biofuel production and carbon capture. *Energy & Environmental Science*, **2**(5), 535-541.

Tommaso, G., Chen, W.-T., Li, P., Schideman, L., Zhang, Y. 2014. Chemical Characterization and Anaerobic Biodegradability of Hydrothermal Liquefaction Aqueous Products from Mixed-culture Wastewater Algae. *Bioresource Technology*.

UN. 2015. *World Population Prospects: The 2015 Revision, Volume I: Comprehensive Tables.* United Nations Publications.

USDA. 2010. Manure and Byproduct Utilization.

USDA. 2011. U.S. consumption of nitrogen, phosphate, and potash, (1960-2011).

Woertz, I., Feffer, A., Lundquist, T., Nelson, Y. 2009. Algae grown on dairy and municipal wastewater for simultaneous nutrient removal and lipid production for biofuel feedstock. *Journal of Environmental Engineering*.

Zhou, Y., Schideman, L., Yu, G., Zhang, Y. 2013. A synergistic combination of algal wastewater treatment and hydrothermal biofuel production maximized by nutrient and carbon recycling. *Energy & Environmental Science*, **6**(12), 3765-3779.

476331 Energy-Dense Liquids from Renewable Energy

Abstract:

Those in the global chemical enterprise now agree that the future of the chemical industry depends on becoming more sustainable. Such a goal includes developing energy sources that are not based on fossil fuels, and that do not release large quantities of climate-changing gases. This goal implies developing technologies that effectively collect stranded renewable energy - solar and wind - and enable long-term energy storage and long-distance energy delivery from remote areas. Solar and wind energy are continuing to get cheaper, but they are far from population centers and storage/transmission is difficult. Hence, small-scale distributed manufacturing processes that directly convert the renewable energy to transportable chemicals are important.

Research Interests:

At the University of Minnesota, I have been working on designing small scale Haber-Bosch process powered by stranded wind-generated electricity. The electricity generated by windmills is used to split water to make hydrogen, and to separate nitrogen from air by pressure swing adsorption. The hydrogen and nitrogen are then reacted to make ammonia. This ammonia is made without the use of natural gas or coal, and hence without emission of climate-changing carbon dioxide. Our analysis of this small process shows that its performance is governed by three characteristic times, those for chemical reaction, for ammonia removal by condensation, and for unreacted gas recycle. To date, our proposed process is limited by the time for ammonia removal, that is, by how long it takes the condenser to liquefy the ammonia product. While this acceleration of ammonia removal should be straightforward, we believe it will require considerable capital investment. As a result, we have been trying to simplify and intensify the Haber-Bosch process. We have not tried to improve the catalytic steps in the reactor at all, because we believe that the existing reaction, optimized by a century of intensive effort, has produced a very rapid potential reaction as good as can be hoped. We have focused instead on reducing the operating pressure and rapidly separating the ammonia produced. In my research, I have showed that the conversion could be increased from 20% to over 90% by coupling reaction and absorption. I have also showed how the operating pressure could be cut from 130 bar to 20 bar, while increasing the specific rate of reaction by cutting the product concentration in the reactor.

My future research has its origin in efforts to develop novel small-scale distributed production processes from stranded renewable energy, for production of carbon-neutral liquid fuels. In particular, I am interested in developing: devices that enable low-pressure ammonia synthesis via reactive-separation in a single vessel; processes for enhanced synthesis of renewable methanol for sustainable chemical industry; and novel supported adsorbents for facile mass and heat transfer for chemical synthesis.

Teaching Interests:

My philosophy of teaching was shaped and developed from my experience as a teaching assistant and guest lecturer in various undergraduate and graduate level courses. Whether I am teaching or learning, I am strongly committed and strive to be student-centered, engaging with students for active learning, and to support diversity as well as cooperation among everyone.

Prior Teaching Experience

My passion for academia and learning culminated into a longing to teach - a way to help others by imparting my knowledge or method of understanding. As a result, my aspirations to teach have pressed me to seek the most unique and challenging positions related to teaching. I have taught as a guest lecturer and teaching assistant in multiple undergraduate and graduate courses in the following subjects: Chemical Engineering Reactions, Mass Transfer, Senior Chemical Engineering Design, and Advanced Membrane Separations.

Teaching in academia requires continual development of one's teaching ability. This aptitude to constantly improve has been my motivation to stay informed not only with the latest practices in chemical engineering, but also the art of teaching. In regards to chemical engineering practices, I am continually interested in exploring and pursuing different strategies for teaching chemical engineering principles. Additionally, my enthusiasm propelled me to take part in the Senior Chemical Engineering Product Design course at the University of Minnesota. In this activity, we are piloting new undergraduate product design projects using chemical engineering principles familiar to students, but have unique design objectives that are innovative for students. The outcomes of our effort was presented at the 2016 American Society of Educational Engineering in New Orleans. In addition, a manuscript based on this work is currently under preparation that will explain the lessons we as faculty members learned, practices we recommend, and objectives that we accomplished.

My fervent desire for teaching extends beyond the classroom as I incorporate teaching principles in my daily life when working with family members, friends, and to people I have never met. [Add More]

Future Teaching Plans

I am prepared to teach the following courses:

- Reaction Kinetics and Reaction Engineering

- Chemical Engineering Design

- Mass Transfer

- Separation Processes

- Junior and Senior Chemical Engineering Lab

As a well-trained chemical engineer, I am more than capable and willing to teach other courses based on the department's needs. I am also personally interested in developing an elective course called "Novel Separation Processes for Sustainability," that I plan to offer to graduate students and upper level undergraduate students.

479779 Upgrading Lignocellulosic Biomass to Biofuels and Value-Added Materials for a Sustainable Future

Abstract:

Research Interests:

With the progressive depletion of fossil fuel resources and the necessity to combat climate change through the reduction of greenhouse gas emissions, there is an emergent need to identify alternative environmentally sustainable energy sources. A potential green energy source is derived from a variety of biomass and industrial waste streams, which can be converted to fungible fuels

Lignin, after cellulose, is the second most abundant biopolymer on Earth, comprising 30% of the organic carbon in the biosphere. Although lignin exhibits a high-energy content, it is typically destined for combustion in biorefineries due to its heterogenic and recalcitrant nature. However, cellulose and hemicellulose can be converted to renewable fuels and value-add products.

Currently, biorefineries under value the potential of using lignin for addressing the worldâ€™s need for fuel and materials. I am interested in genetically modifying bacteria and fungi that are able to utilize lignin and industrial waste streams as their sole carbohydrate source to increased lipid titers for biofuels, generate useful platform chemicals and value-added products for materials, and lignin degradation. I am also interested in developing techniques for enhancing the sustainability of biofuel production, efficiency of carbon conversion, and economic viability.

Teaching Interests:

Coupled with my research interests in sustainability, I am interested in teaching green engineering, bioprocess engineering, engineering economics, and life cycle analysis. I am also interested in applying an entrepreneurial aspect to process and product design.

474453 Highly Active and Durable Extended Surface Electrocatalysts

Abstract:

Highly Active and Durable Extended Surface Electrocatalysts

Shaun M. Alia, Staff Scientist, Chemical and Materials Sciences Center, National Renewable Energy Laboratory

Research Interests:

In the near future, fuel cell vehicles will become cost competitive with other drive train and fuel platforms. The commercial impact of proton exchange membrane fuel cells (PEMFCs), however, is still limited in part by catalyst cost, where the catalyst layer accounts for roughly half of fuel cell costs. Although platinum nanoparticles are commonly used due to high surface area and moderate activity, a reduction in catalyst loading or cost can improve PEMFC prospects and potentially accelerate its progress as a commercial product. In addition to catalyst cost, concerns have arisen due to the durability of the catalyst type.

Extended thin films have been studied as oxygen reduction catalysts since extended platinum surfaces generally offer site-specific activities an order of magnitude greater than nanoparticles. Extended nanostructures also avoid carbon as a support and may offer a long-term durability benefit. The catalyst type, however, has traditionally been limited by low surface area (10 20 m^2 g^{1}), and improvements are typically accomplished through an alloying effect improving site-specific activity. In this talk, progress in developing PEMFC catalysts through various deposition techniques will be presented. This work includes platinum-nickel nanowires, which have surface areas in excess of 90 m^2 g^{1} and mass activities an order of magnitude greater than platinum nanoparticles in half-cell testing.

Recently, aspects of fuel cell catalyst development have been applied to hydrogen-producing electrolysis technologies.

Although the cost of electrolyzers is currently dominated by feedstock costs, capital and catalyst costs will become increasingly vital as the power input shifts to renewable, intermittent sources of energy. Advancement in water splitting will be presented, including the development of extended iridium nanostructures in the oxygen evolution reaction. In cases, these materials produce activities an order of magnitude greater than commercial nanoparticles while offering a potentially significant durability advantage.

Teaching Interests:

My research is in electrochemistry and the development of nanomaterials for energy applications and my teaching experience was attained in chemistry and chemical engineering departments. The breadth of my experiences has provided preparation for teaching a variety of courses. While I am capable of teaching any course in the core curriculum, I am particularly interested in teaching kinetics and transport. Teaching undergraduate or graduate electives is also an interest, on areas aligned with my research, including electrochemistry as it relates to catalyst development. The inclusion of courses related to electrochemistry can inspire students to contribute to the advancement of the field and embraces the spirit of chemical engineering, an ever-evolving discipline at the forefront of technological advancement.

In previous experience as a teaching assistant, I emphasized student participation by having them periodically instruct each other. By running discussion sections and being a guest lecturer, experience was also attained in the practical aspects of teaching, including preparing lectures, lesson plans, and exams. As a faculty candidate, I am interested in promoting student involvement and providing practical applications to support concepts in the form of video or demonstrations. In addition to teaching core chemical engineering courses, I also am interested in bringing energy and nanomaterial related courses into the curriculum.

475433 Electrokinetic Transport in Porous Media for Energy and Environmental Applications

Abstract:

Research Interests: Electrokinetics, Transport Phenomena, Computational Science and Engineering, High Performance Computing, Random Materials

Teaching Interests: Fluid Mechanics, Thermodynamics, Numerical Analysis, Mathematical Physics

Understanding the transport phenomena and electrochemical processes in porous media is of great importance in many applications. Examples include energy applications such as in batteries and supercapacitors, electrochemical processes such as electro-catalysis, as well as industrial scale water desalination or geological scale fluid flow problems in underground reservoirs such as in oil recovery. Although these examples span several orders of magnitude in both length and time scales, the fundamental picture of electrokinetic phenomena is remarkably relevant in almost all of them. Here, we present some recent works on two aspects of electrokinetic transport in porous media.

First, we discuss the surface conduction phenomena and its effect on shortening the charging kinetics in porous electrodes. In most models of electrokinetic transport in porous media, the role of pore microstructure is often neglected and the transport is descried via homogenized, spatially averaged quantities. Inspired by detailed Direct Numerical Simulation (DNS) of the Poisson-Nernst-Planck (PNP) equations at the pore-scale, we show that the pore microstructure can have significant effect on the charging kinetics. Specifically, we observe that in the presence of large applied potentials, the surface conductivity of the Electric Double Layer (EDL) becomes considerably large, competing with that of the bulk of electrolyte. This allows for passage of large surface current through the EDL, which effectively acts to â€œshort-circuitâ€ the bulk and significantly enhance the charging kinetics. We believe this discovery could be utilized to engineer better preforming electrodes for high power density applications.

Second, we describe the effects of electric field on the viscous fingering instability in Hele-Shaw cells. Large scale flow problems in porous media, such as those encountered in underground oil reservoirs, are typically described via Darcyâ€™s law. However, it is well known that many underground rock formations contain surface groups and minerals that dissociate in the presence of water. Convection of these charges by the pressure driven flow can then set up streaming current and streaming potential that affects the flow. Furthermore, electric fields are sometimes used to enhance the oil recovery, e.g. by reducing the oil viscosity through electro-thermal heating or enhancing flow rate via electro-osmotic flow. Thus, a full description of fluid flow in the presence of electric field must also include the electrokinetic phenomena.

It is well known that the moving interface between two fluids in a porous medium becomes unstable if pushed by the less viscous fluid. Here, we report on the role of electrokinetic phenomena on stability of these viscous fronts in Hele-Shaw cells using analytic as well as numerical approaches. Interestingly, we find that instability could be suppressed if the right physical conditions are met or otherwise enhanced, leading to greater mixing of two fluids. These findings might be of interest to researchers working in the area of Electrically Enhanced Oil Recovery.

472746 Thermodynamic Properties and Transport of Fluids in Presence of Cross-Linked Polymers and Confinement: Modeling, Simulations and Experiments

Abstract:

Research Interests: Thermodynamics, Fluid Dynamics, Nanofluidics, Fuels, Molecular Simulations, Oil and Gas, CO2 sequestration

Teaching Interests: Thermodynamics, Fluid Dynamics, Reservoir Characterization and Engineering, Molecular Simulations

Thermodynamic properties of fluids change in presence of cross linked polymers and in confined spaces. Correct prediction of deviation in these properties is valuable to many industries including the oil and gas industry. In last decade, the shale play resources have played a key role in increasing oil production in the United States. However, the average hydrocarbon recovery factors from many of these formations are usually less than 10%. There is a considerable interest in understanding the production of liquids from shale rock plays such as the Eagle Ford Play in Texas. The sizes of pores in shales holding the oil are believed to be of the order of nano meters. The fluids present in such small nano pores have different properties compared to properties measured in bulk. Fluid saturation pressures at given temperatures – bubble points for oils and dew points for condensates in the nano pores are affected by the influence of pore walls in the vicinity of the fluid molecules. Moreover, kerogen, a cross linked polymer like organic matter present in the shale rocks, absorbs a part of oil present in these rocks. Such interactions with kerogen changes the pressure-volume-temperature (PVT) properties of oil in association with kerogen. Approach to bubble point or dew point influences the proportion of liquid or gas produced from a given well, and thus impacts the economic viability. Hence an accurate measure of saturation pressures is important. The traditional equation of state fails to add the complexity of pore wall interaction in calculating thermodynamics properties of confined fluids. It is important to understand how liquids are produced from ultra-low permeability rocks so that production rates and recovery can be optimized. Apart from the fluid properties, the understanding of transport of fluids through the organic and inorganic matrix in rocks is valuable for economic exploitation of oil and gas from shales.

Current research includes study of thermodynamics of solvent (oil) – kerogen (polymer) interaction, Molecular Dynamics simulations for assessment of hydrocarbon storage in kerogen, Monte Carlo simulations in Gibbs and Grand Canonical ensemble for understanding of flow and phase behavior of confined fluids, experiments conducted on isolated kerogen using Differential Scanning Calorimetry (DSC) to understand swelling of kerogen and associated affect on pressure-volume-properties of oil and fabrication of nano fluidics channels for the study of fluid flow in nano channels of tight rocks. In this paper we describe thermodynamic models, molecular simulations and experiments using DSC and in well characterized fabricated nano channels to understand effect of presence of cross linked polymer and confinement on flow and phase behavior of fluids. The results from Molecular Dynamics Simulations show that the kerogen interacts with hydrocarbons. The subsequent thermodynamic modeling demonstrate that kerogen-oil interaction suppresses the bubble point of oil in shales when compared to the bubble point of oil in the bulk. It was observed that for a type II kerogen with hydrogen index (HI) in range of 450-500 mg/g TOC, a suppression of over 1700 Psia in bubble point pressure (at 400 ^{o}F) was observed for in-situ oil from the original bubble point of 3950 Psia (at 400 ^{o}F) of produced 'free' oil in Eagle Ford Shale oil play.

The possibility of enhanced oil recovery through CO2 sequestration is also studied in this research. The self diffusivities of CO_2 were found to be in order of 10^{-9} m^2/s in kerogen. The research demonstrates that CO_2 has more affinity for adsorption in kerogen like organic macromolecules compared to CH_4. The skill sets developed during this research include but are not limited to thermodynamic modeling, molecular simulations of cross-linked polymers and protein like complex macromolecules with different fluids and fabrication of nano – microfluidics devices. Such skills are easily transferrable to the fields of bio-medical engineering, polymer science, material engineering and semi-conductors.

475796 Modeling Multiphysics Transport Phenomena and Device Design for Solar-Fuel Technologies

Abstract:

2ndyear Postdoctoral Fellow

Research Motivation and Interests:

Imagine if we could take a ***leaf*** from nature's book to convert and store available sunlight, water and carbon dioxide in stable chemical bonds of hydrogen and hydrocarbon compounds. These solar-fuels could provide carbon-neutral alternatives to the fossil fuel dominated energy sector with enormous potential for large-scale storage and long-term transportation. Ingenious fundamental science and device-scale engineering advancements are needed to efficiently harness, convert and store solar energy along with favorable economic drivers to facilitate global technological adoption. My research interests and strengths are to develop mathematical models and high fidelity computational simulations closely connected with experimental investigations to design scalable, stable and cost-effective integrated solar energy conversion and storage devices. Specifically, I am interested in probing the complex interactions of the various physical phenomena – optical absorption, heat and mass transfer, heterogeneous chemical reactions and interfacial electron transport to optimize the materials-to-device scale performance of high- and low- temperature solar fuel generators.

Awards: Doctoral Dissertation Fellowship awarded by the Graduate School at the University of Minnesota for 175 Ph.D. candidates from a campus wide competition.

Postdoctoral Project: ***"Design of a Particle-Suspension Reactor for Photoelectrochemical Solar Water Splitting"***

Supervised by Dr. Adam Z. Weber, Lawrence Berkeley National Laboratory, and Prof. Shane Ardo, University of California, Irvine.

PhD Dissertation: ***"Transport and Chemical Phenomena in a Solar Reactor to Split Water and Carbon Dioxide to Produce Synthesis Gas"***

Supervised by Prof. Jane H. Davidson, Christenson Chair in Renewable Energy, University of Minnesota, Twin Cities

Research Interests:

My graduate and postdoctoral research work has helped me acquire an interdisciplinary background in the areas of thermal sciences and fluid mechanics, semiconductor device physics and electrochemistry. I have gained substantial expertise in developing numerical transport models using in-house codes as well as commercial softwares to simulate interdependent transport processes for device-scale performance optimization. I am adept at parallelizing computational codes to implement them on high performance computers (HPCs), which is especially beneficial for three-dimensional and transient problems. Beyond continuum modeling approaches, I am also experienced with Monte Carlo simulations for radiative transport and meso-scale Lattice-Boltzmann approaches to simulate sedimentation of bacterial swimmers. As part of the reactor design team at the University of Minnesota, we designed and tested a 4 kW, high-temperature solar reactor integrated with gas phase heat recovery to operate at 1500 °C. The reactor implements an isothermal ceria redox cycle to split water and carbon dioxide to produce synthesis gas. The prototype development and reactor operation has given me an opportunity to view a heat and mass transfer, reaction-engineering problem from the eyes of materials scientists and structural mechanics researchers. In my current research, I am trying to understand the intricacies involved in semiconductor physics and interfacial electrochemical reactions in particulate suspensions. All my research projects so far have had a significant experimental and prototyping component, which has taught me the value of exploiting combined numerical-experimental approaches to connect the dots in multi-physics problems with varying length and time scales. This experience has also provided me with a versatile knowledge of experimental toolkits – thermocouple instrumentation and control, gas analysis equipment (RLGA and mass spectrometer), UV-VIS spectrophotometer, dynamic light scattering and cyclic voltammetry, to name a few.

Teaching Interests:

I have been a Teaching Assistant for over three years of my graduate study in undergraduate physics (statics and mechanics, electricity and magnetism) and mechanical engineering core courses (fluid mechanics, heat transfer, thermodynamics and measurements laboratory). I have particularly benefited from my experiences of running laboratory classes that involved designing experimental test stations and control devices for teaching fundamental transport phenomena concepts. I've also had the opportunity to mentor graduate students in my research lab to train them on performing numerical simulations and have worked closely with them to write and orally present scientific results. Outside of professional work, I strive to encourage younger students' participation in science and mathematics by volunteering in high schools to help students with basic algebra, geometry and physical sciences.

Future Direction:

As faculty I would like to apply my expertise in transport modeling for designing solar-fuel generators and other energy conversion systems to further the fundamental understanding of the interactions between various physical phenomena, and use this knowledge to offer device-scale engineering solutions. I hope to integrate my experiences from my PhD and postdoctoral work to more closely investigate full-spectrum utilization and heat management in integrated photoelectrochemical devices subjected to concentrated solar radiation. Concentrated solar power provides the benefit of higher power input per unit area of the device but also causes excess heating that could deteriorate the performance of the semiconductor materials. However, intelligent material selection and device design could potentially redirect the ultraviolet and visible portions of the solar spectrum towards water and carbon-dioxide splitting and channel the high-quality heat in the infrared wavelengths for heat storage.

Besides renewable solar fuel applications, I am also interested in solar gasification of biomass and other-waste materials, redox flow batteries, water treatment and pollution abatement with photocatalysts, thermal energy storage in phase change materials and modeling heat and air flow to optimize building energy requirements. I envision that developing numerical models of varying complexity and length scales along with carefully designed experiments can result in optimized performances, irrespective of the end application.

Selected Publications

[1] **Bala Chandran R.**, and Davidson J. H., 2016, "Model of transport and chemical kinetics in a solar thermochemical reactor to split carbon dioxide," Chem. Eng. Sci., **146**, pp. 302–315.

[2] **Bala Chandran R.,** De Smith R. M., and Davidson J. H., 2015, "Model of an integrated solar thermochemical reactor/reticulated ceramic foam heat exchanger for gas-phase heat recovery," Int. J. Heat Mass Transf., **81**, pp. 404–414.

[3] Hathaway B. J., **Bala Chandran R.**, Sedler S., Thomas D., Gladen A., Chase T., and Davidson J. H., 2015, "Effect of Flow Rates on Operation of a Solar Thermochemical Reactor for Splitting CO_2 Via the Isothermal Ceria Redox Cycle," J. Sol. Energy Eng., **138**(1), p. 011007.

[4] **Bala Chandran R.,** Bader R., and Lipiński W., 2015, "Transient heat and mass transfer analysis in a porous ceria structure of a novel solar redox reactor," Int. J. Therm. Sci., **92**, pp. 138–149.

[5] Bader R., **Bala Chandran R.**, Venstrom L. J., Sedler S. J., Krenzke P. T., De Smith R. M., Banerjee A., Chase T. R., Davidson J. H., and Lipinski W., 2015, "Design of a Solar Reactor to Split CO_2 Via Isothermal Redox Cycling of Ceria," J. Sol. Energy Eng., **137**(3), p. 031007.

476218 Understanding and Optimizing Energy and Mass Transport in Porous Materials for Water, Energy, and Thermal Management Applications

Abstract:

Research Interests:

Porous materials are ubiquitous and vital in a wide range of applications. Transport of energy, mass, and charge to the large fluid/solid interfacial area provided by these materials often controls the behavior of such systems. My research examines these transport phenomena as well as the functionalization of the surfaces themselves to optimize performance in several applications including toxin removal from water, energy storage in electrochemical capacitors, and dissipation of extreme heat fluxes from next generation electronic systems.

Part of my ongoing research addresses the challenge of water purification via capacitive deionization (CDI), an emerging technology which removes ionic contaminants from water via electrostatic interactions with the porous electrodes of a capacitor. Specifically, we focus on inverted CDI approaches using surface functionalization of porous electrodes to produce enhanced affinity for a variety of toxins. The porous electrodes act as adsorbents with selectivity for certain toxins. These adsorbents may then be regenerated by the action of an applied electric potential, which drives off the charged toxin species (as compared to traditional CDI where ions are attracted to the electrode surface by an applied potential). We have shown the ability to remove common contaminants from impaired input water streams to levels well below safe drinking water limits and to later expel these contaminants from the porous media without the need for the attendant waste disposal and cost concerns of chemical regeneration. My future research aims toward developing novel materials for better selectivity of difficult to treat impurities and the design of cells to maximize throughput and efficiency.

Another application closely related to CDI is energy storage in electrochemical capacitors. Recent advances in electrode materials provide great potential to enhance the energy density of these systems based on increases in material capacitance and pseudo-capacitance. However, due to the resulting potential for electrolyte depletion based on intrinsic solubility limits or desired operational regimes, the response of these systems involves a complex interplay of electromigration and diffusive transport of electrolyte species throughout the porous structures of the electrodes and spacer. We have developed parametric transport models including electrolyte depletion, capable of accounting for variations in cell geometry, electrolyte composition, and electrode material capacitance, resistivity, and structure. I will extend these models to provide guidance for cell design, electrode material choice, and electrolyte composition leading to electrochemical capacitors capable of addressing arising demands in grid energy storage and electric vehicle applications.

Beyond their use in electrochemical systems, porous materials have great potential in thermal management systems, such as heat pipes. We have studied energy and mass transport in two-phase cooling systems allowing dissipation of extreme heat fluxes, which is key to enabling dramatic performance improvements in a wide variety of electronic components including microwave amplifiers, power transistors in inverters for renewable power generation and electric vehicle applications, and 3D integrated microelectronics. We have produced thin conformal porous copper films with precisely defined microstructure using templated electrodeposition techniques. These structures with high surface area and thermal conductivity support heat dissipation via boiling of fluxes in excess of 1200 W/cm^2 at superheats less than 10 K from hot spots in electronic devices. We have also integrated these films with extended surfaces formed from laser micromachined diamond to create heat sinks capable of extending this performance to large areas. In future, I will focus on improving the mechanistic understanding of two phase mass and energy transport during boiling in porous media and using this knowledge to design two-phase cooling systems employing porous media to solve current and emerging thermal management challenges.

Teaching Interests:

I am prepared to teach a variety of courses in the chemical and mechanical engineering and materials science curriculums. I would be excited to teach materials related courses as well as thermodynamics, heat and mass transfer, and fluid dynamics. I am also interested in developing a new graduate course or seminar providing an integrated treatment of fluid, thermal, charge, and chemical transport in porous media along with relevant surface science.

476141 Harnessing Micro-Scale Transport Processes for Advanced Healthcare, Personalized Medicine and Point of Care Diagnostics

Abstract:

Research Interests:

Recent outbreaks like Zika and Ebola highlights the challenges associated with pathogen diagnostics in the developing world. With the outbreak in Africa, and isolated cases on other continents, the need for an affordable, rapid and portable diagnostic solution has been repeatedly stressed and is one of the most critical issues confronting global health. Unfortunately, the current conventional PCR instrumentation needed to perform "gold standard" DNA-based diagnostic tests is bulky, slow, and expensive, making it unsuitable for resource limited settings in developing countries where dedicated laboratory facilities are not available. Advances in micro-fluidics and smartphone based technology has paved way for novel implementations of traditional molecular diagnostic techniques.

Recently, it has been shown that convective flow fields can be harnessed to perform PCR. My research led me to investigate the underlying physics of these thermally driven micro-scale Rayleigh-Bénard convective flows. This new understanding revealed how a subset of these flow fields exhibit chaotic advection that greatly accelerate temperature driven biochemical reactions such as PCR [1-4]. Having found the optimal design parameters for convective PCR, I coupled this novel isothermal setup with a versatile smartphone based detection unit and an integrated image analysis app [5-7]. This greatly simplifies the design, enabling us to integrate such biochemical analysis platforms with consumer-class quadcopter drones for rapid deployment of nucleic acid based diagnostics[8]. The ability to perform rapid in-flight assays with smartphone connectivity eliminates delays between sample collection and analysis so that test results can be delivered in minutes, suggesting new possibilities for drone-based systems to function in broader and more sophisticated roles beyond cargo transport and imaging.

The surprising interplay between reactions and micro-scale convective flows led me to consider adaptations beyond PCR. Specifically, we demonstrate that such flows, naturally established over a broad range of hydrothermally relevant pore sizes, function as highly efficient conveyors to continually shuttle molecular precursors from the bulk fluid to targeted locations on the solid boundaries, enabling greatly accelerated chemical synthesis. Insights from this study has the potential to provide a breakthrough in our understanding of the fundamental biochemical processes underlying the origin of life [9, 10]. Another aspect of my research involves the hydrodynamical interaction between micron sized particles in intricate flow fields. My computational fluid dynamics models have thrown light on how the micrluidic flow fields can be tune fluid-particle interactions[9, 11].

At Sandia National laboratories, I have continued the development novel molecular diagnotics techniques for point of care applications, which involves an innovative approach to pathogen detection via reverse transcription loop-mediated isothermal amplification (RT-LAMP). A new kinetic model for LAMP reactions is formulated in order to shine light on the underlying reaction pathways. The new insights gained from the model are then used for efficient primer design and determining optimal reaction parameters. The LAMP technique is then coupled with an extremely simple and user-friendly smartphone enabled detection platform capable of detecting bright, unambiguous fluorescence signals produced. Finally, I have made significant contribution to advance the development of "Smart traps" – a next generation of rugged yet inexpensive robots to perform autonomous field surveillance of arboviruses. These traps are able to analyze saliva from vector-borne mosquitoes with daily LAMP assays and transmit data between each other over a wireless XBEE network. The ability to deliver performance comparable to that of current generation systems while simultaneously providing an order of magnitude reduction in cost and turnaround time has the potential to greatly expand the use of nucleic acid based detection assays by moving them out of the laboratory and into settings where they are needed most.

Teaching Interests:

One of the most fascinating subjects I have ever come across would have to be fluid mechanics and transport processes as it is conceptually quite rich and at the same time hosts a tremendous opportunity to incorporate hands on learning experience. There is a need for innovative educational experiences that unify and reinforce fundamental principles at the interface between the physical, chemical, and life sciences. These experiences empower and excite students by helping them recognize how interdisciplinary knowledge can be applied to develop new products and technologies that benefit society. Microfluidics offers an incredibly versatile tool to address this need. In one of my educational paper[1, 12], we describe our efforts to create innovative hands-on activities that introduce chemical engineering students to molecular biology by challenging them to harness microscale natural convection phenomena to perform DNA replication via the polymerase chain reaction (PCR). Experimentally, we have constructed convective PCR stations incorporating a simple design for loading and mounting cylindrical microfluidic reactors between independently controlled thermal plates. A portable motion analysis microscope enables flow patterns inside the convective reactors to be directly visualized using fluorescent bead tracers. We have also developed a hands-on computational fluid dynamics (CFD) exercise based on modeling microscale thermal convection to identify optimal geometries for DNA replication. A cognitive assessment reveals that these activities strongly impact student learning in a positive way.

1. Priye, A., Y.A. Hassan, and V.M. Ugaz, *Education: DNA replication using microscale natural convection.* Lab on a chip, 2012.

2. Priye, A., Y.A. Hassan, and V.M. Ugaz, *Microscale chaotic advection enables robust convective DNA replication.* Analytical chemistry, 2013. **85**(21): p. 10536-10541.

3. Priye, A., et al., *Selecting 3D Chaotic flow states for Accelerated DNA replication in micro-scale*

convective PCR. 15th International Conference on Miniaturized Systems for Chemistry and Life Sciences, 2011. **15**.

4. Priye, A. and V.M. Ugaz, *Convective PCR Thermocycling with Smartphone-Based Detection: A Versatile Platform for Rapid, Inexpensive, and Robust Mobile Diagnostics*, in *Microfluidic Methods for Molecular Biology*. 2016, Springer International Publishing. p. 55-69.

5. Jensen, M., et al., *ANALYTICAL INSTRUMENT SYSTEMS*. 2016, US Patent 20,160,025,630.

6. Priye, A. and V. Ugaz, *PCR to Go*. 2014, Apple iOS appstore https://itunes.apple.com/us/app/pcr-to-go/id909227041?mt=8: Apple iOS appstore.

7. Priye, A. and V. Ugaz, *DNA-to-go: A portable smartphone-enabled PCR assay platform.* arXiv preprint arXiv:1606.02252, 2016.

8. Priye, A., et al., *Lab-on-a-drone: toward pinpoint deployment of smartphone-enabled nucleic acid-based diagnostics for mobile health care.* Analytical chemistry, 2016. **88**(9): p. 4651-4660.

9. Priye, A., *Micro-Scale Complex Flows Enables Robust DNA Replication, Enhanced Transport and Tunable Fluid-Particle Interactions.* 2015.

10. Priye, A., Y.A. Hassan, and V.M. Ugaz, *Thermally-targeted adsorption and enrichment in microscale hydrothermal pore environments.* 17th International Conference on Miniaturized Systems for Chemistry and Life Sciences, 2013. **17**.

11. Priye, A. and W.H. Marlow, *Computations of Lifshitz–van der Waals interaction energies between irregular particles and surfaces at all separations for resuspension modelling.* Journal of Physics D: Applied Physics, 2013. **46**(42): p. 425306.

12. Ugaz, V.M., A. Priye, and Y.A. Hassan, *DNA to Go: A Do-it-Yourself PCR Thermocycler Lab.* American Society for Engineering Education, 2012.

476145 Leveraging Systems-Based Assessment to Understand Research and Development Needs for Emerging Energy Processes

Abstract:

The evaluation of an emerging energy process is often driven by the development of models which utilize process science and engineering fundamentals. The objective of process models is to facilitate an understanding of research and development opportunities, and to prepare a framework for analyzing experimental information. The evaluation scenarios are often characterized by limited data availability owing to their exploratory nature, and leveraging synergies from analogous processes is often valuable in formulating the first steps. In research and development organizations these activities are often categorized as research guidance, process evaluation and planning, or as exploratory research.

Research Interests:

Conceptual process design in combination with techno-economic and infrastructure integration analysis is a robust methodology which helps in understanding critical initiatives for fostering research and development. At the National Renewable Energy Laboratory's National Bioenergy Center, I have worked on the development of process models catering to various bioenergy processes of interest enabled by funding from the U.S. Department of Energy's Bioenergy Technologies Office. These include co-conversion of natural gas and biomass to fuels, catalytic fast pyrolysis of biomass, co-processing of petroleum refinery intermediates and bio-based feedstocks in hydroprocessing, and blending of petroleum and bio-based intermediates. These efforts are significantly important as the bioenergy industry is focusing on the near-term development of conversion pathways from biomass feedstocks to infrastructure-compatible fuel blendstocks. One of the inspirations to contribute towards on-going infrastructure integration analysis efforts emerged from my industrial research and development experience in India (Bharat Petroleum Corporation Limited).

My doctoral research at the University of Utah was focused on the development of conceptual process design principles for a 200 kWth process development unit for Chemical-looping with Uncoupling (CLOU). CLOU is a high temperature gas-solid fluidized bed process for solid fuel combustion to facilitate carbon capture utilization and sequestration (CCUS). The results of my research served as one of the foundational steps in facilitating an understanding of CLOU process fundamentals. The research was funded through the U.S. Department of Energy's Office of Fossil Energy (under Award No. DE-NT0005015).

Teaching Interests:

In addition to my interests in research in energy and sustainable transportation (bioenergy, petroleum and solid fuels), I am highly interested in chemical engineering education. During my post-doctoral research at NREL, I have mentored students, and was awarded an Outstanding Mentor Award by the U.S. Department of Energy, Office of Science's Science Undergraduate Laboratory Internships (SULI) program in 2014. During my doctoral degree, I had served as a Teaching Assistant Scholar towards developing a framework for undergraduate students titled "Inspirations for Innovation" – (http://ctle.utah.edu/tas/projects/asad-sahir.php) with a perspective on product and process design, and have mentored undergraduate students for research. I have also organized workshops at the University of Utah namely "Exploring the Human Dimension as a Teaching Assistant (TA) in the STEM Classroom" in 2013 and "International Graduate Learning, Teaching and Research (IGLTR) Workshop" in 2012.

This poster is a showcase of my research and academic interests which are driven by formulating an understanding the initial knowledge deficiencies associated with the design of an emerging energy process, and subsequently developing bridges through process modeling and chemical engineering principles. As a prospective faculty candidate, I intend to leverage my unique research experience at a national laboratory, industry and during my doctoral degree to help students appreciate core chemical engineering principles, and subsequently inspire and enable them to chart out a course in energy innovation for their community.

478649 Nanoscale Materials for Energy Storage and Conversion

Abstract:

Research Interests:

High efficiency energy storage and conversion are among the main challenges facing governments and industries that supply the energy needs of modern societies. Progress in the development and manufacturing of new low-dimensional and nanostructured materials, which can efficiently store and convert energy, may help us overcome these challenges – if we can design them correctly. In this regard, looking into the molecular structure and dynamics of new materials and characterizing their properties for such applications is an advantage of atomistic modeling.

My research goal is to adopt and utilize molecular modeling-based methods to characterize and predict the properties/behavior of novel materials/systems, and ultimately improve the efficiency of energy storage and conversion systems. I propose to create a multidisciplinary research program focusing on characterizing and predicting thermo-physical, transport and thermo-electrical properties of newly developed materials for energy applications.

Postdoctoral Project: "*Thermal transport in Metal Organic Frameworks (MOFs)*"

Under supervision of Professor C. Wilmer and Professor A. McGaughey

Visiting Scholar: "*Thermoelectric properties of single layer MoS_2 by using ab initio calculations*"

Under supervision of Professor S. Sinha

PhD Dissertation: "*Molecular-Level Modeling of Thermal Transport Mechanisms within Carbon Nanotube/Graphene-based Nanostructure-Enhanced Phase Change Materials*"

Under supervision of Professor J. M. Khodadadi and Professor P. Keblinski, RPI

MS Dissertation: "*Simulation of Turbulent Vortex Shedding Past a Square Cylinder near a Wall*"

Under supervision of Professor M. Raisee

BS Dissertation: "*Analysis of Turbulent Flow through a Channel using 2nd and 3rd Grade Fluids*"

Under supervision of Professor G. Atefi

Past and Current Research:

My academic career path has been a blend of several fields of science and engineering. My formal training was in simulation, which I have learned on many different physical scales: Density Functional Theory (femto), Molecular Dynamics (nano), and Computational Fluid Dynamics (continuum).

My PhD research project was devoted to non-continuum-based modeling of the effects of adding nanoparticles to fluids/phase change materials (PCM) and the heat transport within the resulting nanoparticle suspensions/composites. I adopted the methodology within molecular dynamics simulation approach to calculate the thermal conductivity of multi-component systems. Using the established method and other advanced molecular dynamics-based methods, I studied heat transfer in nanofluids and nanostructure-enhanced PCM (NePCM). I found that among different reported mechanisms of heat transfer in nanofluids, only clustering of nanoparticles can result in high enhancement in thermal transport. For PCM (here, paraffin), introducing CNT and graphene sheets promotes aligning of the molecules in the direction parallel to the CNT axis and graphene surfaces, respectively and consequently leads to considerable enhancements in its thermal conductivity along those directions. Moreover, in collaboration with Dr. Jackson's group at Auburn University, by using molecular modeling, I found that presence of nanoparticles enhances the thin film elasto-hydrodynamic lubrication.

During my one-year stay at UIUC, by utilizing first principle and phonon calculations, I investigated the thermoelectric properties of single-layer molybdenum disulfide (SL-MoS_2) (a 2-D material) and found an anomalously large thermoelectric power factor. Moreover, I developed a new methodology in finding phonon relaxation times.

During the past two years at Pitt and CMU, I have been studying the mechanisms of heat transfer in Metal-organic frameworks (MOFs) during gas adsorption using molecular dynamics (MD) and grand canonical Monte Carlo (GCMC) simulations and spectral energy density (SED) calculations. All these research projects have led to important findings so far and have opened up room for new ideas to the nanoscale heat transfer community which call for more research work, some of which will be outlined in the following section.

Furthermore, over these years I have gained deep knowledge of various aspects of different modeling techniques including MD simulation, Density Functional Theory (DFT), Density Functional Perturbation Theory (DFPT), MC, SED, lattice dynamics, thermal conductivity calculations and quantum modeling that are suitable to be used in studying other interdisciplinary nanoscale research problems as well.

Future research plans

Project 1. Low-dimensional thermoelectric materials

Thermoelectric materials provide a direct way of converting waste heat and solar energy into electricity. For instance, they can be used to convert waste heat in vehicles into electricity. However, lack of a scalable source of thermoelectric materials having the required conversion efficiency has prevented commercializing these devices. One way to overcome the problem of low efficiency is to use low-dimensional materials. Recent developments in nanotechnology have made it possible to adjust the electrical and thermal properties of materials in such a way that they provide a higher amount of thermoelectric figure of merit.

I will devote one part of my research to study thermoelectric properties of 1- and 2-D materials (e.g., nanotubes, nanowires, single and few-layer materials) whose electronic structure and transport properties potentially allow designing more efficient thermoelectric systems. In this general research project, I will focus my research on three properties of materials required for evaluating their thermoelectric efficiency: (1) Seebeck coefficient, (2) electrical conductivity and (3) thermal conductivity. For evaluating the first two properties one needs information on electron relaxation times. I will use ab initio calculations to obtain the relaxation times for phonons and electrons, which are the parameters needed to evaluate the transport properties involved in characterizing the efficiency of thermoelectric materials.

Project 2. Thermal properties of metal-organic frameworks

Metal-organic frameworks (MOFs), porous materials with extremely high internal surface areas, can adsorb large amount of gases and can be used for natural gas storage (e.g. in passenger vehicles). However, one challenge with MOFs is the heat generated during gas adsorption which leads to high temperatures, and in turn, prevents gas adsorption and reduces storage capacity. In principle, designing systems that could dissipate heat more quickly can allow for faster filling times. However, designing porous materials with better thermal properties requires a detailed understanding of the atomic-scale heat transfer mechanisms, which is still lacking in the literature. I propose a research project aiming to further our understanding of heat transfer in these systems. In this project I will use the adopted state-of-the-art molecular modeling approaches including MD, GCMC and SED calculations to study these systems and predict the needed thermal properties.

Project 3. Thermal transport in nanostructure-enhanced energy storage systems

One of the most convenient ways to store thermal energy is using phase change materials (PCM). PCM offer their sizeable latent heat for storing thermal energy at a constant temperature. However, a weakness of such materials is their relatively low thermal conductivity, which strongly suppresses the energy charge/discharge rates. A challenging option is to suspend highly-conductive particles into PCM, which produces "free-form" mixtures/composites5. This challenging problem is one of my research interests on which I will assign one part of my research group to work. I will use non-equilibrium MD method to characterize the required thermal properties of PCM composites. MD is an excellent candidate for handling the length and time scales that are expected in this problem.

Teaching Interests:

For my future academic career, I am prepared to teach general undergraduate and advanced graduate level courses in chemical Engineering, mechanical engineering and materials science, such as heat transfer, fluid mechanics and thermodynamics. Moreover, I would also like to teach interdisciplinary courses such as "statistical physics" and "transport phenomena".

I believe that with the recent growing demands for smaller scale technologies (e.g. nanotechnology), leading engineering departments need to develop courses related to these areas. During the past years dealing with projects in nanoscience, I have always been thinking of developing a course or courses that can effectively address both basic physics and techniques required in the burgeoning areas for engineering students. I think "nanoscale transport phenomena" and "molecular modeling" could be two of such new proper courses within the department of mechanical engineering.

Selected Publications:

H. Babaei, A. J. H. McGaughey and C. E. Wilmer, *Effect of pore size and shape on the thermal conductivity of metal-organic frameworks,* Chemical Science (2016). DOI: 10.1039/c6sc03704f.

H. Babaei and C. E. Wilmer, *Mechanisms of Heat Transfer in Porous Crystals Containing Adsorbed Gases: Applications to Metal-Organic Frameworks,* Physical Review Letters **116**, 025902 (2016). **PRL Editors' Suggestion**.

H. Babaei, J. M. Khodadadi, and S. Sinha, *Large theoretical thermoelectric power factor of suspended single-layer* MoS_2, Applied Physics Letter **105**, 193901 (2014).

H. Ghaednia, **H. Babaei**, R. L. Jackson, M. J. Bozack, J. M. Khodadadi, *The effect of nanoparticles on thin film elasto-hydrodynamic lubrication*, Applied Physics Letters, **103**(26), 263111 (2013). ***The first two authors contributed equally to this work***.

H. Babaei, P. Keblinski, and J. M. Khodadadi, *Improvement in thermal conductivity of paraffin by adding high aspect-ratio carbon-based nano-fillers,* Physics Letters A, **377**, 1358–1361 (2013).

H. Babaei, P. Keblinski, and J. M. Khodadadi, *A Proof for Insignificant Effect of Brownian Motion-Induced Micro-Convection on Thermal Conductivity of Nanofluids by Utilizing Molecular Dynamics Simulations,* J. Appl. Phys, **113**, 084302 (2013).

J. M. Khodadadi, Liwu Fan and **Hasan Babaei,** *Thermal Conductivity Enhancement of Nanostructure-Based Colloidal Suspensions Utilized as Phase Change Materials for Thermal Energy Storage: A Review,* Renewable & Sustainable Energy Reviews, **24**, 418 (2013).

H. Babaei, P. Keblinski, and J. M. Khodadadi, *Thermal conductivity enhancement of paraffin by increasing the alignment of molecules through adding CNT/Graphene,* International Journal of Heat and Mass Transfer, **58**, 209 (2013).

H. Babaei, P. Keblinski, and J. M. Khodadadi, *Equilibrium Molecular Dynamics Determination of Thermal Conductivity for Multi-Component Systems,* J. Appl. Phys. **112**, 054310 (2012).

M. Raisee, A. Jafari, **H. Babaei**, H. Iacovides, *Two-Dimensional Prediction of Time-Dependent, Turbulent Flow around a Square Cylinder Confined in a Channel*, International Journal for Numerical Methods in Fluids, **62**, 1232 (2010).

Successful Proposals: ACS Petroleum Research Fund, 2015-2018; AL-EPSCoR GRSP Fellowship 2010-2014 ($75,000)

465282 Why John Is Right Not to Lecture

Abstract:

The learning styles of the Millennials are quite different than the previous generations, e.g., the Xers, who relied on the lecture-inclass problem-home problem format.. This presentation will discuss some of the characteristics of the Millennials and how the materials that Professor John Falconer and his team at the University of Colorado have developed can better address the Millennials learning styles.

474307 Utilizing Classroom Management Software to Handle Large Classes

Abstract:

A major problem facing instructors of large classes is handling the accompanying increased organizational details and at the same time, trying to be as accessible to the students as possible. This encompasses keeping a grade book, entering grades, and forming and grading group assignments, among others. In addition, the instructor needs to hold office hours, respond to emails, and present reviews. One method that we have adopted with great success is incorporating little-used features of a classroom or learning management system to not only save time, but also to offer our students the same atmosphere and learning environment provided in smaller classes. This is particularly important in the lower-level core courses where faculty availability and classroom resources are often essential for student success.

A classroom management system is an on-line compilation often used solely for keeping grades. There are a number of systems that are currently employed by different universities. Those used by and/or considered by the University of Colorado include Desire2Learn, Blackboard and WebCT. Currently we are working with Desire2Learn and the methods described in this presentation have been used in that system. However classroom management systems are similar enough that these methods can easily be incorporated into others. In fact, we developed many of these practices in one system and easily transferred them to our current system.

Based on both research and experience, our preferred method of teaching is the use of the "flipped classroom" approach where information delivery is outside the classroom. Faculty adopting this approach modify their role from a front-of-the-class lecturer to a collaborative and cooperative facilitator of learning. The flipped classroom results in a shift from covering curriculum to a collaborative effort to master concepts and problem-solving techniques. However, many instructors find it difficult to change their teaching styles because of a lack of time or resources. Some faculty have expressed concerns that devoting class time to active learning might not provide students with sufficient examples of problem solving or sufficient time to introduce material. A well-organized classroom management system supports the flipped classroom approach, even for those large classes.

Components of this approach include on-line quizzes that test the students' knowledge of assigned reading material; screencasts that provide supplementary guidance on particularly difficult concepts; interactive simulations that promote understanding by making abstract concepts tangible; and ConcepTests that are used in class to challenge the students' knowledge and generate discussion. Most of these materials can be found at www.learncheme.com.

The features that a classroom management system provides to insure that the instructor can retain this method of teaching for large classes without increasing the burden include:

- Modular class design with links for "information delivery"
- Group formation tools
- Dropbox, rubrics and grade items for homework assignments
- Clicker integration
- Class email
- On-line office hours capability

In this presentation, we will describe both the implementation and effectiveness of these integrated tools within an active learning environment.

466087 Enabling Numerical Analysis Calculations within Educational Materials Delivered Via PCs and Smartphones

Abstract:

A major educational project called LearnChemE has been underway at the University of Colorado during the last 10 years under the direction of Prof. John L. Falconer involving Chemical Engineering and associated engineering disciplines (www.learncheme.com). The group of faculty members staff involved in this NSF supported project has recently received the 2015 HImmelblau Award of the American Institute of Chemical Engineers for new and novel contributions to computer aids for chemical engineering education. Integration of numerical problem solving has been utilized in many of the concept tests, screencasts and simulations.

User-friendly PC and Android software enabling general numerical problem solving has been under continuing development with PC software first introduced in 1984 under the name of Polymath with the efforts of the first two authors of this paper (www.polymath-software.com). This effort was initially conducted at the University of Connecticut. This overall work received the Himmelblau Award in 2010. Current Polymath related software development continues for Windows PCs and recently for the Android OS is also underway. There is increasing cooperation between the LearnChemE and PolyMath efforts.

This paper will consider current and future opportunities of interaction between current LearnChemE and PolyMath Software groups. Initial considerations are outlined here utilizing current LearnChemE educational materials and available PolyMath software. Additional future possible uses of newly developing software will also be discussed. The application of both efforts to engineering education world-wide will be highlighted. It is also hoped that this work will also be of interest to those interested in developing STEM educational materials (Science, Technology, Engineering and Math) involving PC's and Android smartphones and tablets that can provide resources such as testing, screencasts and simulations that may involve numerical problem solving.

PC Example – Use of Materials from LearnChemE with PolyMath Windows Software

The LearnChemE website has a collection of Student Resources that has a listing for POLYMATH (http://www.learncheme.com/student-resources/polymath). This site provides Introductory Videos that will run on Windows PCs and Android Smartphones and Tablets. Additionally there are currently 13 Example Problems in Kinetics/Reactor Design and four Example Problems in Thermodynamics that will work under Windows and Android devices. Most of these have been created, developed and narrated by Prof. John Falconer.

The example problem in the Kinetics/Reactor Design area titled *Multiple Reactions in a CSTR* will be onsidered. The use of a split Windows screen will be demonstrated/discussed where the video will be available and executable on the left side and the PolyMath 6.2 software will be available for use on the right side Windows screen. The ability to easily enter the program to enter and solve the material balances for the CSTR will be demonstrated and the advantages of this approach will be discussed.

An additional student challenge to the original problem is will be modify the model for the steady-state problem to accountl for the dynamics of the startup of the CSTR system to steady state. This will use the differential equation solving capabilities of PolyMath. Steady-state and the unsteady-state results will be discussed in this presentation/paper. The option to automatically generate Excel spreadsheets and MATLAB m-files for these two cases will be briefly mentioned.

Smartphone Example – Use of Materials from LearnChemE with PolyMathLite Android Software

The LearnChemE Screencast videos can be used directly on Smartphones. However, PolyMath 6.2 runs under Windows. Therefore Polymath Software has created a PolyMathLite app that has most of the capabilities of PolyMath 6.2, and executes directly on most Android smartphone and tablets. The only major changes in the problem solution code from that used in the PCs is the use of vectors to input data and some special funcions to deal with the various types of regressions.

The use of a simple smartphone with the PolyMathLite app will be demonstrated during the presentation by live use (or screen copies) of of a midsized phone display projection. The exact same set of equations as developed in the Screencast video of the *Multiple Reactions in a CSTR* problem will be ussed. The major difference between the smartphone and the PC is that one must switch between the paused video and the PolyMathLite problem entry as only one screen can be displayed at a time. The results between the PolyMathLite solution to the steady-stare CSTR and the PolyMath 6.2 solution are exactly the same. If fact, the algorithms in the Android Smartphone app and Windows PC coding are identical.

Use of the smartphone will be demonstrated in the modification of the code to the ordinary differential equations for the student challenge aspect to model the dynamic startup of the CSTR. The dynamics of the concentration within the CSTR are plotted with a separate plotting Android app or using the new Microsoft Excel Android app. All of this will be demonstrated and documented.

Summary

This presentation/paper identifies the continuing deliverables from two major educational projects coming from chemical engineering faculty and other academic departments that can combine to promote very positive enrichment activities for our students. Many additional possibilities can be considered for related subject areas. In this era of large class sizes, the educational experiences and capabilities of the individual students can continue to be enhanced. These resources from the two projects that have been discussed can also serve to encourage other STEM disciplines to provide similar educational resources, numerical calculations and other activities in support of classwork or individualized learning. E-textbooks can also be supported by PolyMath and/or PolyMathLite and other calculational software. As Android smartphones become very common among students in developing countries, the educational and problem solving capabilities and products of these two projects can be very important in international chemical engineering education.

466827 Spreadsheet Template to Aid in Creating Individualized Version of Objective Homework Problems

Abstract:

Individualized homework problems ensure that each student has at least a minimal level of understanding and problem solving ability. Online programs have been developed that provide individualized problems with guided feedback. However, these services are not available for all courses, generally come at an extra cost and may tend to isolate students, losing the beneficial aspects of group collaboration. The proposed approach strikes a balance between group collaboration and individual accountability and retains the instructor's ability to create novel homework problems.

An Excel spreadsheet template has been developed to aid instructors in developing individualized versions of problem statements. The template's functions and macros make it relatively easy to develop individualized versions of problems. The spreadsheet is also flexible enough to accommodate different types of objective questions. The problem developer decides which variables to alter and the range of those variables. The program then generates 200 random sets of data for those variables. These sets are then utilized in creating the student's version of the problem in the form of a separate spreadsheet. Therefore, each student receives a homework problem that is similar in structure but different in details from the other students.

The student version contains the problem's statement and a place for the student to respond. After submitting their responses, the student receives immediate, right or wrong, feedback. Multiple submissions can be made up to a pre-set limit. Once the limit is reached, the problem is set to base-case conditions and the student is provided with detailed feedback for the base-case problem. The student can then repeat the problem with a new set of variables.

The template was used to create dozens of problems that were integrated into the homework assignments

for an "off-semester" Material Balances course with eight students in the spring semester of 2016. The students both preferred (4.5/5.0) and thought they learned more (4.5/5.0) from the spreadsheet based problems over the traditional homework problems. The spreadsheet problems will be utilized in the fall 2016 Material Balance class at Trine University and compared to similar instructor generated and traditional textbook problems.

469334 How Students "Learn ChemE": The Impact of Departmental Culture on Student Success

Abstract:

Each engineering department has a unique departmental culture based on its history, faculty makeup, geography, faculty teaching styles, and a myriad of other factors. While some departments have a long history with a well-established culture and traditions, other departments that have been recently formed are faced with the challenge of creating a culture from scratch. The success of a department in creating and sustaining a desirable culture can have a significant impact on recruiting, retention, and general satisfaction of its undergraduate and graduate students. These types of activities also serve to shape students' expectations about what it means to enter the engineering profession and shapes how new students identify themselves as engineers. This presentation will discuss John Falconer's contribution to and influence on the chemical engineering culture, not only at the University of Colorado at Boulder, but also in chemical engineering departments world-wide.

470670 Rapid Analysis of Written Responses to Reveal Student Misconceptions in Thermodynamics

Abstract:

Traditionally, engineering courses utilize computations to evaluate student competence. Problem solving typically involves identification of the governing principles and proper application of engineering models. Many of todayâ€™s students have challenges applying models due to misconceptions about the underlying principles. Collection and assessment of short answer written (constructed) responses to conceptual questions can provide insights into why students have difficulties solving numerical problems. Such assessments are seldom used because the evaluation process is time consuming, especially when assessing the percentages of students with certain misconceptions. Accurate automated assessment of constructed responses can provide rapid feedback to instructors about the distribution of understanding/misconceptions even in large classes, providing opportunities for the instructor to reemphasize correct application of principles. In the future, such assessments may also be valuable for self-instruction. The Automated Analysis of Constructed Response (AACR) research group (www.msu.edu/~aacr) at Michigan State University is leading a multi-institutional collaboration (TUES 3 and WIDER funding) with several connected goals: 1) create a national web portal to provide automated analysis of constructed responses (AACR) to conceptual questions in a variety of STEM areas; 2) use the resulting reports to focus community collaborations between STEM education researchers and instructors; 3) transport AACR innovations through ongoing faculty professional development; 4) expand the range of STEM disciplines in which we pursue this research from biology into chemistry, chemical engineering, physics/astronomy, and statistics; 5) engage in ongoing project evaluation for continuous quality improvement; and 6) lay the foundation for sustainability.

This presentation provides a background on the progress in automated assessment of thermodynamics concepts, initially focusing on internal energy. The presentation will provide an overview of the criteria used for question development, the methods used to refine questions, the techniques applied for automated assessment, the reports provided to the instructor and how the reports are helpful in understanding the distribution of misconceptions. The presentation will describe opportunities for participation in the question development process and the status of our current set of questions.

474009 No More Lecturing! What John Falconer Has Done for Student Learning

Abstract:

This talk will provide the author's perspective of the value John Falconer has provided to enable faculty and students to engage in Active Learning of Chemical Engineering. We will discuss how such strategies and tools coincide with research from the learning sciences, as well as Prof. Falconer's contributions to our project, the AIChE Concept Warehouse. In true Falconerian spirit, it will be interactive.

464047 Active Learning in Chemical Engineering

Abstract:

Freeman et al. [1] analyzed 225 studies that compared STEM courses that used active learning to those that used traditional lecturing. The studentsâ€™ performance on exams and on the concept inventory increased by 0.47 standard deviation (when active learning was used, whereas students in the traditional lecture courses were 1.5 times more likely to fail. These studies provide convincing evidence that active learning in the classroom leads to improved student performance. During a lecture, students may think they understand a concept, and only realize they donâ€™t when they try to apply it. An important objective of active learning is that essentially all students are engaged, not just a small fraction.

We have attempted to provide the resources to make it easier for new faculty to use active learning while devoting less time to preparing course materials. Complete course packages have been developed for three chemical engineering courses: material and energy balances, thermodynamics, and kinetics/reactor design. These course packages utilize active learning in class with ConcepTests , interactive simulations, and problems that students work on in groups. They also include screencasts for students to watch before class, reading quizzes, exams, and assignments. The course packages are available in Microsoft OneNote so that they can be readily modified by the user. Access to the course packages is available at www.LearnChemE.com. 1. Freeman, S, Eddy, SL, McDonough, M, Smith, MK, Okoroafor, N, Jordt, H, Wenderoth, MP, Active Learning Increases Student Performance in Science, Engineering, and Mathematics. Proc. Nat. Acad. Sci. (2014) 111, 8410-8415.

468466 Engaging High School Students and Their Parents in STEM through Engineering Outreach

Abstract:

In 1985, women received 29% of the bachelorâ€™s degrees awarded in chemical engineering nationwide. By 2014, nearly thirty years later, this number has improved to 36%, but still remains a far cry away from representing the diversity of the general population. This disparity is stronger in rural communities, such as the large rural population surrounding Cornell University located in upstate New York. The Chemical and Biomolecular Engineering Womenâ€™s group at Cornell University holds a day-long outreach event annually to bring STEM to the rural community of New York. We target female high school students and their parents who are starting to think about their future and considering pursuing an undergraduate education. We invite up to fifty students and their parents who live within a 2+ hour radius of campus to visit Cornell and the chemical engineering department. These students are identified by councilors or teachers as students interested in science, and our task is to keep them interested and engage with them and their parents, to provide them with resources to plan a successful application for an undergraduate degree in engineering.

The event is broken into two parts. In the morning, students carry out three different hands-on lab modules. These lab modules evolve every year but focus on chemical engineering concepts such as process engineering to make cosmetics, DNA extraction, and properties of polymers such as silly putty rheology. Parents participate in workshops and seminars to understand the college application process, STEM careers, and college life. After lunch, students collaborate with their parents in a team lab, typically relating to surface chemistry and transport processes. This lab gives the parents to experience engineering themselves and see their daughters excelling in the activity. To quantify the impact our event has on the participants and garner feedback, we collect extensive survey data before and after the day-long event. Our first classes of students (2010-2012) have finally reached college-age and recent follow-up surveys tell us the longer term impact our event has had on their career paths, which has been substantial. In this talk, I will go over some of the past modules we designed and explain how the outreach event is organized. I will conclude with presenting some of the survey data that quantifies the impact our event has with promoting engineering with the surrounding rural community.

474291 Towards Inclusive Engineering Education and Practice: Examining Engineering Culture

Abstract:

The lack of diversity in engineering education and practice is a durable problem, and while a lot of effort and money has been directed towards "broadening participation," little progress has been made. It is clear that new framings and new strategies need to be designed and implemented in order to move the engineering profession toward greater inclusion of marginalized groups. Several years ago NSF began introducing a number of initiatives calling for generation of strategies that would confront the systems and structures that reproduce social inequality in engineering education and practice, including manifestations of racism, sexism, and ableism in engineering, and also classism, heteronormativity, ageism, and obstacles faced by Veterans and other non-traditional groups (NSF Dear Colleague Letter, 2014). This necessarily requires the focus of efforts to shift from the goal of achieving diversity within the profession to one of cultivating equity, inclusion and justice for engineering students and practitioners, as well as cultivating a culture that promotes just engineering practice.

However, it has been argued that particular ideologies central to engineering culture hinder authentic professional discussion and reflection on the political, social and ethical dimensions of science and technology. Examples of such ideologies include belief in meritocracy, the need for neutrality and objectivity of engineering work with regards to social and political concerns, and the technical/social dualism that allows differential valuation of engineering activities and knowledge in such a way that "technical" aspects are valued above "social" aspects. These underpinnings of engineering culture run deep, yet are often invisible to or unexamined by engineering students, educators and practitioners alike.

This study focuses on uncovering student perceptions with respect to these issues. It was conducted at a land-grant university with a large undergraduate engineering population. Incoming and soon-to-be graduating engineering students were invited to participate in focus groups designed to explore perceptions of engineering culture, especially in terms of the ideologies mentioned above. We also probed the influences of students' intersecting social identities (race, gender identity, sexual identity, ability status, among others) on their experiences in engineering programs. The results will be used to provide a baseline understanding of how students perceive engineering culture, but more importantly will be used as a means to initiate deliberate conversations about potentially problematic beliefs and norms embedded in engineering culture that contribute to the engineering profession's slow progress towards equity, inclusivity and justice.

This work is supported by the National Science Foundation pilot program *Revolutionizing Engineering Departments* (RED) that is aligned with the NSF Engineering (ENG) Directorate's multi-year initiative, the *Professional Formation of Engineers*, to create and support an innovative and inclusive engineering profession for the 21st Century.

474901 Growing a Multi-Cultural Engineering Workforce: An Aiche and Nsbe Strategic Partnership

Abstract:

The adequacy of the United States science and engineering workforce has been an ongoing concern for more than 60 years. And yet, the nationâ€™s colleges and universities are failing to produce a sufficient number of engineering graduates to meet the burgeoning demand for technical talent, a critical labor force gap that is exacerbated by a retiring Boomer technical workforce and changing demographics. This imbalance, particularly in key sectors threatens the future economic competitiveness of the U.S., our quality of life and our national security.

One obvious opportunity to increase the number of engineering graduates from every segment of the domestic population to maximum participation levels in all areas of this burgeoning sector. AIChE and the National Society of Black Engineers (NSBE) have answered the call by developing a strategic partnership to increase the number of graduates from underrepresented populations by supporting collegiate chapters to collaborate to improve academic outcomes of diverse chemical and biological engineering students in college.

This talk outlines the elements of the partnership, and suggests engagement strategies colleges of engineering can leverage NSBE and AIChE collegiate chapters to improve their ability to recruit, educate, retain and graduate engineers who are currently underrepresented among all graduates. This presentation also introduces a learning ecosystem that demonstrates how NSBE chapters are successful at fostering academic self-confidence, institutional integration, positive identity development and effective learning strategies.

474260 Incorporating Interdisciplinarity and Diversity to Maximize Â Engineering Learning Objectives

Abstract:

The student body of undergraduate engineering programs typically has a monolithic demographic. Women and underrepresented group still make up a fraction of engineering programs, which can result in feelings of isolation and lack of acceptance among peer groups. There is a desperate need for a more pluralistic learning experience within engineering classes. For example, many capstone design projects focus on projects that only require technical expertise in one engineering discipline. We hypothesize that multidisciplinary teams of engineers can team together to work on projects based in the arts. In our pilot study, a team of engineers partnered with ballerinas from the Dance Theatre of Harlem to create a technology that would assist dancer effectiveness. The team was comprised of 4 students, half of which were women and were a combination of biomedical engineering and computer science majors. The project sponsor, the Dance Theatre of Harlem, is a predominately African American dance troop. After almost 6 months, the multi-disciplinary engineering team, in collaboration with the ballerinas, developed a wearable technology and smart phone application that aids a dancer in training and injury prevention. The technical expertise required to develop such a technology necessitated undergraduate teams communicate effectively across technical disciplines, as well as engaging with a non-technical project sponsor to achieve a product that would be effective. This study provides best practices for creating these cross-disciplinary, diverse teams and presents success metrics that meet pre-determined learning objectives and ensures a successful project outcome.

474262 The Minority Faculty Forum: A Vehicle for Increased Diversity in Chemical Engineering

Abstract:

The link between diverse faculty in the classroom and a diverse student body is well documented. Yet, racial/ethnic minorities remain underrepresented among chemical engineering faculty and their alarmingly low numbers is a threat to training a diverse chemical engineering workforce. AIChE as a professional society is uniquely positioned to provide an infrastructure to support networking, mentoring, and other opportunities to attract and enhance the career development of minority faculty. To that end, the Minority Faculty Forum (MFF) of the Minority Affairs Committee (MAC) serves as a vehicle to engage and enhance the careers of existing and aspiring minority faculty. MFF was formed in 1995 as an outgrowth of informal gatherings of underrepresented minority faculty at annual meetings. It has evolved over time to become institutionalized as a subcommittee of MAC with regular activities to include networking events and the Janice Lumpkin Travel Grant. MFF plays a critical role in the mentoring and socialization of minority chemical engineering faculty and in advancing the mission and goals of AIChE as a diversity leader in linking academia, industry and the broader community.

465195 Impact of HBCUs on the Number of Minorities Receiving Degrees in Chemical Engineering

Abstract:

A goal of AIChE is to increase the diversity in the Institute by increasing the membership of underrepresented groups. A review of graduation data by the Minority Affairs Committee (MAC) revealed that African Americans receiving engineering degrees has remained at a level of 4.0 – 4.5% over the last several years. A recent study by ASEE also showed a decline in the amount of undergraduate degrees awarded to African American females. To increase diversity in the chemical engineering profession, the number of minorities earning degrees must increase.

Historically Black Colleges and Universities (HBCUs) have historically focused on the education of African Americans. Although the universities have historically served blacks, all ethnicities can enroll. Many HBCUs serve several minority groups including Hispanic communities. Six HBCUs have chemical engineering programs. Around 25 - 30% of American undergraduate degrees awarded in chemical engineering are conferred at the six HBCUs. This report analyzes the relevance of HBCUs on minorities – in particular African Americans – earning degrees and entering the chemical engineering profession.

473874 Steps Toward Improving the Experience of Lgbtq Students in Our Chemical Engineering Departments

Abstract:

The inclusion and retention of students who are part of a gender or sexual minority can involve some unique challenges for engineering faculty and department staff. However, these challenges are not well understood, and have only relatively recently became the subject of engineering education research.

Our LGBTQ students are often growing into their minority identities just as they begin to develop their engineering identities in our classrooms. They may feel forced to conceal their true selves, fearing professional penalties for speaking casually and honestly about their lives, as their non-LGBTQ peers and professors do. They may not themselves be comfortable with their orientation or gender identities and, while attempting to cope with the standard challenges of a chemical engineering degree, they may also be waging an intense internal struggle. This struggle may go on to result in their sudden loss of the love and support of their family and friends, or in hostility in our classrooms by peers or even faculty. Highly stressful experiences combined with an unwelcoming classroom environment have a negative impact on learning.

These challenges for the student are often met by an engineering culture that encourages hetero- and cis-normativity, and simultaneously poses topics of sex, orientation, and gender as taboo for a classroom or laboratory environment.Â In such an environment, it can be difficult to create a more inclusive chemical engineering department or even begin to address the hurdles faced by our LGBTQ students.

ASEE, with support from NSF, launched a Virtual Community of Practice that aims to (1) increase awareness of LGBTQ issues in engineering departments and classrooms (2) build knowledge and skills to disrupt discrimination and promote LGBTQ equality in engineering departments on college campuses and (3) to identify best practices for promoting LGBTQ equality in engineering. To advance these goals, twenty members of the VCP have created STEM-specific â€œSafe-Zoneâ€ (LGBTQ Ally-training) workshops for STEM educators to create more LGBTQ-inclusive classrooms and in departments. Through conference workshops at ASEE and AIChE and through webinars, about 400 engineering faculty have participated in the training to date. This presentation will discuss the results of our VCP to date, including best practices for creating an LGBTQ-inclusive learning environment in chemical engineering classrooms.

467918 Tapping into Diverse Student Assets to Enhance Design Framing Ability and Professional Identity Formation

Abstract:

Students entering engineering schools generally have a very limited understanding of what engineers do in their careers. Students often do not know the interesting work that lies ahead, hence they may drop out of engineering when they encounter the difficult math and science courses in their early years in college. This dropout rate disproportionately impacts those underrepresented in engineering because they may not have family members who are engineers and who can help them gain an appreciation for engineering careers. To address this problem, the Department of Chemical and Biological Engineering (CBE) at the University of New Mexico (UNM), which serves a very diverse population, is introducing 'Design Challenge Modules' in two core classes. A tailored approach for repeated exposure to engineering design practices that seeks to connect diverse student assets with engineering identity is presented in this study. Preliminary evidence for characterizing student assets was provided by means of surveys, videos, interviews, artifacts of design work, and measures of design framing ability in the first two core undergraduate chemical engineering courses at UNM. The research focuses on the engineering design experience by which students engage in a continuous habit of thought followed by action, scaffolding towards their professional identity. A design challenge was implemented whereby students participated in framing the problem and generating solutions to the task that was then assessed and open-coded using mixed methods research to track the framing ability.

Women were considerably more likely than men to view design as a co-evolutionary process, $t\ (122) = 2.69, p<0.01$. It was also found that first generation college students were significantly more likely to agree that design is a learning activity than their traditional peers, $t\ (113)= 2.50, p<0.05$. A comparison of urban and rural student population groups revealed that students from rural communities were more likely to frame the problem using chemical engineering principles. Together, these findings support the premise that constraints in engineering design can extend beyond the technical to student assets. Regression analysis was also used to assess variance in student design solutions. Students

who lack pre-college engineering design exposure and who rated their academic ability lower in introductory courses tended to generate solutions based on divergent problem framing if they viewed constraints as part of the creative endeavor rather than an obstacle. This finding suggests a critical need for engineering faculty to tap into student assets indiscriminately as undergraduates form their professional identity. Success in engineering is impacted by student background characteristics which in turn influence self-efficacy beliefs and their desire to persist along an engineering career pathway.

456662 Global Culture Exchange

Abstract:

At a public engineering school in the Midwest, the undergraduate students are frequently faced with their first introduction to people from other cultures and races. The current political climate and the recent events at the University of Missouri have polarized students and many have chosen to embrace their prejudices rather than learn to overcome them. At my university, the majority of the clashes and negative attitudes have been strongly tied to country of origin. After hearing studentsâ€™ harshly prejudicial comments about classmates and faculty and seeing these negative attitudes in evaluations of teammates on group projects, I developed a plan for a program to help our students see each other in a new way. The department quickly adopted this program.

The Global Culture Exchange program is a monthly program where a student or faculty member presents information about their home. This frequently includes a bit about the history of their country, the geography, and the educational system. Presenters usually include photos of their homes and families, favorite recreational activities, favorite music, and the food. After the presentation, the group is invited to a light lunch featuring foods from that area. The presenter is typically bombarded with questions during the lunch. In each case, the students that present have become, instantly, a more integrated part of the student body. This new attitude persists long after the presentation.

So far, presentations have been skipped the globe from Iran to Russia to Bangladesh to Honduras and more. At the request of the students, we plan to incorporate a few regions of the U.S. as well. The Global Culture Exchange program has been a useful tool to open up communications within the department and encourage the student to be more accepting of those who are different from them.

475976 Conversion Quantification in Epoxides Annealed afte Cationic Photopolymerization

Abstract:

Annealing, a baking process, is widely used in industry because of its ability to improve the properties of polymer films. This research looks into comparing the effects of annealing time and temperature on light cured films made from different epoxide monomers. Each film was held at a temperature below, at, or above the polymerâ€™s glass transition temperature for a specific time period. In addition to the various annealing temperatures and annealing times, films were annealed either immediately after polymerization, 24 hours later, or 48 hours post-polymerization. This time delay before annealing was done to determine whether dark cure occurs and how it affects the annealing process. Dark cure is commonly seen in cationically polymerizing systems, such as those containing epoxides, and can result in conversion increases in the absence of additional illumination. A differential scanning calorimeter was used to anneal the polymer, and changes in monomer conversion due to annealing were measured using Raman spectroscopy at each experimental condition. The influences of annealing on conversion for different epoxide monomers were then compared and the optimal annealing time and temperature for each monomer was determined.

Direct Encapsulation of Pluripotent Stem Cells in Methacrylated Gelatin to Form Engineered Cardiac Tissues

Jennifer Kaczmarek

Sara Head, Petra Kerscher, Dr. Elizabeth A. Lipke

Department of Chemical Engineering
Auburn University, Auburn, AL, USA

hiPSCs for Generation of Cardiomyocytes

Human induced pluripotent stem cells (hiPSCs) are:

- Patient-specific
- Differentiate into all three germ layers (ectoderm, endoderm, mesoderm)
- Able to efficiently differentiate into cardiomyocytes
- Able to be cultured in colonies *in vitro*

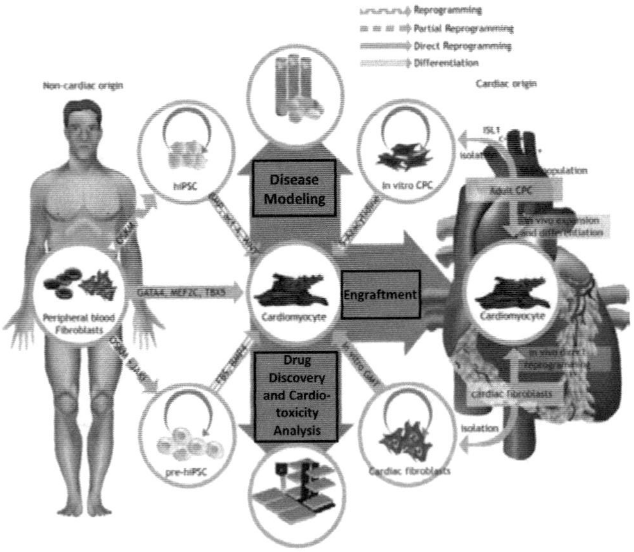

Burridge PW et al. Production of de novo cardiomyocytes: human pluripotent stem cell differentiation and direct reprogramming. Cell stem cell (2012)

hiPSCs for Generation of Cardiomyocytes

Human induced pluripotent stem cells (hiPSCs) are:

- Patient-specific
- Differentiate into all three germ layers (ectoderm, endoderm, mesoderm)
- Able to efficiently differentiate into cardiomyocytes
- Able to be cultured in colonies *in vitro*

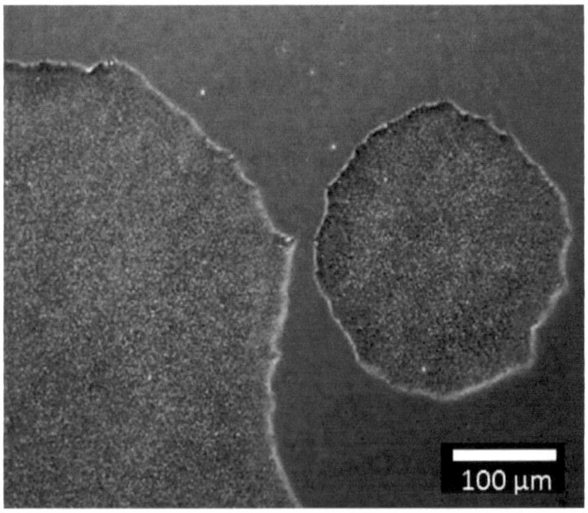

Lipke Lab
Auburn University

3D Biomimetic Materials to Mimic Developing Heart Microenvironment

Biomimetic materials provide a 3D microenvironment

Bioactivity of natural materials aids tissue development

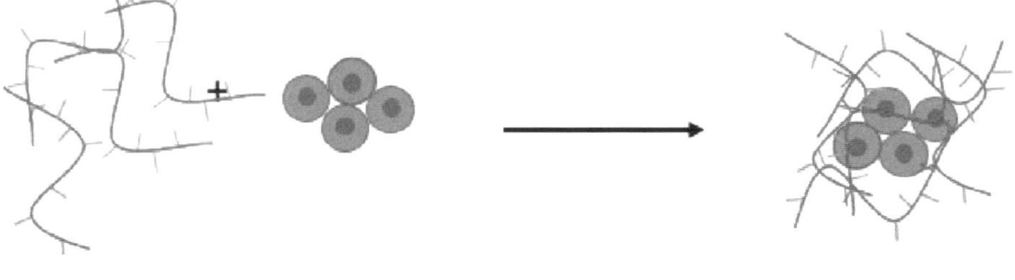

Lipke Lab
Auburn University

Engineered Cardiac Tissue: a 3D *in vitro* model

Multiple Cell-handling Step Approach:

Stem Cells → Cardiomyocytes → Singularized Cardiomyocytes → (+ Biomaterial) → Engineered Cardiac Tissue

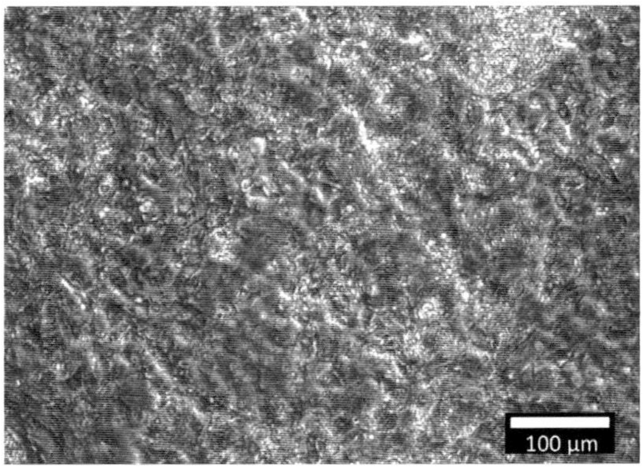

Lipke Lab
Auburn University

Engineered Cardiac Tissue: a 3D *in vitro* model

Multiple Cell-handling Step Approach:

Stem Cells → Cardiomyocytes → Singularized Cardiomyocytes → (+ Biomaterial) → Engineered Cardiac Tissue

Single Cell-Handling Approach:
Stem Cells + Biomaterial
↓
Developing human Engineered Cardiac Tissue (dhECT)

Lipke Lab
Auburn University

Biomimetic Materials Produce Contracting Cardiac Tissues

Polyethylene glycol (PEG) fibrinogen has previously been found to be a suitable biomimetic material for producing functional cardiac tissues.

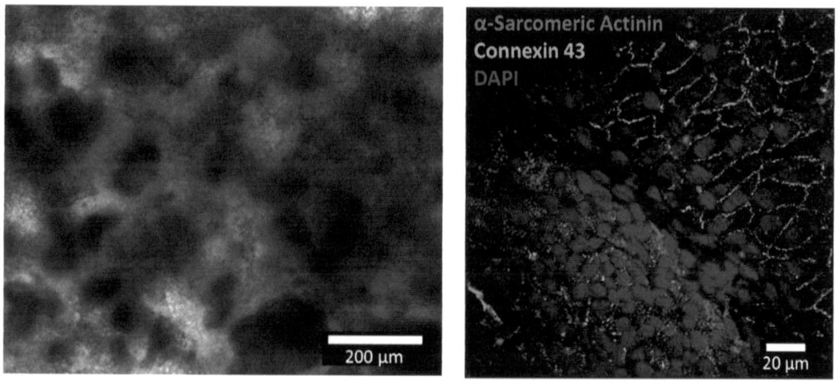

Kerscher, P.

Lipke Lab
Auburn University

Engineered Cardiac Tissue: a 3D *in vitro* model

Multiple Cell-handling Step Approach:

Stem Cells → Cardiomyocytes → Singularized Cardiomyocytes → (+ Biomaterial) → Engineered Cardiac Tissue

<u>Single Cell-Handling Approach:</u>

Stem Cells + Biomaterial

↓

GelMA developing human Engineered Cardiac Tissue (GEhECT)

Lipke Lab
Auburn University

Gelatin Methacrylate (GelMA)

- Natural polymer synthesized by addition of methacrylic anhydride (MA) to gelatin
- Gelatin is derived from collagen, which is naturally abundant in the heart.
- Degree of Methacrylation: 22%

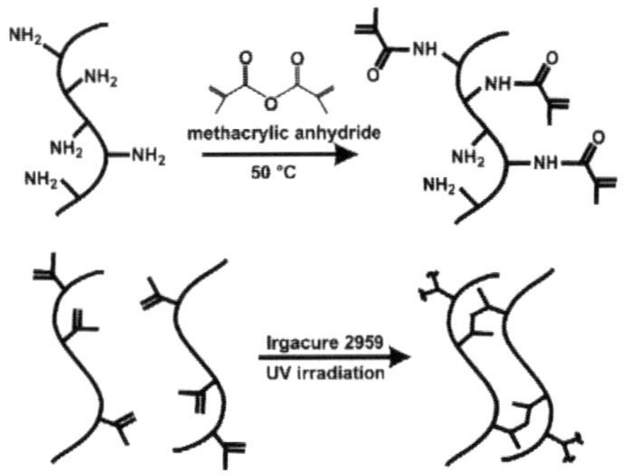

Nichol, J., et. al., (n.d.). Cell-laden Microengineered Gelatin Methacrylate Hydrogels. *Biomaterials*, (31),

Gelatin

GelMA

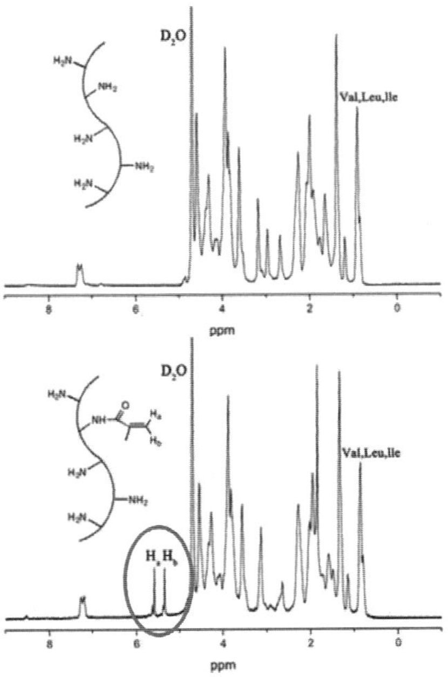

Literature NMR

http://pubs.rsc.org/en/content/articlehtml/2014/RA/C4RA02271H

Lipke Lab
Auburn University

Gelatin Methacrylate (GelMA)

- Natural polymer synthesized by addition of methacrylic anhydride (MA) to gelatin
- Gelatin is derived from collagen, which is naturally abundant in the heart.
- Degree of Methacrylation: 22%

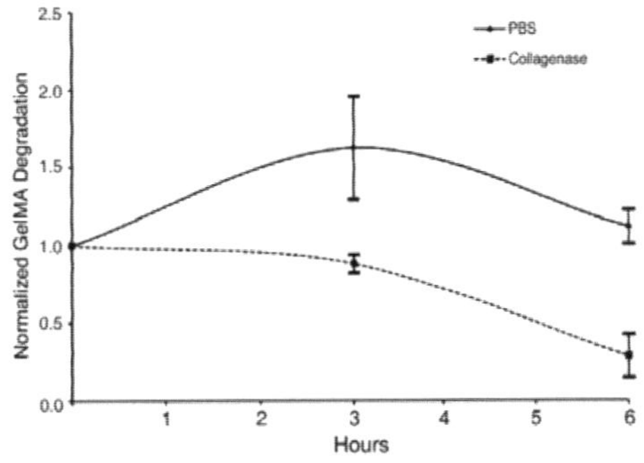

Gelatin

elMA

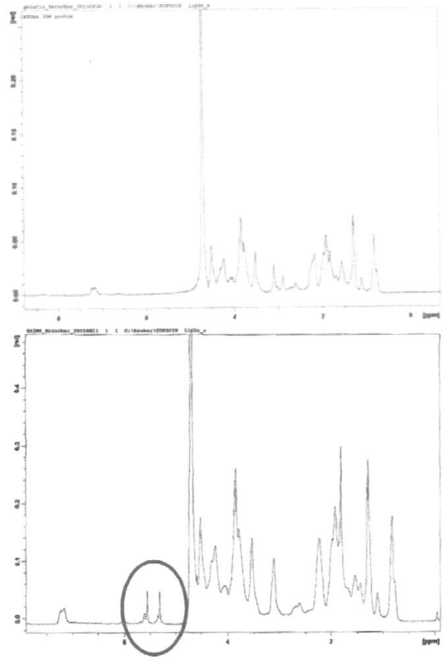

Observed NMR

Lipke Lab
Auburn University

Fabrication of GelMA developing human Engineered Cardiac Tissue

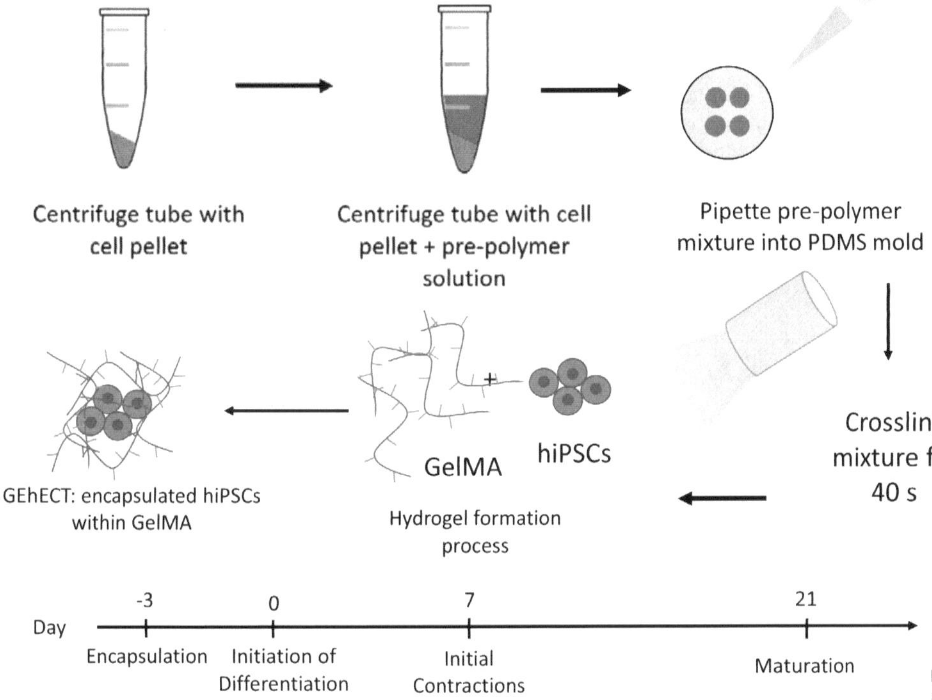

Mechanical Properties of GelMA

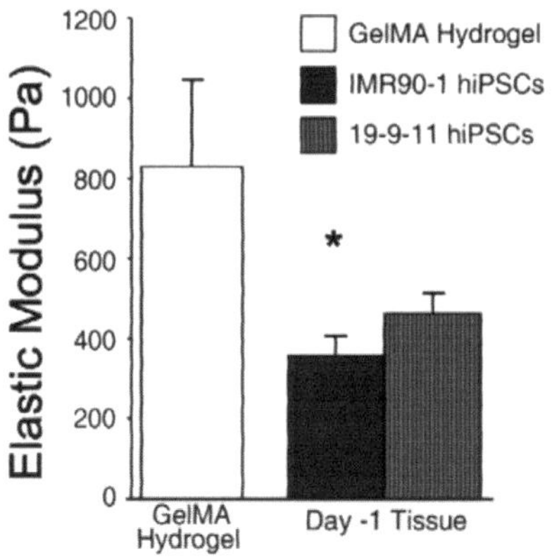

Elastic Modulus of GelMA With and Without hiPSCs

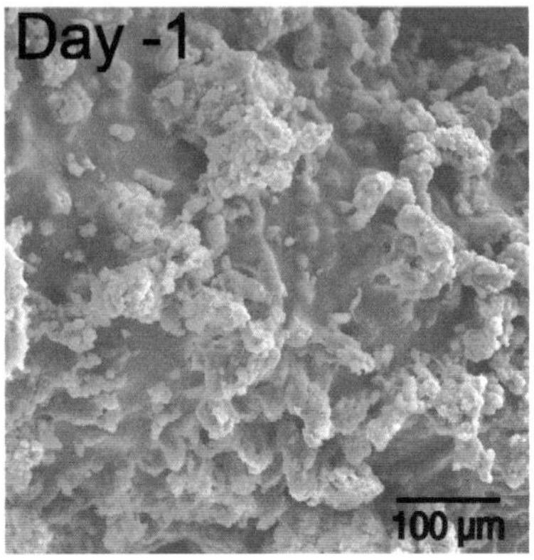

SEM of hiPSC-laden GelMA

Mean ± s.d.
ANOVA $P < 0.05$
* vs. GelMA hydrogel

Cells Remain Viable and Grow Following GelMA Encapsulation

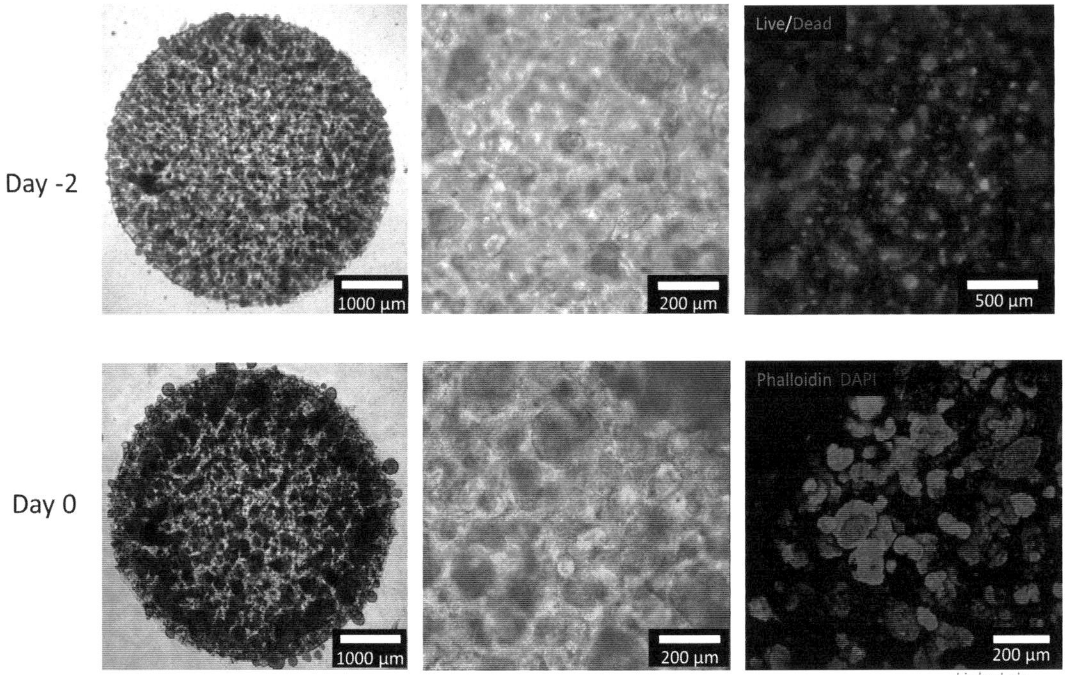

Lipke Lab
Auburn University

hiPSCs Differentiate and Form a Continuous Cardiac Tissue

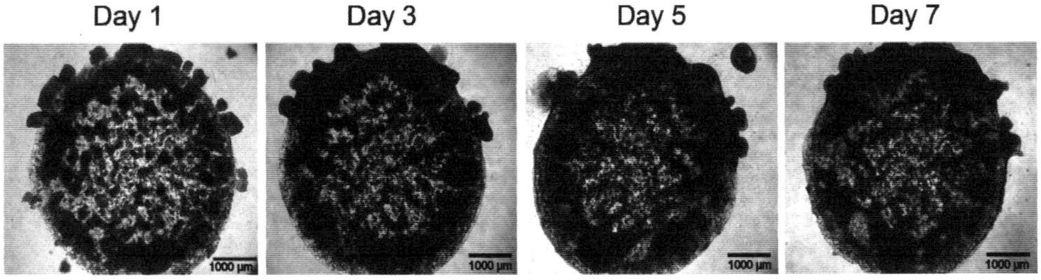

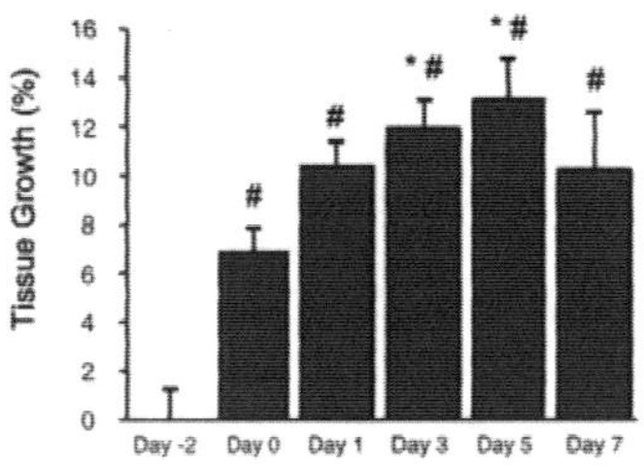

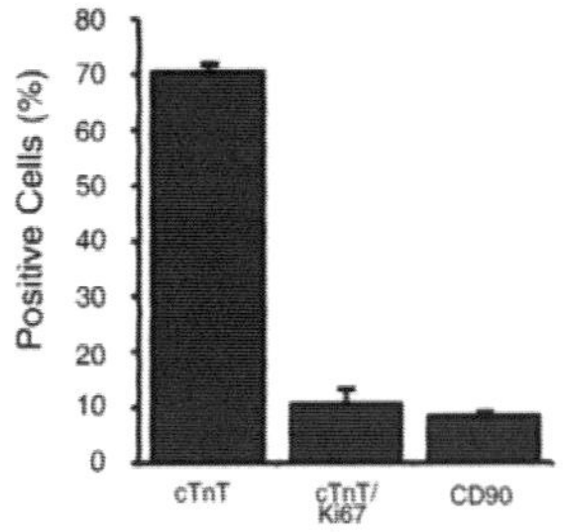

Mean ± s.d.
ANOVA P< 0.05
vs. Day -2
* vs. Day 0

Lipke Lab
Auburn University

Progression of Synchronous Contractions

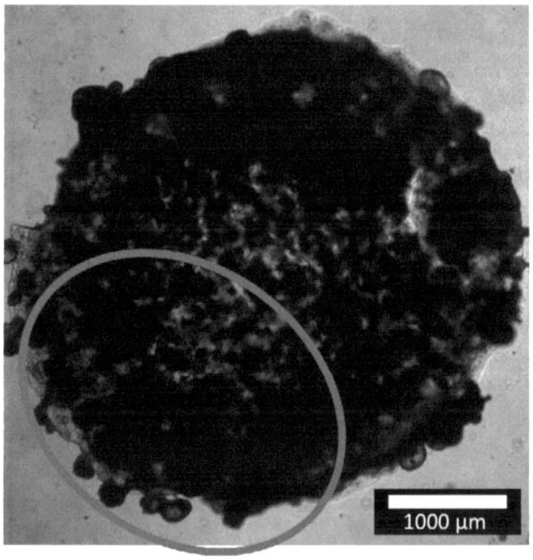

Day 10

Day 20

Lipke Lab
Auburn University

Progression of Synchronous Contractions

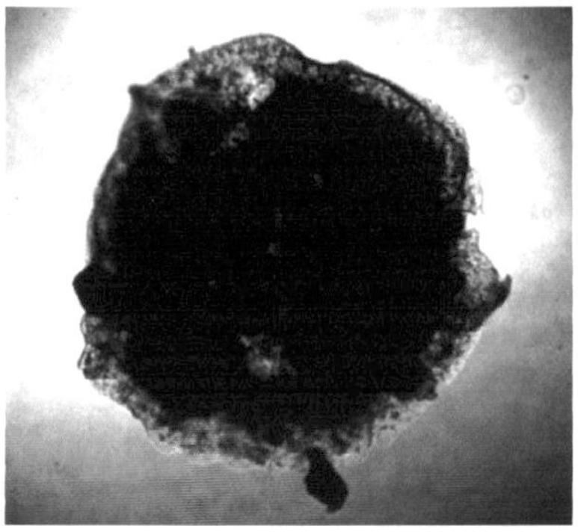

Day 40

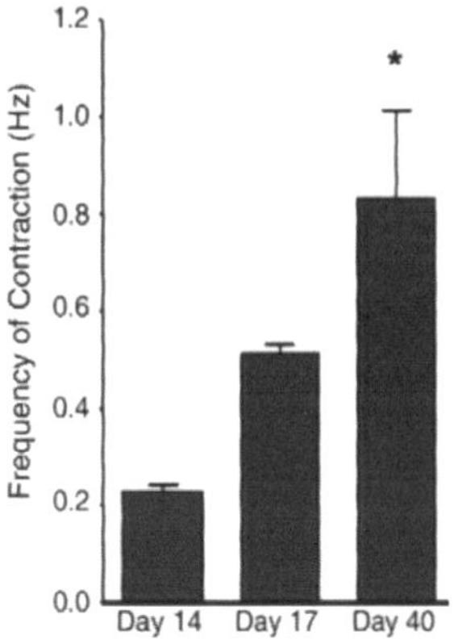

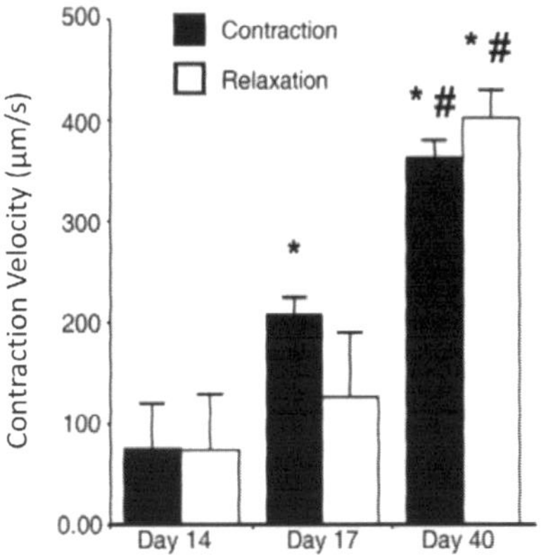

Mean ± s.d.
ANOVA P< 0.05
* vs. Day 14
# vs. Day 17

Lipke Lab
Auburn University

Analysis of Contraction Velocity in GEhECT

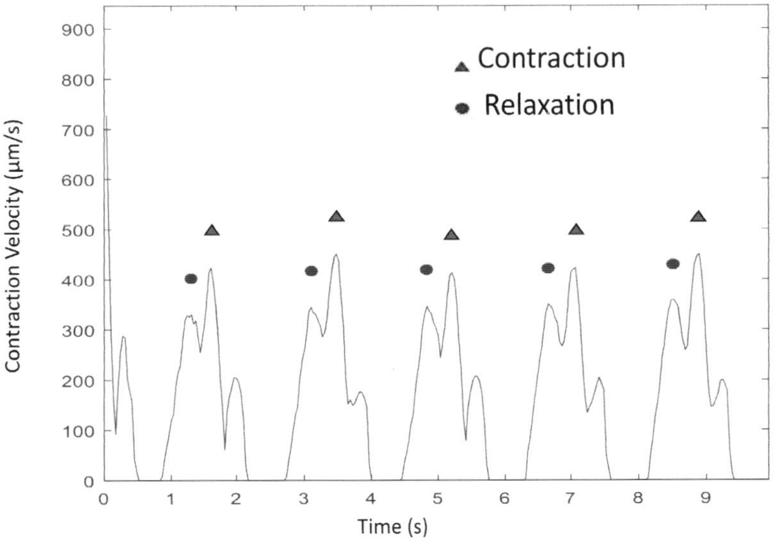

Huebsch, N., et al. (2015).

Lipke Lab
Auburn University

Microelectrode Array Analysis

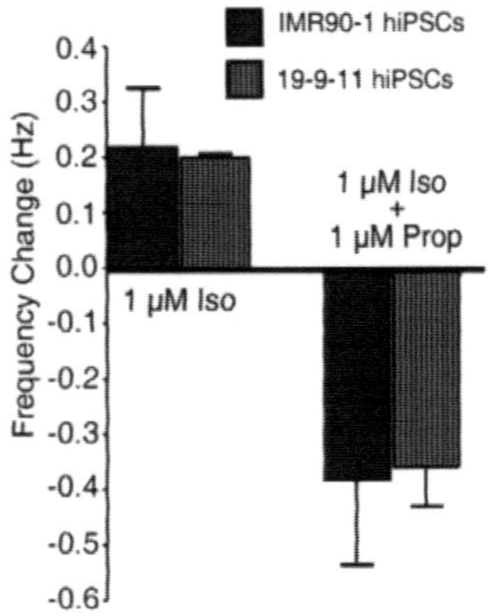

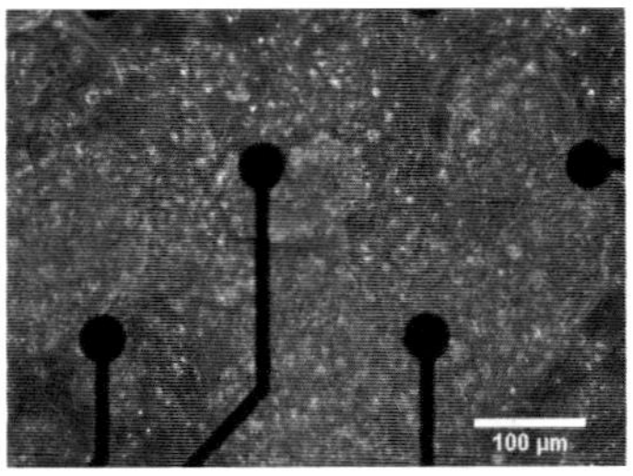

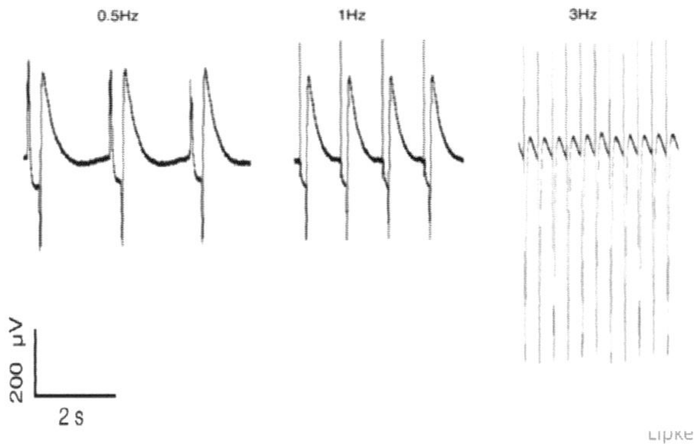

Lipke Lab
Auburn University

GelMA Tissues Show Cell Morphology Indicative of Cardiomyocytes

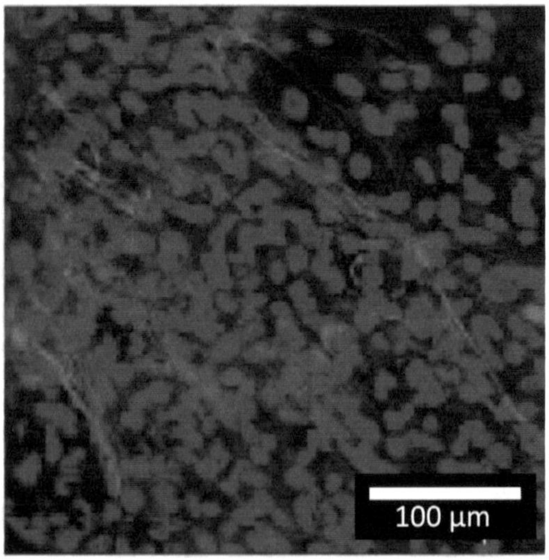

Day 0 Day 20

Phalloidin DAPI

Lipke Lab
Auburn University

GelMA Tissues Show Cellular Features Indicative of Cardiomyocytes

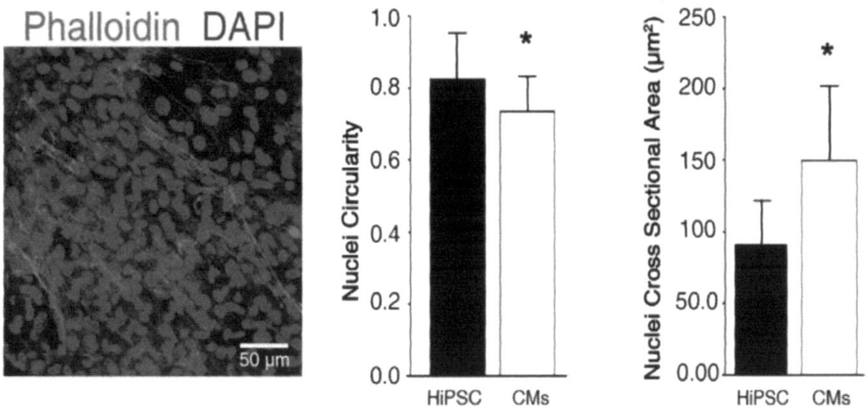

Mean ± s.d.
ANOVA P< 0.05
* vs. HiPSC

Lipke Lab
Auburn University

GelMA-dhECTs Display Cardiac Markers

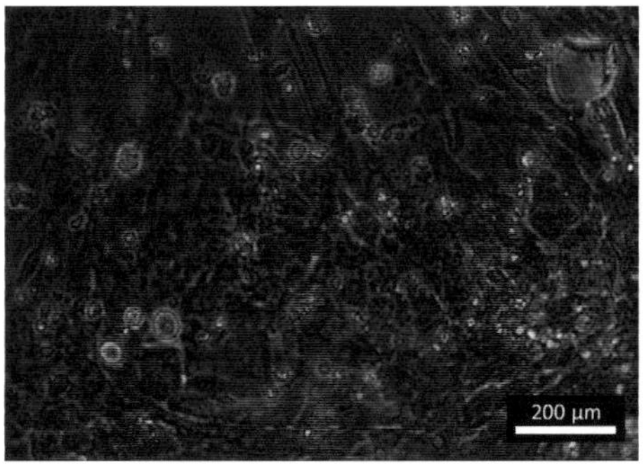

αSA DAPI

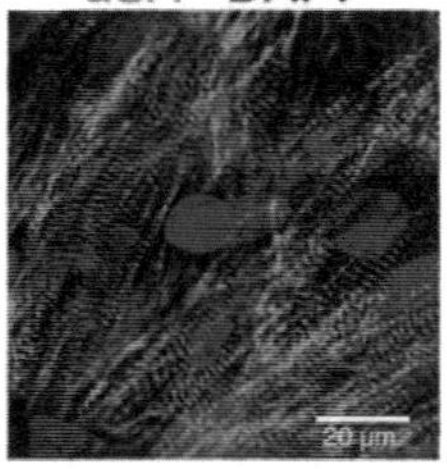

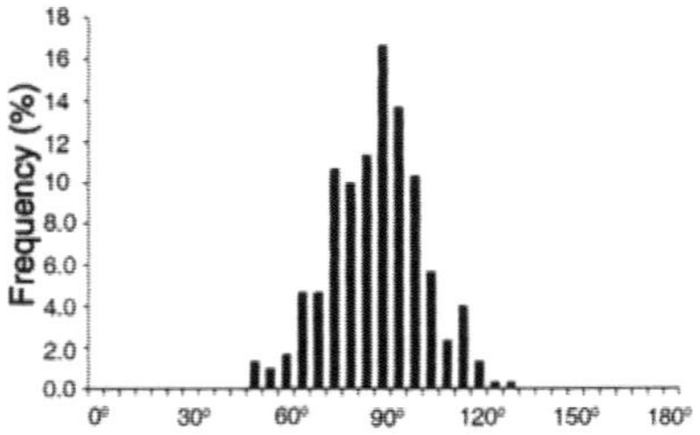

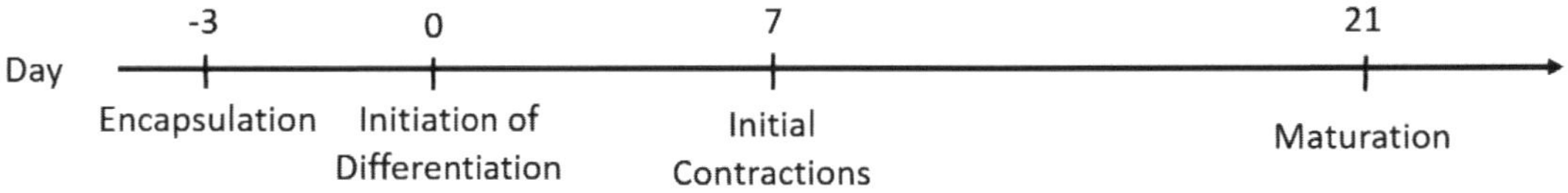

Lipke Lab
Auburn University

Conclusions

Successful synthesis of GelMA with a methacrylation degree of 22%

Encapsulated hiPSCs survive and grow in GelMA hydrogels

Successful cardiac differentiation in GelMA creates uniformly contracting cardiac tissues

The method is a straight forward procedure to create human cardiac tissues *in vitro*

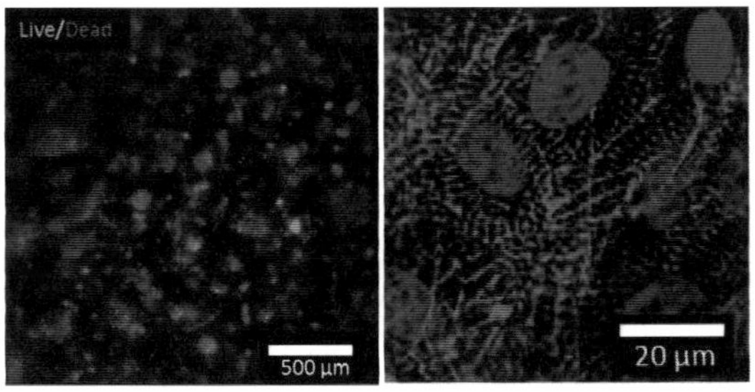

Day 0 Day 7

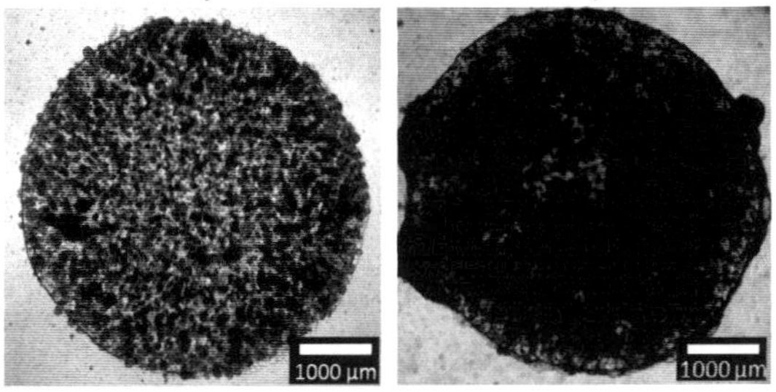

Lipke Lab
Auburn University

Acknowledgements

Lipke Lab Members:

Dr. Elizabeth A. Lipke

Postdoctoral fellow

Dr. Aaron Seeto

Graduate students

Iman Hassani

Alex Hodge

Petra Kerscher

Shantanu Pradhan

Yuan Tian

Undergraduates

Nidhi Goel

Sara Head

Kelsey Henderson

Jennifer Kaczmarek

Funding:

Auburn University
Undergraduate
Research Fellowship

NSF

Lipke Lab
Auburn University

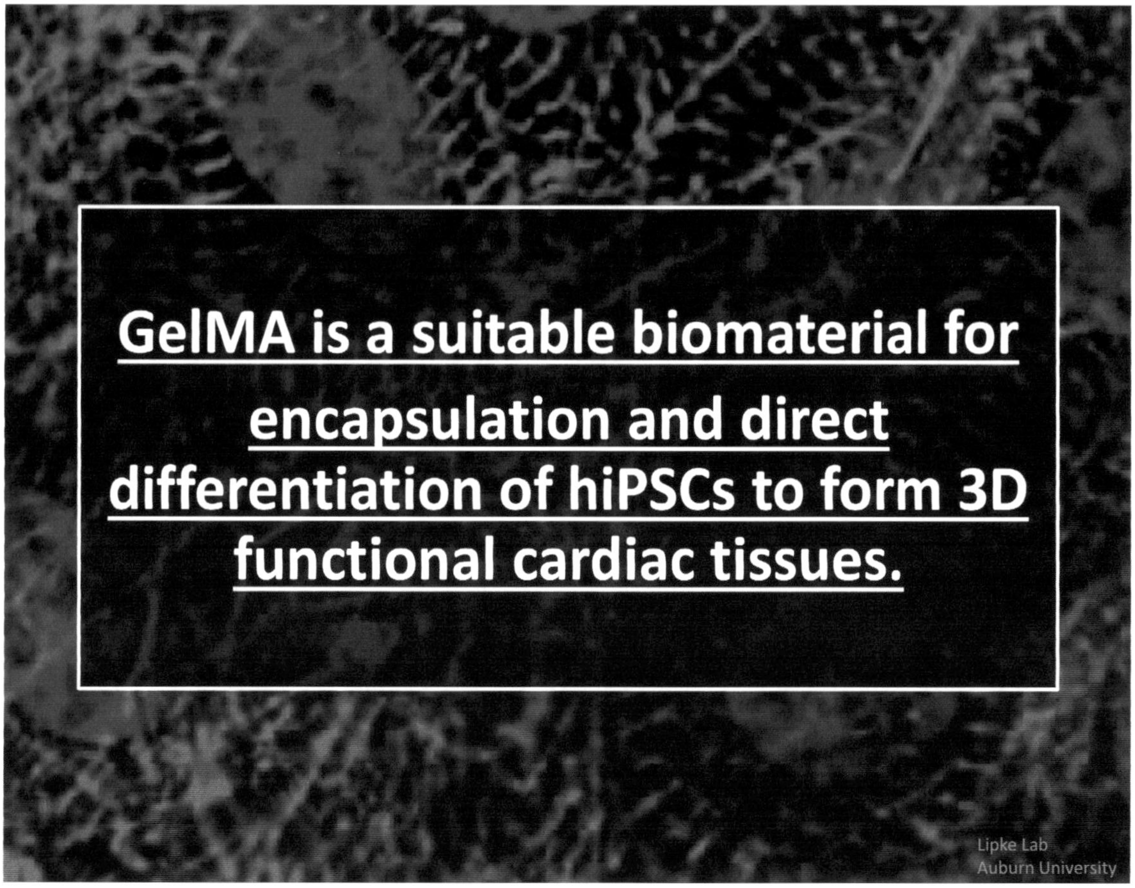

475978 Electrodeposition As a Method to Prepare Thin Film Electrocatalysts for Renewable Energy Application

Abstract:

Reducing anthropogenic carbon emissions requires effective energy conversion and storage devices to support intermittent alternative energy sources and make them economically competitive. The electrochemical reactions used in these devices, namely the oxygen reduction reaction (ORR), oxygen evolution reaction (OER), and hydrogen evolution reaction (HER), require an effective catalyst to improve poor kinetics and generally high overpotential. Precious metals such as Pt, Ir, and Ru are used commercially to catalyze these reactions because of their high activity; however, high cost, rarity and poor selectivity limit their practicality for widespread use. Thus, materials from non-precious metal sources with high catalytic activity are highly sought after. From a device/engineering standpoint, an electrocatalyst material should be scalable for large devices and easily applicable to a gas diffusion or solid electrode surface.

Electrodeposition is a simple, scalable method to prepare materials on many substrates, and is highly tunable as the electrolytic conditions and electrochemical technique affect film growth. In addition to being scalable and versatile, electrodeposition is beneficial to electrocatalysis as the intimate substrate connection allows for more rapid charge transfer, and no pretreatment (*e.g.* binders that can decrease surface adsorption) is necessary to ensure adhesion or stability. Thus, we have developed thin film electrodepositions for ORR, OER, and HER electrocatalysts with promising fundamental results and scalable possibilities for device fabrication. Our $Ni_xCo_{3-x}O_4$ bifunctional oxygen electrocatalysts, a CoP bifunctional water splitting electrocatalyst, and a MnO_x/PEDOT ORR electrocatalyst will be highlighted and examined for their promise as catalyst materials for next-generation energy conversion and storage devices.

This work was supported by the Laboratory Directed Research and Development program at Sandia National

Laboratories, a multi-program laboratory managed and operated by Sandia Corporation, a wholly owned subsidiary of Lockheed Martin Corporation, for the U.S. Department of Energyâ€™s National Nuclear Security Administration under contract DE-AC04-94AL85000.

475979 Synthesis of Cyclic Organic Carbonates Via Catalytic Oxidative Carboxylation of Olefins in Flow Reactors

Abstract:

A methodology for direct catalytic conversion of olefins into cyclic carbonates using peroxide and carbon dioxide under relatively mild conditions is developed. The protocol utilizes packed bed reactors in continuous flow to couple rhenium catalyzed olefin epoxidation and aluminum catalyzed epoxide carboxylation in a single sequence. The use of an in-line membrane separator enables the compartmentalization of mutually incompatible peroxide and tetrabutylammonium iodide. Both aliphatic and aromatic terminal olefins were shown to be viable substrates for the developed method.

475980 Infra-Red Spectroscopic Study of NO Adsorption Mechanism on a Pd Ion-Exchanged Passive NOx Adsorber

Abstract:

The adsorption mechanism of NO at 50â,,ƒ on a H-ZSM-5 zeolite exchanged with $Pd(NH_3)_4^{2+}$ has been investigated utilizing DRIFTs. Two ion-exchanged zeolites with different weight percentages of Pd as well as the zeolite without ion-exchange pretreatment were compared to highlight the changes of the reaction mechanism due to the introduction of Pd. NO adsorptions after different gaseous pretreatments were examined to help identify the chemical state of Pd on the surface and the surface intermediates. The subsequent temperature programmed desorption experiments were also conducted to qualitatively determine the binding energy of the surface intermediates. After the O_2 pretreatment at 500â,,ƒ, Pd were dispersed into the zeolite framework and were mainly in the state of Pd ions, specifically, Pd^+, Pd^{2+} and Pd^{3+}.A surface reaction between NO and Pd^{2+} were assumed to explain the oxidation of NO and the existence of Pd^+. The absorption bands of NO, NO_2 adsorbed on Pd^+, Pd^{2+} and the acid active sites of zeolite were observed and further proved. The results of three different samples were compared to highlight the benefits of Pd as an active metal in the passive NOx adsorber.

475981 Reaction Engineering Routes to Waste Gasification for Sustainable Living Environments

Abstract:

Concern over "green," or environmentally friendly, technologies has risen considerably over the past decade. Thus, over the past 50 years, metropolitan areas have realized a need for waste disposal alternatives as landfills approach capacity and take up valuable land space. Although newer technology has made incinerators more efficient, there is an increasing interest in formulating more 'green' gasification options as alternatives to incinerators.

Gasification converts organic and carbonaceous materials into a combination of gaseous products known as "syngas," or synthetic gas. This research focuses on advancing the current knowledge of a catalytic gasification process as a potential in-situ resource utilization and waste management system. With significance in a variety of engineering applications, this research is of particular relevance towards municipal waste management and as in-situ resource utilization alternative for advancing space exploration.

The catalytic gasification reaction mechanism evolves through a multi-phase reaction mechanism. The process is initiated by a liquid phase oxidation reactions of long chain polymers. The liquid phase oxidation reactions produce carbon monoxide, carbon dioxide, and water. Current long chain polymers being studied in the catalytic gasification mechanism are focused on cellulose and polyethylene as model substrates. The oxidation reactions are complemented by two gas phase reactions: the Water Gas Shift and the Sabatier reaction, the main stages in the pathway of producing hydrogen and methane.

Gasification experiments examine gasification dynamics for different substrates promoted by different catalysts. It utilizes a high pressure batch scale reactor which allows examining controlled multi-phase catalytic isothermal oxidation of solid and melting substrates in the presence high purity air.

The studied substrates, Polyethylene and Cellulose, are both long chain organic polymers, and make up a substantial portion of both space and municipal waste composition. Although similar in nature, these substrates exhibit marked differences as it pertains to gasification, and were therefore selected as model substrates.

Analysis of the liquid and solid reaction residues, along with the gaseous products, enabled the characterization of the gasification kinetic pathway. The results are used in conjunction with a kinetic model for the gas-phase to formulate a methodological model for the gasification of solid waste.

Preliminary results suggest that catalytic gasification provides potential control for the liquid-phase oxidations. Under appropriate conditions, oxidation was found to follow the incomplete oxidation pathway which resulted in significant yields of hydrogen and methane, which are potentially high value products as they can be used for energy recovery.

The results will be presented along with a preliminary extension for a continuous gasification process, which clearly illustrate the potential of catalytic gasification as a sustainable waste management alternative.

475982 A Way for Wafer Washing: Study of Rinsing Flow Dynamics on Rotating Silicon Wafers

Abstract:

During the manufacture of silicon circuits it is important that the surface of the wafer is extremely clean. It has been found that procedures for rinsing contaminants from the surface can be made faster and more effective when the wafer is rotating. However, when a jet of liquid is applied to a rotating surface, centrifugal forces can pull rivulets of liquid beyond the radially advancing wave-front. This is problematic because the rinsing flow is uneven and particles or other contaminants can build up along the edges of these rivulets, increasing product failure at these locations.
A better fundamental understanding of the flow in these systems is essential to improve the effectiveness of spin-rinsing processes while reducing yield losses due to rivulet formation. In the present work, we investigated the effects of wafer rotation rate, jet flow rate, and the presence of a thin liquid film on the wafer on the dynamics of the wave-front of the rinsing flow. It was found that reducing the rotation rate reduced rivulet formation but slowed the wave-front. Also, a higher jet flow rate reduced rivulet formation and increased wave-front speed, but this reduction of rivulet formation diminished at larger radial distances. Coating the wafer with a thin film of water also both reduced the extent of rivulet formation and increased wave-front speed when compared to a clean, dry wafer. Lastly, rivulet formation was studied on wafers coated with thin films of surfactant solutions to understand the effect of surface-tension gradients on the wave-front.

475983 Quantifying Variations in the Biomass Recalcitrance of Douglas Fir

Abstract:

To assist in the development of an economically effective industry for the production of biofuels from forestry residues, the study in this presentation aims to identify and characterize variations between individual Douglas fir trees. Characterizing these variations is intended to provide a foundation for further research into the nature of biomass recalcitrance and to determine the feasibility of potential tree improvement programs aimed at reducing recalcitrance of D. fir. Discussed is a method devised to screen highly recalcitrant woody biomass samples for properties related to biomass recalcitrance.

UV-Solvent Annealing for Morphology and Orientation Control in Self-Assembled PS-PDMS Thin Films

Melissa Kreider
Co-authors: Wubin Bai, Caroline Ross

AIChE Annual Meeting Student Paper Competition

November 14, 2016

Introduction - **Background** - **Experimental** - **Results** - **Conclusion**

Why Block Copolymers?

Semiconductor Industry

Nanofabrication

Pattern Transfer

2

BCP Morphology & Solvent Annealing

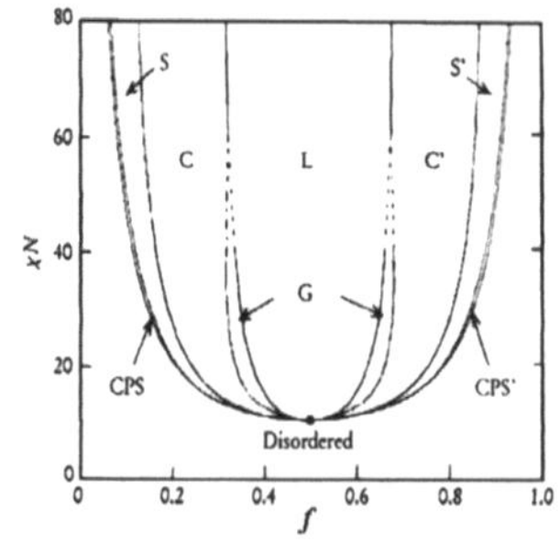

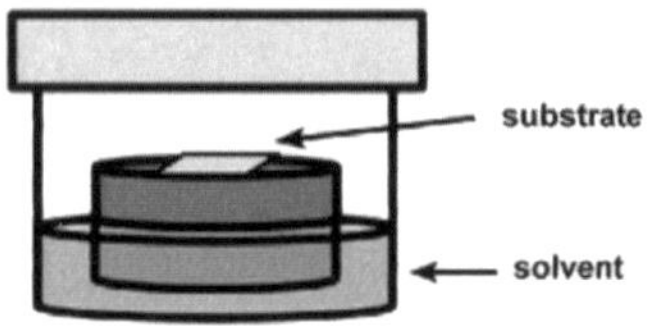

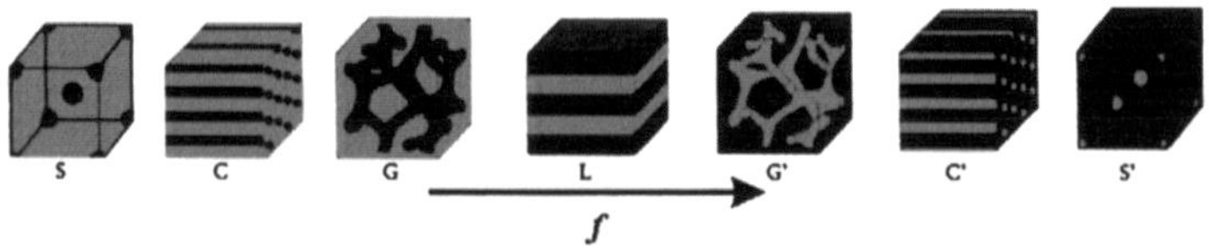

3

Effects of UV irradiation

Polymer Crosslinking

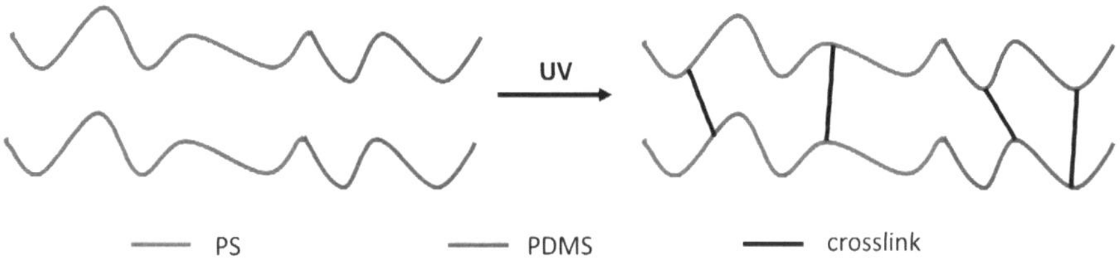

4

Experimental Detail

Spin cast 80 nm SD53

Anneal with toluene & heptane

Apply UV light

Fast Quench

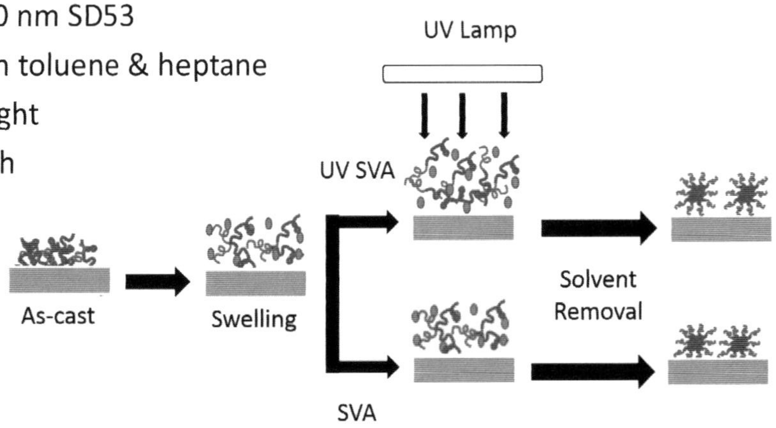

5

UV Absorption

PS-PDMS Light Absorption

UV Lamp: 254 nm emission

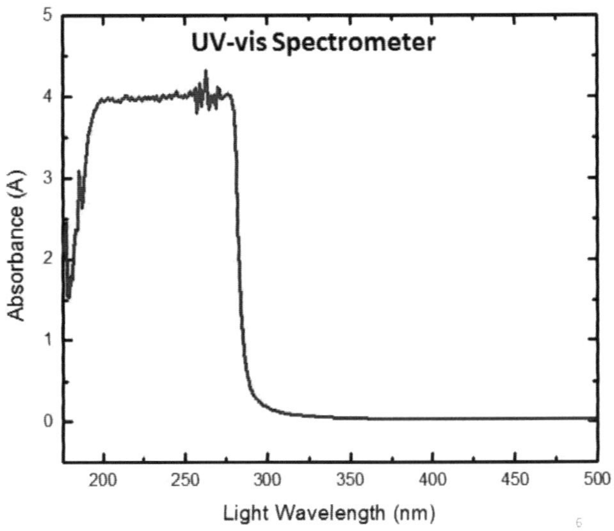

6

Solvent Annealing Swelling Ratio

- Swelling ratio controls the effects of solvent annealing
- UV irradiation induces a change in swelling ratio
- UV irradiated film does not fully deswell

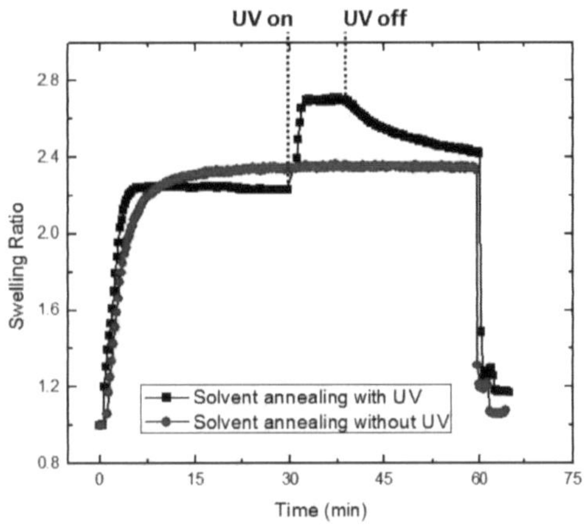

Morphology Changes on Bare Silicon

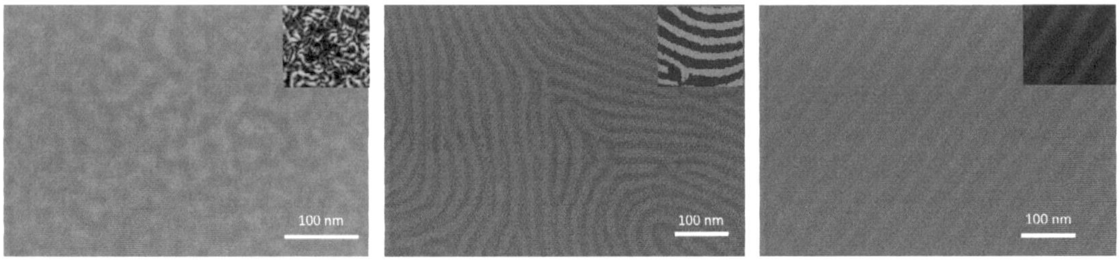

8

Morphology Changes on Bare Silicon

As-cast

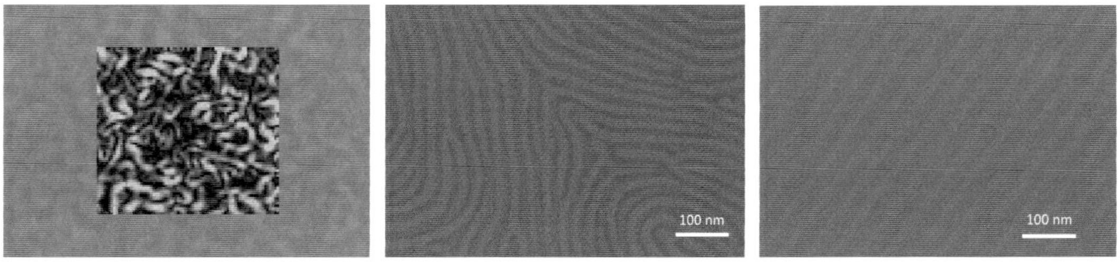

Morphology Changes on Bare Silicon

Solvent Vapor Annealed

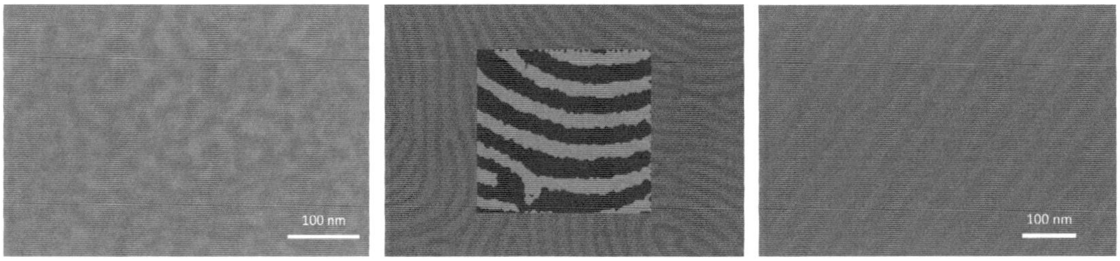

Morphology Changes on Bare Silicon

UV Solvent Vapor Annealed

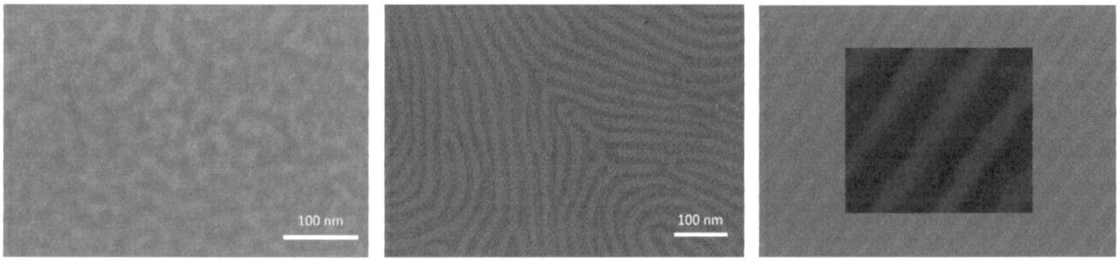

8

Morphology Changes on Trench Substrate

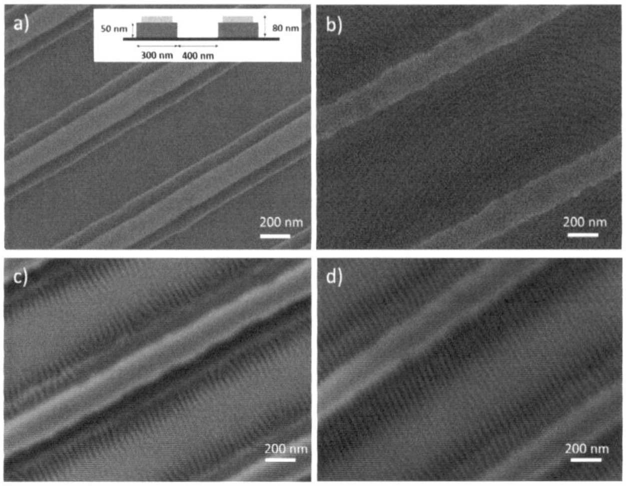

9

Morphology Changes on Trench Substrate

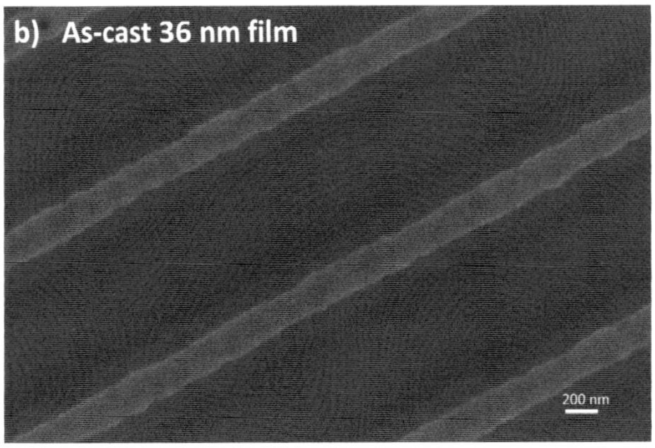

Morphology Changes on Trench Substrate

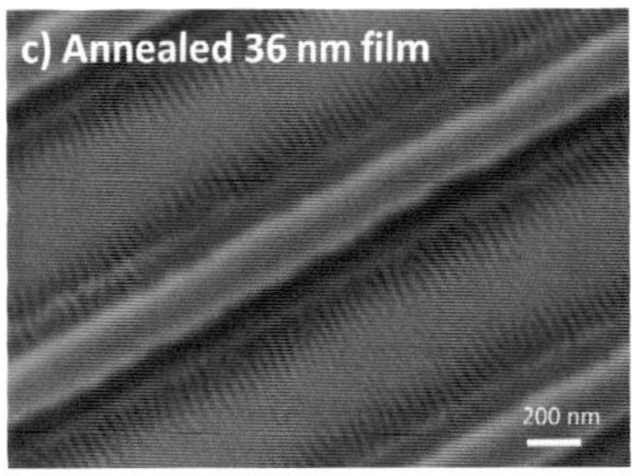

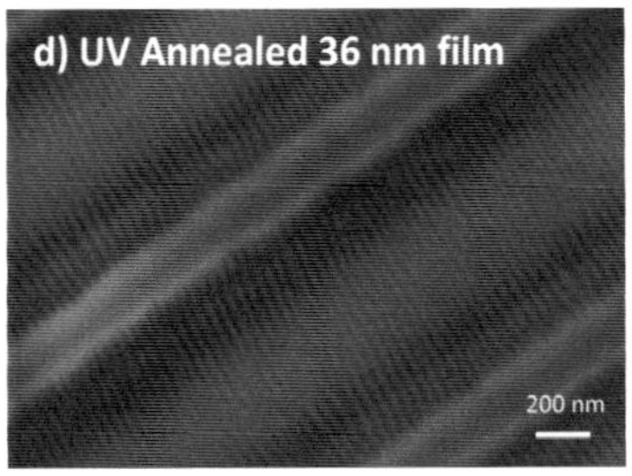

Conclusions

PS-PDMS absorbs UV radiation
UV irradiation affects polymer swelling behavior
UV solvent annealing improved the post-anneal morphology

Acknowledgments

Thank you to:
- Prof. Caroline Ross
- Wubin Bai
- Keehong Lee
- Saman Safari
- George Liontos, Konstantinos Ntetsikas, Apostolos Avgeropolous
- MIT Department of Materials Science and Engineering
- MIT Department of Chemical Engineering
- MIT Undergraduate Research Opportunities Program

468875 "for Here or to Go: Maintaining Mid-Career Faculty Momentum"

Abstract:

Midcareer faculty in engineering often are at crossroads of their careers in terms of opportunities. Whether tenure or non-tenure track, faculty are often presented with a plethora of opportunities in leadership, service, research and international realms. A critical aspect of making decisions at this stage is the application of a well thought out long term academic career roadmap. In the absence of this, faculty can flounder, become discouraged and even become "stuck". This presentation draws on the presenter's extensive expertise mentoring junior faculty, coaching midcareer faculty and designing initiatives to provide a robust roadmap to faculty. Faculty currently searching for that next "big thing" to work on will benefit, as will those that are continuing on a solid upward trajectory to the next promotion.

474873 A Sabbatical Abroad

Abstract:

For many faculty members, tenure allows them the opportunity to apply for a sabbatical. A sabbatical is a means for faculty members to achieve professional growth and development that will simultaneously enhance the intellectual capacity of their institution when they return. In this talk, suggestions for pursuing international sabbatical opportunities will be discussed. In particular, an overview of the Fulbright Scholar Program and tips on how to successfully apply for these fellowships will be provided.

474513 What Comes after Tenure: Experiences of a Faculty Member and Perspective from a Dean

Abstract:

Achieving tenure is a critical point in the career of a faculty member. However, it shouldnâ€™t be seen as an endpoint but rather a stepping point to future success. In this presentation, the author will describe his experiences in his post-tenure life, adding the perspective of best practices he has seen in his role as a university administrator.

474601 From Plasmas to Learners: Opportunities and Challenges in a Post-Tenure Shift in Inquiry

Abstract:

The AAC&U declares that academic freedom is â€œthe indispensable requisite for unfettered teaching and research in institutions of higher education.â€ Such freedom fully blooms after the tenure process. It not only provides faculty with latitude to seek independent inquiry within a field but also enables faculty to change pursuits of inquiry into areas where they identify a need and where their passion drives them. In this presentation, I will reflect on my path that led to a mid-career transition from being a materials processing researcher into pursuing inquiry in the emerging field of engineering education research. Some challenges and rewards along the way will be described.

463429 And Now What? Dedicate Your Life to Helping the Students and Citizens of This World

Abstract:

And now what? Dedicate your Life to Helping the Students and Citizens of this World.

The hard decisions for faculty members come after tenure. What do I do now? Do I continue my research? Do I teach and mentor students? Both? Do I spend more time with my family? Do I start preparing for administration? ÒTenureÓ is a mark only in terms of security of life. But tenure should not change your goals in life. In my career I have found that helping young students, helping people and my community, developing new products and new processes that improve the quality of life of our patients are the most rewarding and satisfying goals. I have dedicated my post-tenure life to helping students, to becoming a catalyst for their successful lives, to addressing their major decisions in life (see also The Peppas Factor: Educating Future Leaders http://www.engr.utexas.edu/features/peppas-2013-teaching-award). At the same time I have dedicated my life to doing what I like to do best (teaching and research) but doing it to improve the quality of life of our patients around the world. I believe you can do the same and I will give examples how you can achieve this most rewarding experience.

461922 Preparing Female Graduate Students for Leadership in Academic and Industrial Careers

Abstract:

In 1985, women received 23% of the bachelorâ€™s degrees awarded in chemical engineering nationwide. By 2014, nearly 30 years later, this number has improved to 32%. Female leadership in chemical engineering remains far from reflecting gender equality as well, with the number of female chemical engineering faculty and practicing chemical engineers hovering around 15%. Many decades of outreach have improved the success of girls in science and math in K-12 and given them the confidence to choose technical majors in college, which has led to the steady increase in females in STEM at the undergraduate level. The next area of focus needs to be on female graduate students. However, challenges remain in supporting female graduate students to advance from graduate school into higher career ranks as faculty and industry leaders. Chemical and Biomolecular Engineering (CBE) has made significant efforts to address recruitment and retention of female graduate students by shifting department culture that increased women to 40% averaged over the last 8 years. This shift is due in large part to a female professional development focus, substantial outreach leadership activities to high school girls, and increased faculty engagement. Programming is designed to improve graduate studentsâ€™ leadership skills and understanding of the potential career opportunities available to them. This programming consists of seminars and workshops planned by students on topics ranging from patents and entrepreneurship to grant writing. Alumnae from the CBE PhD program have returned to speak about their career paths and offer insight to graduating students. With these workshops and seminars, we are increasing the preparation of female graduate students for the challenges of leadership that are often not addressed in graduate school education to empower female students to become leaders in their fields. In this talk, I will discuss the program in place at CBE and share insights into how this program has shifted department and graduate student culture.

466453 Nature Inspired Chemical Engineering: A New Elective Course for Senior Undergraduate and Graduate Students

Abstract:

In 2013, UCL was awarded an EPSRC (United Kingdomâ€™s Engineering and Physical Sciences Research Council) â€œFrontier Engineeringâ€ Grant to form a multi-disciplinary Centre for Nature Inspired Engineering, directed by Marc-Olivier Coppens. The overarching vision of the center is to use nature as a guiding platform to seek potentially transformative solutions to engineering grand challenges, such as sustainable energy, scalable manufacturing methods and clean water. Beyond biomimicry, this nature-inspired approach seeks to reveal fundamental mechanisms in the natural world that underlie desirable properties such as scalability, efficiency or robustness, and can be applied in a broader context to solve similar problems in engineering.

To complement the new research center, a new senior undergraduate / MSc-level elective course on Nature Inspired Chemical Engineering was developed and taught by us. One of the main learning objectives of the course was to stimulate creative thought in leveraging natural phenomena to solve chemical engineering problems. This was achieved by using a variety of active learning and pedagogical techniques, such as, annotated textbook readings of current journal publications, oral presentations highlighting the balance between nature and technology, laboratory demonstrations, and a semester-long group project motivated by student interest in learning from nature to create innovative solutions to chemical engineering problems. Examples of all these activities, including highlights from the final student projects, will be presented.

468052 Transport Lite: Teaching Transport Phenomena through Application-Driven Inquiry in the Brewing Process

Abstract:

Transport Phenomena, CHE-5132 at Villanova University has historically been an upper-level elective designed for students who have an interest in advanced study in the discipline, taught in the spirit of Bird, Stewart and Lightfootâ €™s seminal text of the same name. The fraction of students taking the course for its stated objectives is small. Rather, students often take the course for other reasons: its offered at a convenient time, they need a CHE elective, or are blocked out of other classes. In this talk, the author will describe how he abandoned the traditional approach, and adopted an applications-based approach. Students are introduced to the brewing process in some detail. The principles of transport phenomena are introduced as needed to build a deeper understanding of selected operations in the brewing process; for example, we consider diffusion of CO_2 and O_2 in a bed of malting barley, enzyme diffusion during mashing, flow through a packed bed during sparging, and heat transfer during the wort boil. Metrics comparing the two approaches are thin, but student feed back to the revised offering will be discussed.

451060 The Loped: No Need to Steal This Course, You Could Probably Just Run Along Side It!

Abstract:

A significant amount of recent attention in the institute focused on the need for core chemical engineering skills to feed the pipeline to industry and the call for innovation, equally emphasized. While reading a recent news article by Fareed Zakaria, he quotes billionaire Peter Thiel arguing that we don't live in innovative times by saying: "We wanted flying cars, instead we got 140 characters."[1] Lets consider one description of innovation education from a leading funding organization in the United States (i.e., National Science Foundation). Here we have an excerpt from a web page describing the Integrated Graduate Education Research Training (a.k.a. IGERT) program.

"The program is intended to catalyze a cultural change in graduate education, for students, faculty, and institutions, by establishing innovative new models for graduate education and training in a fertile environment for collaborative research that transcends traditional disciplinary boundaries. It is also intended to facilitate greater diversity in student participation and preparation, and to contribute to the development of a diverse, globally-engaged science and engineering workforce."[2]

Back to the thought on cars, consider that less than 10% of the worlds population owns a car (my informal research yielded a range from 5% to 9%, in the United States vehicle ownership is ~88%). Crude estimates of global bicycle ownership are around 50% with high bike to population ratios in NE Europe and China, while motorcycles or mopeds dominate SE Asia means of transport. If we consider cars, motorcycles and bicycles together, don't they have some essential features that can be applied to many aspects of engineering[?] including chemical engineering: reactions, controls, catalysts, materials, physics, safety features...and elbow grease!

This talk will highlight aspects of a low powered transport – lab class that in an initial offering, I propose cast in a choose your own innovation fashion, where base lab tasks are required but specialization in creative development and systems concepts are leveraged to catalyze cultural change in collaborative research that transcends traditional disciplinary boundaries for a globally-engaged engineering workforce.

---

[1] https://www.washingtonpost.com/opinions/fareed-zakaria-the-need-to-renew-american-innovation/2015/01/01/b0f0d864-913b-11e4-a900-9960214d4cd7_story.html, accessed March, 28th, 2016.

[2] https://www.nsf.gov/crssprgm/igert/intro.jsp, accessed March, 28th, 2016.
http://blogs.worldbank.org/publicsphere/cycling-everyone-s-business, accessed March, 28th, 2016.

- Abstract Text

470372 Molecular Engineering

Abstract:

This paper reviews core curriculum developments for a newly developed Ph.D. program in Molecular Engineering. Molecular Engineering is a dynamic and evolving field and its target problems are complex; breakthroughs require sophisticated and creative engineers who are conversant across disciplines. It deals with material development efforts in emerging technologies that require rigorous rational molecular design approaches towards systems of high complexity.Â A rational engineering methodology that is based on molecular principles is in contrast to widely spread and common trial-and-error approaches. Molecular Engineering steps in where fortuitous discoveries become fewer with increased sophistication of technology.

MolEâ€™s core educational program entails a tri-faceted interdisciplinary approach in molecular design and synthesis, theory/computation, and molecular material characterization. The goal is to bring forward PhDs that possess a familiarity with the fundamental aspects of molecular engineering that involves biological, chemical and physical sciences, and system integration aspects in emerging and established technologies. In particular, MolEâ€™s objectives are to promote scientist and engineers that are experts in either

- molecular synthesis towards specific system assembly and functionality,

- the design of complex systems built up from molecular building blocks, or

- the theoretical engineering of complex molecular systems through predictive molecular modelling.

473076 Crowdsourcing Statistical Phenomena Using an Ipad-Based Brownian Dynamics Simulator

Abstract:

Scientists routinely use Brownian dynamics simulations to understand the nanoscale details governing how polymers respond to forces and deformations. But this important toolbox is generally only accessible only to a small and exclusive community of researchers possessing special expertise and high-level computing resources. Here we show how this barrier can be overcome by taking advantage of the same physics engine that powers popular games like *Angry Birds* to achieve significantly improved computational efficiency, making it possible to perform quantitatively accurate Brownian dynamics simulations of coarse grained (bead-spring model) polymer transport on ordinary tablet computing devices like the Apple iPad. Our app provides a visual representation of polymer coil size, relaxation phenomena, Brownian motion, and transport under an external driving force (i.e., electrophoresis). Key parameters continuously reported during the simulation include radius of gyration Rg and end-to-end length L. The tablet-based format also introduces the opportunity to re-imagine how users interact with molecular simulations by employing familiar touch screen gestures. We show how lessons based on this app can be scaled toward a range of student populations and class sizes via a crowdsourcing approach where students run independent simulations, after which the results are pooled to obtain statistics that govern collective phenomena. Accuracy is improved with more realizations, making this approach ideal for stimulating engagement in large classes.

469627 Tutorial on the SMART-CN Education Modules for Incorporation in the Advanced Undergraduate or Graduate Engineering Curriculum

Abstract:

The Sustainable Manufacturing Advances in Research and Technology (SMART) project funded by National Science Foundation aims to bridge the gap between academic knowledge discovery and industrial technology innovation for sustainable manufacturing. The SMART project involves a multidisciplinary team which has created an interdisciplinary and international Research Coordination Network (RCN) through the joint effort among a number of leading academic laboratories, centers, non-government organizations, and major manufacturing industries.

Meeting the educational goals of the SMART CN has been achieved through the creation of coordination among member universities and developing online modules specific to sustainable manufacturing. These modules are highly structured classroom-classroom-ready materials for use by instructors at the undergraduate or graduate level. The modules can also be used by industrial entities aiming to further their knowledge in sustainable manufacturing through self-study.

The presentation will provide an overview of this educational component of the project with emphasis on the dissemination of modules supported by computer tools for incorporation into the engineering curriculum. These modules include topics like life cycle assessment, green design and engineering, green chemistry, process intensification, process integration, optimization, and process safety among the core module elements. Test-use data, classroom use experience and general student feedback will be shared. These can be adopted by chemical, mechanical, civil, environmental and other engineering disciplines. Extensive case studies demonstrating these concepts are also provided for classroom teaching, homework assignments and term projects. The modules already available through the CACHE website have been reviewed by corporate sustainability teams, as well as academic instructors.

Reference: http://cache.org/super-store

454769 Flow through Packed Beds an Inconvenient Truth

Abstract:

Flow Through Packed Bedsâ€¦An Inconvenient Truth

Jacob Arredondo, Jonathan Worstell, and Meng Zhang

Packed beds are a staple of the chemical industry, and as such, are a critical topic in undergraduate education. Students encounter packed beds in assignments for separations, kinetics, and laboratory courses; and they are usually drawn directly to the Ergun equation for answers.

The Ergun equation, while well known, generally gives poor results, especially for non-spherical packing. Furthermore, the Ergun equation provides no available mechanism for scaling up or scaling down of chemical processes.

In this presentation, we outline a simple undergraduate experiment that demonstrates to students the ineffectiveness of the Ergun equation and facilitates the development of a better solution using dimensional analysis. This assignment is designed to promote critical thinking and shows students that textbooks and academic papers are not always the definitive authority. Additionally, with no moving parts, the experimental apparatus is very robust, i.e. student resistant, in high volume teaching laboratories.

459994 Steal This Project: Entrepreneurial-Based Learning By Building Your Own Reactor

Abstract:

Quite often in reaction engineering education, students are rigorously trained in applying material and energy balances and reaction kinetics for the purposes of reactor selection, sizing, and operating parameters. In chemical engineering laboratory instruction, students are often given the chance to operate and test various reactors or are given the opportunity to experimentally measure reaction kinetics. However, due to a variety of factors, students are rarely afforded the opportunity to build a chemical reactor, which can be an exercise in creativity and safety. At Kettering University, a Keen partner, the reaction engineering laboratory curriculum was revised to incorporate an entrepreneurial-based learning project where student teams selected a chemical reaction of industrial significance and through a guided process designed and demonstrated the reactor. Keen provided grant funds to develop the curriculum and to procure kits of parts which the students can use to build tubular reactors. By the end of 2016 this project will be offered for four quarters. The purpose of this project is open-ended and an exercise in creativity and safe practices as a reaction engineer.

The student teams for this project are expected to research and identify a chemical reaction of potential commercial relevance they wish to evaluate. Due to safety, time, and cost certain limitations such as operating temperature, flammability, toxicity, and material compatibility were be considered while researching chemical reactions. Once a reaction is identified, the student team will propose it to the professor for a safety review and approval. The team will then design a chemical prototype system to evaluate the chemical reaction for possible production to market. The team will propose the design to the class and the professor for approval. The team will then construct the chemical reactor system and evaluate their chemical reaction. It also is required that the team determines a method to determine conversion of the reactants and quantify the products. The final deliverable for this project is an operating chemical reactor system and a PowerPoint presentation. Stretch goals for this project is the determination of the conversion and rate law.

How the project was implemented in the curriculum, rubrics, sample student work, lessons learned, and student survey data will be presented.

462100 Steal This Project: Solar-Thermal Conversion of Biomass to Methanol As an Introduction to Chemical Engineering Design

Abstract:

This project introduces the fundamentals of engineering design over the course of a semester by asking students to design a solar-thermal plant for the production of methanol from biomass. The project involves the gasification of lignocellulosic biomass at high-temperature to produce synthesis gas, which is converted and purified downstream. The high-temperatures are achieved via solarthermal heating during the day, and natural gas combustion at night, requiring students to design a hybrid reactor system.

This project introduces several topics in chemical engineering design that students have likely not encountered before such as life-cycle analysis and solids handling, in addition to common concepts such as reactor design and process heat integration. Students individually perform work on the project in three phases throughout the semester in the form of take-home exams as pertinent topics are introduced. During the final 6 weeks of the semester students break off into teams of 4 or 5 and complete the project. Teams are required to build three different simulations, two in Aspen HYSYS^{TM} and one in Aspen Plus^{TM}, thus familiarizing the students with both software packages. Students must also incorporate textbook design calculations, kinetic modeling and cash flow analysis.

This presentation will describe the project, the unique way in which it was taught, and show sample results from students.

465960 The Virtual Sandbox: Particle Flow Physics Taught with Interactive Tools

Abstract:

Processes in the chemical industry involving particles and powders are (i) difficult to describe, and (ii) often pose a risk for process stability and safety. To guarantee process and product safety, process simulation, as well as the simulation of particle flow, became a main trend over the past decades. Accordingly, there is a strong need to cover the simulation of particulate systems in basic and advanced courses of particle technology and multiphase reactors. The virtual sandbox project focuses on the adoption of the augmented reality sandbox presented by Reed et.al. (2014) for teaching the basics of particle flow physics. It comprises a sandbox, a Kinect camera, and a simulation program. The sand surface is scanned, and loaded into the particle simulator LIGGGHTS®. The simulation results can be projected onto the sand to close to loop from reality to simulation, and back to reality.

The deeper scientific meaning of the project can be summarized as follows: the first key element is to teach the students the importance and comfort to simulate particle based processes, rather than performing experiments which suffer from the opaque nature of sand. The second element is to introduce the complexity and the effect of liquid bridges that exist between individual grains. Our idea is to guide the students through a set of models that are needed to describe these bridges: starting with models that describe the genesis of a pendular bridge (Wu et al., 2016), over the seminal work of Halsey and Levine (1998) on roughness effects, to the most recent developments in force models. Third, basics concepts of 3D vision are introduced, in order to make students aware of the enormous capabilities of modern camera systems. Last but not least, new teaching methods, e.g., inquiry-based learning, are developed and applied to the above topics: via additional small-scale and easy to reproduce experiments we make an attempt to illustrate the above scientific concepts even to high school students (age 16 to 18), as well as primary school pupils (age 6 to 10).

Acknowledgement
The project is founded by the Austrian Science Fund FWF WKP67.

References
T. Halsey, A. Levine, "How Sandcastles Fall", Physical Review Letters 1998, 80, 3141-3144.

S.E. Reed, L. Kreylos, S. Hsi, L.H. Kellogg, G. Schladow, M.B. Yikilmaz, H. Segale, J. Silverman, S. Yalowitz, and E. Sato, "Shaping Watersheds Exhibit: An Interactive, Augmented Reality Sandbox for Advancing Earth Science Education", Americal Geophysical Union, Fall Meeting 2014.

M. Wu, S. Radl, J.G. Khinast, A model to predict liquid bridge formation between wet particles based on direct numerical simulations, AIChE Journal 2016,62, 1877-1897.

471828 Process Controls Final Projects Inspired By Real Unit Operations Laboratory Modules

Abstract:

Northeastern University specializes in opportunities for experiential learning, both through a co-op program as well as through direct coursework. The style of experiential and inquiry-based learning have been particularly reinforced through two intermediate-level Unit Operations courses, in which students have been tasked with developing experiments, selecting measurement parameters, and designing analysis. Studentsâ€™ previous experiences have been utilized in the development of final projects for their Process Controls course.

In order to provide a culmination of the entire Process Controls course and to challenge students to design a complete controls system relying on some combination of feedback, feedforward, and/or cascade control, a final project was designed to connect the concept of Controls to a real-world system. Student teams were given the option of selecting from six different modules they had observed and analyzed in a previous laboratory course, including distillation, liquid-liquid extraction, heat exchangers, reverse osmosis, and water remediation. These modules had required complete manual control, and had a range of parameters that could be adjusted during operation. For their final project, students were challenged to design a complete control system that required them to select control variables, handle multiple disturbances, and operate within acceptable limits. Students derived overall balances, designed block diagrams, calculated tuning parameters both by hand and through simulation software, and analyzed their proposed control system for stability and economic viability.

The project culminated in both a technical written report and a presentation. Across three semesters and multiple sections, every team has designed a different system resulting in different quality of control. Requiring the students to defend their design decisions and forcing them to account for a range of disturbances has helped students develop a much more complete grasp of Process Controls and given them greater confidence exiting the course.

This presentation will provide examples of the modules and assessment of the project to date.

472364 Developing Professional Skills through a Service Learning Project on Aluminum Recycling

Abstract:

Given the volume of scientific and technical material that must be covered and assessed in most standard core chemical engineering courses, addressing the softer educational objectives required for ABET accreditation often poses a challenge for instructors. These objectives include teamwork experience, presentation skills, and an awareness of social responsibility and the global implications of engineering decisions. In this project, students were provided an opportunity to develop those professional skills by applying basic technical knowledge to understand and influence a lifestyle choice: recycling aluminum. The students were enrolled in a sophomore-level material and energy balances course. In a semester-long project, the students worked on teams to learn about aluminum refining and recycling technology from a chemical engineering perspective. Using skills learned in the course material, they were able to perform basic calculations that gave them a grasp of the difference in energy requirements between the two processes. They also participated in a service project collecting and recycling cans during tailgate events at football games, which provided them with practical experience of local recycling infrastructure. Finally, they implemented their learning to prepare a presentation for lay audiences regarding the energy requirements and environmental effects of aluminum refining. Each group delivered their presentation to a different external audience. Student surveys reveal that their teamwork, management, and communication skills were enhanced through the project, as was their cognizance of how engineering can relate to societal challenges. Students also reported that the project helped them better connect to the technical course material by relating it to a real process whose broader implications they could understand.

472416 Case Studies for Vertically Integrated Curriculum Projects

Abstract:

Often, students are unable or unwilling to â€œconnect the dotsâ€ when it comes to various courses in the chemical engineering curriculum. To combat courses being interpreted as isolated or unrelated to one another, U.C. San Diego has begun experimenting with a curriculum-wide course project. In this project, students are exposed to a single industrial-scale chemical process and repeatedly analyze it from different perspectives in each class. Starting with material balances, expanding to complex energy/entropy/exergy balances in thermodynamics, to designing the reactors, and so on. In this discussion, several case studies already implemented as UCSD are discussed, including: methanol synthesis, the Haber-Bosch process, vinyl chloride production, and hydrodealkylation (HDA) of toluene. Specific focus will be attached to how the projects progressed and expanded from course to course.

472951 It's Not Stealing If It's Free: Practical Process Instrumentation, Data Analysis, and Control with Open Source Hardware and Software

Abstract:

Data acquisition, analysis, interpretation, and visualization skills aren't just for "data scientists"—all chemical engineers benefit from a strong practical foundation in data. Unfortunately, many undergraduates have a hard time connecting the abstract concepts from their computer science course requirements back to the core chemical engineering curriculum. I'll present a hands-on project-based course concept that reinforces chemical engineering fundamentals while emphasizing the incredible utility of computational tools in the practical context of PHYSICAL COMPUTING—coding motivated by the task of working with sensors and actuators and the challenge of gleaning insights from the messy world of real data.

472384 Promoting Analogical Reasoning in a Numerical Methods Class

Abstract:

The biggest challenge for engineering classes in numerical methods or modeling is illustrating the application of numerical methods to engineering systems. In chemical engineering, these courses are often taught shortly after the introductory mass and energy balance class so students are lacking in advanced applications needed to illustrate numerical methods for solving dynamic systems. To overcome this limitation, we apply these numerical methods to an analogous problem that is not technical. This approach removes the cognitive load associated with learning the new application alongside learning a new numerical method. If the non-technical examples are sufficiently relatable, and analogous to engineering examples we hypothesize that transfer will occur to more technical applications. Further, by choosing analogous applications that are humorous and engaging, it can increase instructor immediacy and student confidence. Results from applying this approach to several large numerical methods classes are presented. Best practices and the importance of active learning to this process is discussed.

455605 Chemical Engineering - Biology Crossover Project - Customer Client Interactions

Abstract:

The freshman level Introduction to Chemical Engineering CHME100 course was run in the spring 2015 term. This course was modified to include modules that familiarized the students with key chemical engineering principles. Traditional topics were covered that included fluid flow, heat transfer, electrical systems and controls. A specific focus of each topic was modules or case studies that dealt with applications to industry, potentail biological hazards or contaminants, and process safety. Within each of these modules, the presence of microorganisms was incuded in the discussion from a safety, economic and environmental standpoint. These freshman level chemical engineering students were given a project to help bridge the traditional engineering focus to an all-encompassing project. The first step was to work within a chemical engineering team to assess a problem at a facility. After the chemical engineering students worked together to collect data, assess the problem and work on a possible solution, they met with a junior level molecular biology student group. The CHME 100 students communicated and collaborated with the molecular biology students as a customer or client would interact to solve the problems associated with microorganisms in industrial settings. They presented their work in a written report and an oral presentation. The interactions and experiences of this module will be discussed.

472602 A Longitudinal Study of the Epistemological Beliefs of Chemical Engineering Students

Abstract:

There are classes in which a textbook is followed closely, some in which it is not followed at all, and anything in between. However a textbook is always required. Studies show that students feel more comfortable in "textbook based" classes than in classes without one, in spite of the high cost of textbooks. Unfortunately, in today's technological word it is impossible to cover current knowledge in a single monograph and thus it is becoming more important to explore alternatives to a textbook in engineering classes. Three years ago, we started a study to determine the effect of textbook free classes on the development of lifelong learners. It became evident that it was necessary first to evaluate the epistemological beliefs of the students and their ability to understand primary sources. A three-years-long study was started and an online survey for alumni was developed. We have now collected data for two and half years from sophomore, junior and senior classes. A survey borrowed (and modified) from C.R.E.A.T.E. was used. This survey is based in the previous work by Schommer (1990) but it has been modified to adapt to a different discipline. This survey was modified further to include questions about the use of textbooks and lifelong learning perceptions. The classes affected were Thermodynamics 1 (first semester sophomore), Thermodynamics 2 (Second semester sophomore), Bioseparations (second semester junior), and Bioreactors (second semester senior). The surveys were administered to current students in class (20 minutes) or as a "homework". More than 95% of the students participated in the surveys. A separate on-line survey addressing primary sources reading, use of textbooks and epistemological believes was prepared for alumni (approximately 300) of the Biochemical Engineering Emphasis Program. The longitudinal study will be completed in one more years.

A few results from the surveys can be highlighted. The difference between the same group of students in a textbook-free environment at the beginning and at the end of the semester are striking and several factors and aspects related to their epistemological sophistication show differences that are statistically significant.

However, their ability to decode primary sources is still quite limited at the end of the semester. This group of students will be tested again in Fall 2016. There is a drastic (and positive) change in their epistemological beliefs after one semester of being exposed to a class that uses primary sources as the main teaching material. Their scientific curiosity has improved from 62% to 73% ($P(T<=t)= 0.04$). In conclusion we noticed a drastic change in the epistemological believes and in the ability of reading primary sources in one group of students as they were introduced to a text free course for first time in their lives.

457954 Scenario-Based Learning in a UK Chemical Engineering Programme

Abstract:

The Department of Chemical Engineering at UCL in London, UK has, as part of its Integrated Engineering Programme, introduced six week-long scenarios in the first and second year of study. The scenarios run twice per term, involving 130â€“150 undergraduate students and a number of teaching staff.

The topics for each scenario is chosen to reflect the material taught in the regular modules running that term, thus supporting the studentsâ€™ learning and providing an opportunity for them to test out their new knowledge. The problems are also chosen to reflect different aspects of the chemical industries, for instance oil and gas transportation, air separation, and equipment design, to give the students an awareness of the types of problems they may encounter as practicing engineers.

The assessment of each scenario varies depending on the problem statement, and involves either oral presentations, poster sessions or videos, as well as written reports of different formats, e.g. newsletter, a technical bid etc, again aiming to reflect the aspects of technical presentation and writing the students may encounter later in professional life.

The main responsibility for the development and the delivery of the scenarios rests with two full-time teaching fellows, who in addition to the scenarios also support the other aspects of the Integrated Engineering Programme. For each scenario, they work with module leads to develop the technical brief, the assessment and the programme for the week.

This presentation will give an overview of our work and will summarise our main experiences so far including extensive student feedback.

459127 Using Science Comics to Improve Undergraduate Chemical Engineering Education

Abstract:

In order to meet the demands of a growing student population, with an increasing number of sections and instructors for every core chemical engineering course, new curricula tools have been developed to help supplement current instruction and provide a stronger and more consistent educational experience for all undergraduates at Northeastern University. Of particular concern has been the need to improve student understanding of more complex, theoretical course concepts that do not have an obvious real-world application or are not traditionally represented in visual form.

Science comics have been developed by integrating written lectures and instruction with visual depictions to provide new curricula to help students better grasp more abstract and complex topics. Specifically, comics have been implemented in Unit Operations Laboratory, Thermodynamics, Transport, and Process Control courses to help explain concepts of data analysis, fugacity, heat exchangers, and PID controllers. The comics have provided students with â €˜visual notesâ€™ that allow them to revisit the lectures where the concepts were originally taught; instead of relying on the plain text in their notes, the comics depict a lecture or a discussion between professor and students to represent the original instruction and contribute to a greater student understanding.

In the first two semesters of implementing the comics in chemical engineering courses, students have shown moderate improvement on exams and technical papers addressing the specific concepts addressed by the comics. Further, students have indicated significant improvement in their confidence in addressing the topics. Over 20 other universities and institutions have now also implemented the comics in their courses to varying degrees of success. Based on the student interest and response, further comics are being developed with Northeasternâ€™s College of Art, Media and Design as Capstone projects as well as in collaboration with other colleges. This presentation will discuss current efforts in the research and further assessment.

DESIGN OF DIFFERENTIATED CURRICULA IN FRESHMAN CHEMICAL ENGINEERING COURSES

Dr. Brenda Hutton-Prager, University of Mississippi, Oxford, MS
bhprager@olemiss.edu

Introduction

Students typically enter their first year of college with differing levels of college-preparedness, attitudes towards study, study skills, metacognitive skill, academic performance, grasp and comprehension of the English language, and level of independence. One possible influencing factor on these varying maturity levels is the level of cognitive development currently practiced by each student, best described by Piaget and post-Piagetian theory (*1, 2, 3, 4*). This wide disparity is particularly noticeable in lower-performing states with respect to K-12 education. The State of Mississippi, for example, ranked 50th in education indicators in 2014 (*5*), while in 2015, Mississippi ranked 50th and 49th in general mathematics and reading testing respectively (*6*).

The student population at the University of Mississippi (UM) currently contains 60% resident from Mississippi (*7*), and hence wide variation of the identifiers listed above is observed at freshman level. This disparity is even broader when one factors in particularly high performing students also enrolling in UM via the Sally McDonnell Barksdale Honors College and Center for Intelligence and Security Studies, or students participating in specialist programs such as service projects overseas, Engineers Without Borders, the Center for Manufacturing Excellence, the Mississippi Excellence in Teaching Program, and the Ole Miss' Chinese Language Flagship Program (*8*). In an Ole Miss (UM) news blog discussing 2014 enrollment statistics (*9*), it was stated that 'many' students were the first in their families to attend college; and that the overall student population was diverse in both culture (over 90 nationalities represented) and in age (ranging from 15 – 87 years).

The ease with which educators successfully deal with these differences partly influences the overall learning progression for each student, as well as student retention rates in their original STEM choice of course. Most professors are not formally trained in teaching methodology, and develop their skills via a range of activities such as teaching as they were taught (lectures with little student involvement), self-learning, and/or participating in meetings / short courses / conferences on new teaching methodologies through organizations such as the American Society of Engineering Education (ASEE) or First Year Engineering Education (FYEE). Many active learning and engaging student strategies are presented in such gatherings to assist professors in developing their courses, however few strategies presented deal specifically with overcoming issues of disparity amongst students in a class.

Differentiated learning (DL), commonplace in many K-12 educational settings, is frequently used to effectively address differences within a class such that *all* students can learn effectively, be adequately challenged, and significantly improve their critical thinking skills, amongst other benefits. However, it is not well-practiced in most post-secondary educational institutions. The current educational research is investigating how to adapt this technique to college level, implement it into large classes (> 30 students), and gain momentum in widespread implementation across all freshmen engineering classes. In addition to the benefits mentioned, it is hoped that longer term retention of knowledge, higher engagement, and enhanced learning independence of students will be achieved.

The context of this work is primarily introductory STEM courses in freshman year at UM.

Background

Summary of Differentiation and its Characteristics

Many learning strategies, pedagogies or theories such as Active Learning (AL), project-based learning (PBL), Gardner's multiple intelligences and learning modes, Vygotsky's constructivist theory, and Bloom's taxonomy, amongst others, contribute positively to enhanced student learning. Most benefit is gained however when these strategies are considered under an umbrella of a differentiation framework (DF) (*10, 11, 12*). This is because within a DF, student need is identified early into the course, and activities utilizing the various strategies outlined are tailored for the student to progress their learning from their current level of understanding.

Characteristics of a DF are well defined in the literature for K-12 education (*10, 12*). The DF commonly described has the following five key differentiation principles (DP) (*12*):

1. Understand **student needs** and preferred learning modes
2. Focus on **key concepts** and provide **multiple approaches** to learning
3. Provide **challenging learning experiences** within each student's Zone of Proximal Development (ZPD)
4. Foster **collaboration** between student and student, and student and faculty
5. Create **independent learners** and student ownership of learning

DL has gained wide acceptance in the K-12 education sector, where it is generally believed to be the best tool currently available to address differences within a class (*10, 11, 12, 13, 14, 15, 16, 17, 18, 19, 20, 21, 22*). These many examples have shown improvements of student outcomes in a particular class as a result of implementing differentiated techniques. While differentiation is commonly believed to assist struggling learners, it can also be used to extend the more advanced students by giving them appropriate challenging tasks to maintain their learning growth as well (*17*). This tool is not without its critics, where some believe it is an idealized theory unable to be practically implemented (*23*).

Closely linked to DL is the necessity to conduct formative assessment, whereby teachers identify the current level of student knowledge and issues during the course / topic delivery. This 'assessment for learning' guides teachers into making necessary changes based on their students' needs part-way through a topic and hence improves their learning and understanding in readiness for summative assessment. Black and Wiliam (*24, 25*) argued emphatically that in order to raise standards in a classroom, it was necessary for teachers to regularly make use of such formative assessment tasks. They showed overwhelmingly through more than 250 publication reviews that formative assessment improved student outcomes. Stiggins (*26*), also backed these views by calling for increased usage of 'assessment for learning' tools to help diagnose student needs during the learning phase, in order to promote a more successful outcome. More recently, Popham (*27*) again supported the need for formative assessment, but questioned why so few teachers regularly used such strategies to target student needs and hence improve their learning. These works demonstrate the need to train faculty on existing techniques that have already been well-proven.

In the higher education sector DL has been less utilized, however there are some studies that have demonstrated similar student benefits when differentiated techniques have been implemented into higher education classes. Chamberlin and Powers (*28*) implemented a differentiated curriculum for first year mathematics students in the Rocky Mountain region, USA, by introducing the main principles of differentiation from K-12 settings. They found clear improvements in student content knowledge and their perception of learning compared with a control group. Konastantinou-Katzi,

Tsolaki, Meletiou-Mavrotheris and Koutselini (*29*) also created a differentiated program for 27 first year engineering mathematics students. Importantly, they reported 13 additional A-grades and 15 less 'C-or-below' grades upon completion of their differentiated program. There are significantly more studies in higher education STEM settings reporting the benefits of strategies such as PBL, design-based learning, experiential learning, and clicker usage in post-secondary settings (*30, 31, 32, 33, 34*), which can be used as tools within a DF.

In a study several decades ago, up to 50% of students entering college were identified as still displaying concrete operational thinking, based on Piaget's cognitive developmental stages (*1*). Growing evidence was cited describing the development of formal operations (Piaget's final cognitive development stage) well into adult years. In the 1970s in Illinois, a federally funded DOORS project (*4*) emphasized freshmen learning 'thinking skills' relevant to their respective disciplines. Through targeted thinking-skill activities, it was shown that most students participating in the program achieved similar intellectual progress as their peers, and retention rates improved. While not officially a 'differentiated' study, this work does nevertheless demonstrate improved student outcome when student need is formally addressed.

Assuming that many students entering college today are still performing at concrete operational thinking, higher educational instruction must cultivate the formation of this final stage of cognitive development as well as post-Piagetian cognitive thinking. Appropriate strategies could be incorporated into a differentiated program such as that currently being pursued. Given the widespread success apparent in many classrooms around the world as a result of DL, this method will assist educators in dealing with wide disparity in a class.

Differentiation Framework for Higher Education

The differentiation principles described earlier are also relevant at college level, and indicate a progression of learning from understanding student needs through to creating independent learners. Despite the applicability of these principles to higher education, additional focus is required on later-acquired cognitive thinking. The relevant theories discussing cognitive development include the final cognitive steps in Piaget's theory (*1, 4*) and also the fostering of creative thinking in post-Piagetian theory (*2, 3*). Another useful text specifically developing creativity in engineers is that by Walesh (*35*), which will also be used by the authors to develop the framework further. These theories will be incorporated practically within the framework in DP3 and DP5, with learning activities targeting formal operational thinking and scientific creativity. These integrate different perspectives and contradictions into what is known as dialectical thinking (*2, 3*).

Methodology

An iterative approach to the development of differentiated STEM curricula for freshman classes was adopted. Using this technique, various activities matching the five differentiation principles were prepared and delivered to ChE101: Introduction to Chemical Engineering, at UM. With careful data collection of each activity, data analysis and reflection on the degree of success, such activities and overall curricula will be modified iteratively until a robust program is fully developed.

Key activities included formative testing; differentiated homework sheets; knowledge-retention testing; experiential learning activities; independent study assignments; and study skills enhancement based on the same five principles. Data collection and analysis ranged from student incoming characteristics, skills and learning preferences, through to feedback regarding the key activities. Pre-topic tests were compared with post-topic tests to quantify the learning achieved, and regular quizzes throughout the semester testing earlier information were used to judge retention of knowledge. Independence of students was assessed at the beginning of the semester, and then reviewed later on with more robust, independent study assignments. All information gathered was used reflectively to continue to build or modify the course in order to fully tailor the learning for each student.

Results and Discussion to Date

Differentiation Framework

A framework based on the five differentiation principles described above was developed by Dr. Hutton-Prager. A summary of this framework appears in Table 1, where a brief description of each differentiation principle and examples of class / learning activities are presented. To be proficient in implementing this framework, an educator needs to develop a 'teacher tool-kit' with a wide range of activities from which to draw upon for the class. Identifying characteristics and learning needs of a class at the beginning of a semester will to a large extent dictate the selection of appropriately-matched tasks for learning. Activities within this 'tool-kit' need to range from those that teach the basic concepts through to more challenging assignments to develop critical thinking skills and depth of learning; collaborative projects; and activities that lead to independent learners.

Table 1. Brief summary of the differentiation principles (DP) and their applications

DP	What does this DP mean?	Examples / characteristics of DP
1	Identify knowledge of entry-level student ability for course; determine preferred learning modes as well as needs / issues throughout each topic	Pre-topic tests; questionnaires on preferred learning modes; formative testing within a topic to identify potential issues.
2	Identify key concepts of the course and the minimum requirements; deliver learning via a range of learning activities based on feedback from students' preferred learning modes	Lectures with gaps to encourage engagement; student development of movie clips / talks; group participation with Google Apps; thinking experiments; actual experiments; think-pair share activities; flipped classrooms.
3	Expand the concepts by providing challenging and in-depth activities to promote student growth, critical thinking, and problem-solving skills. Sufficient variety of tasks required to target each students' zone of proximal development (ZPD).	Differentiated homework sheets; longer-term assignments such as computer simulations, design projects, or problem-based learning activities; individual and/or team work.
4	Provide collaborative activities to prevent student isolation, foster trust amongst students, and develop collaborative learning approaches between students and faculty.	Team work assignments as discussed in DP3; class activities done in specific groups for specific outcomes; explicit teaching of team dynamics.
5	Encourage students to become independent learners through explicit teaching of learning habits, also based on the five DP. The aim is to develop student confidence in approaching new tasks successfully.	Explicit teaching of study skills that foster independent learning; independent study topics (such as contentious or ethical topics) for students to practice their study skills to learn independently.

Gathering Initial Data (DP1)

Students in the ChE101 class were asked several questions to elicit information regarding their 'starting' characteristics, such as learning preferences, confidence with mathematics and chemistry, and existing knowledge. This information was used to help inform future planning of the classes.

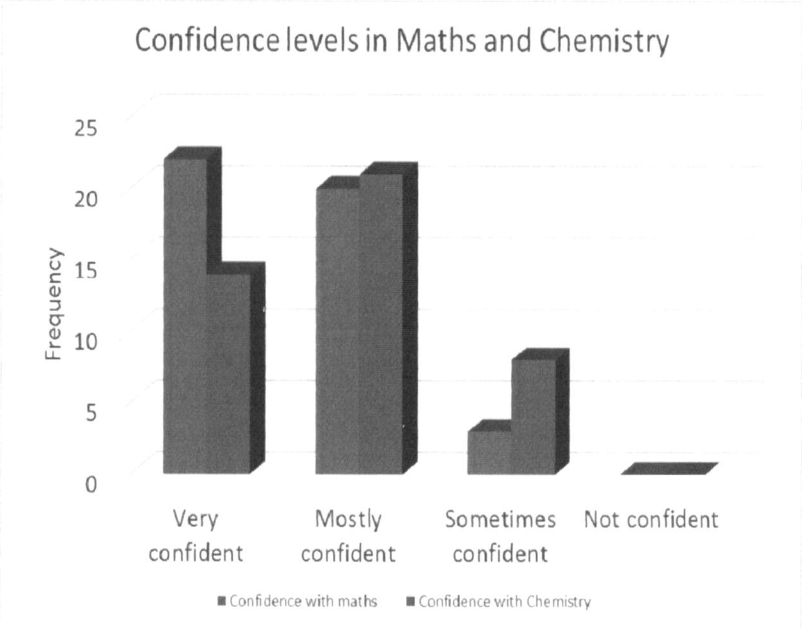

Figure 1: Student responses to the question about their confidence levels in mathematics and Chemistry

Figure 1 shows the students' perception of their confidence in mathematics and chemistry at the beginning of the course. Virtually all of the class recorded "very" or "mostly" confident, with only small numbers indicating less confidence. However, a pre-topic test issued also at the beginning of the course portrayed a different result. An aggregate of five questions asking straightforward statistical, interpolative and linearizing of equations found that only 60% of the class obtained correct answers, while four chemistry questions asking basic mole, density and temperature calculations resulted in an average of 64% of the class obtaining correct answers. These questions were designed to identify common misconceptions with both mathematics and chemistry, and showed that this was the case for many students.

Figure 2 shows the breakdown in preference of typical learning activities within the class, showing a fairly even division between group and individual work. Interestingly, 'traditional teaching' of copying down lecture notes, while not the most common response, still had a credible rating in the overall statistics. An 'Index of Learning Styles Questionnaire' by Felder (*36*), was completed by all students, and averages showed strong preferences in "sensing" and "visual" learning (72 and 82% respectively); approximately even distribution between "active" and "reflective" learning modes; and a slight preference of "sequential" over "global" learning (62%). The strong preference for sensing learning matches the strong dependence of concrete operational skills from Piaget's theory noted by several educational researchers regarding incoming freshmen students (*1, 2, 3, 4*). Typical characteristics of sensing learners include being able to solving problems using routine methods, and difficulty in proceeding with a problem that slightly deviates from a routine solution. This is indicative of performing at the concrete operational level of cognition, and therefore progression to formal operational thinking and post-Piagetian creative thinking is a necessary goal for this class.

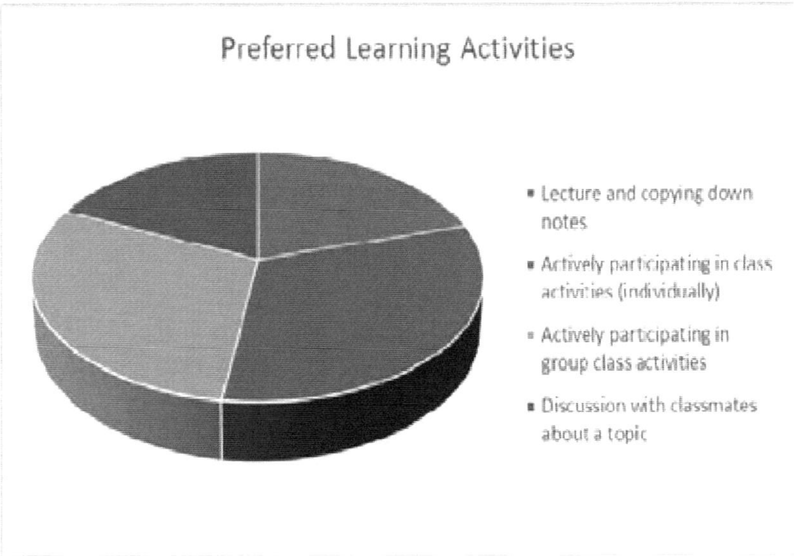

Figure 2: Breakdown of preferred learning activities within the class

Challenging Learning Activities (DP3)

A series of 'differentiated homework sheets' were planned for the class. These were designed with three questions in 'beginning' level (just below minimum standard); three at 'intermediate' level (at standard) and three at 'advanced' level (above standard level to score an A in the subject). The exam will also consist of questions from all three levels, with the majority of students being able to achieve an average grade, and the more capable ones able to achieve an A grade.

Students were required to answer any four questions by choosing the category in which they felt most comfortable, plus one additional question to push them to the next level. Advanced students chose an easier question first to confirm their knowledge before attempting the three more difficult questions. Analysis of questions chosen in the first homework sheet found that 42% of responses were at the beginning level; 45% at the intermediate level, and 13% at the advanced level. These figures indicated that students were making use of the extra support questions to learn the concepts, with some also aiming higher at the advanced level. Choice of response will continue to be monitored over the remaining homework sheets to look for trends and also compare with tests and exams.

Conclusions and Future Work

Implementation of DL in ChE101 is in progress, and preliminary data collected to date has exposed some important characteristics of the class, such as differences between perceived and actual ability in mathematics and chemistry, learning preferences within the class, and a large number of students cognitively functioning at the 'concrete operational thinker' level. The first differentiated homework sheet revealed that 42% of the class chose least one homework problem from the 'beginning' question bank, with only 13% choosing one or more problems from the advanced level.

Learning will continue to be delivered to the class in a differentiated manner, and data / feedback collected to inform future lessons and refine the differentiation framework for wider dissemination into the future.

References

1. Tomlinson-Keasey, C. A. (1978). Chapter 1: Piaget's Theory and College Teaching, *Essays from and about the ADAPT program, Paper 29*, pp 1-6.
2. Yi-Chi Chang, Y., Chiou, W-B. (2014). Diversity believs and postformal thinking in late adolescence: a cognitive basis of multicultural literacy, *Asia Pacific Educational Review, 15*, pp. 585-592.
3. Wu, P-L, Chiou, W-B. (2008). Postformal thinking and creativity among late adolescents: a post-piagetian approach, *Adolescence, 43(170)*, pp 237-251.
4. Campbell, T., Zucco, K., McGill, P., Taylor, K., Thompson, R. (1979). DOORS Project: Development of Operational Reasoning Skills, *Illinois Central College, East Peoria*, retrieved from: http://files.eric.ed.gov/fulltext/ED203775.pdf
5. Chalk, S. (2015). Quality Counts Introduces new State Report Card; U.S. Earns C, and Massachusetts Ranks First in Nation, *Education Week Research Center*, retrieved from: http://www.edweek.org/media/qualitycounts2015_release.pdf
6. Bernardo, R. (2016). 2015's States with the Best and Worst School Systems, *WalletHub*, retrieved from: https://wallethub.com/edu/states-with-the-best-schools/5335/
7. The University of Mississippi office of admissions (2016a). University of Mississippi quick facts 2015, retrieved from: http://www.olemiss.edu/aboutum/quickfacts/
8. The University of Mississippi office of admissions (2016b). The View Book 2016-2017, retrieved from: http://www.olemiss.edu/info/admissions/admissions-guides/index-viewbook.html
9. Diggs, M. (2014). UM Enrollment Tops 23,000 Students for Fall Semester, *The University of Mississippi News*, retrieved from: http://news.olemiss.edu/um-enrollment-tops-23000-students-for-fall-semester/
10. Rock, M. L., Gregg, M., Ellis, E., Gable, R. A. (2008). REACH: A Framework for Differentiating Classroom Instruction. *Preventing School Failure,* 52(2): 31 – 47.
11. Tomlinson, C. A. (1999). *The Differentiated Classroom, Responding to the Needs of All Learners.* Alexandria, VA: Association for Supervision and Curriculum Development.
12. Prager, B. H. (2013*). An analysis of the Current VVCE Mathematics Study Design and Comparison with other Equivalent Designs with respect to the Principles of Differentiation.* Masters Thesis, The University of Melbourne.
13. Watts-Taffe, S., Laster, B. P., Broach, L., Marinak, B., McDonald Conner, C., Walker-Dalhouse, D. (2012). Differentiated Instruction, Making Informed Teacher Decisions, *The Reading Teacher, 66(4)*, pp. 303-314.
14. Valiandes, S. (2015). Evaluating the impact of differentiated instruction on literacy and reading in mixed ability classrooms: Quality and equity dimensions of education effectiveness, *Studies in Educational Evaluation, 45*, pp. 17-26.
15. Nicolae, M. (2014). Teachers' beliefs as the differentiated instruction starting point: research basis, *Procedia Social and Behavioral Sciences, 128*, pp. 426-431.
16. Wu, E. H. (2013). The Path Leading to Differentiation: An Interview With Carol Tomlinson. *Journal of Advanced Academics,* 24(2): 125 – 133.
17. Manning, S., Stanford, B., Reeves, S. (2010). Valuing the Advanced Learner: Differentiating Up. *The Clearing House,* 83(4): 145 – 149.
18. Herold, B. (2016). Literacy Gets a personalized Makeover, *Education Week Spotlight On Differentiated Instruction, Jan*, pp. 2-4.
19. Sparks, S. D. (2016). Differentiated Instruction: A Primer, *Education Week Spotlight On*

Differentiated Instruction, Jan, pp. 7-8.

20. Bullock, J. (2016). Differentiated Instruction, One Size Does Not Fit All, *Education Week Spotlight On Differentiated Instruction, Jan*, pp. 9-11.
21. Davis, M. R. (2016). District's Ambitious Effort Fuels Personalized Learning, *Education Week Spotlight On Differentiated Instruction, Jan*, pp. 12-15.
22. Tomlinson, C. A. (2016). Commentary: To the Contrary: Differentiation Does Work, *Education Week Spotlight On Differentiated Instruction, Jan*, pp. 16-17.
23. Delisle, J. R. (2016). Commentary: Differentiation Doesn't Work, *Education Week Spotlight On Differentiated Instruction, Jan*, pp. 15-16.
24. Black, P., Wiliam, D. (1998). Inside the Black Box: Raising Standards Through Classroom Assessment, *Phi Delta Kappan, 80(2)*, pp. 139-144, 146-148.
25. Black, P., Wiliam, D. (2010). Inside the Black Box: Raising Standards Through Classroom Assessment, *Kappan, V92(N1)*, pp. 81-90.
26. Stiggins, R. J. (2002). Assessment Crisis: The Absence Of Assessment FOR Learning, *phi Delta Kappan,* pp. 758-765.
27. Popham, W. J. (2013). Waving the Flag for Formative Assessment, *Education Week*, retrieved from: http://www.edweek.org/ew/articles/2013/01/09/15popham.h32.html?r=1954966171
28. Chamberlin, M. & Powers, R. (2010). The promise of differentiated instruction for enhancing the mathematical understandings of college students, *Teaching Mathematics and Its Applications*, Oxford University Press, 29: 113-139.
29. Konstantinou-Katzi, P., Tsolaki, E., Meletiou-Mavrotheris, M., Koutselini, M. (2013). Differentiation of teaching and learning mathematics: an action research study in tertiary education. *International Journal of Mathematical Education in Science and Technology,* 44(3): 332 – 349.
30. Han, J. H., Finkelstein, A. (2013). Understanding the effects of professors' pedagogical development with Clicker Assessment and Feedback technologies and the impact on students' engagement and learning in higher education, *Computers & Education, 65*, pp. 64-76.
31. Farrell, S., Cavanagh, E. (2014). Biodiesel production, characterization, and performance: A hands-on project for first-year students. *Education for Chemical Engineers*, 9: e21 – e31.
32. Mantri, A. (2014). Working towards a scalable model of problem-based learning instruction in undergraduate engineering education. *European Journal of Engineering Education,* 39(3): 282 – 299.
33. Gomez Puente, S. M., van Eijck, M., Jochems, W. (2011). Towards charactersing design-based learning in engineering education: a review of the literature, *European Journal of Engineering Education, 36(2)*, pp. 137-149.
34. Golter, P. B., Thiessen, D. B., Van Wie, B. J., Brown, G. R. (2012). Adoption of a Non-Lecture Pedagogy in Chemical Engineering: Insights Gained from Observing an Adopter. *Journal of STEM Education,* 13(5): 52 – 61.
35. Walesh, S. G. (2017). *Introduction to Creativity and Innovation for Engineers*, Pearson Education Ltd, London.
36. Soloman, B. A., Felder, R. M. (n.d.) Index of Learning Styles Questionnaire, NCSU, retrieved from: http://www.engr.ncsu.edu/learningstyles/ilsweb.html

465446 Rapid Prototyping: Impeller Design Using 3-D Printing and Â Data Analysis to Enhance Creativity and Critical Thinking in the Chemical Engineering Curriculum

Abstract:

Integration of educational modules in core chemical engineering courses is an effective way to familiarize students with new technologies within the standard four-year undergraduate curriculum. This work involves the development of educational modules and laboratory experiments to introduce 3-D printing rapid prototyping and advanced data analysis to students. The modules and experiments are self-contained and flexible allowing for their use in a wide range of courses at every level of the curriculum. Rapid prototyping is becoming essential for efficient and cost-effective equipment and process development in many industries. For many applications, ranging from biomedical devices to industrial scale impeller design, 3-D printing is an important element in the rapid prototyping tool box of the chemical industry. The educational modules presented here familiarize students with the rapid prototyping design process. Students design and 3-D print impellers and use experimental design techniques to test these in laboratory scale tank-type mixing equipment. Solid/liquid system cloud height and off-bottom velocity are used to characterize the process, and to develop models for mixing efficiency as a function of process parameters (liquid height and impeller distance from tank bottom), fluid viscosity and particle properties. The mixing characteristics of polylactic acid (PLA) 3-D printed axial and radial impellers were studied as well as mixing characteristics of more complex designs such as jet and lily impellers. The effects of 3-D printing parameters, impeller material of construction, and water/glycerin solution viscosities (1 cP to 12 cP) on mixing efficiency for solid (40 mesh sand)/liquid systems were investigated using a Box-Behnken experimental design and will be discussed. In addition, flow visualization experiments using complex impellers will be presented. The focus of this work is the development of 3-D printing rapid prototyping educational modules and laboratories that include design, manufacturing, statistical experimental design and data analysis to enhance creativity and critical thinking in the engineering curriculum while familiarizing

students with new technologies and processes. Educational modules for sophomore level courses and laboratory experiments for senior level unit operations courses will be presented.

467831 Stealing a Freshman-Level Separations Project

Abstract:

At California Baptist University, students from all engineering majors take Introduction to Engineering Design during their freshman year. For spring 2014, the faculty wanted to change the nature of the design project and create a project/competition that would touch on a variety of engineering disciplines. The challenge was to find a component of the project that could relate to chemical engineering while being part of a larger project with components from other disciplines and avoiding the need for students to handle chemicals. The faculty discovered and "stole" a project from Ohio University that involved separations on a macro-scale using beads and other objects.

Use of "the Separation Project" in the Ohio University chemical engineering program was motivated by a desire to include a hands-on project early in the curriculum that could be safely conducted outside of a laboratory class, and was inspired by a project described in a 1996 Chemical Engineering Education article by Carol McConica. [Carol McConica, "Freshman Design Course for Chemical Engineers", Chemical Engineering Education, 30(1), pp. 76-80, 1996.] The core goal of the project is to design, build, and test an automatic system to separate a mixture of small objects. It provides an allegory for chemical separations, in which the chemical engineer identifies and exploits differences in the properties of the components in order to separate them from a mixture. It illustrates technical principles such as recovery and purity and the difficulty of achieving high levels of both, the effect of changing the order of separation steps, and the reality of variability in operation. It can also be used as a vehicle to develop teamwork and communication skills, to teach the iterative nature of the design process, and to introduce principles of economics or of statistical analysis.

The core concept of the Separation Project can be modified by changing the number and types of objects and by setting different milestones and deliverables to be a useful learning vehicle in a variety of settings. The project was initially assigned by Young at Ohio University in the sophomore "Material Balances" course, but has moved to the freshman Introduction to Chemical Engineering course to provide an early design experience in the curriculum. Young also used this project for three years in a class of STEM education majors, to give them experience with the engineering design process and how they might use it in their own K-12 classrooms. A teacher in Texas has adapted it for use in a summer engineering program for rising high school freshmen.

At CBU the separations project was incorporated as the third and final module of a team competition. Teams would begin with a container filled with objects. The first module needs to acquire and transfer the objects to a transport unit. The transport unit moves the objects to the separations unit. Then the separations unit exploits property differences to try to achieve high recovery and purity (as was done with the project at Ohio University). Teams are made up for students from various majors and broken up into three groups, one for each module. Students are exposed to principles of chemical, mechanical, electrical, civil, software, and biomedical engineering. Students learn about the basic design process and develop teamwork and communication skills. Students also encounter the difficult task of interfacing with other groups on the teams, where they need to find a balance between optimizing performance of their module with meeting the needs of the other modules and the team as a whole. The final competition occurs during a large Design Showcase where the public is invited and design projects from juniors and seniors are also put on display.

472687 Impact of Peer Mentoring on Undergraduate Student Mentees and Their Peer-Mentors in Both the Near and Long Term

Abstract:

The Partnership for Retention Improvement in Math, Engineering, and Science (PRIMES) is an NSF STEP program implemented across 8 different departments in 3 different schools (Engineering, Education, and Arts and Sciences) at the University of Louisville. The cornerstone of this program is the development of learning communities and supplemental instruction teams based on the Colorado Learning Assistant model to support several of the foundation courses for each of the participating departments. In Chemical Engineering, these courses include Introduction to Chemical Engineering, Introduction to Materials Science and Material and Energy Balances. In the 5 years of the programs existence, there have been over 50 Chemical Engineering peer-mentors who have impacted hundreds of undergraduate students. In this talk we will discuss the impact of the PRIMES program on both the mentors and their mentees in terms of both classroom performance as well as longer term impacts as they matriculate through the program and into the job market.

473046 Teaching and Assessing Critical Thinking in a Course-Based Senior Research Laboratory in Chemical Engineering

Abstract:

We have redesigned our senior two-quarter, required laboratory sequence with an overarching emphasis on critical thinking and decision-making based on data. We believe that the need for such quantitative critical thinking is universal for all of our chemical engineering students, and although it is an important goal of many classes, it is rarely achieved. As such, we have redesigned our two-quarter sequence using a scaffolded approach. In the first quarter, students are given an experimental system (an engineered E. coli strain that produces fatty acid ethyl esters (FAEEs) as a biofuel) and asked to optimize the timing of the induction step to produce the biofuel product. Students engage in iterative cycles of deciding what data to collect, analyzing their data, and making decisions on what experiments to conduct next. They are given an opportunity to cycle through this process several times throughout the quarter, while also presenting their analysis and conclusions with their colleagues.

In parallel to these activities, teams of 3-4 students develop proposals for research projects to conduct in the second quarter, which are all aimed at improving renewable energy opportunities. These student teams refine their proposals over the duration of the first quarter and produce a final proposal with clear research goals and experimental plans to address those goals. During the implementation of their student-designed research projects in the second quarter, student teams continue to develop the skills that were the focus of the first quarter, while also learning how to develop and troubleshoot their own experimental progress. This culminates in a public poster session where students present their research projects and findings to a broad audience including faculty, staff, graduate students and undergraduate students in the department.

The effectiveness in achieving the student learning goals was measured using both informal student interviews as well as direct evidence of specific activities in student work examples that were adapted from the Wieman's Experimental

physics "cognitive task analysis" (*The Physics Teacher*, 2015).

471759 Beyond Gut Feeling: Teaching "Engineering Intuition" in the Technology Age

Abstract:

Technology is an essential component of modern engineering problem-solving. In both industry and academia, computer programs allow us to solve more complicated problems in a fraction of the time it would take to solve manually. Modeling software allows us to simulate entire plants before physically building a single unit.

With the rise of software as a tool in engineering problem-solving, comes the need to teach our students the difference between 'making the machine work for you' and 'working for the machine.' Often, the convenience of software solutions manifests in complacency—eager students accepting solutions the software provides without question. The ability to recognize whether a software's solution (or assumptions used) is reasonable or ridiculous—engineering intuition—has become an important skill to teach.

To better understand what factors may contribute to a students' engineering intuition, we gave students open-ended simulation problems that were always solvable by the software, but the scenario was not always physically realistic. Students were pressed to interpret the solution and comment on the scenario in the problem—for example, would they recommend the distillation column the software designed, and why? We then considered potential correlations between students' ability to assess a simulation-based problem's practicality and a number of factors: student internship experience, program of study, ideal job, GPA, and learning styles.

This multi-disciplinary study is ongoing at two U.S. non-PhD-granting institutions. Required junior-level courses (chemical engineering at one institution, aeronautical at the other) with a heavy simulation component are being evaluated at both institutions. Preliminary results suggest correlations may exist between practical experience and ability to assess simulation problems. The one correlation found at both institutions suggests that high-achievers (characterized from GPA) are the most likely to take extra steps (such as perform additional calculations or look up real-world scenarios) when required to thoroughly assess a solution.

Continuing work will shed more light on what factors are most involved in student ability to assess simulations, and allow us to refine our instruction of engineering intuition through the use of computer simulations in the classroom.

470396 Vertical Integration of the Chemical Engineering Curriculum By Expansive, Progressive Projects

Abstract:

Too often, students identify each course in their curriculum as isolated and unrelated to any other course, before or after. This can lead to significant confusion or anxiety when the student is â€œsuddenlyâ€ forced to incorporate all these topics into their senior design and senior lab courses. Often, these students are left at a severe disadvantage by being forced to augment their understanding and means of analysis of a system - all because of a lack of understanding of course-to-course cohesion. A project-based approach to the entire curriculum can be utilized to avoid this situation. Years prior to the senior design project, students can be organized into groups and encouraged to analyze an overall industrial process from day one of their engineering education. They start with the most basic analysis of material and energy balances, but then expand those analyses as they progress. Instead of simply observing a whole new project with each subsequent course, they continue with the same process, utilizing material with each new course and correcting/optimizing their analysis. By the time they reach their senior year and their senior design course(s), they will have fully analyzed a single process from all perspectives and are better prepared for the detailed design criteria expected of them. Not only are they better prepared for their senior level courses, they are also more practiced for their post-academic careers.

471518 Revisiting the Tubular Flow Reactor Experiment

Abstract:

Revisiting the Tubular Flow Reactor Experiment

Yanir Maidenberg and J. Patrick Abulenicia

Abstract

The tubular flow reactor experiment is a common reaction kinetics example in the unit operations laboratory sequence. In this experiment, students are typically tasked to examine reaction conversion as a function of both volumetric flow rate and residence time. More specifically, students evaluate if the fluid-flow regime (i.e. laminar or turbulent flow) or the time in the reactor plays a more effective role in generating the desired products.

The chemical reaction that has been traditionally used at our institution, as well as many other institutions, is the saponification of ethyl acetate to form ethanol. Briefly, solutions of sodium hydroxide and ethyl acetate are prepared in separate vessels and then introduced simultaneously at prescribed flow rates into the tubular flow reactor. Students can calculate the degree of conversion for the reaction by either monitoring the conductivity - either at the inlet or outlet - or by performing a titration against hydrochloric acid to calculate uncreacted sodium hydroxide. This process would be repeated over several volumetric flow rates to cover the laminar and turbulent regimes. Although this is a robust reaction, the authors began exploring ways to improve some areas of the tubular flow reactor experiment. First, the conductivity measurements from this setup exhibited great fluctuation as the reactants flowed, which made it difficult for students to reliably collect data. Second, variations among different groups were found due to the bias inherent in determining start point and end point of titrations. Third, sodium hydroxide is a strong base, and can be dangerous if handled improperly (it has a health rating of 3). Last, it was found the reaction proceeded too slowly to discern any statistically meaningful variations.

To this end, the authors began to explore other possible chemical reactions for this experiment, with the following criteria: 1) it had to be safe, 2) conversion should be easily measured, and 3) relatively inexpensive. A fourth (but less important) criterion is that the reaction undergo a color change, so that visual confirmation of the reaction conversion could be observed in the clear glass reactor at the institution. In this presentation, the authors will outline several alternative reactions for the tubular flow reaction experiment, as well as a summary of past work by chemical engineering educators on this topic.

"DESIGN-IFYING" BIOCHEMICAL ENGINEERING SEPARATIONS

Incorporating Small Design Projects into a Traditional Lecture Course

Jason R. White
Department of Chemical Engineering
University of California Davis
2016 AIChE Annual Meeting, San Francisco, CA

Objective/Motivation

- To bring design problems into traditional lecture courses
  - Get students more engaged
  - Improve course-level student outcomes
- To better prepare students for their senior plant design project experience
- Start to fill in the gap in design-related activities between the freshman year design experience and the senior year plant design project

Biochemical Engineering UG Curriculum @ UCD

Freshman			Sophomore		
Fall	Winter	Spring	Fall	Winter	Spring
Mathematics					
Chemistry					
Ethics	**Coffee Lab**	GE	Chemistry Lab	Material Balances	Matlab
Writing	GE	Physics			Biology

Junior			Senior		
Fall	Winter	Spring	Fall	Winter	Spring
Applied Math	Fluids	Heat Transfer	Economics	Biochm Kinetics	**Plant Design**
Thermo 1	Thermo 2	Mass Transfer	Kinetics	Biochm Seps	Biochm Lab
Biology	Thermo Lab	GE	Controls	Biotech Facility	Biochm Elect
GE	GE	Transport Lab	Biology	Biochm Elect	GE

Implementation of Mini-Design Projects in the Bioseparations Course

Day	Week 1	Week 2	Week 3
1	Lecture on Module Begins	Lecture	
2	Lecture	Project Assigned In-class group time	Project Due
3	Lecture	Lecture	
4	Assessment from Previous Module Homework #1 Assigned	Homework #1 Due Homework #2 Assigned In-class group time	Homework #2 Due Assessment on the Module

Mini-Design Projects

Module	Centrifugation	Filtration	Chromatography	Precipitation
Key Design Parameters	Rotation speed, ratio of desired to cells to contaminant cells in product	Combined area of 2-stage feed-and-bleed unit	Height equivalent of a theoretical plate, properties of the column packing material	Average velocity, length of tubular flow precipitator, pressure drop
Objective	To maximize the profitability of the process	To meet an outlet purity specification	To achieve an effective separation of product from contaminants	To determine if scaled-up process was realistic

5 9/24/2016 UCDAVIS

Post-Project Survey

Please rate the following questions on a scale of 1-5. Check the appropriate box.

	1 (none)	2 (little)	3 (some)	4 (good)	5 (excellent)
How would you rate your understanding of the material required for this assessment prior to this series of lectures?					
How would you rate your understanding of the material required for this assessment at this point in time?					
How would you rate the educational value of the group mini-project?					
	1 (poor)	**2 (fair)**	**3 (satisfactory)**	**4 (very good)**	**5 (excellent)**
How satisfied are you with how the instructor taught this part of the course?					

What was one thing about this topic that you found particularly interesting?

What about this series of lectures do you feel needs the most improvement?

What feedback (positive or negative) do you have regarding the mini-project?

6 9/24/2016 UCDAVIS

Raw Survey Data

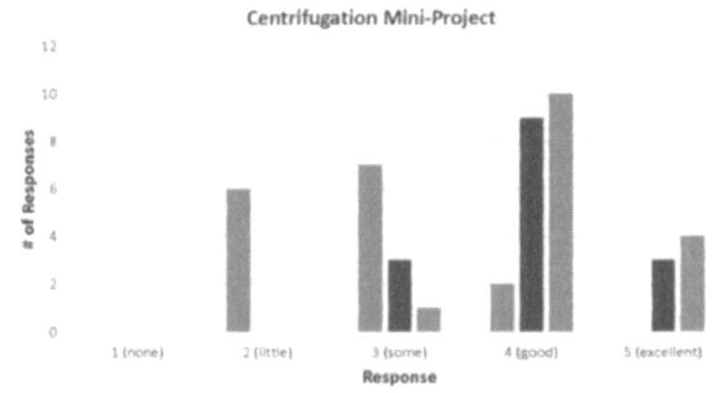

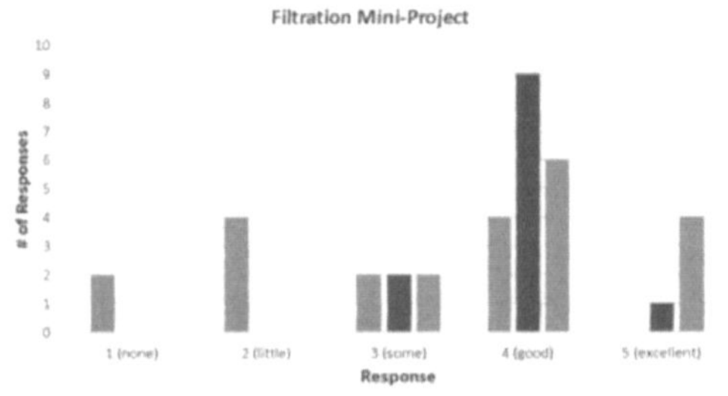

How would you rate your understanding of the material required for this assessment prior to this series of lectures?

How would you rate your understanding of the material required for this assessment at this point in time?

How would your rate the educational value of the group mini-project?

Raw Survey Data

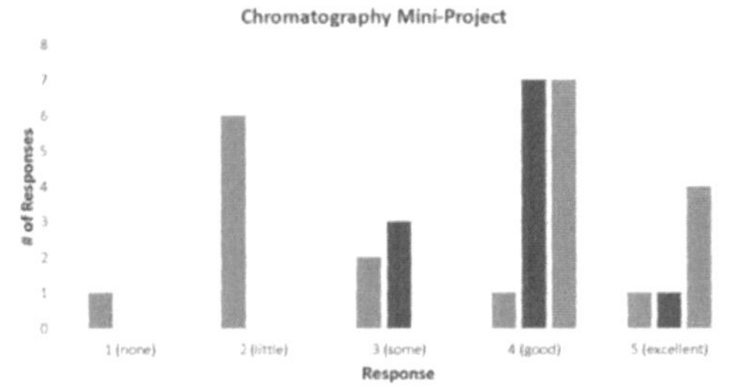

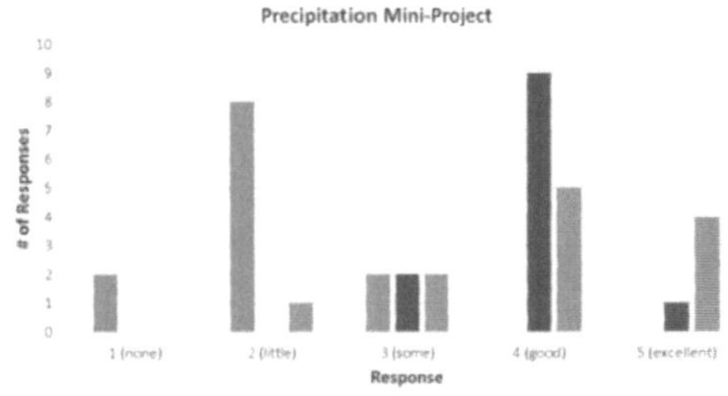

How would you rate your understanding of the material required for this assessment prior to this series of lectures?

How would you rate your understanding of the material required for this assessment at this point in time?

How would your rate the educational value of the group mini-project?

Analysis

Measure	CF (Av ± SD)	F (Av ± SD)	CH (Av ± SD)	P (Av ± SD)
Pre-module perceived understanding*	2.7 ± 0.7	2.7 ± 1.2	2.5 ± 1.1	2.0 ± 0.6
Post-module perceived understanding*	4.0 ± 0.7	3.9 ± 0.5	3.8 ± 0.6	3.9 ± 0.5
Factor increase in perceived understanding	1.5 ± 1.4	1.5 ± 1.9	1.5 ± 1.9	2.0 ± 1.6
Education value of mini-project*	4.2 ± 0.6	4.2 ± 0.7	4.4 ± 0.5	4.0 ± 1.0
Satisfaction with instruction**	3.9 ± 0.6	3.5 ± 1.3	3.6 ± 1.3	3.9 ± 0.8

* Responses: 1(none), 2(little), 3(some), 4(good), 5(excellent)

** Responses: 1(poor), 2(fair), 3(satisfactory), 4(very good), 5(excellent)

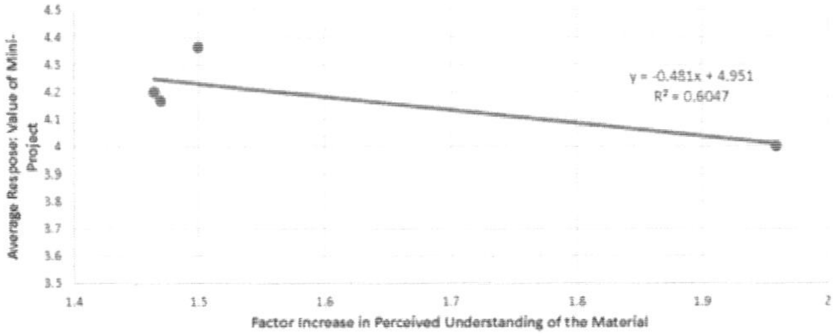

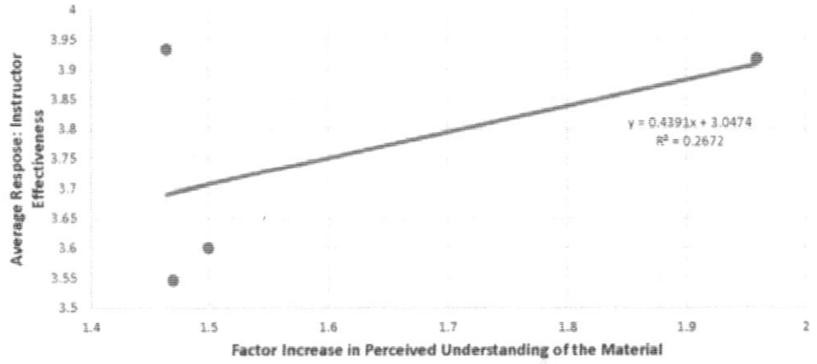

Student Comments

"I really enjoyed the mini project. It pushed us to use prior problem solving skills and tools (MATLAB) to solve a very realistic problem in industry - maximizing profit when dealing with your product and contaminants. We should be able to solve theses types of problems when working in industry and having the opportunity to practice solving these problems now is helping us apply our engineering knowledge."

- In response to the centrifugation mini-project

Student Comments

"The project was a good overall assignment that challenged our understanding of the material."

- In response to the filtration mini-project

"The mini-project was good. That probably doesn't help too much, but overall, the mini-projects have been a big help in understanding the material."

- In response to the precipitation mini-project

Student Comments

Negative Comments:

- Some thought the problems were too easy, others too hard/not enough information provided

- Didn't enjoy group work, flaky teammates

- Not enough class time was provided for groups to meet

13 9/24/2016 UCDAVIS

Conclusions and Future Directions

- Students generally appreciated the design focus of the mini-projects
- Determining if the projects are impacting student outcomes for this course is still a work in progress
    - No control group
- Changes will be made in regards to the timing of when the project is assigned with respect to the module
- Study to be formalized for future offerings of the course
    - Future surveys to include questions relating to ABET students outcomes

14 9/24/2016 UCDAVIS

QUESTIONS?

Thank you for your attendance!

469569 Don't Take the Difficult Path in Solving Systems of Nonlinear Equations

Abstract:

The LearnChemE group at the University of Colorado has produced a voluminous library of Screencasts, ConcepTests and Interactive Simulations that is widely used by the chemical and biochemical engineering community. We became interested in particular with the PolyMath Screencasts and the associated examples (see link below[1]), as we found that these examples can be used for additional educational purposes to those that were originally envisioned by the developers.

For example there are courses associated with the use of the computer for problem solving in ChE and other engineering disciplines with titles and content such as Numerical Methods and Programming. In such courses, artificial examples and assignments are often used for demonstration and training. We have found that using ChE related problems for demonstration and training can increase considerably a student's motivation to learn such subjects. Our textbook (Cutlip and Shacham[2]) provides such examples, and we have found that with the help of the LearnChemE examples we can effectively demonstrate and apply additional principles.

During the screencast where the mathematical model of a particular phenomenon is explained, the user can enter the model equations simultaneously into the PolyMathLite[3] package using an Android Tablet or Smartphone. The screencast can be paused as necessary to complete the problem solution entry. This allows the PolyMathLite software to solve the problem, and the user can compare the results with those obtained by the instructor in the screencast.

Consider for example the "VLE: Wilson's Equation" screencast which details the modeling of the vapor-liquid equilibrium for a binary mixture using Wilson's equation. The equations can be entered into the

PolyMathLIte app. as shown in Figure 1. Note that PolyMathLite automatically detects that the problem involves a system of nonlinear equations when the RUN button is pressed. Note that a HELP is provided within the PolyMathLite app.

Execution of this program yields determination of the 10 unknowns where the critical unknowns are the bubble point temperature (T = 344.226 K), the gas phase mole fraction of component 1,(y_1 = 0.86) and gas phase mole fraction of component 2, (y_2= 0.139).

In addition to possible use in a "Thermodynamics" course, this problem can be of interest for a course that teaches "Numerical Methods". The inexperienced user would think of generating a solution involving 10 implicit algebraic equations with 10 unknowns. This would require initial estimates for all variables. An additional concern is that the presence of logarithmic terms in some equations might cause some of the functions to become undefined for non-positive values of some of the unknowns. The practical solution process can be considerably improved by rearrangement of the equations so that some can be written in explicit form of single variables. Also reformulation of the equations to replace the logarithmic terms by exponential terms. This screencast demonstrates that a systematic method for carrying out such rearrangements and reformulations can be an important contribution to a successful problem solution. This concept is a good practical addition to a "Numerical Methods" course. The revised form of the equation set is shown in Figure 2.

Note that there is only one implicit equation that remains in the revised form (f(T) = 1- y_1 – y_2) with T as the only unknown that requires an initial estimate. A practical interval to bracket the temperature solution can easily be explored. The remaining equations were converted to explicit forms and the logarithmic expressions were converted to exponential expressions. The resulting set of equations was easily solved with PolyMathLite, and the resulting solution agrees nicely with the solution obtained within the screencast video.

Additionally the PolyMathLite solution report also contains a ready to run MATLAB[4] m-file program for solving the same problem when MATLAB is available. The MATLAB program generated for solving the revised form of the VLE Wilson's Equation problem is shown in Figure 3. The PolyMathLite generated MATLAB program can serve as a basis for MATLAB programming exercises of various difficulty levels.

References

1. "LearnChemE – Educational Resources for Chemical Engineering" website of the University of Colorado, Boulder (http://www.learncheme.com/student-resources/polymath)
2. Cutlip, M. B., and M. Shacham, Problem solving in chemical and biochemical engineering with POLYMATH, Excel and MATLAB. 2nd Ed., Prentice Hall, Upper Saddle River, N. J. (2008)
3. PolyMathLite is a product of Polymath Software (http://www.PolyMathLite.com.).
4. MATLAB is a trademark of The Math Works, Inc. (http://www.mathworks.com).

Figures - The figures for this paper did not transfer to AIChE with the paper so they are available for download at the download link:

http://www.polymathlite.com/aiche2016

Figure 1. Input of the VLE Wilson's Equation problem in to PolyMathLite (Partial)

Figure 2. Revised form of this example problem (Partial)

Figure 3. MATLAB program generated by PolyMathLite for solving the revised form of this example problem

470402 Teaching Agglomeration through Simulation in Undergraduate Education

Abstract:

Particle size enlargement by agglomeration is common in a variety of industries, particularly in pharmaceutical and agricultural processes. Therefore, it is frequently taught in particle technology courses. To follow the evolution of particle size with time, population balance equations (PBEs) are commonly used. The problem is that these PBEs are often intractable for undergraduate students due to the fact that these partial integral differential equations only have analytical solutions for a very limited number of cases, and are challenging to solve for other cases. To address this issue, a simulation model was developed for student use. The goal of the software is to allow students to perform simulations for size independent agglomeration kernels and size dependent agglomeration kernels and compare the results. This model can also be used for comparison with experimental results. This preliminary version of the software will be presented and discussed.

Simulation Solutions, Inc.

Minds-On/Hands-On Training for Operators™

"Evaluating Performance in Chemical Engineering Process Safety Education using Simulators"

Mr. Donald C. Glaser
President

Mr. Matthew Garvey
Project Engineer

Simulation Solutions, Inc.
179 Avenue at the Common, Suite 1
Shrewsbury, NJ 07702-4804
dgalser@simulation-solutions.com

Robert G. Bozic, Ph.D.

Lieutenant Colonel U.S. Army(Retired)
Lecturer in Chemical Engineering Design and
Chemical Engineering Process Safety
Department of Chemical Engineering
Columbia University in the City of New York
robert.bozic@columbia.edu

COLUMBIA ENGINEERING
The Fu Foundation School of Engineering and Applied Science

Outline

- Background/Motivation
- Safety Recommendations for ABET
- CHEN E4501 Chemical Engineering Process Safety
- SAChE Integration
- Learning from History (Case Study & Design)
- Using Simulations to Teach and **Evaluate Process Safety Education**
- Acknowledgements and References
- Conclusion and Future Work

2

Background and Motivation

"SAChE Recommendations for ABET Safety Content in Chemical Engineering February 18, 2010"
"The Safety and Chemical Engineering Education (SAChE) Committee makes the following recommendations with regard to student understanding to meet ABET curriculum requirements.

1. The graduate must understand the importance of process safety and the resources and commitment required. This should include the important incidents that define process safety, and how these incidents affected the practice of chemical engineering.

2. The graduate must be able to characterize the hazards associated with chemicals and other agents. This must include toxic, flammable, and reactive hazards.

3. The graduate must understand and be able to apply concepts of inherently safer design.

4. The graduate must understand how to control and mitigate hazards to prevent accidents. This should include generally accepted management systems, plant procedures and designs to prevent accidents.

5. The graduate should be familiar with the major regulations that impact the safety of chemical plants.

6. The graduate should understand the consequences of chemical plant incidents due to acute and chronic chemical releases and exposures.

7. The graduate should be reasonably proficient with at least one hazard identification procedure.

8. The graduate should have an introduction to the process of hazard evaluation and risk assessment. Resource materials to assist in meeting these requirements can be found on the SAChE web."

3

(www.sache.org accessed 13 Apr 2015)

Background and Motivation

"ABET Curriculum Statement approved in October 2011"

"1. Curriculum
The curriculum must provide a thorough grounding in the basic sciences including chemistry, physics, and biology, with some content at an advanced level, as appropriate to the objectives of the program. The curriculum must include the engineering application of these basic sciences to the design, analysis, and control of chemical, physical, and/or biological processes, including the hazards associated with these processes."

4

(www.abet.org accessed 13 Apr 2015)

CHEN E4501 Chemical Engineering Process Safety Course Description

"Aimed at seniors and graduate students. Provides classroom experience on chemical engineering process safety as well as Safety in Chemical Engineering certification. Process safety and process control emphasized. Application of basic chemical engineering concepts to chemical reactivity hazards, industrial hygiene, risk assessment, inherently safer design, hazard operability analysis, and engineering ethics. Application of safety to full spectrum of chemical engineering operations."

((http://www.columbia.edu/cu/bulletin/uwb accessed 27 Sep 2015)

CHEN E4501 Chemical Engineering Process Safety Course Significance

"Chemical Engineering Process Safety will be used anywhere in the world across the complete range of chemical engineering operations. Process Safety is current recommended ABET curriculum content by the Safety in Chemical Engineering Safety Committee. Academia and industry contributed to the content of applying chemical engineering concepts to chemical reactivity hazards, industrial hygiene, risk assessment, inherently safer design, and hazard operability analysis. With the ultimate aim of safer operations in the chemical engineering profession, the applications of the Chemical Engineering Process Safety will make the world safer."

5

(Fall 2015 Course Proposal)

CHEN E4501 Chemical Engineering Process Safety Course Objectives

1. Learn about the importance of process safety and the resources and the commitment required. This should include the important incidents that define process safety, and how these incidents affected the practice of chemical engineering.

2. Learn how to characterize the hazards associated with chemicals and other agents to include toxic, flammable, and reactive hazards.

3. Learn how to apply concepts of inherently safer design.

4. <u>Learn how to control and mitigate hazards to prevent accidents.</u>

5. Become familiar with the major regulations that impact the safety of chemical plants.

6. Learn about the consequences of chemical plant incidents due to acute and chronic chemical releases and exposures.

7. Become familiar with the process of hazard identification and evaluation, and risk assessment

8. Become familiar with engineering ethics.

6

(Spring 2015 CHEN 4501 Syllabus)

Course Design and SAChE Integration

Week 1: Process Safety and Course overview
Intro to Process Safety (Chp 1) and Intro to SAChE (www.sache.org)
Week 2: Chemical Reactivity Hazards
Chemical Reactivity (Chp 8) and Hazard Identification (Chp 11)
(SAChE Chemical Reactivity Hazards completion certificate)
Week 3: Industrial Chemical Hygiene (Chp 3) and Toxicology (Chp 2)
(SAChE Chemical Process Safety in the Chemical Process Industries (Crowl) completion certificate)
Week 4: Risk Assessment (Chp 12)
(SAChE Risk Assessment completion certificate)
Week 5: Prevention of Fires and Explosions (Chp 6 and Chp 7)
(SAChE Dust Explosion Control (Louvar) completion certificate)
Week 6: Chemical Reactor Safety
(SAChE Runaway Reactions (Theis) completion certificate)
Week 7: Safety Procedure and Designs (Chp 13)
(SAChE Process Safety 101 (Bernardi) completion certificate)
Week 8: Inherently Safer Design
Intro to Reliefs (Chp 9), Relief Sizing (Chp 10)
Week 9: Inherently Safer Design (Continued) and Engineering Ethics
Intro to Reliefs (Chp 9), Relief Sizing (Chp 10), Order of Engineer.
(SAChE Inherently Safer Design completion certificate)

Situational Awareness

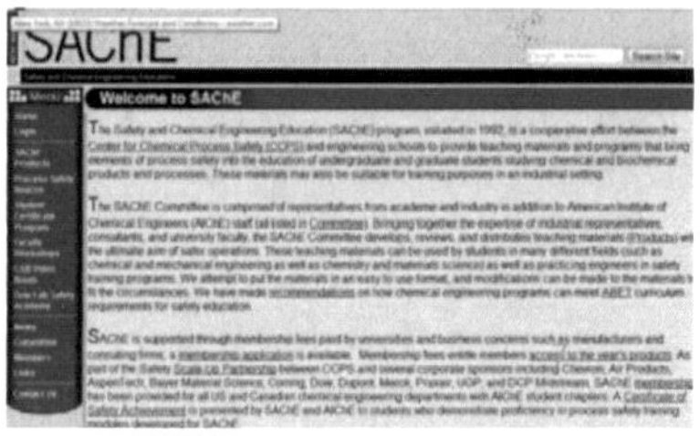

(www.sache.org 27 Sep 2015)

Course Design and SAChE Integration

Week 10: Guest lecture: **Simulation Solutions, Inc.**
Week 11: Intro to HAZOPs and Layer or Protection Analysis
(SAChE Process Safety Lessons Taught from Experience (Wiley) completion certificate)
Week 12: HAZOPS Project Progress Review
Week 13: HAZOPS Project Progress Review
Week 14: Presentation of Accident Investigation Case Studies
Week 15: Study Days
Week 16: Final HAZOPS Project Due

Situational Understanding

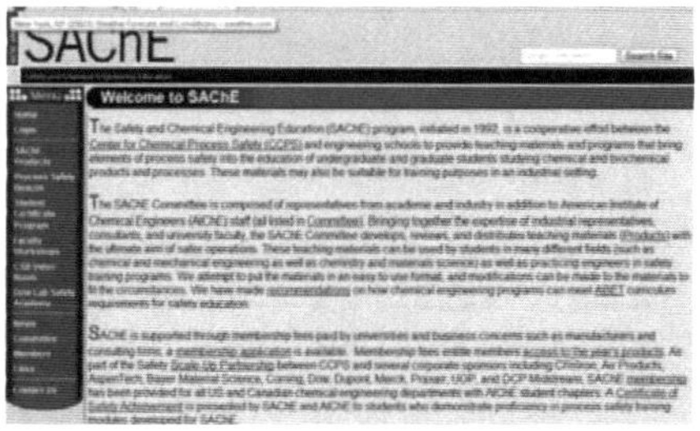

(www.sache.org 27 Sep 2015)

Learning from History
Case Study and Design

1. Emphasis on Block Flow Diagram to develop a common operating picture

Case study presentations Spring 2016

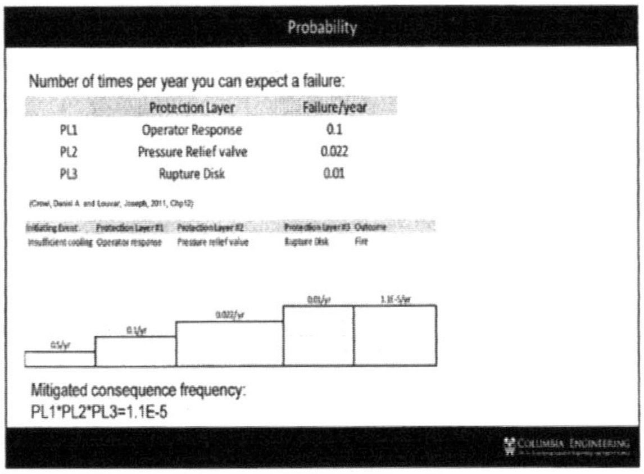

2. Emphasis on Probability of Failure

9

Learning from History
Case Study Format

Oral Communication Skills- (Briefing During Class)

Written Communication Skills- (Report turned in to the instructor)

Process Flow Diagrams

Case History (the accident, the event that produces injury/death/damage, and the hazard, the chemical or physical condition that can cause damage. Identify and evaluate the hazards involved. Determine what controls were used, and what should or could have been used. Define the scenario, a description of the events that result in an accident; the indecent, the loss of cantonment of material/energy, the consequence, result of accident, and the likelihood, expected probability of the event. Remember to address, risk- as defined, Risk = f(frequency, consequence) of the accident and determine what aspects of inherently safer design, moderate, substitute, minimize, and simplify, could have been used in order avoid such an accident)

Lessons Learned

Documentation/References

(Crowl, Daniel A. and Louvar, Joseph, 2011, Chp1)

10

Learning from History
Design

Oral Communication Skills - (In Class Progress Review)

Written Communication Skills- (Report turned in to the instructor)

Process Flow and Process Control Block Diagram Diagrams

Design Results (Outline what you recommend as a replacement and what your major conclusions are from your design and safety analysis. Design Results include summary of ASPEN simulation results, your safety analysis results from the Process Hazard Analysis Risk Assessment, Event Tree Analysis, Layer Of Protection Analysis, HAZOPS Analysis. Make sure you address aspects of Inherently Safer Design: Moderate, Substitute, Minimize, and Simplify. In this project, use Aspen Plus 8.6 for simulating your material and energy balances. As appropriate, make use of good engineering assumptions in order to complete this design project. Remember to assume, check, and validate assumptions. Keep a record of your assumptions and include them in your presentation.)

Discussion (Discussion of major decisions for your design results and your safety analysis)

Documentation/References

11

(Crowl, Daniel A. and Louvar, Joseph, 2011, Chp1)

Learning from History
8 Significant Disasters (Case Studies)

Morganton, NC Jan 2006 (explosion)

Anacortes, Washington Apr 2010 (explosion)

Flixborough, England, June 1974 (pipe rupture)

Jacksonville, Florida, Dec 2007 (explosion)

Texas City, Texas May, 2005 (explosion)

Patterson, NJ Apr 1998 (explosion)

Bhopal, India, Dec 1984 (toxic release)

Seveso, Italy, Jun 1976 (runaway reactor, toxic release)

12

(www.csb.gov accessed 31 Jan 2016)

(Crowl, Daniel A. and Louvar, Joseph, 2011, Chp1)

Evaluating Chemical Engineering Safety Training Results

Donald C. Glaser & Matthew Garvey
Simulation Solutions, Inc.

29 Mar 2016 and 31 Mar 2016

13

Objectives

- Learn more about the importance of process safety and the resources & commitment required.
- Learn how to control and mitigate hazards to prevent accidents.
- Learn about operator training through the use of simulators.
- Review process control and chemical engineering design concepts.

14

Simulation Solutions, Inc.
Outline

- Overview of Operator Training
- Review of Operator Safety & INSTO™ Methodology
- Introduction to P&ID / DCS / Outside Operator
- SSI "What-If" Exercises
- SSI Troubleshooting Exercises
- Q&A for Students

15

Simulators provide life-like Training for New and Experienced Operators

Laboratory Experience

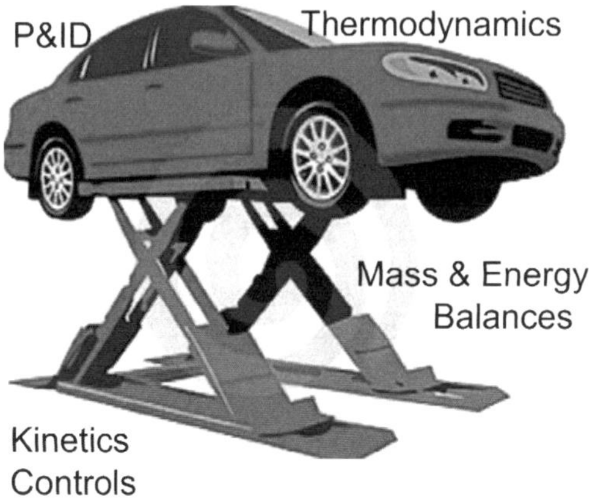

Real World Experience

16

Minds-On/Hands-On Troubleshooting Process

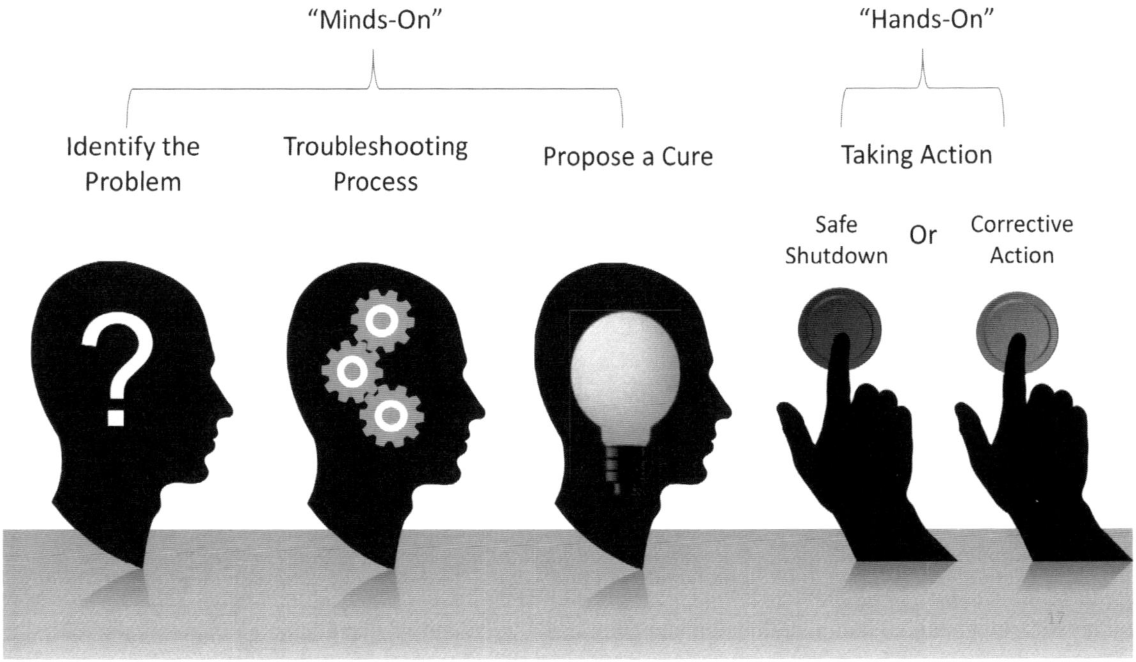

Grafting an *Operations Mindset*™ to New Operators

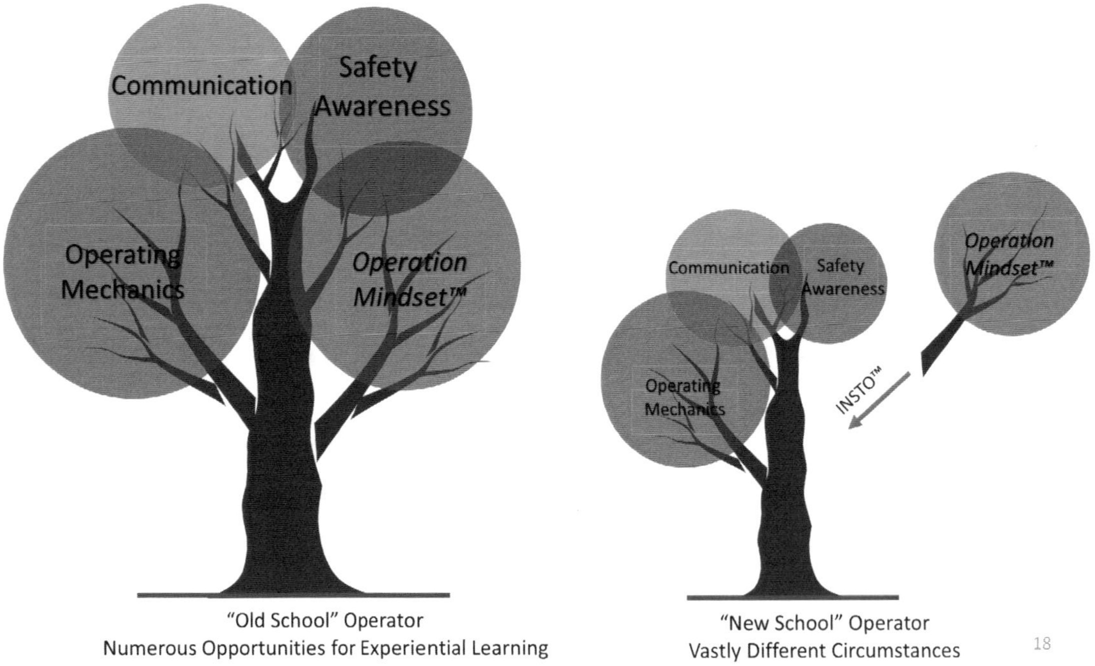

Coursework Flow

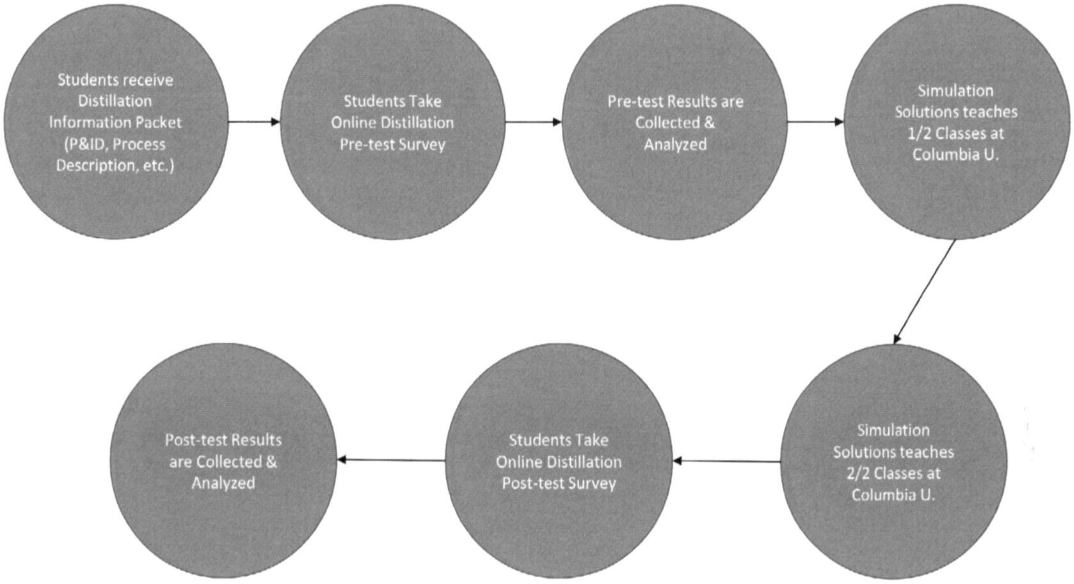

19

Coursework Flow

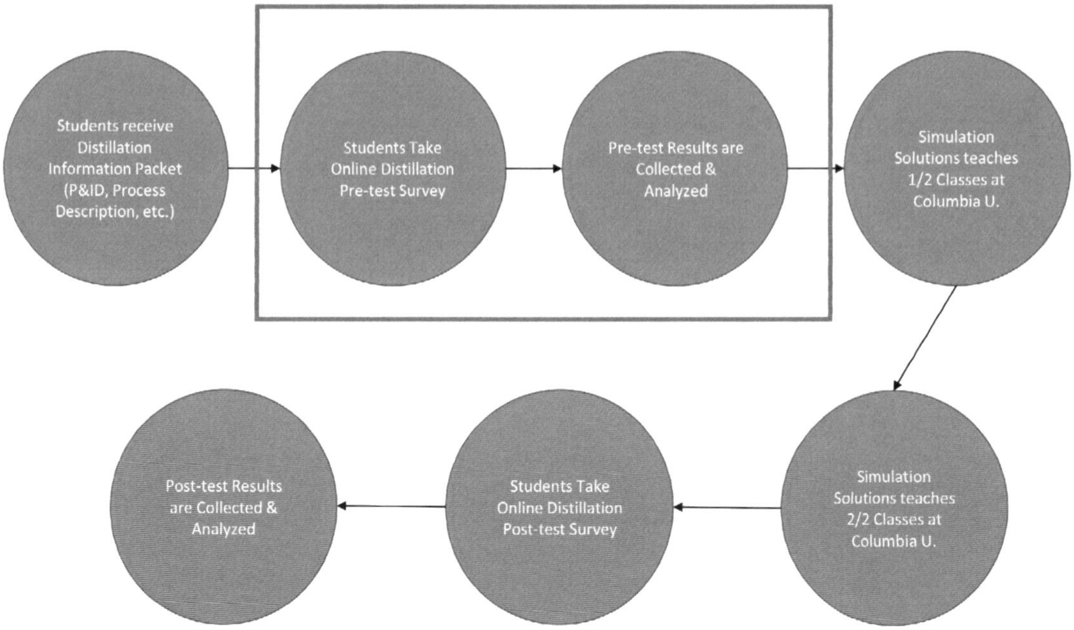

20

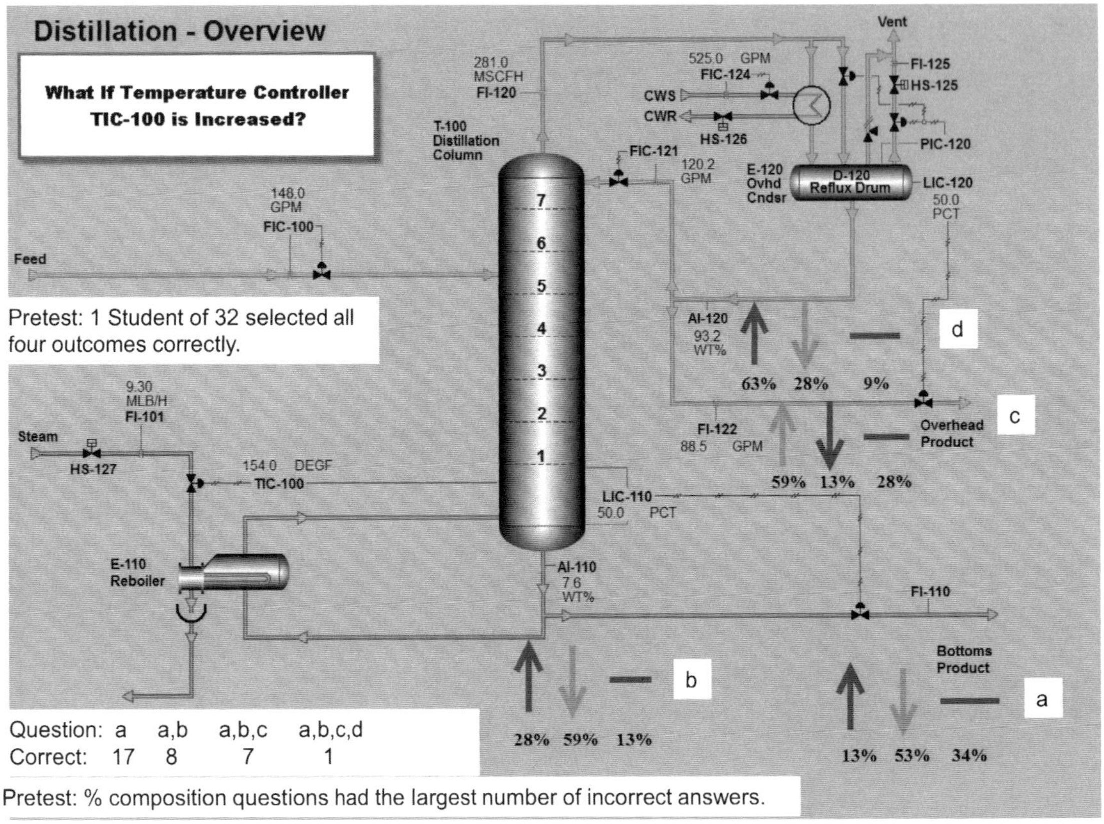

SAChE/ Course Objective/ Lesson Objective "Learn how to control and mitigate hazards to prevent accidents."

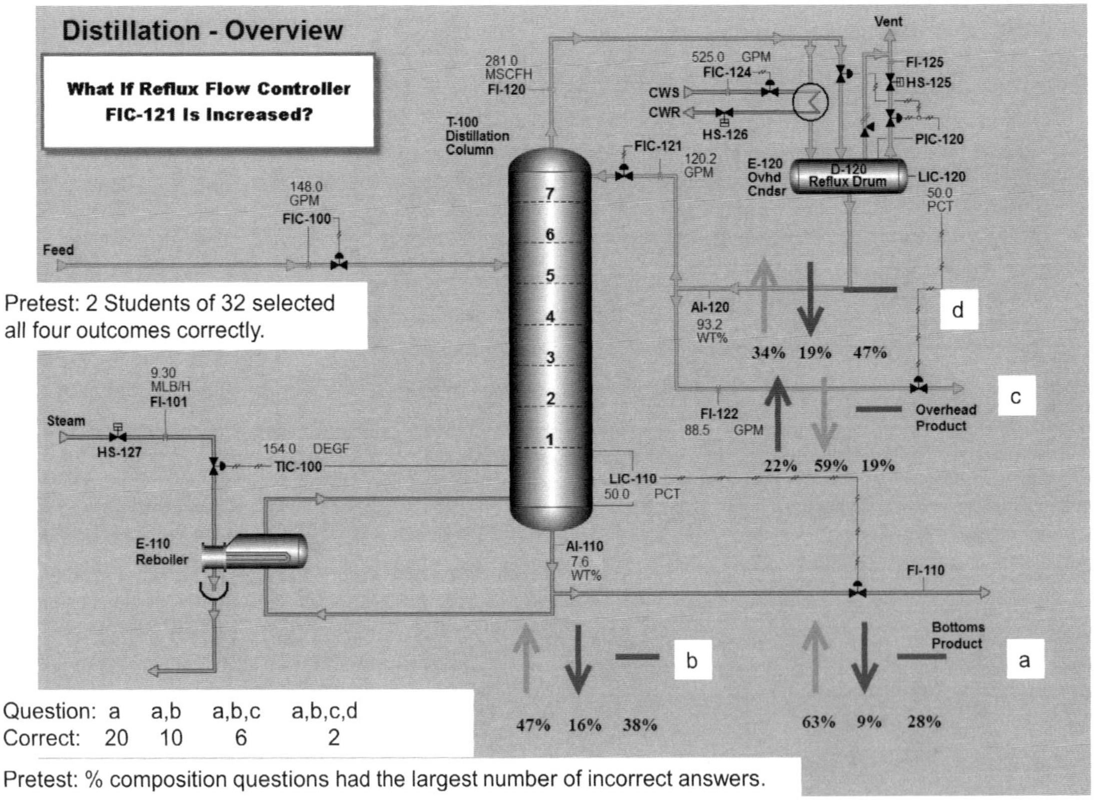

SAChE/ Course Objective/ Lesson Objective "Learn how to control and mitigate hazards to prevent accidents."

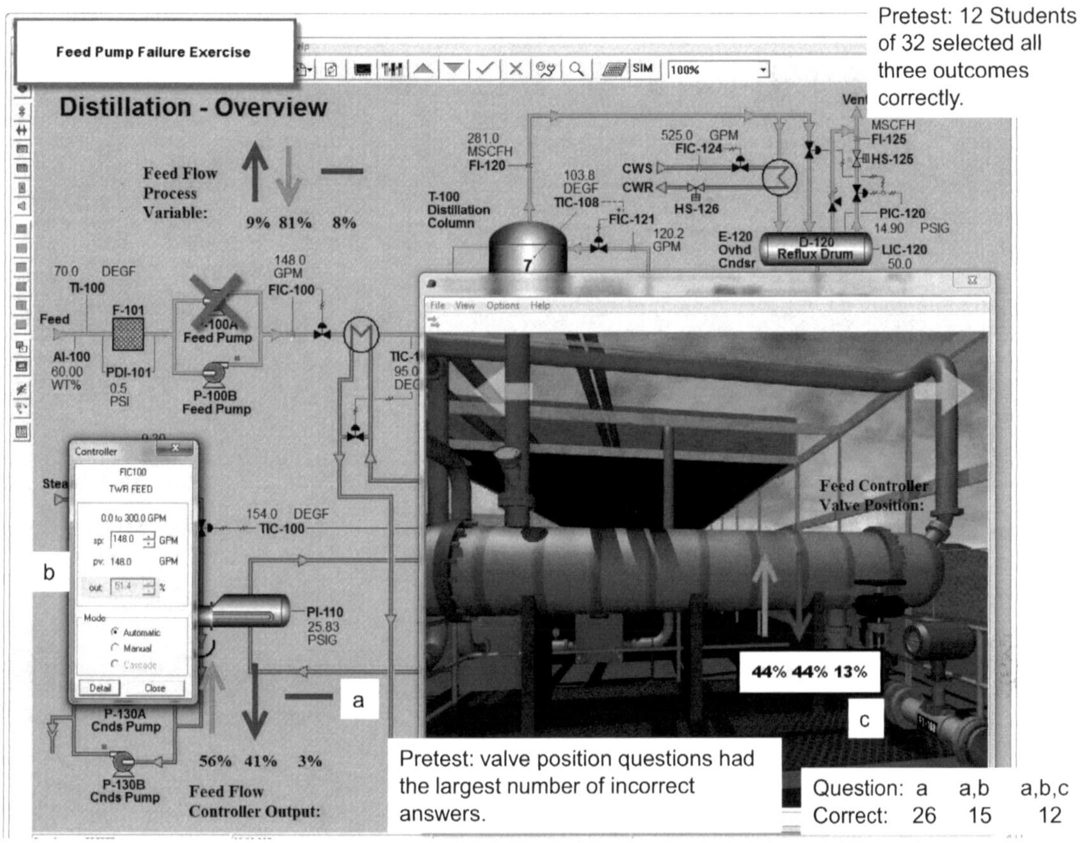

SAChE/ Course Objective/ Lesson Objective "Learn how to control and mitigate hazards to prevent accidents."

Feed Pump P-100A Failure Recovery Steps

Step No.	Operator Action	Number of Students Remaining with Proper Sequence (32 Total)
1.	Place Feed Flow Controller in Manual	16
2.	Lower Feed Controller Output to 10%	10
3.	Start Spare Feed Pump, P-100B	8
4.	Increase Feed Flow Output until Process Variable is at the Design Flow Rate	4
5.	Place Feed Flow Controller into Automatic Mode	4

24

SAChE/ Course Objective/ Lesson Objective "Learn how to control and mitigate hazards to prevent accidents."

Feed Pump P-100A Failure Recovery Steps

Step No.	Operator Action	Number of Students Remaining with Proper Sequence (32 Total)
1.	Place Feed Flow Controller in Manual	9
2.	Lower Feed Controller Output to 10%	7
3.	Start Spare Feed Pump, P-100B	7
4.	Increase Feed Flow Output until Process Variable is at the Design Flow Rate	3
5.	Place Feed Flow Controller into Automatic Mode	3

25

SAChE/ Course Objective/ Lesson Objective "Learn how to control and mitigate hazards to prevent accidents."

Coursework Flow

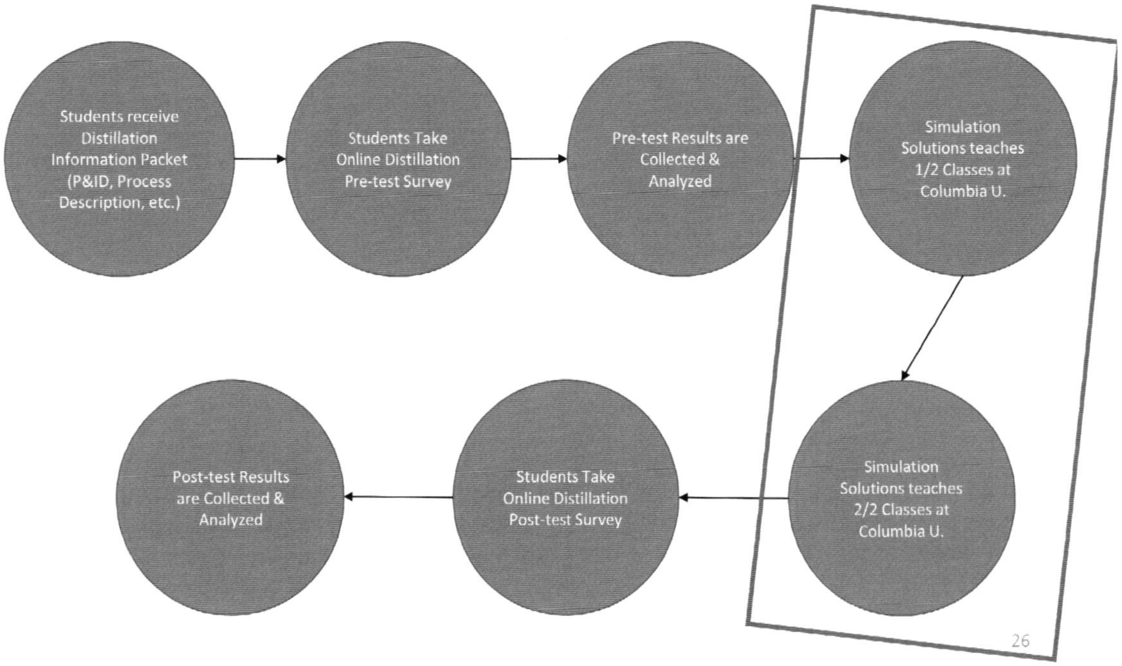

Simulator Demonstrations

Distillation

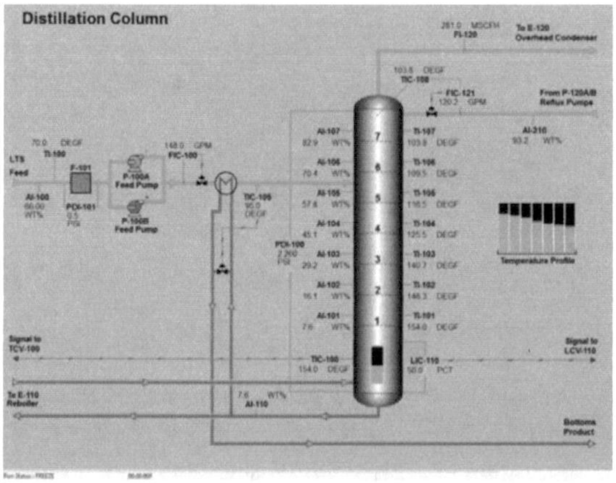

27

"Minds-On/Hands-On" Training™

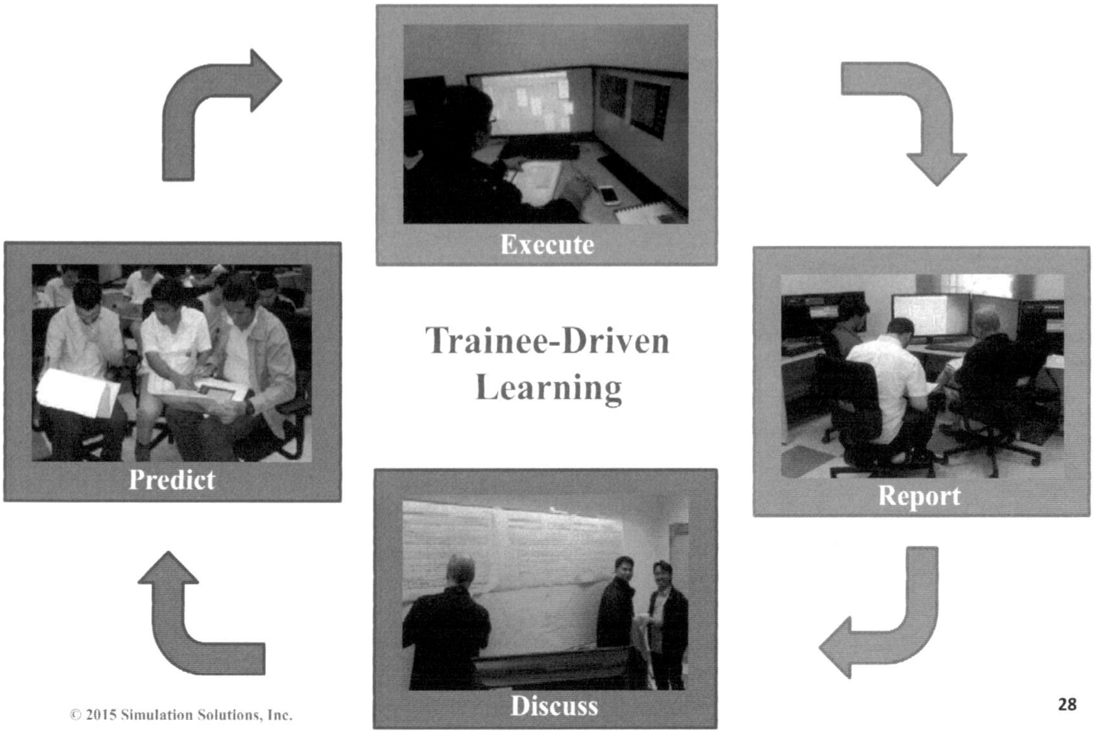

28

31

Distillation
"What-If" Flip Charts

How will the Bottom of the Tower react if Feed Flow is increased?

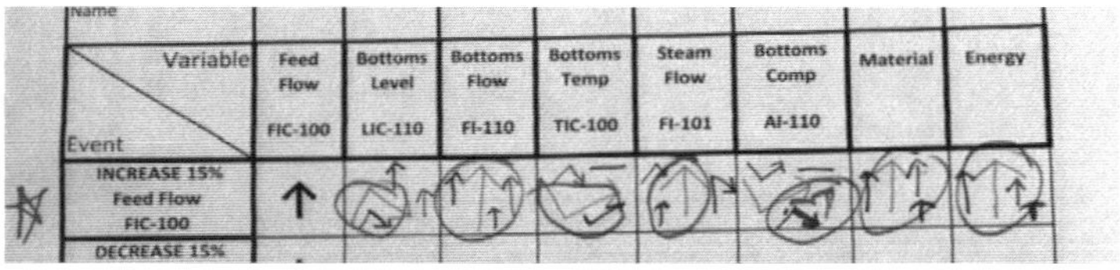

How will the Top of the Tower react if Reboiler Steam is Increased?

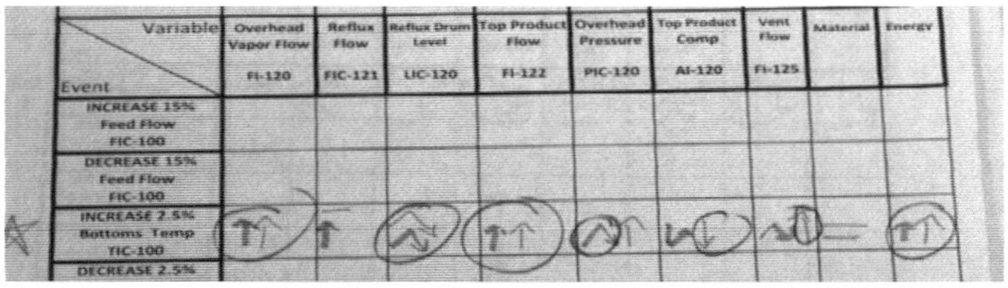

How will Product Flows & Compositions react if Overhead Pressure Increases?

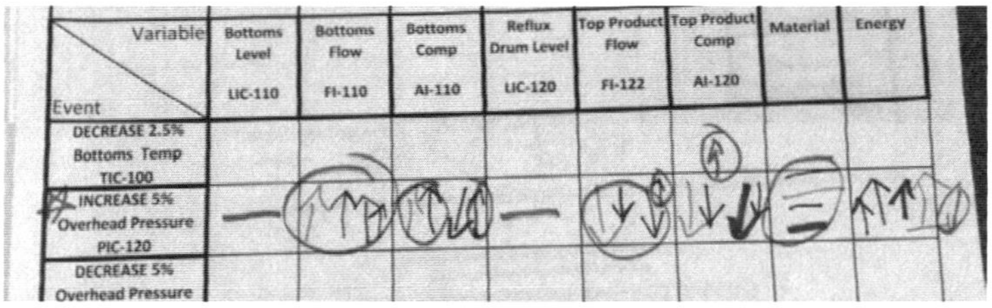

34

Distillation
"What-If" Discussion

- Students were asked to predict how Controllers and Indicators would respond to a Setpoint change in a Distillation Simulator. (i.e. Increasing Feed Flow Rate, Tower Temperature, Reflux Flow Rate, Overhead Pressure).
- Students did not always correctly predict controller responses. For example: When more Feed is put down a Distillation Column the bottoms level will rise, but eventually come back to set point given the controller in this system. This is achieved by the Bottoms Flow increasing. Some students only predicted that this level would rise, and not return to setpoint.
- Another area of trouble was predicting Product Stream purities. For example when the Overhead Pressure was increased in the system, many students predicted that the Top Product would have less "lights" in the stream. An increase in Pressure will actually sweeten the Top Product Stream.

35

Troubleshooting Exercise Results – Loss of Steam to the Tower

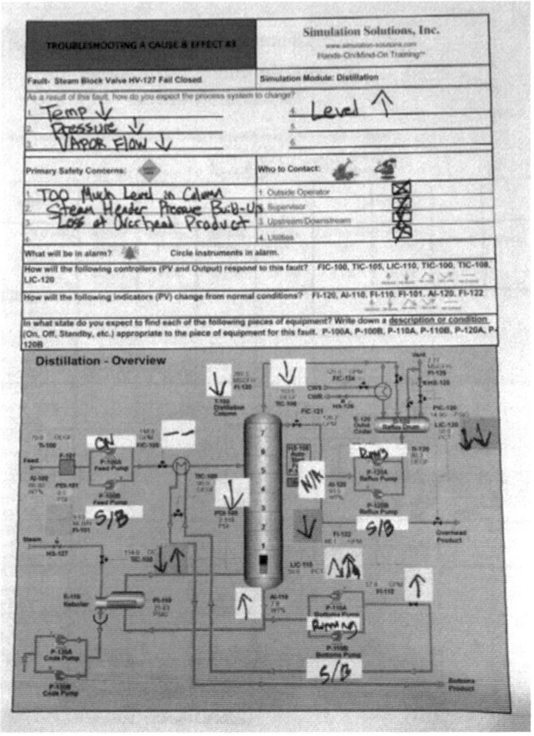
TROUBLESHOOTING A CAUSE & EFFECT #3

Simulation Solutions, Inc.

Hands-On/Mind-On Training

Fault- Steam Block Valve HV-127 Fail Closed

Simulation Module: Distillation

As a result of this fault, how do you expect the process system to change?

1. Temp ↓
2. Pressure ↓
3. VAPOR Flow ↓
4. Level ↑
5.
6.

Primary Safety Concerns:

1. TOO Much Level in Column
2. Steam Header Pressure Build-Up
3. Loss of Overhead Product
4.

Who to Contact:

1. Outside Operator ☒
2. Supervisor ☒
3. Upstream/Downstream ☒
4. Utilities ☒

What will be in alarm? Circle instruments in alarm.

How will the following controllers (PV and Output) respond to this fault? FIC-100, TIC-105, LIC-110, TIC-100, TIC-108, LIC-120

How will the following indicators (PV) change from normal conditions? FI-120, AI-110, FI-110, FI-101, AI-120, FI-122

In what state do you expect to find each of the following pieces of equipment? Write down a description or condition (On, Off, Standby, etc.) appropriate to the piece of equipment for this fault. P-100A, P-100B, P-110A, P-110B, P-120A, P-120B

Distillation - Overview

36

Troubleshooting Exercise Discussion

- Students were asked to predict Controller and Indicator Responses if Steam Flow to the Tower went to zero, due to a closed block valve.
- A focus was put on the "big picture" safety concerns of this fault, specifically how this malfunction will affect downstream/upstream units.
  - For example, the unit that is receiving the Overhead Product as their input will no longer have any material.
  - Compositions of both product streams will be affected and could be a concern to downstream units.
- Some students had trouble predicting Controller Output responses, which is instrumental to the "cure" to the malfunction.
  - In this example, if the Steam Block Valve fails closed, then the Temperature Controller will be wide open at a 100% Output. Before manually opening the steam Block Valve, the Operator will want to put the Temperature Controller in Manual and lower the Output, so that the tower is not hit with an excess flow of Steam.

37

Student Survey Results 2016 (Graded on Scale 1-5)

	For each statement below, rank from 1-5, where 5 indicates strong agreement and 1 indicates strong disagreement.	CHEN E4501 Lessons 19 and 20	Standard Deviation
1	As a result of CHEN E4501 lessons 19 and 20 my ability to understand the importance of process safety and the resources and commitment required has improved. (course objective 1)	4.67	0.62
2	As a result of CHEN E4501 lessons 19 and 20 my ability to understand how to control and mitigate hazards to prevent accidents has improved. (course objective 4)	4.33	0.73
3	As a result of lessons 19 and 20 my understanding of operator training through the use of simulators has improved.	4.63	0.63
4	As a result of CHEN E4501 lessons 19 and 20 my understanding of process control and chemical engineering design concepts has improved.	4.48	0.89

29 of 31 students responded.

38

Student Survey Results 2016 (1 of 2)

General Comments about Simulation Solutions Safety Course Spring 2016 at Columbia University
CHEN E4501 Chemical Engineering Lessons 19 and 20

- good as expected
- I'm still confused about the some basic concepts regarding controllers. The speech may have went too quickly
- I hope that we have more reading material on this since the lecture is a bit fast and overwhelming
- Good
- I thought it was engaging and fun! I don't feel like an expert now but it was a good introduction to operator controls and applied concepts. I feel with further training this would really prepare someone to work as an operator
- very interesting and informative
- This experience has greatly improved from last year. Through expanding the lesson to two classes it was much easier to absorb the material and it does not feel as rushed. Additionally, it is very interesting to experience a practical application of chemical engineering rather than theoretical. (most likely the grader who took the course in 2015)
- Personally, the first lesson felt hard to follow but the second lesson I caught on and felt I had gained and improved my knowledge of process control
- These 75 minutes felt like 15 mins. Very interesting. But I think if this can be 2 weeks, it would help understand controls even better. Thanks Dr. Bozic
- Zero complaints. Great presenters, practical applicability was very helpful, only suggestion would be to have the exercises be a little more guided to hear more of the proper mindset in finding solutions and allowing us to do more exercises.
- Really likes working with simulator. Was a lot of fun, wish we were able to do more, but it covered a good amount given the time. Definitely Recommend!
- Really enjoyed and learned much from them!
- It was better than last time. Do it again.
- It was really fun to have an interactive class. These lessons were a great way to see the inner though process of a future engineer.

39

Student Survey Results 2016 (2 of 2)

General Comments about Simulation Solutions Safety Course Spring 2016 at Columbia University
CHEN E4501 Chemical Engineering Lessons 19 and 20

- Helpful for basic understanding of chemical engineering simulating. I can't thoroughly understand details but it's overall good.
- The contents on What-if problem was great. Hope to learn more on hazard control topics.
- It was great to learn the operating process by using simulation software, which helped a lot to improve my understanding in terms of various situations.
- was really informative, interactive and fun.
- it helped me have a global view of control system and the entire chemical reactor process.
- It's really interesting to learn about the engineering design and simulators. Definitely should keep the lessons in the future!
- useful information would prefer all answers to worksheet so I can review on my own & a copy of the simulation program
- These lessons were useful but it would be better if we can run the software on macbook too for everyone to get the experience for using the simulator
- I wasn't sure about the presentations/simulations at first but ended up having a great time learning about process control/design concepts. My understanding has definitely improved, even though I don't quite understand everything yet. Before the lessons I had no idea how to read the provided diagram but now I feel like I have a better understanding.
- very helpful to learn the process control
- Process was very cool. It may be worthwhile to mention that the software for the demo isn't very demanding- i.e. I was initially concerned that the software being loaded would be a huge file size, something like Chem CAD or ASPEN, that it wouldn't be able to run on my (deleted word) computer. It worked fine though.

40

Coursework Flow

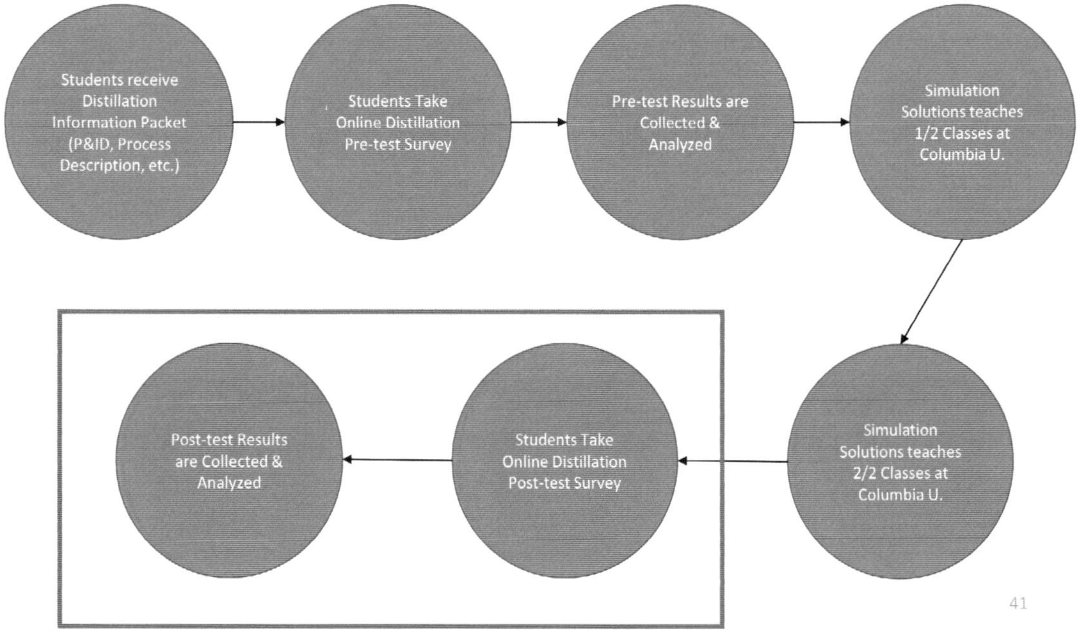

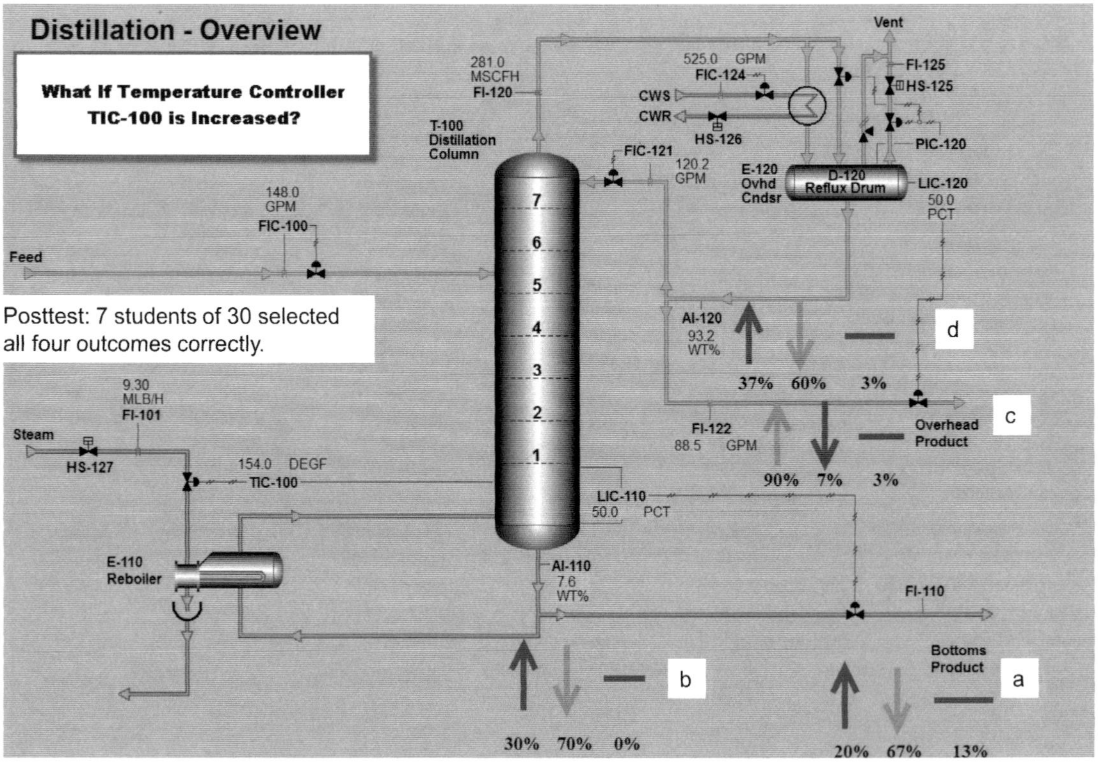

Question:	a	a,b	a,b,c	a,b,c,d
Correct:	20	14	7	7

Post-test: % composition questions had the largest number of incorrect answers.

SAChE/ Course Objective/ Lesson Objective "Learn how to control and mitigate hazards to prevent accidents."

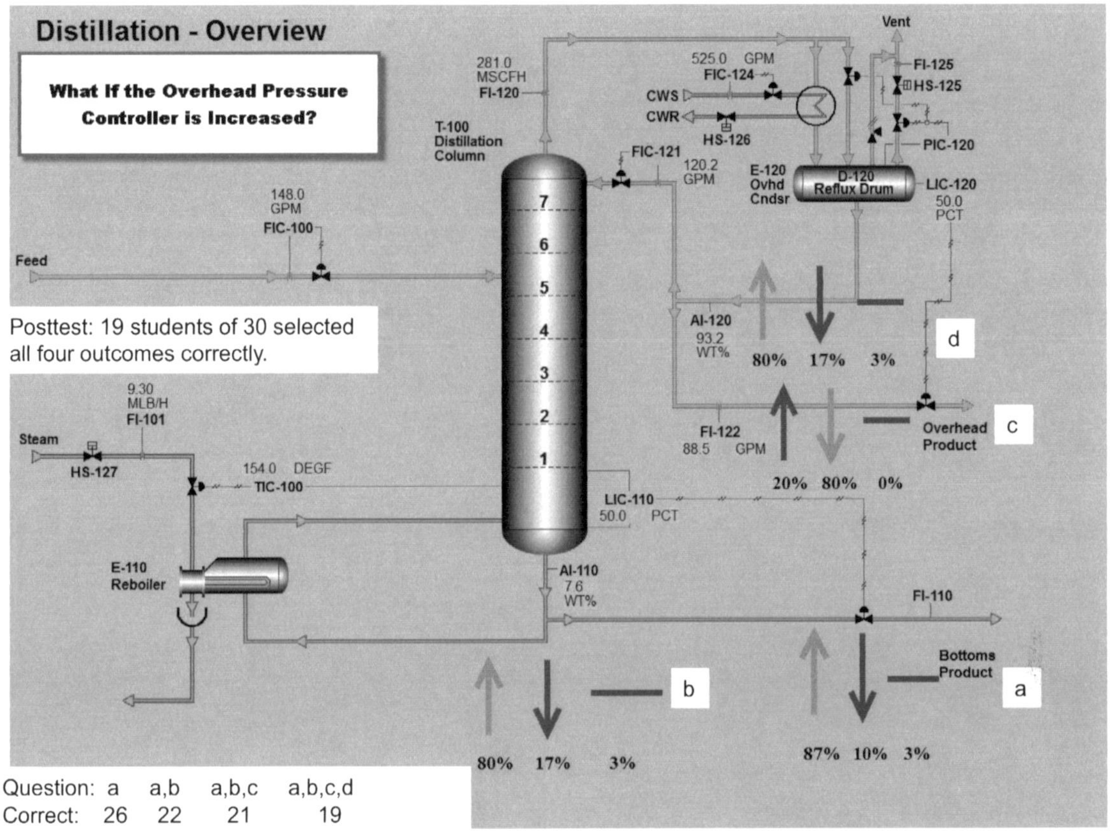

SAChE/ Course Objective/ Lesson Objective "Learn how to control and mitigate hazards to prevent accidents."

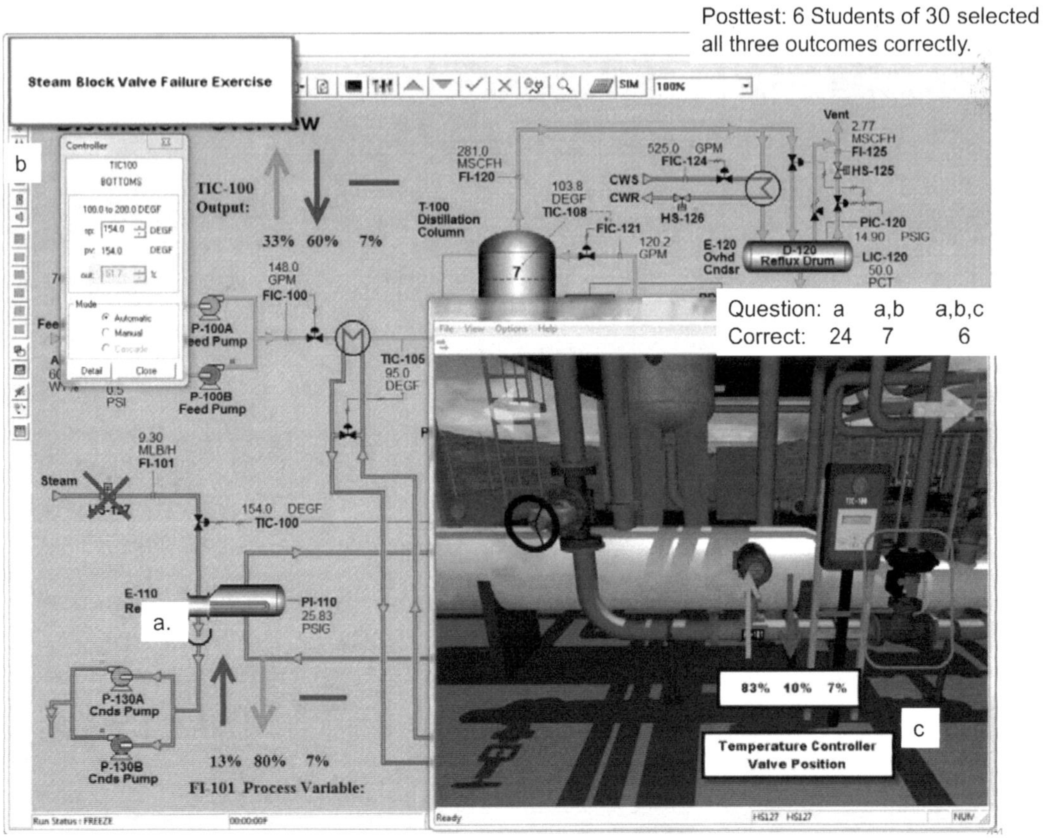

SAChE/ Course Objective/ Lesson Objective "Learn how to control and mitigate hazards to prevent accidents."

Condensate Pump Failure Recovery Steps

Step No.	Operator Action	Number of Students Remaining with Proper Sequence (30 Total)
1.	Place Temperature Controller in Manual	27
2.	Lower Temperature Controller Output from 100% to the Design Rate of 50%	25
3.	Start Spare Condensate Pump, P-130B	24
4.	Adjust Temperature Controller Output until the Design Temperature of 154 degF is achieved.	23
5.	Place Temperature Controller in Automatic Mode	23

45

SAChE/ Course Objective/ Lesson Objective "Learn how to control and mitigate hazards to prevent accidents."

Comparing Pre-test and Post-test Data

- Students showed substantial improvement in recovering from a plant upset in a safe manner in order to mitigate and control a hazard.
- Students showed improvement in predicting Product Stream Flows when a Setpoint change is made in a Distillation Unit.
- Predicting Controller Output responses and effects on composition can be viewed as areas that needs improvement, as understanding control schemes is vital to safe operations in a process unit.

46

Lessons Learned

Lessons learned

By scheduling the Project Progress Review Prior to the case study, students were encouraged to go in a direction that required them to do the case study work prior to the Progress Review. This seemed to help decrease procrastination.

Students were inspired and enjoyed opportunities to use computer software that simulated process equipment.

Student performance on the pretest and post test showed a marked improvement. With only two 75 min classes, and a small amount of individual "stick time" on the simulator, student ability to predict outcomes improved considerably.

Having a pre-test and post-test seemed to increase interest in the event as well as provide the major means by which performance was evaluated.

47

Acknowledgements

Thanks to the Columbia Students of CHEN E4501!
Big Thanks to Don Glaser and Matthew Garvey!
Go Lions!

Simulation Solutions, Inc.
Minds-On/Hands-On Training for Operators™

48

Conclusion and Future Work

Continued success with initial integration of SaChE Recommendations into curriculum.

Continue a "Back to Basics" approach with Process Flow and Process Control Diagrams

Continue the approach related to operator training and evaluation of process safety education and examining ways to incorporate more predictive style exercises with the use of simulators.

Refinement of process simulation in the classroom with CHEN E4500 Principles of Process Design Course

Future work: on process simulation and Operations Center Role Playing Scenarios in the classroom with CHEN E4501 Chemical Engineering Process Safety Class

49

References

ASPENTECH V8.8 Software (used for design work)

Beer, David and David McMurrey, *A Guide to Writing as an Engineer 2nd Edition*. Wiley, New York, 2005

Crowl, Daniel A. and Joseph F. Louvar, *Chemical Process Safety" Fundamentals with Applications 3rd Edition*, Pearson, New York, 2011

Felder, Richard M. and Ronald W. Rousseau, *Elementary Principles of Chemical Processes 3rd Edition*, 2000, John Wiley and Sons, New York, New York. Section 4.3a Flowcharts, p90-92

Graham, Nasby, "Using Process Flowsheets as Communication Tools" *Chemical Engineering Progress*, Volume 108, No. 10.Oct 2012. p36-44, Available in Chemical Engineering Progress on line

i.ytimg.com/vi/8A1xSCUtB-M/hqdefault.jpg accessed 22 Jan 2015

Perry, Robert H. and Don W Green, *Perry's Chemical Engineering Handbook*, 7th Edition, 1997, McGraw Hill, New York, New York

Seborg, Dale E., Thomas F. Edgar, and Duncan A. Mellichamp, *Process Dynamics and Control*. Wiley Inc, New York, New York (1989) p7-9

Walker, Vornel, "Designing a Process Flowsheet" *Chemical Engineering Progress*, Volume 105, No.5. May 2009. p15-21, Available in Chemical Engineering Progress on line

www.aiche.org/academy/courses/sache/sache-certificate-program accessed 19 Jan 2015

www.aiche.org/ccps/community/students-teachers-and-communities/certificate-safety-achievement accessed 19 Jan 2015

www.bing.com/images/search?q=jacksonville%2C%20florida%20chemical%20plant%20dec%202007&qs=n&form=QBIRMH&pq=jacksonville%2C%20florida%20chemical%20plant%20dec%202007&sc=0-0&sp=-1&sk=#view=detail&id=757F70F7B398E519B65D6CACB55A59E57E50CB72&selectedIndex=5 accessed 22 Jan 2015

www.bing.com/images/search?q=pasadena%20texas%20oct%201989&qs=n&form=QBIR&pq=pasadena%20texas%20oct%201989&sc=0-19&sp=-1&sk=#view=detail&id=1A52930A92CCEC5F28D374DBCC19EA634FE9A02F&selectedIndex=1 22 Jan 2015

www.csb.gov accessed 3 Feb 2015 used for case study videos

www.csb.gov/assets/1/7/Tesoro_Anacortes_2014-May-01.pdf accessed 28 Jan 2015

www.csb.gov/assets/1/19/Final_Report_small.pdf accessed 28 Jan 2015

www.sache.org accessed 2 Jan 2015, used for SaChE

www.srai.org/wp-content/uploads/2010/06/pic2.jpg accessed 22 Jan 2015

www.toxipedia.org/download/attachments/9353/Seveso.chem.jpg?version=1&modificationDate=1238498605000&api=v2 andhttp://www.bing.com/images/search?q=Seveso+Disaster+Affected+Area&Form=IQFRDR#view=detail&id=A9CEBE175583891AADF628A0E49D369903208B4E&selectedIndex=0 accessed 22 Jan 2015

www.bing.com/images/search?q=port%20wentworth%20georgia%20feb%202008&qs=n&form=QBIR&pq=port%20wentworth%20georgia%20feb%202008&sc=0-24&sp=-1&sk=#a accessed 22 Jan 2015

www.bing.com/images/search?q=texas%20city%20mar%202005&qs=n&form=QBIR&pq=texas%20city%20mar%202005&sc=8-11&sp=-1&sk=#view=detail&id=F2A258E4AD124686AFE250ED468FBB03FA6C7066&selectedIndex=6

50

453486 Credentialing Students in Their First Separations Course-ChE360 Using a Problem Based Approach

Abstract:

Problem-based Learning (PBL) is an approach that challenges students to "learn how to learn," working in groups to seek solutions to real world problems. The process replicates the commonly used systemic approach to resolving problems or meeting challenges that are encountered in practice, and will help prepare students for their careers. In the spring 2016 semester offering of ChE360 at New Jersey Institute of Technology, we added one new feature – simultaneous use of Aspen Plus®. The course is delivered twice per week for 80 minutes per session. In this semester the plan was to deliver new information on the selected separation process in a workshop format in the first part of a week, then use Aspen Plus® simulations in the second part of the week to validate results obtained through graphical construction. Each significant example calculation that accompanied a new topic area such as using the McCabe-Thiele Method to determine the number of theoretical stages along with exit stream flow rates and compositions were subjected to this new course delivery approach. In addition, the students worked in groups on the simulations and were required to document their results in a structured report format and each report was graded as a mini-project. Formative assessment using Students Assessment of their Learning Gains (SALG) resulted in 100% response rates with at least 67% of the class reporting moderate –great gain in understanding content, impact on their attitude, increase in their skills, integration of their learning and helpfulness of their class activities.

453706 The Partition Function for Student Laboratories

Abstract:

Increasing lecture course size involves locating a larger lecture hall and arranging more teaching assistants to do the grading. However, increasing students in laboratory courses has no such simple solution. Increasing student enrollment in laboratory courses involves

- fire code concerns
- available number of experiments
- chemical inventory, especially with regard to solvents
- evaluation of experimental reports

To resolve these issues, it is necessary to develop more experiments that are robust with regard to usage and to use creative scheduling. This presentation discusses the ways Rice University and the University of Houston have responded to increases in student enrollment.

462666 Mini-Labs That Scale! Activities, Demos, and Simulations for Heat Transfer Concepts

Abstract:

Inquiry-based activities are simple, inexpensive, relatively rapid experiments that have been demonstrated to repair student misconceptions in important areas such as heat transfer and thermodynamics. For large classes with laboratory sections, it's possible to run these experiments. However, for classes with limited lab time or physical resources, we have re-worked the experiments to be used as demonstrations, simulations, or thought experiments. We have also assessed the effectiveness of each of these modes using the Heat and Energy Concept Inventory (HECI). All modes of these experiments, along with data on how resource-intensive they are to implement and how effective they are, will be posted in the AIChE Concept Warehouse. In the talk, we will demonstrate some of these experiments and simulations live, as well as explain how to access the materials for free through the Concept Warehouse.

460252 Active Learning in Senior Process Design and Plantwide Control

Abstract:

This presentation will illustrate how we have engaged students to increase learning outcomes in the chemical engineering senior design course at Lehigh University by using the KEEN (http://engineeringunleashed.com/keen/) framework and active learning.

The Lehigh senior design course is two semesters. Students work in teams of 3. Each team is assigned a different and significant industrial process. The goal of the first semester is to complete the economic optimal steady-state design. The goal of the second semester is to demonstrate a working plantwide control structure. Commercial grade steady-state and dynamic simulation software is used.

The KEEN framework is based on the 3 C's: engaging Curiosity, making Connections, and Creating value. The KEEN framework helps us create challenging projects that ignite student curiosity and critical thinking. The students make connections to their prior course work and work experiences. They engage with multiple faculty and industrial consultants. All design projects require the students to identify the value proposition of their work. We use an active learning process: tell, show, do, report, and question. 150 minute classes are broken into segments. Lectures with examples are followed by in-class student team exercises. Students report their results and then engage with an open discussion of questions and answers. 75 minute recitations are held in the computer laboratory where each student has an Aspen workstation to develop process insight through simulation working individually and in teams. Two oral reports are given during the semester along with multiple group meetings with the faculty. A final all day formal presentation is given by each student team to the faculty, industrial consultants, and invited guests. This formal event is also held by web conference to engage those unable to attend in person or who would like to watch the event at another time.

465383 E-Lessons As Pre-Laboratory Training Tool

Abstract:

To increase cognizance among Chemical Engineering undergraduates of the new technologies brought in by the fast growing pharmaceutical/biopharmaceutical sector, a growing number of Chemical Engineering schools are introducing bioanalytical experiments in their undergraduate laboratory modules. Two difficulties often faced by students when performing bioanalytical experiments are: large amount of information to be processed during the laboratory session and the fact that they often lack the background theory of the experiment they are about to perform. Limited or no background knowledge, in combination with overload of information in a short period, hinders studentsâ€™ learning, often restricting their learning to mere observation. A viable solution to this challenge, especially for large classes, lies in providing pre-laboratory training to the students. This presentation discusses the development and use of e-lessons as a pre-laboratory training tool to augment studentsâ€™ laboratory experience. The pilot study involved creating e-lessons for two bioanalytics experiments: *polymerase chain reaction and high performance liquid chromatography.* Each e-lesson consisted of two screencasts and a self-assessment quiz. Student feedback showed that e-lessons provided the much needed guidance to the students, making learning more personalized and allowing students to have better control over their learning.

469807 Creation of Low-Cost Spectrophotometers for Use By Students

Abstract:

Spectrophotometers are standard equipment in chemical engineering laboratories and industrial processes. They may be used to track reactions which change solution color over time, and determine kinetic parameters or properties of cellular growth. Commercial spectrophotometers are very expensive, often costing thousands of dollars, making them prohibitively costly for use by all students in a large classroom setting.

We have developed several teaching modules around inexpensive spectrophotometers built by students, which we have used in classrooms of approximately 100 students. In our fall introduction to chemical engineering course, our freshmen students are given a package containing a few materials, including an LED and photoresistor, totaling less than a dollar, and instructed to build a photometer using also an Arduino Uno board. The rest of their spectrophotometer can be created with any materials they desire, often consisting of scraps and garbage. In the spring semesterâ€™s freshman design laboratory, students work in groups of three and redesign their apparatus with materials and tools provided in the lab; here they make a more robust product that they will use throughout the semester. These spectrophotometers, depending on the studentâ€™s design, may rival even commercial products in their signal to noise ratio and range of sensitivity.

We have also created many of these inexpensive spectrophotometers for use throughout our lab for various purposes. Additionally, we have used them in our K-12 outreach efforts as portions of high school introduction to engineering projects, during engineering summer camps, and in the Science Olympiad.

471330 Attracting Students to Chemical Engineering with Coffee: A Large Enrollment Lab and Design Experience

Abstract:

We describe a new class developed at U.C. Davis titled "The Design of Coffee," which serves as a non-mathematical introduction to chemical engineering as illustrated by the process of roasting and brewing coffee. Hands-on coffee experiments demonstrate key engineering principles, including material balances, chemical kinetics, mass transfer, fluid mechanics, conservation of energy, and colloidal phenomena. The experiments lead to an engineering design competition where students strive to make the best tasting coffee using the least amount of energy – a classic engineering optimization problem, but one that is both fun and tasty. "The Design of Coffee" started as a freshmen seminar in 2013, and it has exploded in popularity: it now serves 1,533 students per year, and is the largest and most popular elective course at U.C. Davis. In this talk we focus on the class pedagogy, with an emphasis on how coffee serves as an engaging and exciting topic for teaching students about core scientific principles in an approachable, hands-on manner.

473027 Embedding Hands-on Mini Laboratory Experiences in a Core Undergraduate Fluid Mechanics Course: A Pilot Study

Abstract:

Laboratory instruction plays a critical role in engineering education by enabling students to apply theory-oriented concepts to predict behavior in real systems, analyze and interpret data, and critically evaluate the experiment results. But recent enrollment pressures have led to larger class sizes, making it increasingly challenging to achieve the level of engagement needed to effectively deliver hands-on content. Here we report the results of a pilot study aimed at addressing this need by embedding laboratory experiences directly into a core undergraduate fluid mechanics course. Instead of the 1 to 2 semester delay that typically exists between the junior-level theory course and the senior-level dedicated unit operations laboratory, we introduced three self-contained â€œmini-labsâ€ into the theory course that were each designed to be comparable in scope to a homework assignment. Students completed hands-on experiments approximately 30 - 45 minutes in duration that supported learning objectives involving friction losses in pipes, flow measurement, and centrifugal pump analysis. A companion assignment involving data analysis and interpretation of results was developed that was designed to be more focused than a conventional laboratory report. Our pilot program spanned offerings of the course during two different semesters and involved a total of approximately 70 students. We describe our efforts to develop the laboratory experiments and assignments, schedule the lab sessions using an online tool, and discuss our implementation experiences. A post-assessment yielded positive feedback indicating that students gained confidence in solving engineering problems and identifying connections between theory and practice. Results of this pilot will be used to expand the mini-lab format to other core undergraduate courses in support of an innovative new shared instructional facility currently under construction at our institution.

473078 Micro-Fluidics As a Versatile Platform to Promote High Impact Hands-on Learning Experience in Chemical Engineering Classrooms

Abstract:

The pedagogical value of hands on experience that a laboratory provides to aid science and engineering education is well known, but they are often resource and time intensive to execute. Thus, there is a need to seamlessly incorporate innovative experimental and cross disciplinary experiences into classrooms. Microfluidics offers an ideal platform for such educational experiences owing to its inherent low cost and portable implementation, overcoming barriers that have previously hindered the ability to embed sustainable high impact learning in classrooms. Here we introduce a series of â€œDo it yourselfâ€ (DIY) format microfluidic experiments which are simple enough to be executed by undergraduate students while simultaneously exciting them to recognize how interdisciplinary knowledge can be applied to develop new products and technologies that benefit society.

Firstly, we present a capillary based microfluidic viscometer incorporating a simple design where the students can track the motion of colored fluid with their smartphones as it is driven by capillary action. This simple setup not only enables students to measure rheological properties of fluid without dedicated rheometers, but also presents them with a more intuitive interface to visualize the effect of system parameters on the rheological properties of the fluid. Secondly, we experimentally demonstrated the calculation of pressure losses in microfluidic flow systems with the aim to verify the standard pressure loss correlation as a function of flow Reynolds numbers and also measure the pressure drop across various microfluidic flow resistors. Finally, we also introduced chemical engineering students to molecular biology by challenging them to harness microscale natural convection phenomena to perform DNA replication via the polymerase chain reaction (PCR). For this demonstration, we constructed convective PCR stations incorporating a simple design for loading and mounting cylindrical microfluidic reactors between independently controlled thermal plates. A portable motion analysis microscope enables flow patterns inside the convective reactors to be directly visualized using fluorescent bead tracers [1]. The problem of microscale convective PCR provides a particularly

effective vehicle to connect concepts bridging the physical, chemical, and life sciences. A cognitive assessment reveals that these activities strongly impact student learning in a positive way.

[1] Priye, Aashish, Yassin A. Hassan, and Victor M. Ugaz. "Education: DNA replication using microscale natural convection." *Lab on a Chip* 12.23 (2012): 4946-4954.

473435 Chemical Engineering in a Maker Studio for All Engineers

Abstract:

Laboratory courses are a fundamental portion of chemical engineering education but most students only encounter such courses in their junior and senior years. At UC San Diego, a coordinated effort across all engineering programs has been made to develop so-called "Experience Engineering" courses for freshman and sophomore students to increase exposure to the engineering disciplines and improve retention in subsequent courses. These courses utilize a common work area known as a Maker Studio or Maker Space, a flexible learning space that allows students to a great deal of flexibility in tackling hands-on problems. Here, an Experience Chemical Engineering laboratory course utilizing the Maker Studio is described as a low cost, readily implemented, and generally well-received addition to the Experience Engineering program at UC San Diego.

"Renovating and Rejuvenating Chemical Engineering Unit Operations Lab in a Space Limited Environment."

Robert G. Bozic, Ph.D. LTC U.S. Army(Retired) and Scott A. Banta Ph.D.
Department of Chemical Engineering
Columbia University in the City of New York
robert.bozic@columbia.edu
sb2373@columbia.edu

Abstract

In response to requirements to consolidate student educational labs for Chemical Engineering and Earth and Environmental Engineering, the Columbia University Department of Chemical Engineering recently significantly renovated and upgraded the Chemical Engineering Unit Operations Lab through a School of Engineering and Applied Sciences, SEAS, supported project. Space saving adjustments, interoperability, flexibility, connectivity, and maintaining educational goals as well as considerations for future lab adjustments were addressed during this project. The considerations for this type of renovation are similar for many programs, but may come as a non-standard change. Lessons learned from the project as well as a suggested approach to such a change for the chemical engineering unit operations lab of the future will be shared.

Background and Motivation

In a space limited environment, optimizing space in SEAS is a priority. Contributors made necessary space saving changes while upgrading the educational environment and increasing the capability of the CU ChemE Unit Ops lab.

Increase in Outcome Coverage from Old to New Experiments

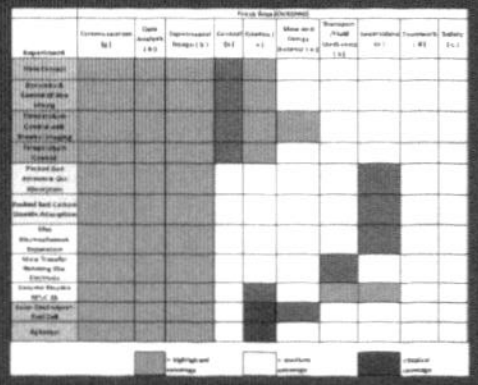

Project Goal

A safe, dynamic, hands on, undergraduate chemical engineering lab experience in support of program outcomes.

Renovation Criteria/Considerations

-Outcome coverage
-Conceptual design tested by marking off the area of the lab to check for suitable space
-Times scales for each experimental run
-Resources required for each experiment
-Connection to electrical, water, utilities
-Ventilation
-Temperature/Climate control
-Waste disposal
-Chemical storage
-Inherently safer design of experiments
-Back up internet connections (cable and wireless)
-Phone connection
-Space for students/instructors and their belongings based on change of seasons
-Flexibility for changes to lab layout

Initial Lab Concept Sketch

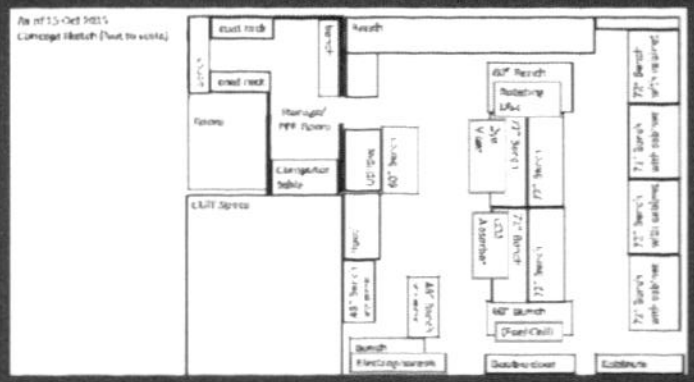

The space required for each bench and experiment was tested in order to determine the configuration prior to purchase.

Before Renovation

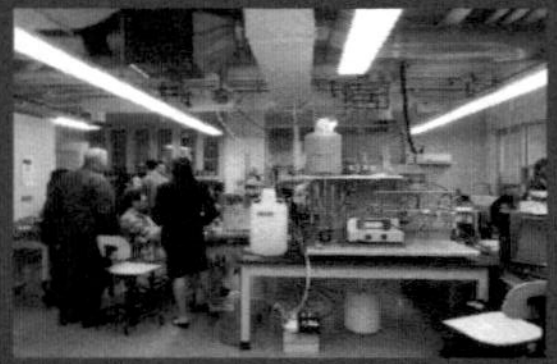

side wall space and vertical space available

old storage space

→

After Renovation

side wall space and vertical space utilized

faculty/student support room

Lab Rep Co Flexible/Mobile Lab Bench Design

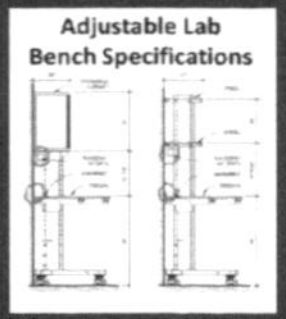

All benches are adjustable and mobile with Internet cabling an electrical wiring fully integrated.

Initial Lab Bench to Room Blue Print

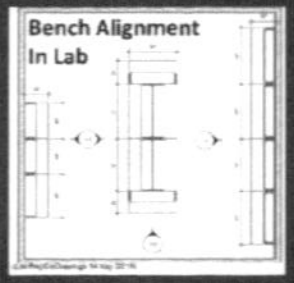

Some Lessons Learned

-Moving equipment too many times leads to damaged/lost equipment.

-Access to utilities must be considered from all points of the room.

-Access to communication resources should be considered.

-Risk must be assessed then re-assessed as things change based on situational awareness and situational understanding.

-Think, then make a move. (i.e. "measure twice, cut once")

-What can go wrong, will go wrong.

-Anyone engineer can make a plan, the good engineer can make a good plan, but the great engineer has a plan for things go wrong.

Risk Assessment and Safety

A Risk Assessment was completed on each new experiment added. Coordination with Environmental Health and Services was conducted throughout each step of the renovation process.

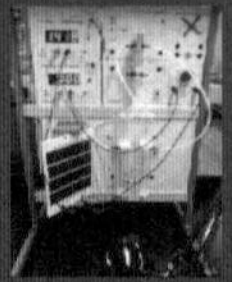

New Solar to Fuel Cell Experiment

New CO_2 Adsorption Experiment

New Flow Control Experiment

New Temperature Control Experiment with Thermal Imaging

Acknowledgement

•Columbia Facilities
•Lab Rep Co.
•Israel Rodriguez, CU SEAS Facility Coordinator

Photo Credits

Dr. Robert G. Bozic, Lieutenant Colonel U.S. Army (Retired)

Educating Chemical Engineers with Interest in the Automotive Industry

Chipman, G.D.; Fuller, T.F.

Introduction

There are now over 1 billion vehicles in the world. The growth of the automotive industry is predicted to accelerate as time continues, with a prediction of reaching 2 billion vehicles in the world by the year 2035.[1] As the number of vehicles increases, the oil reserves in the world decrease. Many governments, realizing that the world's dependence on fossil fuels is unsustainable, have begun to pass laws to curtail fossil fuel use in their countries. One example of this is the agreement reached in 2012 between President Obama and 13 major automotive suppliers to improve the corporate average fuel economy (CAFE) to 54.5 MPG for cars and light duty trucks by the year 2025.[2] This new endeavor has led to many automotive companies increasing their research efforts into electric vehicles and the batteries that power them.

The transition of the automotive industry from conventional internal combustion engine (ICE) vehicles to electric vehicles (EVs) has been set in motion by aggressive fuel economy standards. However, improvements to the power source of EVs are needed to be competitive with current ICE vehicles. Lithium-ion batteries, the predominant power source of electric vehicles, have made great steps towards improving, but further progress is needed in the areas of shelf and cycle life, power and energy density, and the safety and cost of the battery.[3] Prospective researchers of battery systems in the electrochemical field include material scientists, electrical engineers, chemists, and chemical engineers.

Figure 1 Vehicle Electrification

This is only one of the difficulties faced by chemical engineers who have an interest in working on one of the grand challenges of this time by researching batteries for EVs. Other challenges include understanding other aspects of the vehicle operation; including vehicle dynamics, controls, packaging, and other knowledge not generally taught as part of obtaining a degree in chemical engineering. There exists opportunities for chemical engineers to better prepare themselves for careers in the automotive industry. These include the following: 1.) Taking courses on electrochemistry to be able to understand basic principles of batteries and fuel cells, 2.) Electives from other disciplines that teach about vehicle dynamics, controls, packaging, and operation, and 3.) Extracurricular opportunities that allow for hands on experience in the automotive industry. For those with a passion to contribute on an exciting topic, these are some tips that could help them to in finding and succeeding in their desired career.

Electrochemical Education

Some of the major topics covered in chemical engineering courses include thermodynamics, kinetics, and heat and mass transfer. While these are also at the core of electrochemical engineering and the processes involved in battery performance, the courses in chemical engineering are lacking for two reasons. 1.) The basics of electrochemistry are not taught in these courses, and are really needed to have a fundamental understanding of the operation of electrochemical storage devices used to power electric vehicles. 2.) Ideally, the problems in each of the core courses of the chemical engineering major would include example problems dealing with electrochemical systems. However, this is rarely the case. This is perhaps because of the unfamiliarity of students and teachers alike with electrochemical principles that make adapting the curriculum to these circumstances more difficult.

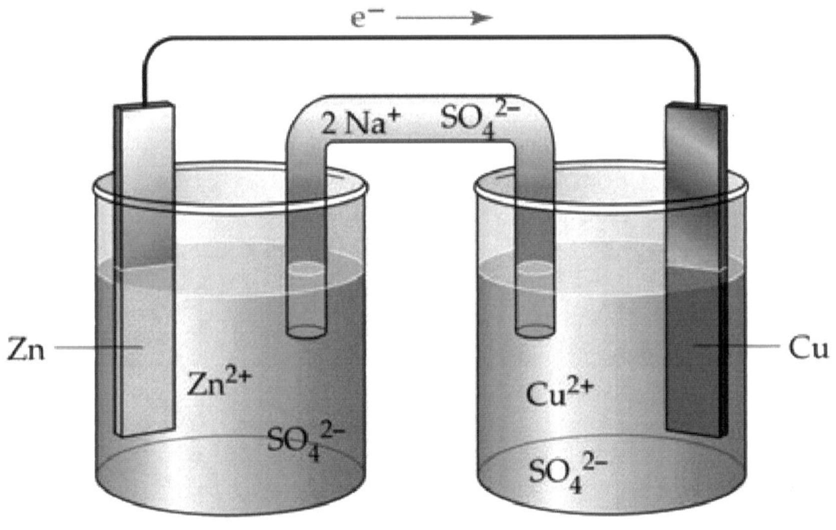

Figure 2 Basic Electrochemical Diagram of a Battery System

In addressing the first area of concern, many chemical engineering departments have instituted an electrochemistry course as an elective for senior level and graduate students that teaches the fundamental principles of fuel cells and batteries. This is a good starting place for chemical engineering students interested in working in the automotive industry and in doing research on EV batteries. The course covers basic electrochemical fundamentals like Faraday's law, how to determine a cell's voltage, and the current that is present in a cell.

Electrochemistry courses also prepare students by educating them about the common structures and characteristics of battery systems. Perhaps an understated fact of an electrochemistry course is that it introduces students to the vocabulary that is used within the field. These include words like capacity, capacitance, coulomb, faradaic reactions, double layer, solid electrolyte interphase, among others. An introductory course in electrochemistry prepares students because of its ability to give students the basic principles and skills needed to begin their own deeper analysis of the field.

Cross-Disciplinary Courses

The world is increasingly becoming one of specialization. That being said, those who possess a diverse skillset offer unique insight into many different problems. While knowledge of electrochemistry is a good starting place for chemical engineers with interest in the automotive industry, other points of knowledge would be useful as well. These could include vehicle dynamics, controls, and computer aided design. Cross-disciplinary work has become a focus point in many industries. Likewise, students who have broader skill-sets are also in high demand. Emadi, a professor at McMaster, encourages electrical engineering students to take courses in chemical engineering to become more multidisciplinary themselves.[4] This multidisciplinary education will enable them to be more effective, and knowledgeable in the cross-disciplinary teams that they will work in.

Many of the requirements for the lithium-ion battery of an EV are determined by the power and energy needed to move the vehicle. These are calculated by knowing the vehicle dynamics. Some examples include the all-electric range, acceleration times, and vehicle efficiency. A basic understanding of the forces that exist allow for a better understanding of the limitations of EVs. These forces include air drag, the weight of the vehicle, and the tractive force. A basic vehicle dynamics course can also cover the procedure to model vehicle performance using the equations learned as part of the course. An EV can be designed in conjunction with understanding the basic functions of the major components of the vehicle. This includes the motor, battery, inverter, and charger to list a few. An introductory course to vehicle dynamics allows students to understand the basic principles at play in vehicle performance, and to extend this to be able to design potential EVs while analyzing the limiting factors of the design.

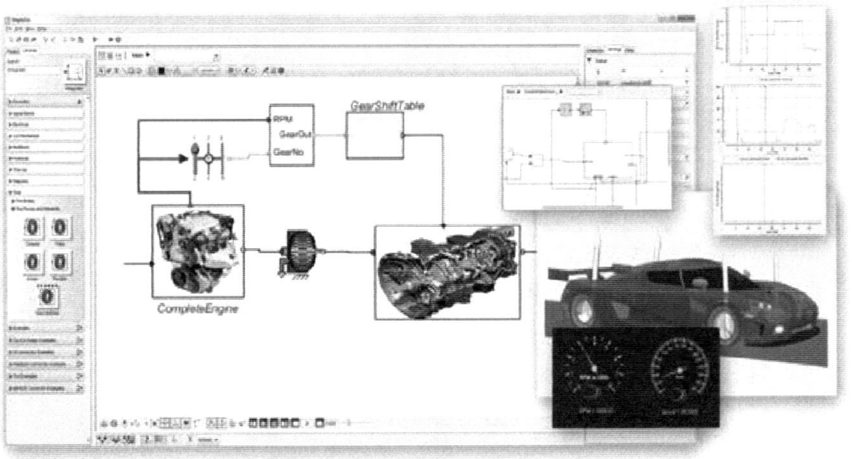

Figure 3 Vehicle Dynamics Simulation

EV efficiency and performance also depend on the ability to regulate the performance of the vehicle. Chemical engineers typically encounter at least one controls course that is focused on controlling chemical processes. While this controls course is good for basic theory and limited application, having a better, deeper understanding of controls theory allows for prospective automotive workers to consider how their designs will be implemented. Controls is a major section of the automotive industry and another important consideration when designing EVs. Controls also is also another area, as opposed to battery research, that could be filled by chemical engineers in the automotive industry. For those interested in the controls sector, a greater focus on controls courses would be advised as part of their course work.

Figure 4 CAD of Electric Vehicle

One major challenge currently faced by EVs is the placement of the battery. A larger battery leads to a more electric range. However, it also leads to more packaging complexity. Having exposure to basic computer packaging techniques allow automotive engineers to be able to understand the difficulty faced in packaging their designs, and to be able to assess the feasibility of their proposed design. Understanding how to use basic CAD programs better enables chemical engineers to understand the limitations in space that are faced in designing a place for the EV battery system.

Pertinent Extracurricular Experience

Another way to prepare chemical engineers for a career in the automotive industry is by participating in activities like SAE, Formula racing, motor sports, or AVTC competitions. The latest AVTC competition is titled EcoCAR 3 and is sponsored by the Department of Energy and General Motors. The competition gives 16 teams in North America the opportunity to convert a conventional Camaro into a hybrid electric vehicle. The goal being to maintain the performance and safety standards while improving the fuel economy and emissions. This team has provided opportunities for some chemical engineering students at Georgia Tech to participate in the vehicle development process and be exposed to experiences not typically encountered in conventional classroom settings.

Figure 5 Georgia Tech EcoCAR 3 Team

EcoCAR 3, while not about building a battery, does provide exposure on how battery packs are assembled, design considerations that need to be accounted for when placing a battery pack in the vehicle, the safety stipulations for working with high voltage, and working on a multidisciplinary team. Participants are also exposed to some of the more practical limitations like soldering, lugs, wiring, and packaging. Getting to not only propose a design, but actually build it, provides invaluable experience in the challenges faced by those that are charged with implementing the designs that are submitted by engineers. This is valuable for any engineer, to realize difficulties that are encountered as designs are taken from the theoretical into reality. This will assist in giving engineers a better sense of how to produce designs that are more easily implemented in the real world.

The design experience also provides students the opportunity to use the knowledge gained in the class room in a real world setting. Creating real problems in the automotive industry that they are tasked with solving using the information they have learned in their college classes. This exposes students to many of the equations that they learned in their classes. The experience also teaches them the properties that are limiting in the proposal of any potential design and implementation. For example, considering questions like where the battery will be placed, how it will be cooled, and how to ensure safe operation. It also exposes students to the wiring that must be placed in the vehicle and its limitations. Students must properly account for the size and length of wire that is needed, and the appropriate fuses for the system, accounting for EMI, and other things that are not taught in the current chemical engineering curriculum.

While chemical engineers are not the conventional choice for automotive teams, they have been some of the more valuable. A steep learning curve is faced because of the lack of direct relation between course work and application, but it has given chemical engineering students the opportunity to be exposed to the automotive world and to gain knowledge they could not have otherwise gained. The experience has been invaluable in helping to prepare them for careers in the automotive industry.

Survey of Chemical Engineers in the Automotive Industry

One valuable perspective in knowing how to best prepare chemical engineers for the automotive industry is to talk to chemical engineers who have already successfully obtained careers in the automotive industry. As they are already working the jobs that automotive companies hire chemical engineers to perform, they would be the best qualified to know how, in retrospect, they would have best prepared for their current careers. A few chemical engineers were surveyed to obtain their advice on what activities are most important to preparing to work in the automotive industry. Their responses are provided in this section.

One chemical engineer who works with fuel cells had the following advice:

> "A graduate course in electrochemistry has proven to be very useful. On some rare occasions, when something has to be tuned on a test stand, I may get to draw upon what I learned in controls (you have to have some awareness of PID control), but not much more than that."

> "In the auto industry, many chemical engineers have to have awareness of catalysis, especially if you work on either fuel cells or aftertreatment (catalytic converters). Therefore, a focus on catalysis and materials science would be good. Chemical engineers in the auto industry usually do have to have some understanding of characterization techniques for materials: surface area measurements (BET, N2 porosimetry); electron microscopy (SEM, TEM, STM, STXM, etc.); X-ray techniques (XRD, XPS, XRF); chemisorption; ICP or ion chromatography analysis for compositions; and many other techniques. Even if you work in paint, many of these material characterization methods are applicable. Often, when you see vehicle program "firefights" occur to identify a problem, you'll see people quickly trying to get SEM data."

> "Chemical engineers in the auto industry often have to figure out how to get micron-scale or even nano-scale layers of something to perform incredible functionality. Automotive chemical engineers have packaging constraints that petrochemical engineers do not. These layers may be catalyst washcoats in a porous aftertreatment monolith, catalyst layers in a fuel cell, electrode and separator layers in a battery, or paint layers. We need to understand how to make inks or paints that are well-dispersed and how to characterize them. We need to understand transport of inks (whether by spraying, coating, etc.) to a substrate, and we need to understand how the processing affects the structure of a layer. The more well-versed students are in colloidal science, the better."

Another chemical engineer had this general advice and offered some bulleted suggestions:

> "Recognize all the processes on-board a vehicle that tie back to chemical engineering. Most or all of the propulsion systems (and many other systems) are based on some chemical process that needs to be well understood and controlled to result in a safe and effective vehicle. So for example: internal combustion, battery electric and fuel cell electric. So it's important to realize chemical engineers are not restricted to the conventional chemical process type industry and the product development end of the automotive industry is well suited for chemical engineers."

- "Not everyone (especially non-engineers) appreciates or realizes the advice given above so I have encountered a slight stigma against chemical engineers in product development. For interviews, Resumes and cover letters must be prepared to reinforce the alignments of a chemical engineering academic background with the field of automotive engineering you are applying for."
- "One of the best paths for chemical engineers in the automotive industry is in the embedded controls domain. So, take all the controls courses you can. For example, MPC (model predictive controls) was developed by chemical engineers and is likely to make its way in to vehicles soon. Also, take some classes in C and if possible a class that would get you an introduction to controller hardware basics."
- "Although this may be a little out of scope, try to get some automotive experience before you graduate. Either through EcoCAR, Formula SAE, internships etc. This really helps."

Two other responses are given below:

> "I worked in automotive industry for five years and two months until October 2015. During graduate school, I participated in the EcoCAR competition, did an internship at Ford for three months and above all worked with a thesis advisor, Prof. Tom Fuller, who motivated you by example to look at the big picture while working on the fundamentals at say microstructure or nanoscale. A place like Georgia Tech also offered me the opportunity to take up courses offered in various departments. I took electrochemistry courses as a refresher during graduate school. When I joined full time at Ford after my PhD, I continued to focus on fundamental research to answer important questions from the management such as bottlenecks to fast charge and published papers in peer reviewed journals and presented at technical conferences as these would enable the automotive and battery industries to come up with solutions faster."

> "The first thought that comes to me is to accept and embrace new opportunities, even if you are not sure you are capable. For me, I went from being a researcher to a chief engineer without much time to adjust. That "jump" helped me understand my future collaborators and customers within GM, which was great preparation for being an R&D director, my current role."

Some of the common themes include an electrochemical course, pertinent extracurricular activities (like EcoCAR), taking other courses, varying from controls to catalysis to material characterization, and obtaining research experience. Being open to opportunities and realizing that chemical engineers are capable of doing anything they set their mind to helps those interested in the automotive engineering industry, or any other industry, to succeed.

Conclusion

One of the biggest challenges facing the current generation is to create sustainable transportation. Chemical engineers are uniquely equipped to assist in finding the best solution. However, chemical engineers are not conventionally seen in the automotive industry perhaps because of a lack of preparation. For chemical engineers who are interested in working in the automotive industry, advice has been given on how to best prepare for their desired career. This advice included taking an electrochemical course that would help them to learn the basics of batteries, fuel cells, and other electrochemical storage devices that are becoming more commonly used in the transportation sector. Other relevant courses were suggested, ranging from controls courses to vehicle dynamics courses to learning basic CAD techniques to help with packaging and routing in the vehicle. Pertinent experience on a automotive team goes a long way in giving chemical engineers relevant experience that will prepare them for their future careers. Advice was also given from chemical engineers who are currently working in the automotive industry. While many of the same points were highlighted, they also pointed out the importance of being willing to embrace challenges in order to be successful. As long as a chemical engineer is willing to put in the work necessary, they will be successful in any industry. The advice given above, however, will prove useful in better preparing any prospective automotive engineer for their career.

References

1.) *Transportation Forecast: Light Duty Vehicles Light Duty Stop-Start, Hybrid Electric, Plug-In Hybrid Electric, Battery Electric, Natural Gas, Fuel Cell, Propane Autogas, and Conventional Vehicles: Global Market Forecasts, 2015-2035*. Navigant Research. Web. 10 Nov. 2016.

2.) "Obama Adminsitration Finalizes Histroic 54.5 MPG Fuel Efficiency Standards." The White House, 28 Aug. 2010. Web. 10 Nov. 2016

3.) Aurbach, D., K. Gamolsky, B. Markovsky, Y. Gofer, M. Schmidt, and U. Heider. "On the Use of Vinylene Carbonate (VC) as an Additive to Electrolyte Solutions for Li-ion Batteries." *Electrochimica Acta* 47.9 (2002): 1423-439.

4.) Platt, John R. "How to Get Hired in the Automotive Industry." - *IEEE*. 7 Apr. 2014. Web. 10 Nov. 2016.

454605 Shared Spaces and Schedule Optimization for Undergraduate Teaching Laboratories

Abstract:

Singapore Institute of Technology is a newest autonomous university in Singapore. Since the establishment of SIT in 2009, studentsâ€™ enrolment in the university has been increasing year on year. This has led to challenging space limitation in SIT due to its rapid expansion.Â This presentation will discuss how SIT resolves these constrain through shared spaces and optimization of class schedules with due considerations given to safety and resources.

454777 Distillation: The Forgotten Art

Abstract:

Distillation: The Forgotten Art

Jacob Arredondo, and Jonathan Worstell

Distillation may not be a glitzy, glamorous unit operation these days, but is still remains important with regard to profit generation in the chemical industry. For that reason, it should be taught in undergraduate laboratory courses. However, distillation experiments have a size limitation, which is determined by the amount of solvent stored in the laboratory.

To relax this size limitation, we have designed three-day laboratory sequence. During the first day, senior students batch distill one of several binary organic mixtures using bench top equipment. They do a flash, empty column, packed column, three-valve column, and six-valve column distillation; then analyze their results using Lord Rayleighâ€™s equation and McCabe-Theile diagrams.

The second day, students operate a three-tray air/water Plexiglas column that has interchangeable sieve, valve, and bubble cap trays. The students determine the flooding and weeping operating regime for each type of tray.

Finally, during the third day, the students distill their assigned binary mixture in a twenty five tray Oldershaw column. They control this column manually, which is a good learning experience.

This presentation includes a video demonstration of the second dayâ€™s experiment using the Plexiglas column.

455231 An Education Module for Natural Gas Storage

Abstract:

Over the course of six years working as a student liaison and with faculty to make sustainability a rigorous education component, feedback from students strongly suggests that a large portion of chemical engineering graduates desire to work in the energy sector, specifically in the oil and natural gas industries. However, petroleum engineering students maintain an advantage to gaining employment in these fields over strict chemical engineers. As a participant of a Center for Sustainable Engineering faculty development workshop, an expected outcome includes the dissemination of an education module relevant to the topic of sustainability. The outcome in my case involves the development of a series of modules related to Natural Gas upon the realization of the so called *Natural Gas Revolution*. Synergistic tight shale gas-hydraulic fracturing and directional drilling technologies act synergistically to provide for the recovery of energy resources touted as a bridge to overall cleaner energy production and use. 2015 marks a point in time where the majority of electricity generation in the U.S. originates from natural gas as a primary energy source instead of coal. This poster outlines in detail, the first module that will be shared as part of the CSE repository, focusing on the topic of Natural Gas Storage, providing chemical engineers with a succinct technical exposure to natural gas storage related to energy sustainability. As a whole, modules presented together illustrate Natural Gas practice from an interdisciplinary perspective including elements from; chemical engineering; health sciences; environmental engineering, and contains significant links to policy-regulation.

457212 Applied Learning in Biopharmaceutical Laboratories

Abstract:

The biopharmaceutical industry has a need for competent chemical engineers who are familiar with good manufacturing practice regulations and the hands-on practical knowledge of using specialized equipment such as stainless steel bioreactors and single-use bioreactors. Additionally, the chemical engineers also need to have an understanding of cell culturing techniques and sterility concepts such as clean-in-place and steam-in-place.

This presentation will share approaches on the development of an applied learning biofermentation laboratory course in SIT to train chemical engineers for the biopharmaceutical industry.

458580 Learning Tools in an Interactive Textbook for Material and Energy Balances

Abstract:

Since textbooks are experiencing a 21st century makeover, chemical engineering textbooks should follow other larger markets, like freshman chemistry and physics, to create student-centered and interactive tools. One textbook has recently been released by zyBooks that provides an interactive, web-based experience for material and energy balances. While studentâ€™s response to the new textbook will be covered in a talk, the rationale of different features of the book will be explored. First, question sets with learning questions are scaffolded from easier to more difficult as well as explain why each answer is correct or incorrect. Animations turn figures into interactive constructions of new concepts (e.g., phase diagrams), operating principals of common equipment (e.g., a distillation column), and developing useful equations (e.g., the energy balance for a steady state system with negligible kinetic and potential energy). Finally, students receive feedback in real time as they read/participate in each section. Feedback is also provided to the instructor to verify that the reading was fully completed, partially completed, or incomplete.

460068 Chemical Engineering Comics to Teach Assumption Making

Abstract:

One of the most difficult aspects of chemical engineering for undergraduate students to comprehend, and therefore one of the most difficult to teach, is the concept of assumptions. How to make them correctly, when to apply assumptions, and how to use them in line with mathematical concepts and definitions are a requirement for any successful engineer, yet it is something that has proven especially challenging for undergraduate students to grasp. Providing students with not only a grasp on the concept but also a willingness and confidence to apply assumptions when possible is necessary in undergraduate education. In an effort to create supplemental curriculum to help support students in their development with this concept, the authors have prepared a Chemical Engineering science comic to be implemented in their course work. The comic should make the concept of assumptions more accessible to engineering students, as well as act as a way to introduce the topic to students in a useful and thought provoking manner. The comic utilizes visual depictions of applied assumptions, supported by an dialogue between a professor and student, to create an additional learning tool. With future comics to come in the series, it is the hope that these chemical engineering comics can act not only as a starting point for discussions of assumption making to undergraduates but also be applied to other engineering concepts in the future.

460347 Nmsu Guide to Laboratory Safety

Abstract:

The chemical laboratory safety guidelines were nonexistent at NMSU in 1981 when I arrived. After I published the"Guide to Writing and Problem Solving for Chemical Engineers" used at 8 universities, it was apparent to me that we needed a laboratory guide to laboratory safety. Because I read in CEP that George Whitmyre, Laboratory Coordinator at University of Delaware, had drafted a safety guide, I wrote to him asking if I could use it. He graciously allowed me to use it in writing,and allowing me to modify it as requiredin New Mexico and so I drafted the "Guide to Safety in the Laboratory for Chemical Engineers" and it was apporoved by the general counsel at NMSU in writing for use in our department at NMSU. Some time later, the Vice President for Economic Development, AverettTombes, asked me if I would draft a Laboratory Safety Manual for the University. I agreed to do so and I put together a committee for the Environmental Safety and Health that was a broad committee of expertise across the university.The meetings held subseqently drafted the composite document as appropriate, though it was strongly based on the Departmental Laboratory Safety Guide in Chemical Engineeringdrafted by me. It did include guidelines on laser safety an a very brief section on Radiation safety(because radiation safety was already covered in the radiation safety manual through the the environmental safety office drafted by Prof. A. Burr.The committee completed the task and it was completed and distributed to all Deans and Department Heads in October 1995. The manual was adopted by the Board of Regents and adopted online in 1995 an it remained online until 2015. The Environmental Safety Training Office then removed it as "Obsolete". It is my contention that it should be updated and left online because the first item in the Code of Ethics for Chemical Engineers is that the health and safety of the public should be held paramount in the practice of chemical engineering. The point is that when the laboratory safety manual is removed the tendency naturally becomes one of complacency. And when complacency develops, the hazards tend to come back.

461566 Enhancement of Lab Experiments Using CAD and Prototyping

Abstract:

Finding specific instrumentation and/or specialized equipment for laboratory experimentation can often be challenging or event cost prohibitive. Rapid Prototyping can be used to customize laboratory parts to effectively build and implement an experiment. Laboratories with active participation of the students, acquiring hands-on experience, are of great importance in engineering schools as practical knowledge gained in the area of expertise can significantly impact the quality of student education. By designing equipment in collaboration with students, instructors have more input throughout the design process and subsequently students get involved early on the design of experiment aspects of laboratory practices. The use of Computer Aided Design (CAD) and Rapid Prototyping are excellent tools to customize laboratory and research instrumentation. Over the past year, more than one hundred projects have been implemented materializing prototyped parts using a 3D printer or a Laser cutting machine. Among other things, this has kept costs down while at the same time producing very unique experimental equipment.

This contribution present a show case of examples than can be easily implemented in courses dealing with fluid mechanics, heat and mass transport. In addition, this work also includes examples of students led learning objective writing as well as examples of studentsâ€™ outcomes improvement.

461568 Integrating Concepts of Social Responsibility and Cultural Awareness into a Global Service Learning Study Abroad Course

Abstract:

Engineers generally tend to be technically oriented and results focused. Meeting technical challenges is what motivates many engineers. However, it is important for engineers to realize that they are part of a global community and that the processes and products they design can have far reaching impacts. This is especially true in developing countries where even technically well-conceived projects may not fit the needs of the communities for which they are intended. Additionally, cultural differences complicate the work environment. Students engaged in global service learning projects need to understand the role engineers play in society. Solving technical challenges is only one aspect of the engineerâ€™s job. Increasingly, engineers must consider the broader impacts of their design choices. This poster presentation will include a case study that documents the organization of a Global Service Learning course that introduces the students to the concepts of social responsibility and cultural awareness.

In the summer of 2015, a group of chemical and mechanical engineering undergraduate students from the University of Kentucky Paducah Extended Campus Program participated in a Global Service Learning and Sustainability course focused on implementing a project to convert waste plastic into a sustainable liquid fuel in rural southern India. Service learning projects conducted in other countries often face cultural hurdles that are quite different than in traditional study abroad programs. Working side-by-side with local people requires adapting to cultural practices that may be quite different from the studentsâ€™ expectations. The learning outcomes for this course included providing students with clear understanding of the key issues of sustainability and sustainable community development in rural and underdeveloped regions and educating them as to how engineers must consider cultural factors when working in underdeveloped regions.

This course included both a traditional classroom lecture component and a hands on service learning project focused on applying the principles of green chemistry, appropriate technology and sustainable engineering to address the problems of waste plastic accumulation in underdeveloped regions. This course was offered as a six-week summer term course in the Indian state of Tamil Nadu. The Global Service Learning and Sustainability course was developed in collaboration with the Organization of Development Action and Maintenance (ODAM), an NGO headquartered in Tiruchuli, India. The case study presented here documents the successful implementation of this technology in Tamil Nadu, India and the cultural hurdles that must be addressed in an international service learning project.

465973 Application of Low-Cost Components to the Chemical Engineering Practices

Abstract:

Since the Bolonia process started in 1999, the focus in education changes from a professor centered process to a student centered process. It took shape in the concept of "capabilities" or some specific skills to be worked during the teaching process at universities. Some of them are creativity, oral communication or autonomous learning. That change in the educational process implies that it is necessary a change during classroom or laboratory activities. Traditionally, laboratory practices in chemical engineering were done by using pre-built equipement calibrated to properly work during laboratory lessons. It was done so because of the complexity and high cost of the chemical equipement used. It would be a nonsense to think that students could assemble and start up a distillation column or a CSTR. Instead of that, students find a totally assembled and perfectly working equipment were they only take some data ignoring the work required and the problems arised during mounting and setting. If we join that with the increasing concern about capabilities like autonomous learning or team working, we concluded that it would be great if we could desing an activity were students desing, build, operate and improve a simple chemical process. One could think that it is an utopy because of the great cost of the components needed. However, in the last years have emerged some low-cost hardware communities that can help us to achieve our aim. Specifically, Arduino (and others like Raspberry Pi) microcontrollers can be bought nowadays for less than $25. They have both analogical and digital inputs and outputs, and can be easily programmed. These features make them right to use in PSE teaching, for example, to test different control algorithms for a specific controller. Besides, not only cheap microcontrollers can be found nowadays, but also a wide amount of cheap sensors and transducers (temperature, pressure, flow...) are commercially available too. All together encouraged us to design a meta-activity were two small size (plate and shell tube) heat exchangers were provided to the student with pumps, valves, deposits, tubing, sensors, microcrontollers and all electronic components required. That operation was chosen mainly because of safety aspects (only water is used at temperatures below 60 ºC). In this work, we present all technical aspects of the meta-activity predesigned as well as the educational tasks employed. We also

present students feedback as well as the rubrics employed to measure the capabilities tested.

470474 Addressing Social Challenges through Creativity, Engineering, Nanotechnology & Diversity

Abstract:

Fully capitalizing on the investment in nanoscience research will require a technical workforce educated to innovate, manage, and evaluate the socially-responsible implementation of nanotechnology. This project created interdisciplinary undergraduate engineering courses at the University of Connecticut where nanotechnology was used as the context to teach creative thinking, project management skills, and evaluation of social outcomes of novel technology. "Nanoscience and Society" is an introductory course on nanoscience with an emphasis on social issues. Students evaluate social impacts of existing technologies through case studies, guest lectures, and group work. Contrary to most early engineering courses, this class is explicitly designed to appeal to non-engineering majors and those who are particularly altruistic-minded, with recruitment a prime goal. "Managing Creativity and Innovation" targets junior engineering students and offers specific training in being creative and in managing creativity in diverse teams using nanoscience as the technological context. The objective is to shape a more adaptable workforce able to embrace and evaluate evolving technologies to meet future challenges. In "Capstone Design Marketplace" senior undergraduates create their own design challenge, and respond to the design challenge of a peer from a different engineering major. Learning goals include increased confidence and competence employing technical knowledge and skills, enhanced interdisciplinary knowledge, improved project management skills, and improved technical communication skills. Overall, this "ASCCEND" program has increased interdisciplinary engineering education at our institution and has enhanced awareness for societal impacts of technology. However, while we expected that social benefits would serve as an effective vehicle for promoting a more diverse engineering pipeline, extensive surveys revealed no significant differences among majors or gender that the "potential to make a difference" plays in a student's choice of major. Interestingly, parents played a significant role in the selection of engineering as a major, especially for female students, thus providing a strategy for broadening participation of women in engineering. By providing students with specific training in managing creativity, evaluating technology in interdisciplinary settings, and working across traditional disciplinary boundaries, this program helps participants to become more agile and productive members of the evolving global workplace.

465655 A Joint Chemical and Mechanical Heat Exchanger Project, Trying to Emulate Industrial Experiences

Abstract:

Creating "real world" experiences in the classroom can be challenging. To do this in heat transfer, we developed a project that required students from different courses and different majors to work together. The heat exchanger design project was given to the chemical engineering students, whom must design a shell and tube heat exchanger, but they must also hire mechanical engineering consulting firms to design more efficient and compact heat exchangers. Each chemical engineering group was paired with two mechanical engineering groups. The meeting consisted of the clients presenting their problem to both of the consultant firms followed by a short one on one meeting with each firm. Following this, all communication was to be conducted via technical memo, including a progress report containing preliminary designs. The students then presented to everyone in both courses and graded the presentations.

This project sought to mimic the interactions of chemical and mechanical engineers in the design of a heat exchanger. Mechanical engineering students cover heat exchangers in great detail and learn about compact heat exchanger designs whereas chemical engineering students are more interested in the overall operation of the system and are taught the m ore classical heat exchangers. Many times in industry, chemical engineers will calculate the overall heat transfer coefficient times the area of heat exchange (UA) and then pass this information on to a mechanical engineer with other design constraints.

Surveys were conducted to determine the students' perspectives on the project, skills gained, and attitudes towards the other engineering discipline. Here we present the project details, our perspectives of the project, outcomes, students' perspectives, and our revisions going forward.

472122 Optimizing External Coursework in a Flipped Chemical Reaction Engineering Classroom

Abstract:

Starting in spring 2014, the Chemical Reaction Engineering Class (called Process Kinetics) at the University of Connecticut incorporated a flipped classroom approach. The course includes online daily questions (based on reading assignments), online video lectures, and intentionally short reading assignments combined with only very brief (typically <10 min) of lecturing in class and extensive in-class examples worked by the students. In addition, the class included traditional homework assignments and exams. During the first half of spring 2014, this course was taught as a hybrid lecture/flipped classroom. Based on student midcourse feedback, the course was pushed to a fully flipped classroom and the entirety of the spring 2015 was taught was a flipped classroom.

This presentation will include a description of the course layout, extensive lessons learned from this experience from the perspective on the professor, and an analysis of student feedback from each semester. Overall, the students responded very favorably to the course as a whole. Among the favorite components were the online videos and in-class examples. A large fraction of the class took advantage of the opportunity to rewatch online lectures, review online problems, and revisit the in-class examples in preparation of exams and when completing homework. The flipped classroom set-up also provides an ideal platform for applying just-in-time teaching approaches, particularly if there are specific subtopics with which a large group of students are struggling.

This poster will focus on the relationship between student performance in the course and how well they kept up on the daily course materials and expectations. Specifically, I will compare how far before class meetings assignments and videos are typically completed and compare this to performance in other aspects in the course. I will also discuss future plans to implement gamification to encourage students to complete these assignments in a more timely manner based on preliminary results that demonstrate that students who are successful in the course typically complete

assignments further in advance on the class session where the information is covered.

473144 An Ontological Approach Towards Curriculum Alignment in Chemical Engineering

Abstract:

The curriculum has been identified as a major aspect of teaching in higher education (Byrne, 2006; Cassel, Davies, LeBlanc, Snyder, & Topi, 2008; Dexter & Davies, 2009; Mälkki & Paatero, 2015; Morsi, Ibrahim, & Williams, 2007) and its complexity has also been identified (Byrne, 2006), especially for the domain of Chemical Engineering. Solving problems relating to the curriculum and teaching and learning in general, is becoming more complicated. In an effort to address the problems that arise in relation to curriculum, educationalists and technologists employ advanced technologies. The technologies used range from simple concept maps (Morsi, Ibrahim, & Williams, 2007) to software solutions (Dexter & Davies, 2009) and semantic solutions (Cassel, Davies, LeBlanc, Snyder, & Topi, 2008). One of the latest attempts in employing semantic technologies and more specifically knowledge modelling has been presented in (Bussemaker et al., 2016), where the authors presented the CheEdo ontology. The ontology represents topics in Chemical Engineering and Learning outcomes.

In this paper we extend CheEdO ontology to facilitate vertical alignment of the curriculum. The ontology is modelled in Web Ontology Language (OWL) and SWRL language has been employed to create the rules that guide the alignment process.

The updated ontology is validated with a case study taking into account recent developments in UK curricula.

References

Bussemaker, M., Trokanas, N., Cecelja, F. 2016 "An Ontological Approach to Chemical Engineering Curriculum Development ", Proceedings of the 26th European Symposium on Computer Aided Process

Engineering, June 2016.

Byrne, E.P. 2006, "The Role of Specialization in the Chemical Engineering Curriculum", Education for Chemical Engineers, vol. 1, no. 1, pp. 3-15.

Cassel, L.N., Davies, G., LeBlanc, R., Snyder, L. & Topi, H. 2008, "Using a computing ontology as a foundation for curriculum development", Proceedings of the Sixth International Workshop on Ontologies & Semantic Web for E-Learning, pp. 21.

Dexter, H. & Davies, I. 2009, "An ontology-based curriculum knowledgebase for managing complexity and change", Advanced Learning Technologies, 2009. ICALT 2009. Ninth IEEE International Conference onIEEE, , pp. 136.

Mälkki, H. & Paatero, J.V. 2015, "Curriculum planning in energy engineering education", Journal of Cleaner Production, vol. 106, pp. 292-299.

Morsi, R., Ibrahim, W. & Williams, F. 2007, "Concept maps: Development and validation of engineering curricula", Frontiers In Education Conference-Global Engineering: Knowledge Without Borders, Opportunities Without Passports, 2007. FIE'07. 37th AnnualIEEE, , pp. T3H.

473225 Beyond the Nuts and Bolts of Energy Technologies: Multidisciplinarity in Chemical Engineering Education

Abstract:

Introductory knowledge to energy has been incorporated in chemical engineering curricula as a result of the close connection between energy technologies and the chemical engineering discipline. Courses in introductory energy education typically expose students to the entire spectrum of energy technologies by surveying the technological landscape of the industry (e.g. through a sequence of topics including fossil energy, hydropower, wind, solar), and by examining each energy technology within its respective developmental and regulatory context. This approach, although intuitive and well-developed, is flawed in its portrayal of energy technologies as independent and mutually exclusive choices. Such an approach frustrates students with the complexity of mundane energy-related factual knowledge, and renders the multidisciplinary essence of engineering education elusive. We tested a new framework for the teaching and learning of energy-related topics by designing courses for chemical engineering majors, focusing on multidisciplinary energy education rather than dissemination of factual energy knowledge. Our innovative coursework highlights the impact of energy technologies, both renewable and non-renewable, within a number of socioeconomic and environmental dimensions. In our latest offering of a newly developed introductory energy course, Energy Technology and Social Change, we present energy as a key driving force for social, economic, and technological evolution within human society. Students analyze energy technologies from various eras of human history, and identify the key role energy plays in shaping human society as a whole. Under this new framework, chemical engineering students learned how to conduct value judgements based on characteristics of energy technologies, and became able to perform holistic assessment of emerging energy technologies that are augmenting or replacing existing technologies.

471427 Intersection of Active Learning Experiences and Learning Preferences in the Gateway Course

Abstract:

In this investigation, we sought to link student learning preferences with class performance in an introductory chemical engineering course with roughly 160 students split over three sections. In addition to the professors' lectures using "notes with gaps" and active learning opportunities, all students were also required to attend a one-hour weekly recitation-style Problem Session led by teaching assistants (TA's) who reviewed lectured content and covered homework and example problems. The weekly Problem Sessions occurred in a classroom setting similar to lecture, with students sitting in rows facing the TA. To review a homework or example problem, TA's asked students how to proceed step-by-step in solving a part of the problem. Based on feedback from the class, the TA's helped the students ultimately arrive at a solution to the problem. To create more active learning opportunities, a "Creativity Space" in the Engineering library with moveable whiteboard walls was utilized twice during the semester to allow students to work in groups, solve problems in steps, and critique other groups' solutions, with the TA's providing guidance as the students worked together. We gauged the effectiveness of normal Problem Sessions, as well as the active learning exercises, via a student survey where they identified their learning preferences, such as more example problems, more lecture content review, more group work, etc. The survey responses were analyzed both by overall class preference as well as by breaking the responses down into smaller groups based on the students' final grades in the class. This survey and analysis elucidated specific student learning preferences and enabled identification of strengths and weakness in the Problem Sessions at each level of performance in the class. Overall, we identified several areas of improvement to address the needs of lower performing students while also revealing that generally, students enjoy and benefit from active learning experiences. Our findings will be used to improve Problem Sessions in future semesters.

474018 Re-Situating Learning at Oregon State

Abstract:

This poster presents our efforts towards *revolutionizing* the undergraduate learning environment in the School of Chemical, Biological, and Environmental Engineering at Oregon State University (CBEE). While we currently provide students many innovative learning opportunities, we are concerned that the extent to which these efforts are marginalized and isolated may deter from their influence. To address this issue, we are in the process of creating a more holistic, inclusive, professionally-based learning environment woven through both curricular requirements and co-curricular opportunities for students. We specifically target social inequality by seeking to create engineering educational systems and interpersonal interactions that are professionally and personally life-affirming for all people across their differences. We seek to catalyze change through construction of a culture of inclusion and a shift in our learning environments from sequestered activities to realistic, consequential work. This requires a fundamental change in the nature of department culture (values, norms and structure).

Project activities include: (1) curricular redesign of nine core sophomore- and junior-level studio classes to include more ***realistic, consequential work*** leveraging research-based pedagogies like ***problem-based learning*** and ***model-eliciting activities***; (2) growing faculty and studentsâ€™ ***capacity to engage issues of inclusivity*** by shifting their cognitive and affective knowledge of power and privilege; (3) planning and implementing ***student professional development pods***, longitudinally mixed student teams where students help one another understand the university experience and how it relates to professional practice; and (4) implementing ***formal changes in governing policies and procedures*** within CBEE.

This work is supported by the National Science Foundation pilot program *Revolutionizing Engineering Departments* (RED) that is aligned with the NSF Engineering (ENG) Directorateâ€™s multi-year initiative, the *Professional Formation of Engineers*, to create and support an innovative and inclusive engineering profession for the 21st Century.

474353 Promoting Systems Thinking for Holistic Undergraduate Engineering Education

Abstract:

Practitioners in many disciplines, including Engineering, are being increasingly called upon to use systems approach to address challenging problems related to sustainability. The systems approach transcends disciplines and is especially required when answers to inherently complex issues are sought. It has been suggested that handling complexity requires working in an intuitive domain not considered in engineering education. The need to improve studentï¾’s ability to understand complex systems necessitates improving their systems thinking skills. We will discuss about preparing undergraduate engineering students to handle challenging interdisciplinary problems and improve their systems thinking skills.

474442 Overview of Chemical, Bioengineering, Environmental, and Transport Systems Division (CBET)

Abstract:

The National Science Foundation (NSF) is an independent federal agency created by Congress in 1950 "to promote the progress of science; to advance the national health, prosperity, and welfare; to secure the national defenseâ€¦" With an annual budget of $7.5 billion (FY 2016), we are the funding source for approximately 24 percent of all federally supported basic research conducted by America's colleges and universities. In many fields such as mathematics, computer science and the social sciences, NSF is the major source of federal backing.

474443 Highlights of CBET Cluster on Chemical and Biochemical Systems

Abstract:

The National Science Foundation (NSF) is an independent federal agency created by Congress in 1950 "to promote the progress of science; to advance the national health, prosperity, and welfare; to secure the national defenseâ€¦" With an annual budget of $7.5 billion (FY 2016), we are the funding source for approximately 24 percent of all federally supported basic research conducted by America's colleges and universities. In many fields such as mathematics, computer science and the social sciences, NSF is the major source of federal backing.

474444 Highlights of CBET Cluster on Bioengineering and Engineering Healthcare

Abstract:

The National Science Foundation (NSF) is an independent federal agency created by Congress in 1950 "to promote the progress of science; to advance the national health, prosperity, and welfare; to secure the national defenseâ€¦" With an annual budget of $7.5 billion (FY 2016), we are the funding source for approximately 24 percent of all federally supported basic research conducted by America's colleges and universities. In many fields such as mathematics, computer science and the social sciences, NSF is the major source of federal backing.

474445 Highlights on CBET Cluster on Environmental Engineering and Sustainability

Abstract:

The National Science Foundation (NSF) is an independent federal agency created by Congress in 1950 "to promote the progress of science; to advance the national health, prosperity, and welfare; to secure the national defenseâ€¦" With an annual budget of $7.5 billion (FY 2016), we are the funding source for approximately 24 percent of all federally supported basic research conducted by America's colleges and universities. In many fields such as mathematics, computer science and the social sciences, NSF is the major source of federal backing.

474446 Highlights of CBET Cluster on Transport, Thermal and Fluid Phenomena

Abstract:

The National Science Foundation (NSF) is an independent federal agency created by Congress in 1950 "to promote the progress of science; to advance the national health, prosperity, and welfare; to secure the national defenseâ€¦" With an annual budget of $7.5 billion (FY 2016), we are the funding source for approximately 24 percent of all federally supported basic research conducted by America's colleges and universities. In many fields such as mathematics, computer science and the social sciences, NSF is the major source of federal backing.

474447 Interactive Question and Answer Session with NSF Program Directors

Abstract:

The National Science Foundation (NSF) is an independent federal agency created by Congress in 1950 "to promote the progress of science; to advance the national health, prosperity, and welfare; to secure the national defenseâ€¦" With an annual budget of $7.5 billion (FY 2016), we are the funding source for approximately 24 percent of all federally supported basic research conducted by America's colleges and universities. In many fields such as mathematics, computer science and the social sciences, NSF is the major source of federal backing.

474619 Experience on a New of Micro-Adaptive Instruction Methodology Developed for the Improvement of Flipped-Classrooms

Abstract:

Implementation of many of the modern student-based teaching techniques such as the flipped classroom, and active learning, consumes significant class time and eventually limits the overall covered material. Something, which could lower their efficiency and limit their applicability. This presentation discusses the combination of the micro-adaptive instruction as a methodology to extend "free" in-class-time in order to introduce the problem-creation-solving learning approach. Here I present the preliminary results and conclusions from this combination of methodologies on teaching "external flows" of a fluid dynamics course. Micro-adaptive instruction is implemented the past four years at the same course.

469079 An Educational Matlab App for Pharmacokinetic/Pharmacodynamic Modeling of ACE-Inhibition

Abstract:

The goal of this project was to introduce engineering design and the discipline of chemical engineering to incoming freshmen through an interactive, hands-on activity. Our project focused on the development and instruction of an interactive MATLAB simulation app inspired by research interests in the hormone Angiotensin II and its effects on kidney cells during diabetes. Angiotensin II raises blood pressure. Several pharmaceuticals known as Angiotensin Converting Enzyme (ACE) inhibitors are on the market to block Angiotensin II production to lower the blood pressure. The chemical reactions involved in natural production of Angiotensin II and the biological response to pharmaceuticals to lower blood pressure in a time- and dose-dependent manner were modeled in MATLAB and packaged into an easy-to-use app. This app can be utilized in other educational contexts such as chemical kinetics courses and introductions to chemical and biomedical engineering applications. We give a brief example of how the app was adapted for use in the chemical reaction engineering junior level chemical kinetics course. The presentation primarily focuses on the use of the app during the 2015 & 2016 Oklahoma State University College of Engineering, Architecture and Technology (CEAT) Summer Bridge Program, a three-week preparatory program designed to guide incoming freshmen through the transition from high school to college. Through this project, the students were exposed to chemical and biomedical engineering design principles and gained problem solving experience.

The mathematical models in this study describe the change in drug concentration over time as well as the Angiotensin II levels over time. The models use pharmacokinetic and pharmacodynamic relationships and were packaged into an interactive Graphical User Interface (GUI) in MATLAB. The students in the Summer Bridge Program used the app to design optimal pharmaceutical dose amounts and frequencies needed for lowering blood pressure using two different ACE inhibitors (Benazapril and Cilazapril) in patients with either normal or impaired kidney function. Plots were generated for the drug and Angiotensin II concentrations over time. These plots were analyzed and evaluated by the students to make an educated decision as to the optimal dosage conditions for their virtual patient. In addition to the use of the simulation, the students participated in activities to introduce basic calculus, differential equations, and coding to give the students a better understanding of the simulation functionality and its connections to their basic engineering and mathematics coursework.

The students were assigned specific drug types and virtual patient kidney functions and asked to design the optimal dosage and frequency for those cases. After they decided on a final design, the students were asked to create a short report and presentation to convey their results and learning outcomes to their peers. Overall, the students seemed to enjoy seeing pharmaceutical applications of chemical and biomedical engineering as well as having the aspect of creative decision making throughout the project.

Through use of the MATLAB app, students practiced open-ended engineering design, connected calculus and coding to solve realistic engineering problems, constructed a scientific presentation, practiced public speaking, and gave feedback on peersâ€™ presentations. This app is a useful tool for students to gain familiarity with chemical and biomedical engineering concepts including design, without having significant background training.

466263 Fluid Networks: From a Science-Based Training to a Learning Centered Pathway, a Collaborative Development Involving a Teacher and a Pedagogical Engineer

Abstract:

Fluid networks: from a science-based training to a learning centered pathway, a collaborative development involving a teacher and a pedagogical engineer

N. Veuillez, C. Gatumel

University of Toulouse, centre RAPSODEE, Ecole des Mines d'Albi, campus Jarlard, 81013 Albi Cedex 09, France

cendrine.gatumel@mines-albi.fr / nathalie.veuillez@mines-albi.fr

The Ecole des Mines d'Albi of Toulouse University offers a master degree in process engineering with a mention on materials and energy in building sector. This communication deals with a course about fluid networks originally designed for the students training, that had to be adapted for apprentices.

Training in French engineering schools is three years long and lead to obtain a master degree. Students classically get into engineering schools after a two years intensive training based on fundamentals and have to pass a competitive entrance examination. They have excellent skills in mathematics and scientific matters. Since a decade, engineering schools offer a different way to obtain that degree under the form of sandwich courses. Engineer-apprentices get into school after a technical degree. They have more technical and practical skills than students but poorest abilities in fundamentals. Their training is three years long too but it combines a working experience in a company. While classical deductive approach is, if not the best but suitable for students, an inductive approach is preferable for the apprentices. We are also in presence

of two populations with different ways of learning and two different-designed training who have to get to an identical master degree.

At this evolution's kick off was a series of lectures about fluid mechanics, graph theory and numerical resolution completed with exercises in the form of tutorial classes. The teacher in charge of the evolution of this course was not the one who had created it and actually neither the new teacher nor the students were totally at ease with this three input subject course and its theoretical and linear approach. The new teacher was supported in this task by a pedagogical engineer, their shared diagnosis was: "The students feel the subject is important but difficult, highly academicals, they don't make the connections between scientific fields, and their results are not that good".

No revolution, but step by step, many changes, through an intensive, very pleasant and "fruitful" collaboration. Two principles underlay this work: the course has to make sense for the students, the "learning outcomes" have to become specific and clear; and the way of learning must support student's involvement, autonomy and self-direction.

Five years after, the course looks quite different:

- It's now a blended course, with lessons at school and e-learning sessions on the school's LMS (learning management system: in this case a Moodle open source platform),
- A contextualized course, with a great witness, Mines Albi's alumni now working on a city's heat network,
- And many activities on the platform: students have to train, they receive automatically feedback trough the platform, they interact on the forum, they self-evaluate their skills and have to give feed-back on their way of learning [1], on their progresses and also on the remaining difficulties.

The new version of this course was proposed first to the apprentices and is now applied to the students' training. The impacts of all the changes are still being analyzed, but some results can be mentioned yet:

- The teacher "feels no longer alone" to re-build the course,
- And she feels, she has a different teaching position: no longer expert above students, rather trainer alongside (passing from one side to the other of "Jean Houssaye's pedagogic triangle [2], the pedagogical engineer said)
- Last but not least, she noticed both apprentices and students are more involved in their learning process: "they even stay after the end of the lesson!".

Those qualitative results must be soon completed by some quantitative data. And the evolution is going on, collaborative and peer-review process will be soon introduced in the course.

References

[1] Benjamin Bloom, *Taxonomy of Educational Objectives*, 1954, New York, Longmans, Green and Co.

[2] Jean Houssaye, *Le triangle pédagogique : les différentes facettes de la pédagogie*, Issy les Moulineaux, ESF éditeur, 2014

Chemical Engineering Technology Program Conversion to Asynchronous, Online Delivery

Chester Little, Ph.D., P.E.
Dept. of Physics and Astronomy, Austin Peay State University

Abstract ID# 467364

Abstract

The abstract has been removed to avoid duplication.

Motivations for Online Conversion at APSU

The information and explanations contained in this paper were originally developed for several presentations delivered over a 2-year period by the ChET faculty members at Austin Peay State University. References for those presentations are included.

The Chemical Engineering Program (ChET) at Austin Peay State University was started in the Fall semester of 2009 to meet the needs of the growing polysilicon processing industry in Tennessee. The major employer of the region requested an accelerated delivery of graduates to meet its expedited operations schedule. In fact, it was necessary for some of the ChET students to leave campus before graduation in order to begin an intensive operator training program. Therefore the ChET faculty agreed to record the lectures of the last semester classes so that these students could finish the coursework away from campus. This first temporary attempt at online, distant course delivery was very problematic and was not viewed as a success. Nevertheless, the ChET faculty did not want to completely abandon the idea of online course delivery but recognized that success would require much more effort. An "after-action" review was conducted by the faculty members and yielded the following key learnings.

- the required preparation time for online course delivery is easily double the preparation time for normal, live delivery
- students need a consistent delivery format for lectures, homework, quizzes and exams as prescribed by an internet-based Course Management System
- a set of recorded lectures with slides was not adequate
- a substitute was desperately needed for the problem-solving sessions that normally occurred in the live course delivery

A second opportunity for consideration of online course delivery occurred a few years later when it was recognized that the potential student population for a Chemical Engineering Technology career did not fit the typical (traditional) student pattern. Furthermore the faculty acknowledged that the Chemical Process Industries often have unusual work schedules. Therefore the typical ChET student could be described as follows:

- older, usually had significant family obligations,
- typically worked 30 (or more) hours per week with episodes of extended overtime, and
- often worked rotational schedules.

In addition, the university hoped to attract students who were outside the university's geographic region or were mobile workers (rotating to isolated worksites). Therefore the ChET faculty at APSU decided to make a second attempt at online course delivery with a new methodology targeting the potential ChET student population and addressing the above key learnings.

Course Preparation Time

I n hindsight, it sounds quite simple but the following statement must be declared to all involved in the delivery of an online course. The starting point for online delivery is the endpoint of course preparation for normal, live delivery. Furthermore one should expect to expend the same amount of time and effort on the online conversion as was expended on the original course development. The "handover" of a live course for online conversion should include:

- a well-used book and course syllabus,
- a full suite of lecture slides in electronic format,
- a full complement of out-of-class exercise material (homework and/or project assignments), and
- a full suite of test material (quizzes and exams).

Development of a Protocol for Consistency

In order to assure that the format of the entire course material is on a consistent basis, it is very helpful to follow a consistent protocol for the conversion of the material. It is extremely beneficial if the online conversion effort for a certain course can actually start during (or before) the last live delivery of the course. If so, the following preliminary steps should be taken to expedite the eventual course conversion.

1. Record the "normal" lectures of the course during the last live delivery. The pauses and breaks for problem-solving discussions or questions become apparent. It is good practice to announce the advancement of slides by stating the slide number.
2. If possible, capture all white-board and document-camera material. Make note of the occasions when confused looks by the students cause the instructor to repeat, emphasize or better explain a topic or concept.
3. Recognize that an online lecture actually requires less runtime than live delivery of the same material. Furthermore online delivery is not constrained by the normal lecture time period (e.g. 55 minutes). Therefore there are opportunities to revisit the Course Breakdown Structure and Plan in order to rearrange the lectures into more logical divisions (according to subject matter rather than elapsed time segments).

An example of a modified Course Breakdown Structure can be seen as Figure 1.

In addition to those preliminary course conversion activities that should be done during the last live delivery, there are a few conversion standards that the faculty members should adopt to assist the consistency effort. Regarding media files (audio and video), the university's Course Management System and/or the Distance Education Department will likely have guidelines, requirements and suggestions. The faculty members should thoroughly review these guidelines and requirements. Although opinions and preferences are very common, it is recommended that faculty members agree on file formats for presentation slides (e.g. MS PowerPoint), media files (e.g. MPEG formats), audio and video resolutions and even the software for rendering videos (e.g. Camtasia Studio). Tutorials for the use of these agreed standards may be useful for some faculty members.

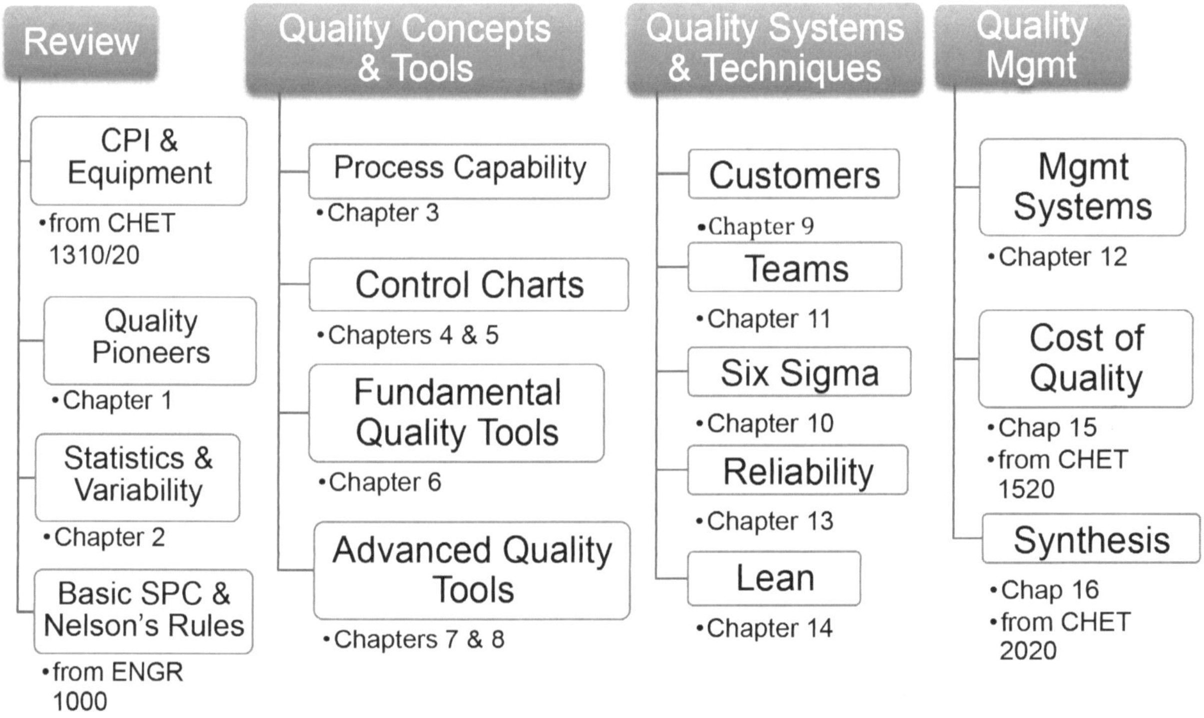

Figure 1 – Course Breakdown Structure for Quality Control Course

With the completion of these preliminary steps and preparations, a faculty member can perform the conversion of an existing live course for online delivery on a lecture-by-lecture, slide-by-slide basis. The following steps will be helpful to assure an efficient, consistent production of online course lectures.

1) If available, listen to/review the live recordings of the last lecture material recognizing that lecture slide order could have been re-arranged. Note: Changes to an individual slide are often very few.
2) If the instructor is unfamiliar with the lecture material and/or live recordings are not available, it is helpful to prepare a script for audio recordings. Presenter's notes are very valuable as guides for an audio script.
3) Record "studio" versions of new lectures as individual slide audio clips. Listen to your recordings to catch/edit mistakes. Note: It is not necessary to add pauses for the students because they will have the ability to pause/rewind/replay lectures as needed.
4) Decide whether slide animations should accompany the audio delivery. Some animations are actually distracting to the students who prefer to view the entire

slide while listening to the audio. If slide animations are warranted, use the following steps:

a) Listen carefully to the slide audio and note elapsed times for key messages (natural groupings of information).
b) Validate the order of the above key messages on the lecture slide and rearrange, if necessary.
c) Parse the slide elements (bullets, graphs, tables, calculations etc.) to honor the above key messages.
d) Develop a storyboard for the parsed slide elements with the elapsed times from the audio file. A storyboard example can be seen in Figure 2.
e) Set animation times (entry) for the slide elements according to the storyboard.

5) Insert audio file and set to start automatically. Verify animation synchronization, if applicable.
6) Set slide transition timing as desired.
7) Set slideshow to play narrations and use timings.
8) Repeat above for subsequent lecture slides.
9) Convert and render video according to the chosen software package and agreed formats and resolutions.

Slide Element	Slide Element Description	elapsed time
0	Title, Problem Description & Audio File	time zero
1	ASME B31.8, t_c equation	2:42 (162 sec)
2	S - SMYS	3:28 (208 sec)
3	X52 etc.	3:40 (220 sec)
4	E – longitudinal factor	3:46 (226 sec)
5	F – construction factor	3:57 (237 sec)
6	T – temperature factor	4:22 (262 sec)
7	t_c calculation	4:42 (282 sec)
8	t_r calculation	4:58 (298 sec)
9	end of audio file	6:17 (317 sec)

Figure 2 – Slide Animation Storyboard Example

In addition to the lectures, the other course material must be converted to a consistent basis. Quiz questions should be formatted consistently as required by the automated quizzing feature of the Course Management System (CMS) so that the students can get timely responses to their quiz answers. Homework and project assignments should be prepared using the same file format for uploading into the CMS. For example, it is convenient to use MS

Excel for homework assignments with the stipulation that students prepare their answers on the same sheet as the problem statement. If possible, exams should be developed using a similar format. Dropboxes should be enabled for homework, project and exam submissions.

Media Enhancements and Problem-Solving Sessions

Science, Engineering and engineering technology lectures are seldom complete without one or more problem-solving exercises to consolidate the learning process. In addition, there are other breaks during any technical lecture when a graphical sketch or picture, a plot of paired variables and/or a mathematical representation (i.e. symbolic equation) are needed to properly convey a concept or relationship. The appropriate opportunities to explain the material in another manner are often naturally determined in a live lecture. Unfortunately, these opportunities are not as apparent during the discontinuous slide-by-slide (lecture bullets and audio file) work progress for online delivery. It should be noted that the student's viewing/listening of a lecture can also be discontinuous in time. Therefore extra effort should be taken to determine when natural pauses in the informational flow are needed during a lecture to supplement with graphical elements or problem solutions. One must recognize, however, that too many lecture breaks cause confusion for the student. It is recommended that figures, tables, sidebars or examples from the textbook should not be pasted and repeated in the lecture material; a reference to or further explanation of the textbook is helpful and sufficient. Honestly the best method for capturing these extra material situations is a review of the last live delivery.

After identifying the opportunities for providing additional explanation, the method of producing the extra material and the corresponding method of delivery need to be determined. The animated lecture slide, as described in the previous section, is the simplest method to add emphasis beyond the static lecture slide with accompanying audio. The animated lecture slide works quite well for a straightforward example. The pace of the information delivery is set by the audio file not the animation.

Another common delivery method involves the preparation of a separate movie file (video & audio) that is then attached in an extra presentation slide and plays inside the lecture presentation. A good rule-of-thumb for choosing the attached media file method is the determination that multiple screens or slides are necessary to complete the exercise. The multiple slide condition suggests that the exercise is more complicated and cannot be adequately delivered in a single animated slide. Then the remaining consideration would be whether the extra material is so complicated (or requires an extended amount of time) such that the original lecture is disrupted. If so, one should consider delivering the extra material as a stand-alone movie and placing a link or reference in the original lecture.
The stand-alone movie should be reserved for the more complicated problem solving or graphical explanation scenarios. For sure, any explanation requiring a combination of many mathematical equations, nomographs, intermediate results, plotting, unit conversions or the use of reference material would be a good candidate for a stand-alone movie. To better describe these more complicated scenarios, the term "recitation" is often used to designate these longer instructional episodes. Other useful determinations for choosing to produce and deliver a separate recitation are:

- the amount of time is beyond a recognized maximum (e.g. 10 minutes),
- problem solution involves the use of multiple concepts,

- problem scope is across multiple lectures, and
- media file could be used for multiple courses (or labs).

Examples for these delivery methods and more-detailed explanations are available in the presentations that are listed as referenced material.

Reconnection of Lecture & Lab

During the 2+ year ChET curriculum conversion effort, it was discovered that the online students became disconnected from the lecture material when undertaking certain lab activities. During the normal live delivery of a course, the instructor would naturally emphasize some pertinent material during the week of a corresponding lab exercise. This natural lab/lecture connection and its associated learning synergy are lost in the online delivery mode of lectures even if the lab courses remain as scheduled, live events. It should be noted that the ChET faculty members intended to combine the existing 3 lab courses with an additional lab course and then deliver all of these lab exercises across two long-weekend campaigns. It became obvious that this observed loss of lab/lecture connection would become more pronounced when a lecture course was separated from the lab exercises by many months. Therefore it was necessary to develop a fit-for-purpose methodology to reconnect the lab and lecture material.

The first task in this effort was a mapping exercise to connect lecture material with the individual lab assignments. Although many of these connections are relatively obvious (control valve operation with lectures about control valves or lockout/tagout assignments with hazardous energy isolation lectures), other important synergistic lab/lecture connections were discovered. Ironically it was also discovered that the synergy works both ways; the lecture material is better understood with reminders of lab exercises.

The accepted solution to reconnect the lab and lectures was the development and use of specialized recitations. These short recitations were developed using the appropriate lecture slides, normal recitation videos, lab protocols, pilot plant process drawings and pre-job safety analyses. The following steps can be used as guidelines to develop these short lab/lecture reconnection recitations:

1) Review the lab protocols and pre-job safety analysis forms for locations to interject technical material from lecture slides and/or pilot plant documentation.
2) Find lecture sequences that coincide with the identified lab protocol locations.
3) Record “sloppy” audio versions of the lab activities that will be described or shown in the reconnection recitation. Why sloppy? Only the natural, comfortable time durations of the explanations are needed at this point
4) Prepare, rehearse and record video segments of the lab activities according to the audio durations from above. It is often helpful to speak during the video capture to assist the dubbing effort of more polished audio clips.
5) Watch/edit the video segments to determine whether frames need to be added or deleted to better match the audio timing. Still shots (especially lecture slides, freeze frames or process drawings) should be added to allow adequate time for the explanations of the lab activities.

6) Prepare a storyboard for combining still shots and video clips honoring the timings of the “sloppy” audio files.
7) Record polished audio clips for the video segments and still shots.
8) Combine all media elements and fine-tune the video clips with the addition or removal of the frames to reasonably match the polished audio clips.
9) Render the above into a movie file according to the prescribed resolutions of the Course Management System.

Examples of lab/lecture reconnection recitations can be found via the referenced material.

Note: It is recommended to use students in the video (and even audio) recordings to give the exercise some credibility with future students. In addition to a small amount of pay, the student-actor can add a nice bullet point to his/her résumé as a possible discussion highlight during a job interview.

Use of Imbedded Quiz Questions

Some organizations have expressed the desire to use the online ChET program work product of lectures, recitations, quizzes, homework assignments and exams without a complicated Course Management System. The lectures and recitations exist as separate movie files while the homework assignments and exams usually exist as MS Excel files; these can be easily handled through a dropbox arrangement. However the quizzes present a more serious challenge in that quizzes are more beneficial with immediate grading and correct answer response. During the 2+ year period of course conversion, a significant upgrade was developed by one of the common software packages for this type of work. Quiz questions can now be imbedded into the lecture movies for the student to answer as he/she progresses through the lecture material. The student answers and quiz score are either uploaded to a web-based server or emailed to the course instructor. Of course, the student receives immediate feedback regarding incorrect and correct answers. The use of “as-you-go” quiz questions requires the faculty member to revisit the course and lecture development to determine the appropriate location for the quiz questions. The techniques for imbedding quiz questions have been demonstrated but the imbedding effort was not undertaken for the entire curriculum and has not been tested with students.

Examples of imbedded quiz questions can be found via the referenced material.

Future Work

Unfortunately the Chemical Engineering Technology program at Austin Peay State University was suspended in May 2016 before the entire ChET curriculum could be converted to online, asynchronous delivery. The conversion methodology and techniques have been developed, tested, modified and accepted but one course (Operations Management) has not been converted. The lab/lecture reconnection recitations have not been continued and the imbedded quizzing was not implemented. The program pilot plant equipment has been removed from campus and placed for bid on an open auction. Therefore it is virtually impossible to recover the ChET program at APSU. Other opportunities have been identified for the capture, re-use and transfer of this information and knowledge.

Complete Transfer of ChET Program - The full body of the work is available for any institution that would like to implement a two-year Chemical Engineering Technology program based upon the North American Process Technology Alliance (NAPTA) model and books. The remaining course conversion and transition of the converted course material could be accomplished in approximately 4 man-months of work; this time estimate includes uploading the existing quizzes to a different Course Management System. The imbedding of the quiz questions into the lectures could be delayed indefinitely. The lab/lecture connection effort would need to be repeated for the new lab/pilot plant arrangement.

Compression of ChET Curriculum into a 1-year Certificate – Using the converted ChET course material with the addition of 4 fit-for-purpose courses, a 30 credit hour certificate has been proposed.

Another possibility for the use of the Chemical Engineering Technology body of knowledge involves expanding the breadth and depth of the subject matter with additional technical and mathematical rigor. Austin Peay State University has several concentrations within its existing Engineering Technology program and there are separate paths within these concentrations. The Manufacturing Engineering Technology concentration (B.S in ENGT) is an appropriate location for much of the ChET material. Two proposals are currently being prepared to accomplish this. The first proposal will directly expand the ChET material as a Process Technology path. A second proposal is being considered in which the ChET material is supplemented with additional Health, Safety & Environmental coursework to provide a Sustainability path. Both of these paths will reside within the Manufacturing Engineering Technology concentration.

Abstracts have been submitted for all of the future work described above. Papers and presentations are currently in preparation to provide greater detail about these possible avenues to capture the ChET program institutional knowledge (and importantly the online delivery work products) for use in other programs and situations.

References will be added to the final version.

458575 An Interactive Textbook for Material and Energy Balances: Student Feedback and Performance

Abstract:

Textbooks are experiencing a 21st century makeover. The utility of the textbook has diminished for the digital natives populating higher education today, including the availability of solutions manuals. Therefore, my goal was to create an interactive, web native textbook for engineering students, specifically for a material and energy balances course that introduces students to chemical engineering and its fundamental concepts. A first version included many interactive features, but self-publishing, which keeps costs to the students lower, was not sustainable. A second iteration is now available with a fully interactive experience.

Several unique features distinguish the new book - available from zybooks.com - from traditional textbooks including question sets with scaffolded learning questions, animations, and real time feedback for the student and instructor. The features of the book will be summarized in the talk, but discussed in detail during the Education Division poster session. This talk will focus on student feedback and student achievement during the Spring 2016 semester. With over 100 students using the zyBook for a material and energy balances course, student surveys about the bookâ€™s function and features will be explored as well as quantitative participation data and correlation with student performance.

472508 A Required, Non-CreditÂ Professional DevelopmentÂ Seminar Series for Chemical Engineering

Abstract:

A professional development seminar series at Bucknell University has been implemented for more than 25 years in the Spring semester to all four years of Chemical Engineering majors. This course is required but also listed as zero credit. It gets an attendance rate of nearly 90% every week as motivated by both the opportunity for "bonus points" in other core classes and due to the value of the course to the students. Faculty organizing the seminar series receive teaching credit. Most speakers are alumni of the department and receive reimbursement only for mileage and occasionally for overnight accommodations. Despite the minimal amount of incentive to obtain new speakers and the lack of credit students received, the series has been well-attended and well-received over the years. Its success lies in the value students receive and the support and importance placed on the seminar series by the faculty of the department. We use the seminar, in part, to address ABET outcomes on lifelong learning and to reinforce this theme to the students.

The seminar series also serves to show students the types of careers and opportunities that will be available to them as practicing chemical engineers. In addition to the external speakers from industry, government, or academia, we also have special seminars each semester that feature such topics as undergraduate research, career development, engineering ethics, AIChE and related activities, and National Engineering Week. The seminar also serves as a common time to get all class years and faculty together in one place which helps to build a strong sense of belonging and camaraderie in our department, even among the first-year students. For the faculty, this series provides a continuous source of information about current industrial practice and directions. Finally, the series helps to build and maintain relationships between alumni and the department.

This talk will provide a larger overview of this seminar series and document its assessment and outcomes. We will present data including historical attendance records, course evaluations and survey results. Discussion will include how we evaluate the success of professional development opportunities for our students as well as what would be involved in setting up a similar program at other institutions.

470409 Successes and Challenges in Teaching Professional Skills to Undergraduates in the 21st Century

Abstract:

As society evolves, so do professional norms. The modern workplace spans generations with vastly different experiences, especially with respect to communication and technology norms. While some may not have experienced computers until well into their careers, many current U.S. undergraduates are unfamiliar with a time before smartphones. With this dichotomy comes a new set of challenges for both the teacher and student as we prepare individuals in the classroom for life after graduation.

Our immense technological capabilities in the 21st century allow us to tackle more problems in rapid succession compared to what was possible 100 years ago. They also allow us to remain in constant contact, often blurring the lines between professional and personal time or space. Thus, teaching professional skills to our undergraduate students not only continues to be an important task, but also an evolving one.

In the third year of teaching Technical and Professional Communication for Engineers (TPC) at the undergraduate level, new modules and activities have been added to better reflect the professional environment of the 21st century, as well as student interests. Electronic communication etiquette, body language/nonverbal communication basics, and developing elevator pitches are among the modules that have been expanded. Time management and prioritization is an example of a module added on account of unanimous student interest.

Reactions to the course have always been largely positive, and the extra attention given to professional skills has proved valuable to the students. Mid-term and end-of-course evaluations highlight the immediate practicality of course material, and students report greater awareness of the communication that surrounds them in their professional lives.

469562 Self-Reflection Assignments for Evaluating Non-Technical Skills and Setting Goals for Professional Development

Abstract:

The presentation will focus on a series of class exercises called "self-reflection assignments" that have been used in two offerings of the undergraduate chemical kinetics course at Oklahoma State University (OSU) and will provide evidence to support the impact of the activities. The self-reflection assignments involve brief essays and goal setting related to professional, non-technical or "soft" skills. The assignments were adapted from a document called the Self-Evaluation Rubric originally developed for a physics laboratory course at Berkeley used for guided reflection assignments. The outcomes of the assignments have not been previously measured and reported in an engineering course. The Self-Evaluation Rubric contains a list of primary and advanced non-technical skills important for scientists and engineers. These skills include persistence, organization, connections, self-compassion, courage, mental resourcefulness, communication, diligent skepticism, collaboration, and reflection. In Spring 2015 & 2016, written assignments were used in the OSU CHE 3123 course to foster development of non-technical, professional skills. Five times during the semester, junior level undergraduate students (63 in the first offering and 81 in the second) were given an essay prompt to use the Self-Evaluation Rubric to identify one or more skills to work on in the following two-week period (first essay) or one-month period (subsequent essays except the last). In their essays, students were asked to assess their current proficiency level (beginning, developing, or succeeding) in the skill according to the detailed description in the Self-Evaluation Rubric, describe their goals related to the skill and their plan for improvement, and share progress in the skill(s) if any had been made since the previous essay submissions. This process required the students to identify the change they want to make, come up with a plan to implement that change, be consistent in their implementation, and frequently assess how their changes worked. The essays were submitted electronically as homework assignments (worth 3 points out of 1000 each for a total of 1.5% of the total course grade) and were graded for completion. The instructor provided electronic feedback through the classroom management system ranging from one sentence to a few paragraphs praising students' efforts, making suggestions, or answering questions that arose in the essays. Besides the intended goal of professional skill development, the assignments also gave the students practice with written communication and forged connections between the students and the faculty instructor for the course.

A classroom research study was conducted in Spring 2016 to assess the impact of the assignments using pre- and post-assignment surveys of student perceptions of their non-technical skill levels (beginning, developing, or succeeding) and their opinions on the impact of the assignments on any changes in their skill levels. The hypothesis is that the self-reflection assignments positively impact student progression toward higher levels of non-technical skills. Further, we hypothesize that skills that students focus on actively and skills that students are simply made aware of through the exercises and do not actively reflect on in course assignments improve during the semester with some level of attribution to the self-reflection assignments. The research study involves surveying students before and after the series of self-reflection assignments to quantitatively assess changes in aggregate skill levels using the Self-Evaluation Rubric and student perceptions of their own proficiencies in the skills. The Self-Evaluation Rubric has been published under a creative commons license, and educational research on the content of student reflections using the rubric in a physics class has been published in the physics education literature. In this study, we quantified the number of students that rank themselves as beginning, developing, or succeeding for each of the 10 skills on the rubric before and after the use of self-reflection assignments in the course. Our surveys also included open-ended questions about potential causes for skill level changes, if any, about skills they focused on throughout the term for the self-reflection assignments, and about their opinions of the self-reflection assignments in general. These surveys were administered anonymously and electronically without inclusion of any identifying information. Student participation in the surveys for the research study is optional. Student self-reflection assignment submissions were not utilized in the study. The data will be presented in aggregate form. This research methodology was approved by the OSU Institutional Research Board.

The research may produce new knowledge on the efficacy of the self-reflection assignments in improving non-technical skills in the context of an engineering course. Testing the hypothesis through the survey assessments will provide quantitative feedback on whether or not this instructional technique should be recommended for use in future course offerings and in other engineering courses at OSU and at other academic institutions. Non-technical skill development is important for engineering undergraduate programs as these skills comprise a significant portion of the required student outcomes for engineering baccalaureate degree program accreditation. The research study data collection is ongoing. Data will be analyzed in Summer/Fall 2016 for inclusion in the presentation. The limited sample of the results available at the time of abstract submission positively support the hypothesis.

456348 *Integration of Academic Coaching Fundamentals into Engineering Classroom Activities*

Abstract:

Chemical engineering education at all accredited institutions meets high technical rigor for student development and competency. Other skills such as interpersonal communication and intrinsic motivation, however, are more difficult to teach and assess. One means of imparting these skills into students is by drawing upon the fundamentals of academic coaching and integrating these ideas into classroom assignments. Academic coaching has been used as an alternative to advising and changes the faculty-student dynamics from an advice giver/receiver pair to two equally invested partners. By changing these dynamics, the student is empowered to develop their own solutions to his/her problem and commit to follow through on the planned solution. To achieve this, academic coaches uses a number of approaches that aid the student in developing his/her own solution including: encourage students to follow curiosity, view the problem differently by changing the perspective, asking powerful questions that provoke self-reflection, and utilizing approaches that require creative response.

While academic coaching has traditionally been applied for student interactions outside the classroom, we have drawn inspiration from the coaching ideals and implemented elements of coaching into classroom assignments. Academic coaching approaches were applied into an interdisciplinary sophomore/junior thermodynamics class, Engineering Thermodynamics, and also two chemical engineering classes Advanced Math for Chemical Engineers and Chemical Engineering Thermodynamics. These approaches included connectivity questions and connectivity projects â€“ questions designed to develop links between the course material and real-world applications. The courses also integrated greater use of software to aid in solving homework problems and develop student confidence and increased computational abilities. Our efforts also incorporated additional collaborative learning techniques to develop inter-personal skills and promote functioning within a team environment.

We hypothesized that students exposed to coaching both as an advising alternative as well as in the classroom would benefit and outperform fellow classmates exposed to just coaching as advising or to no coaching at all. To assess this, a short questionnaire was administered to the students at the beginning of the Fall 2015 semester and a follow-up survey at the end of the Spring 2016 semester. These questions were designed to determine if students had increased levels of personal inquisitiveness, commitment and follow-through, and intrinsic motivation. Results of the survey, academic coaching, and in-class academic coaching techniques will be discussed.

464295 Freshmen As Employees on Senior Capstone Laboratory Projects

Abstract:

Over the past four years we have established a mentor program between each student in our freshman design laboratory and our senior teams in their capstone laboratory. To aid our seniors, this project was meant to enhance their learning through teaching, which is a proven means to deepen student understanding. An additional aim was to give seniors experience hiring and managing "employees" to supplement their experience working on teams made up only of their cohorts, and give them insight into the perspectives of potential employers. For freshmen, the project was meant to give a clearer understanding of their academic trajectory though their interactions with their seniors, to build their engineering identity, and to give them experience in applying for an engineering position.

In our introductory chemical engineering course (fall semester of the freshman year) we instruct our freshmen on resume preparation and iteratively improve their resumes over the semester. In the spring semester of the freshmen year, they enter our design laboratory where they execute several design projects. Through the same period, seniors enter the last semester of our Senior Projects Laboratory series, in which they propose and execute novel projects. About a third of these projects are selected to progress and the list of green-lit senior projects, along with single-page descriptions, are given to the freshman class. The freshmen then select the projects they wish to work on and tailor their resumes for their most desired projects. Seniors then receive all the freshmen resumes and select their top candidates. Freshmen are placed onto teams in rounds, based on senior choices.

For three week, freshmen work with their seniors for a minimum of 8 hours. At the end of this period, freshmen author a report and evaluate the management skills of their senior team mentors. The seniors also evaluate the team working and productivity of their freshmen "employees". Surveys are also used to gauge freshmen perception of the experience in general.

Over 300 freshmen and 200 seniors have participated in this program and the survey results have been greatly positive, but dependent on several important factors. Student experiences in this program depend greatly on freshmen's ability to manage problems of scheduling, logistics, and conflict resolution, which many of them have not yet encountered in a professional environment. Furthermore, we found that seniors must be incentivized to create clear and meaningful expectations for their freshmen teams from the early proposal phases of their projects to ensure the partnership is fruitful for both groups. We describe several tools that have been introduced to address these problems and improve the freshmen-senior collaborative process over the years.

472075 Creating Socially Minded Engineers through Student-Led Air Pollution Assessments of Local Towns

Abstract:

This presentation will introduce a novel methodology to introduce chemical engineering students to service-learning through a directed student project. The overarching objective of this project is to create engineers with higher levels of community and policy-awareness while training them on many of the technical skills required in engineering practice. The specific course is an elective titled *Transport and Transformation of Air Pollutants* with an approximate class size of 20-25 predominantly senior-level chemical engineering students. The course is intended as either an overview of air pollution and atmospheric chemistry for non-specialist graduate and undergraduate students or an introduction to these topics for graduate students specializing in air quality. I cover the source and impacts of air pollutants, physical and chemical processes governing air pollutant concentrations, and prediction of concentrations using computational models. The major topics include: kinetics of chemistry in the atmosphere, pollutant transport, pollutant removal, atmospheric science, aerosol microphysics, and estimation methods for exposure.

For this specific project, a team of undergraduate and graduate students will carry out an air pollution assessment in a local town. This assessment includes (1) estimating pollutant emission amounts and locations, (2) selected monitoring at anticipated hotspots, (3) surveying the impacted community, and (4) compiling existing air pollution data in the area. Students are asked to identify hotspots, assess the current state of air pollution, and provide recommendations for the community moving forward. They share their findings with the partner town via a final report, a brief policy memo, and a presentation to the town (including community members).

This course aims to (1) create more community-minded professionals, (2) provide experience with open-ended problems using incomplete/flawed data, (3) teach communication skills, and (4) train students in handling uncertainty.

456481 Web-Based Simulation Games for the Integration of Engineering and Business Best Practices

Abstract:

Web-Based Simulation Games for the Integration of Engineering and Business Best Practices

Bruno A. Calfa[1] and William F. Banholzer[1]

Universities do an outstanding job providing a technical education. From undergraduates to postdocs, students learn science and engineering fundamentals, how to read the literature, analyze data, and test hypothesis. However, we believe there is a serious need to improve training on industrial/business fundamentals. Therefore, there is a pressing need to introduce business fundamentals into the engineering curriculum. Producing students who have a stronger appreciation of industrial realities and technical evaluation skills will provide a more competent workforce, as well as more knowledgeable government employees.

Case studies and business simulations are proven teaching tools (Sin and Center (2002), Bequette et al. (1998), Seay and Eden (2008), Corsi et al. (2006)). They immerse students in realistic actual or hypothetical situations developing their problem-solving skills. In addition, cases and simulations are a natural way of using and applying open-ended and practical problems, team projects, and written reports and oral presentations.

In this talk, we introduce a new web-based tool, *Chemical Business Simulator* (http://uwchembussim.che.wisc.edu/) that consists of a framework for business simulation games. The games enhance the chemical engineering curriculum by requiring application of both engineering and finance principles to define successful commercial chemical processes. The framework is extendable, i.e., additional cases can be developed and integrated without completely redesigning the website. We will briefly discuss currently implemented cases and share some feedback from students.

References

A. Sin and A. M. Center. Gas Station Pricing Game: A Lesson in Engineering Economics and Business Strategies. Chemical Engineering Education., 36(4):278-280, 2002.

B. W. Bequette, K. D. Schott, V. Prasad, V. Natarajan, and R. R. Rao. Case Study Projects in an Undergraduate Process Control Course. Chemical Engineering Education., 32(3):214-219, 1998.

J. R. Seay and M. R. Eden. Incorporating Risk Assessment and Inherently Safer Design Practices into Chemical Engineering Education. Chemical Engineering Education., 42 (3):141-146, 2008.

T. M. Corsi, S. Boyson, A. Verbraeck, S.-P. van Houten, C. Han, and J. R. Macdonald. The Real-Time Global Supply Chain Game: New Educational Tool for Developing Supply Chain Management Professionals. Transportation Journal., 45(3):61-73, 2006.

---

[1] Department of Chemical and Biological Engineering. University of Wisconsin-Madison. Madison, WI 53706. USA.

459162 Statistical Analysis of Undergraduate ChE Curricula of U.S. Universities

Abstract:

Chemical Engineering (ChE) is a historically relevant degree that is constantly evolving in order to accommodate the emerging research and industrial trends (e.g., biology, nanotechnology, energy). However, adding new courses to the curriculum without removing older ones presents a challenge because it is already overloaded. The problem is further exacerbated by the need to justify any curricular changes to ABET reviewers and various university committees. It is believed that a thorough statistical analysis of the current state of U.S. curricula could eventually be used to guide or justify curricular changes. Therefore, this talk will present results of our recent comprehensive study that is intended to become such a tool. Publically-available undergraduate degree sheets were aggregated from every accredited ChE department in the US. A comprehensive analysis was performed, and subject-specific descriptive statistics were reported for core ChE courses. In addition to the core courses, biology-related data were also accumulated into the analysis, due to the growing popularity of the subject. Besides exposing various trends and commonalities in ChE curricula, this presentation will focus on two different approaches to teaching Transport Phenomena, the significant presence of Bio-related courses, and the apparent scarcity of traditional Statics/Dynamics and Process Safety courses. Ultimately, the identification of commonalities between what departments are teaching is intended to be used as a first step toward making any changes; while a host of other factors, such as regional considerations and student feedback, should be included in the final decision. To close, an example of the use of this tool is provided in the form of a case-study of refining the NJIT curriculum.

459838 Engaging Students through Active Learning

Abstract:

As class sizes continue to grow, many instructors struggle with engaging and assessing their students. Active learning techniques may be used to simultaneously engage students and serve as an assessment tool. In a traditional lecture environment, students can begin to lose their focus after as little as ten minutes. Without intervention, this loss of focus can lead to complete disengagement. Numerous studies have shown that active learning techniques not only engage students, but may lead to deeper understanding, higher grades, and a reduction in the learning gap between high and low performers. In a large classroom, timely and effective assessment of students may be challenging. Without feedback, studentâ€™s perception of their learning may not be reflective of their actual progress. Active learning techniques may be utilized to assist students in formative assessment (including self-assessment), as well as summative assessment. In this seminar, we will review some of the research findings, explore active learning techniques that can be utilized in STEM classrooms, and sample some of the tools/technology that may aid in implementation in large classrooms. Bring your smart phone, tablet, or laptop and learn how you can use these simple techniques in your classroom.

466975 Meaningful, Consequential Learning Using Studios in Large Enrollment Classes

Abstract:

Recently chemical engineering programs in many universities have shown significant growth related to student interest in energy production, sustainability, and innovation chemical processing. With the increasing student populations, many programs have struggled with serving larger number of students while maintaining an effective learning environment. One approach has been to limit enrollment in the program and another has been to deliver the same class multiple times. This presentation reports an alternative approach in which larger lecture classes are punctuated by smaller studios. In the studios, students are afforded the opportunity to actively engage the content presented in lecture. We have incorporated studios in ten core classes beginning in the 2011-2012 academic year.

Studios are activity based (e.g., students solving problems, explaining phenomena from in class demonstrations, answering conceptual questions, working on virtual laboratories, etc.). The enrollment in a single studio is limited typically to about 25-30 students to achieve an instructional model of facilitating learning and rapid feedback. In its original formulation, student activity was driven by concept-based worksheets where students would be directed through a series of qualitative and numerical questions to make meaning of the material presented in the previous lecture or for the coming lecture the following day. However, using this approach, we observed some students in a "worksheeting" mode where they sought to provide answers to "get the points" rather than deeply engage in the content. As a consequence, we are in the process of shifting to a Studio 2.0 design that is based on intentional integration of *realistic, consequential work*. In this approach, students are placed in the role of engineers and provided tasks that are framed to require them to activate disciplinary knowledge and practices to solve real-world problems. The fundamental principle is to *situate learning* in contexts similar to ways that the content is used in practice. In this presentation, we present initial redesign approaches and results.

465190 Educate Matlab in Chemical Process Control Using Flipped-Classroom Modules

Abstract:

Process modeling and control techniques play an essential role in improving process efficiency, product quality and process safety in chemical industry. The education of process modeling and control, on the other hand, is challenging as the related courses like Chemical Process Control are math intensive. MATLAB Simulink has been proved as an effective tool to educate college students with modeling techniques, as it provides a friendly user-interface in which students can use the block modules to build mathematical models and perform simulation. However, no flipped-classroom approach has been developed to educate students MATLAB Simulink in solving process modeling and control problems in chemical engineering. Since flipped-classroom approaches allow students various chances to access the lecture videos, it may facilitate students' learning in both programming skills in MATLAB Simulink and the process control techniques. In this work, we developed three flipped-classroom modules that covered the following skills in process modeling and control: 1) simulation of ODE model in chemical engineering; 2) operation of Laplace transform and transfer functions; and 3) design and analysis of closed-loop feedback systems and PID controllers. Accordingly, eleven examples and thirteen videos were developed to cover the aforementioned skills that are essential to process modeling and control in chemical engineering.

The inverted-classroom approach was implementation in the following steps for each of the three teaching modules: 1) the videos were sent to students to watch one week before the lecture; 2) a quiz was given at the beginning of the lecture to make sure the students did watch the teaching videos; 3) a question & answer section was given to clarify students' questions on the teaching videos; 4) the students were required to reproduce the MATLAB programs shown in the teaching videos in class and turn in their work by the end of the lecture; 5) students was given another lecture time to do homework and the instructor answered students' question in class; 6) the instructor graded students' homework and gave suggestions for students to improve their programs.

At the end of each teaching module, especially after the graded homework was given back to students, one anonymous survey was given to students. Fifty senior students taking the course Chemical Process Control in Spring 2015 at Villanova University participated in the surveys. Each survey included three sections: Section 1 was used to evaluate the improvement of students' knowledge in process control and modeling and their skills in MATLAB Simulink; Section 2 was designed to evaluate the effectiveness of the flipped-classroom approach in facilitating students' learning; and Section 3 included two open-ending questions for students to list the aspects they like the inverted-classroom teaching approach and prove suggestions for further improving the teaching modules. All these surveys were in the form of scale survey questions.

On the basis of students' answers in surveys, the two-sample Kolmogorovv-Smirnov statistic method was used to analyze the data. The results indicated that students' knowledge in process modeling and control taught in this course was improved after they took the training. In particular, the average for student's knowledge levels before taking the teaching modules was around or below 3 out of 5, while the average rose above 4 out of 5 after they finished the teaching modules and homework. This means that the implementation of the process modeling and control knowledge in MATLAB enhanced students' understanding the knowledge taught in the classroom. The results also showed students' ability in using MATLAB to simulate ODE models, perform Laplace transform, and design PID controllers was significantly improved. Specifically, students evaluated their MATLAB modeling skills rose from below 2 to above 4 out of 5 after taking the teaching modules. In Section 2 of the surveys that include six questions on the effectiveness of the inverted-classroom approach to improve students' overall learning and ability to solve process modeling and simulation problems, the degree for students' agreement with the effectiveness of the inverted-classroom approach generally rose from around 3 to 4 after they took the designed teaching modules. As for opening questions in Section 3 of surveys, the aspects students most liked the inverted-classroom approach include: 1) they were able to watch the videos repeatedly at the time and pace convenient to them; 2) the instructor was able to answer their questions in homework immediately in class; 3) the videos offed another chance for them to understand the concepts introduced in the in-class format from the perspective of MATLAB simulation. The aspect students suggested for further improvement was that some of them mentioned they spent more time than the one for the in-class format. These survey results generally indicated the designed flipped-classroom approach did facilitate students' knowledge levels in process modeling and control techniques and improve students' skills of MATLAB Simulink simulation for chemical processes.

466900 Student-Created Youtube Videos, a Multimedia Assignment, to Foster Active Learning in a Large Engineering Classroom

Abstract:

36 groups of 4-5 junior Chemical Engineering students were challenged to create informative and captivating YouTube videos of a transport phenomenon highlighting a mass or heat transfer principle that they learned in the ChE342 classroom or beyond. This innovative ChE342 course projectÂ has not only successfully transformed boring deductive classroom teaching into social sharing of engineering principles, but also led to the creation of a YouTube channel (https://www.youtube.com/channel/UCBpw-AyX_Sw8EVsKObSiUVw) titledÂ â€œ*The Fun of Mass and Heat Transferâ€* as a global outreach platform. Our qualitative and quantitative data have clearly demonstrated the effectiveness and statistically significant impact of this teaching innovation on student learning (Fei Wen and Eshita Khera, *Chemical Engineering Education*, in press). While still under construction, the YouTube channel already has viewers from 68 countries around the world including US, UK, Mexico, India, Israel, Brazil and South Africa. The YouTube channel also has subscribers who are STEM educators interested in replicating the video contents in their own high school classrooms. This innovation has recently been awarded the Provost's Teaching Innovation Prize at the University of Michigan and spurred great interest among faculty to replicate this project in other engineering courses. Taken together, the Student-created YouTube Video represents an effective, scalable and sustainable teaching innovation, which is making a broad impact on engaging students in the learning process, enhancing their ability to communicate, collaborate, innovate and take risks, as well as worldwide outreach.

469557 Chemical Engineering Undergraduate Research at the University of Kansas and the University of Delaware

Abstract:

There are lots of great opportunities for undergraduate students to participate in leading edge research at the University of Kansas and the University of Delaware in Chemical and Biomolecular Engineering. Faculty research projects provide a unique opportunity for undergrduates to become involved as early as their freshman year in serious scientific work with a leading expert in the field. Each year, the Departments circulate a description of faculty projects and the roles that undergraduates would provide in the research. Some projects have funding to hire student assistants during the summer and the academic year. The departments also host frequent seminars and colloquia, open to undergraduates, where scholars from Universities, Industry and National Laboratories present their recent work. This presentation will highlight some of the recent undergraduate projects in chemical engineering at the University of Kansas and the University of Delaware.

471912 Information and Communication Technologies Use in Teaching Physical Chemistry Courses: Combining Virtual Traditional Teaching and Flipped Classroom Modalities

Abstract:

The implementation of Virtual Education Programs (VEP) has gained interest in most higher education institutions in recent years, mainly because of the rapid development of communication technologies and globalization and also because of their versatility, and collaboration opportunities that they generate.

Besides the advantages escribed, VEP are often criticized in terms of the dehumanization of the educational experience, students must also develop self-teaching abilities and in most of the cases, the fear caused by the lack of faculty accompaniment is present.

In Chemical Engineering major, this practice is often derived to administrative or computational based courses. In this work we report the experience of teaching courses of the Physical Chemistry curriculum (Surface Phenomena and Chemical Reaction Kinetics). The lectures were carried out in virtual traditional teaching modality, performing synchronic sessions with the use of a conference broadcasting software. Students were able to interact with the teacher in real time and to solve exercises and equation derivations with the use of an electronic board.

Lectures were uploaded to moodle, which was also the mean by which information was shared i.e. solved problems, homework, exams and grading. The versatility of the experience allowed the teacher to share didactic material like related web-pages, lectures and videos, and to organize activities as a seminar, where students performed oral presentations in non-native language (English) to experts in the field which gave them in return: scientific feedback, advice and encouragement.

Physical chemistry courses face the complexity of the abstract nature of the concepts involved, as well as the derivation of equations and implementation of mathematical solving procedures. Despite seemingly a disadvantage, the virtual environment allowed the teacher to use visual materials for better understanding of concepts. Students were also motivated to do web-surfing which enhanced the self-learning experience. Advisory sessions were carried also online; during the lecture sessions the help of another professor was required to act as a moderator.

The experience received a grade of 9.4/10 from the students and higher authorities by means of the â€œinstitutional teacher evaluationâ€. In terms of educational achievement, grades reflect a normal distribution with similar average grade and fail index in comparison to that obtained in the same courses in previous years where the courses were lectured physically. Teachers of the following courses (Chemical Reactors Design) report similar educational achievement and good understanding of the required concepts.

As far as we know, these were the first semester courses performed completely in a virtual traditional teaching mode in the Chemical Engineering major in Mexico, this experience gives us the opportunity of programming new courses linking higher education institutions and also the ability of promoting more versatile programs for the future Chemical Engineering students.

453150 Leveraging Simulation and Experiment in Process-Control Education and Training: Case Study of Development of Educational Software Suites for Either Stand-Alone Study or in Conjunction with Laboratory Measurements

Abstract:

Laboratory experiments to demonstrate the underlying principles of process control can sometimes challenge scalability for larger classes/sessions, through comparative lack of feasibility given time constraints. The primary aims of the laboratory were to allow students gain an understanding of feedback control, the typical characteristics associated with P, PI and PID controllers and introduce the concept of a transfer function for a system and how one might determine it. An important objective of this project was to rectify this by conducting a critical reappraisal of a control apparatus and experiment at UCD Chemical Engineering. In particular, changes introduced included: (a) reducing the number of set point changes in experiments, (b) increasing the set point change from 10% to 20% in the case of proportional controller experiments to emphasis the difference between different proportional band settings, using disturbances as opposed to set point changes when demonstrating derivative action, and (c) using 30% proportional band as opposed to 5% proportional band when using a PID controller to improve control loop stability. These changes have resulted in total experiment time being reduced from by a quarter. 201 minutes to 153 minutes.

A second major theme lies in the incorporation of process-control simulation into the physical laboratories, to develop a learning tool to supplement a process control laboratory. A code was generated within MATLAB, which could generate a transfer function from an open-loop step response recorded in the laboratory. A simulation model was created in an application within MATLAB called Simulink. This allowed for a simulation of the laboratory control settings and a comparison of real and simulated data. The calibration of this was limited due to a lack of replication of experimental data; however the model itself functions as was intended. Due to a lack of certainty within the laboratory results, a simulation exercise to demonstrate the characteristics of the PID controller was created and shows promising potential as a stand-alone control-simulation environment, or for use with laboratory efforts.

456861 Low Temperature Conversion of Biomass into Activated Biochar for Water Treatment in Resource-Constrained Settings

Abstract:

Access to clean water is a major challenge in many parts of the world, especially in remote rural areas such as rain forest communities in Africa. New approaches are needed to ensure adequate water supply and quality, as they are inextricably intertwined with food security and public health. In this presentation, we will focus on the synthesis, characterization, and application of activated biochar for water purification. Unlike conventional gasification reactor systems designed to operate at large scales, not much has been investigated in the past about small-scale batch pyrolysis. Nevertheless, batch systems offer advantages such as increased char yield and net exothermicity. Various types of cellulosic biomass (eg: sawdust, coconut shells) can be used as carbon precursors via pyrolysis. We will describe a process where catalytic activating agents are used in batch setups to lower the pyrolysis temperature and simultaneously activate the carbon produced to BET surface areas in excess of 500 m2/g . A systematic study of process parameters and characterization of the resulting activated carbons is used to optimize the chemistry and minimize contaminants from the produced filter-carbon.

468039 Small-Scale Multiple Effect Distillation (MED) for Brackish Groundwater Desalination

Abstract:

Multiple effect distillation (MED) is the oldest thermal desalination process and has a typical plant capacity of 600 to 300,000 m^3/day. Because MED units do not use membranes, the capital cost, system complexity, and maintenance costs are usually lower than those for reverse osmosis (RO) systems. The purpose of this study was to engineer a small unit that could be operated in the field using renewable energy-sourced heat, namely, heat from the pyrolysis of waste biomass.

A small-scale MED unit was designed and fabricated at NMSU to produce approximately 0.5 m^3/day of very pure water (<10 ppm total dissolved solids (TDS)) from brackish groundwater. The unit is being used to study scaling behavior as a function of water chemistry and identify optimum operating conditions for each water. The fabricated MED unit consists of two falling film evaporators with horizontal smooth copper tubes in a parallel feed arrangement; the flow rates are approximately 60 kg/hr. brackish feed water and 20 kg/hr. of distilled water based on brine concentration factor of 1.5. The feedwater is sourced from the Brackish Groundwater National Desalination Research Facility (BGNDRF) in Alamogordo, NM and has a TDS concentration of 1,000-5,000 ppm. The MED operates under partial vacuum (0.2 bar absolute) and temperatures of 60-70°C. The MED performance was improved by adding a heat exchanger, water heater, and pump to bring the feedwater temperature closer to its boiling point prior to be sprayed into the effects. The aluminum parts in the MED effects were anodized to prevent corrosion. Remote thermocouples were used to monitor the temperature at the surface of copper tubes to study the effects of dry spots. The temperatures and pressures selected in this design are meant to be compatible with local brackish groundwater chemistry in terms of salt solubility to limit scaling on the heat exchanger surfaces.

472093 Bench Scale Investigation of the Biosand Filter Media As the Fluoride, Turbidity, and *E. coli* Removal

Abstract:

Biosand filter (BSF) is a simple and inexpensive facility that can be built as a community-based water treatment technique using locally-available materials. It has been proven that BSF can remove turbidity, *E. coli*, arsenic in water efficiently by modifying the filtration media. Drinking water fluoridation is prevalent in the world. High fluoride intake would cause dental fluorosis and other toxic health effects, especially for children. Therefore, drinking water de-fluoridation is necessary to safeguard public health. There is little information on the use of BSF for fluoride removal. For instance, some previous works reported that aluminum oxide-coated pumice (AOCP) was effective for fluoride removal and had better removal efficiency compared to activated alumina. Other research works found that pumice media BSF can provide similar water quality as conventional sand BSF but 30% more filtered water. Based on these findings, we have investigated the optimization of using AOCP surfaces to serve as media for BSF to clean contaminant water. Preliminary results have shown that AOCP can remove 93% of fluoride in surface water. In our setting, the traditional sand layer has been replaced by AOCP surfaces. We have found that filtration rate, contaminants removal efficacy are prime parameters that enhance BSF performance. It is expected that our findings will be able to predict the behavior of BSFs and their implementation in low-income communities or as point of use (PoU) systems.

463624 Capacitive Deionization of Brackish Waters Using Biomass-Derived Carbons

Abstract:

Irrigated agriculture water represents the bulk of demand for water in most of the countries suffering from acute water scarcity in the Near East and North Africa, as well as Mexico, Pakistan, South Africa, and large parts of China and India. Greenhouses are emerging solution. Current testing in Mozambique and Sierra Leone has demonstrated significantly reduced water use. Yet even a few gallons per day per greenhouse is challenging because many local water sources are brackish. Capacitive deionization (CDI) is a sustainable, energy efficient, and cost effective technology for removal of low or moderate salinity typical of brackish water as compared to technologies that remove the water instead. Indigenous production of the carbon for the electrodes holds the key to such technology being adopted as a source of irrigation water, simultaneously facilitating the use of such using renewable energy sources. Biochars have recently been explored as electrode materials in CD, and can be easily produced from local plant-based agricultural waste materials.

The goal of this study is thus to identify biomass process conditions to enable the best performance as electrode material in CDI. Studies on CDI so far have been targeted towards achieving the best adsorption capacity, which directs towards the use of highly processed and tailored carbons. This is mostly achieved through chemical activation processes using toxic or corrosive and expensive materials which require deionized water for subsequent washing of the carbon, for achieving better CDI performance. Hence, with the effort directed towards making the process viable for use locally in regions with acute water scarcity, a change in the approach to tailor the electrode material for use in CDI is proposed. This approach is to determine the best possible material utilization by ensuring maximum possible contact between the liquid and the biochar surface (wettability) and by generating the least Ohmic loss thus resulting in highest electrical efficiency.

472455 The Use of Electrodeionization for High Water Recovery in Brackish Water Treatment

Abstract:

Currently reverse osmosis dominates desalination because of low cost, reliability, and choices in the market place. However, there are many new desalination opportunities, especially in brackish water applications, where the size and scale make it possible to consider other technology. One of these technologies is electrodeionization. Electrodeionization is a technique in which we use wafers to remove ions to low concentration through an electricially driven process. Because the cost is related to the amount of ions being moved this technology can be cost competitive at low to medium ion concentration and low to medium scale. The purpose of this presentation is to explore places where electrodeionization can be economically feasible and show the fundamental scientific challenges with the technology. This work will specifically focus on the high water recovery possible with electrodeionization with experimental results to show high current efficiencies at concentration concentrations of greater than 10% salt.

466903 Total Phosphorus Removal from Sewage in Microbial Electrochemically Assisted Septic Tanks

Abstract:

Background

Subsurface sewage treatment systems (SSTS), commonly known as septic systems, typically consist of a septic tank and an absorption field (or drainfield or leachfield), to treat sewage from households. The septic tank as pretreatment step decreases total suspended solids (TSS), biochemical/chemical oxygen demands (BOD/COD), and pathogen levels from sewage influent, while the drainfield or soil absorption field captures nutrients (nitrogen and phosphorus) from the tank effluent so that the water leaving SSTS would have minimal environmental impact. By modifying the conventional septic tank through incorporating electrode materials to make use of microbial electrochemical process, this study assessed this novel design in terms of sewage treatment performance. Since phosphorus becomes increasingly threatening to water environment, it was the goal of this novel design to remove it from liquid phase and potentially capture it at electrode surface.

Methods

The experiments were conducted with 1-L scale simulated septic tanks at four different applied voltage levels (0.50 V, 0.63 V, 0.75 V, and 0.88 V, selection based on previous preliminary results) with control of without applied voltage. Tanks were set up and operated at two temperatures (15 and 25 °C), a hydraulic retention time of 8.3 days, two cycles of feeding and withdrawal during each HRT. The treatment conditions of the reactors are listed in REF _Ref450506281 \h Table 1 08D0C9EA79F9BACE118C8200AA004BA90B02000000080000000E0000005F0052006500660034003500300035003000360032003800310000000 . ANOVA and pairwise multiple comparisons based on Bonferroni test were conducted for water characteristics with OriginPro Version 8 at significance level of α=0.05.

Table SEQ Table * ARABIC 1. Treatments of simulated novel septic tanks for 171-day operation

	Treatment*		
Reactor #	Temp, oC	Voltage, V	Chlortetracycline (CTC)
Reactor #1	25	0	No
Reactor #2	25	0.50	No

Reactor #3	25	0.63	No
Reactor #4	25	0.75	No
Reactor #5	25	0.88	No
Reactor #6	15	0	No
Reactor #7	15	0	1 mg-CTC/L
Reactor #8	15	0.50	No
Reactor #9	15	0.63	No
Reactor #10	15	0.75	No
Reactor #11	15	0.88	No
Reactor #12	15	0.75	1 mg-CTC/L
Reactor #13	15	0.88	1 mg-CTC/L

*Water characteristics data of reactors #6 and #7 were pooled for data analysis, and the same procedure was applied to reactors #10 and #12, and reactors #11 and #13, for the reason that CTC addition was not detected to affect other water characteristics.

Results and Discussion

For total phosphorus, results suggested that septic tanks partially removed it from influent (REF _Ref450507954 \h Table 2 08D0C9EA79F9BACE118C8200AA004BA90B02000000080000000E0000005F005200650066003400350030003500300037003900350034000000), and the novel design of microbial electrochemical further decreased its concentration ($p=0.0001$) with an influence from temperature ($p=0.0003$), while the interaction of voltage and temperature was not significant ($p=0.2723$). Those performance differences in total phosphorus removal were further elucidated by pairwise multiple comparisons given in REF _Ref450508201 \h Table 3 08D0C9EA79F9BACE118C8200AA004BA90B02000000080000000E0000005F005200650066003400350030003500300038003200300031000000 . The results concluded that at 25 ºC, 0.63 V or higher voltage at the selected electrode material is practically effective to remove most total phosphorus (>89.7%) from sewage, resulting in an effluent TP concentration less than 0.82 mg/L over 171-day operation; and that at 15 ºC, phosphorus removal efficiency was smaller and it required 0.88 V to remove over 90% of total phosphorus from sewage.

Table SEQ Table * ARABIC 2. Total phosphorus concentrations in influent and effluents of different reactors

	Ave	SD	TP removal
	mg/L	mg/L	%
Influent	7.95	1.39	-
Control, 25 ºC	6.98	1.72	12.2
0.50 V, 25 ºC	1.81	3.67	77.2
0.63 V, 25 ºC	0.82	2.92	89.7
0.75 V, 25 ºC	0.14	2.30	98.3

0.88 V, 25 °C	0.11	2.31	98.7
Control, 15 °C	7.36	1.60	7.45
0.50 V, 15 °C	6.30	3.84	20.7
0.63 V, 15 °C	4.31	3.90	45.8
0.75 V, 15 °C	2.26	2.48	71.6
0.88 V, 15 °C	0.49	2.36	93.9

Table SEQ Table * ARABIC 3. Pairwise multiple comparisons results for total phosphorus levels in effluents

Comparisons (Bonferroni Test)	Significance at alpha=0.05 or lower
0.50 V, 25 °C vs. Control, 25 °C	*
0.63 V, 25 °C vs. Control, 25 °C	*
0.63 V, 25 °C vs. 0.50 V, 25 °C	NS
0.75 V, 25 °C vs. Control, 25 °C	*
0.75 V, 25 °C vs. 0.50 V, 25 °C	*
0.75 V, 25 °C vs. 0.63 V, 25 °C	NS
0.88 V, 25 °C vs. Control, 25 °C	*
0.88 V, 25 °C vs. 0.50 V, 25 °C	*
0.88 V, 25 °C vs. 0.63 V, 25 °C	NS
0.88 V, 25 °C vs. 0.75 V, 25 °C	NS
Control, 15 °C vs. Control, 25 °C	NS
Control, 15 °C vs. 0.50 V, 25 °C	*
Control, 15 °C vs. 0.63 V, 25 °C	*
Control, 15 °C vs. 0.75 V, 25 °C	*
Control, 15 °C vs. 0.88 V, 25 °C	*

0.50 V, 15 ºC vs. Control, 25 ºC	NS
0.50 V, 15 ºC vs. 0.50 V, 25 ºC	*
0.50 V, 15 ºC vs. 0.63 V, 25 ºC	*
0.50 V, 15 ºC vs. 0.75 V, 25 ºC	*
0.50 V, 15 ºC vs. 0.88 V, 25 ºC	*
0.50 V, 15 ºC vs. Control, 15 ºC	NS
0.63 V, 15 ºC vs. Control, 25 ºC	*
0.63 V, 15 ºC vs. 0.50 V, 25 ºC	*
0.63 V, 15 ºC vs. 0.63 V, 25 ºC	*
0.63 V, 15 ºC vs. 0.75 V, 25 ºC	*
0.63 V, 15 ºC vs. 0.88 V, 25 ºC	*
0.63 V, 15 ºC vs. Control, 15 ºC	*
0.63 V, 15 ºC vs. 0.50 V, 15 ºC	*
0.75 V, 15 ºC vs. Control, 25 ºC	*
0.75 V, 15 ºC vs. 0.50 V, 25 ºC	NS
0.75 V, 15 ºC vs. 0.63 V, 25 ºC	NS
0.75 V, 15 ºC vs. 0.75 V, 25 ºC	*
0.75 V, 15 ºC vs. 0.88 V, 25 ºC	*
0.75 V, 15 ºC vs. Control, 15 ºC	*
0.75 V, 15 ºC vs. 0.50 V, 15 ºC	*
0.75 V, 15 ºC vs. 0.63 V, 15 ºC	*
0.88 V, 15 ºC vs. Control, 25 ºC	*
0.88 V, 15 ºC vs. 0.50 V, 25 ºC	NS

0.88 V, 15 ºC vs. 0.63 V, 25 ºC	NS
0.88 V, 15 ºC vs. 0.75 V, 25 ºC	NS
0.88 V, 15 ºC vs. 0.88 V, 25 ºC	NS
0.88 V, 15 ºC vs. Control, 15 ºC	*
0.88 V, 15 ºC vs. 0.50 V, 15 ºC	*
0.88 V, 15 ºC vs. 0.63 V, 15 ºC	*
0.88 V, 15 ºC vs. 0.75 V, 15 ºC	*

* indicates significant difference, p-values less than 0.05; NS indicates non-significant difference, p-values equal or larger than 0.05.

Sludge solids were collected from simulated septic tanks (reactors at 25 ºC, REF _Ref450512244 \h Figure 1 08D0C9EA79F9BACE118C8200AA004BA90B02000000080000000E0000005F005200650066003400350030003500310032003200340034000000) and quantified for mixed-liquor (ML) total suspended solids (TSS) and total solids (TS), sludge volume index (SVI), and composition. After log transformation, both MLTS and MLTSS displayed a linear increase along with voltage level. At 0.63 V where TP removal achieved close to 90%, the MLTSS and MLTS increased by 2.53 and and 1.17 folds, respectively. However, the corresponding SVI of the sludge was only 52.5% of the control reactor, indicating the formation of denser sludge, which suggested a smaller increase of sludge volume (85.3%) than the weight increase. The composition analysis based on energy dispersive X-ray spectroscopy (EDS) revealed the concentration and amount of phosphorus in sludge samples (REF _Ref450512615 \h Table 4 08D0C9EA79F9BACE118C8200AA004BA90B02000000080000000E0000005F005200650066003400350030003500310032003600310035000000). More phosphorus was captured in sludge sediment when voltage was increased, but the phosphorus balance, assumed to be captured on electrode surface, was higher at voltages levels of 0.50 and 0.63 V than 0.75 and 0.88 V.

Figure SEQ Figure * ARABIC 1. Mixed-liquor total suspended solids (A), mixed-liquor total solids (B), and sludge volume index (C) of the sludge from reactors with different voltage treatments at 25 ºC.

Table SEQ Table * ARABIC 4. Phosphorus balance in different reactors.

	Amt of eff P	Amt P removed	Amt P in Sed	P of balance (on electrodes)
	mg	mg	mg	mg
Inf	178	-	-	-
Control, 25 ºC	143	35	13	21
0.50 V, 25 ºC	37	141	26	114
0.63 V, 25 ºC	17	161	37	125
0.75 V, 25 ºC	3	175	92	83
0.88 V, 25 ºC	2	176	92	83

Conclusions

These results confirmed the phosphorus retention capability of the novel septic tank, suggesting a lower burden on the soil absorption for nutrient removal. The amount of sludge solids was also increased, which may require a more frequent tank cleaning.

467843 What Is the International Impact of Europe's Bologna Process?

Abstract:

In 1999, Education Ministers from 29 European countries signed the Bologna declaration, creating the European Higher Education Area. What is the impact on Chemical Engineering education? What does it mean to internationalization in our field? Official statements, independent assesments, student's views will be compiled. The paper can be discussed with the audience.

460817 Student-Created Assessment: Midterm Test of a Graduate Course, Mathematical Methods in Chemical Engineering

Abstract:

Students were divided into groups, each of which was asked to provide a question paper and suggested solutions like a course instructor, as their midterm assessment. The course, Mathematical Methods in Chemical Engineering, is a graduate core course, mainly covering analytical and numerical techniques for solving differential equations generated by modeling of chemical processes. The class size was normally between 30 and 50. Detailed guidelines and sample product created by the instructor or previous students were provided at the beginning of the semester. The grading was based on: whether the questions in the product can assess student achievements of module learning objectives; whether a "normal" student can complete the paper reasonably well within 4 hours; the clarity of the problem statements; the significance of the physical problems; the reasonability of modelling and the correctness of using techniques to solve; the creativity or critical discussion on the problem, modelling, and/or the solving the model. An oral presentation, highlighting their ideas, was given by each group, for sharing and grading fairness. A survey was conducted at least 6 months after this assessment to get student opinions on this experiment. The product showed that most students achieved and knew how to assess the achievement of the module learning outcomes. Many creative questions and critical discussion on the questions came out, suggesting that students challenged their comfortable zone, and also learned through the assessment procedure. Students' responses to the survey questions showed their interests and acceptance of such type of assessment. They also expressed that they liked this learning experience, and it was fairer than the conventional exam, because students can show their mastery of knowledge in their own way, without being afraid of misunderstanding the questions.

459654 Feedback- and Engagement-Enhanced Education: A Newly-Designed Senior Course, Chemical Product Design

Abstract:

A new technical elective course, *Chemical Product Design*, has recently been offered by the Department of Chemical and Biomolecular Engineering at the National University of Singapore. This course is targeted to senior students (undergraduate Year 4 or masterâ€™s) to equipped them with skill of designing high-value-added specialty chemical products (e.g., devices, medicine, personal-care products), as an supplement of traditional process design skill (for commodity products). It covers the basic methodology with illustrative examples from many areas (based on the textbook: *Chemical Product Design*, by Cussler and Moggridge, 2nd edition, 2011). The course involves active-learning lectures and student teaching (which is recorded for self-reflection and feedback/evaluation from the lecturer, teaching assistant and peers) so that students gain competence of thinking divergently and critically, and confidently solving open-ended problems through group discussion. Besides the technical aspects, the module also emphasizes engineering professionalism (including ethics and sustainability) and communication skills (by supervised group discussion and student teaching with feedback loop). After completion of the course, students are able to apply the basic methodology to design a product in group, write a professional report, teach others based on the learning objectives, and evaluate otherâ€™s teaching.

466727 Student Activation and Interaction through Tutorials and Adequate Scheduling

Abstract:

Introduction

Activating students and interacting with them are hot topics in education. In this respect, blended learning, entailing the combination of a number of pedagogic approaches, irrespective of learning technology used, represents a valuable approach [1]. The idea of blended learning is founded upon the notion that every student has an individual way of processing new theories. By offering a broad range of educational tools, the chance for a student to successfully grasp a new concept drastically increases [2-5]. Currently, e-learning is being hyped as the ultimate form of blended learning. However, this form is not always easily accessible and similar outcomes might be obtained through another form of blended learning, e.g., tutorials. Moreover, a critical assessment and reorganization of the course schedule is expected to already significantly enhance the students' motivation.

Kinetic Modelling and Simulation: starting situation

At Ghent University, Belgium, Kinetic Modelling and Simulation (KMS) is one of the core courses in the chemical engineering program. During this course, 20 to 30 students are confronted with a multidisciplinary field comprising mathematics, statistics, chemistry, engineering and informatics aiming at simulating and, hence, elucidating chemical kinetics. The KMS course spans a single semester, i.e., 12 weeks, and was constructed upon 5 different pillars: theory, examples, exercises, projects and feedback, see Figure 1. The theory, example and exercise sessions were given during the contact hours. In total, 8 ex cathedra theory sessions of 3 clock hours were scheduled in these 12 weeks to cover 5 topics: statistics (A), linear regression (B), non-linear regression (C), experimental design and model discrimination (D) and multi response regression (E). Exercises were only given on the topic of non-linear regression and were situated relatively far in time from the corresponding theory sessions, c.q., the theory on non-linear regression was given in weeks 4 and 5 while the corresponding exercises were held only in weeks 6, 7 and 9. During the latter weeks, the theory corresponding to other concepts such as experimental design was taught which did not promote the students' learning process. Three projects were given on three different topics, i.e., linear regression, non-linear regression and experimental design aiming at model discrimination, after which students got overall feedback. At the end of the course, 4 to 5 PhD students presented their work as an example for different aspects in kinetic modelling.

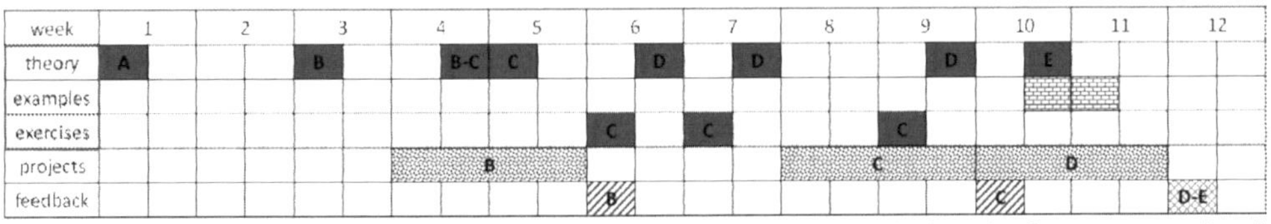

Figure 1: Schedule of the course Kinetic Modelling and Simulation before reorganization

Although the course has been evaluated rather positively over the years, some issues were identified: (1) during the theory sessions, **interaction was limited** due to the rather abstract content, (2) students were **ill-prepared** for the exercise sessions, (3) **not all concepts were put in practice** through exercises and/or projects and (4) throughout the semester, the **workload was concentrated at specific times**.

Reorganizing of the course according to blended learning principles

Aiming at enhancing the students' motivation and interaction with the teachers, a well-considered reorganization of the KMS course has been implemented, see Figure 2. Ex-cathedra theory blocks are, in this scheme, limited to 1.5 hours only and are followed by an equal amount of time for tutorials and hands-on trainings in which the students immediately process and apply the theory covered in the preceding time block. In addition, after a number of theory and tutorial sessions, a project assignment is given as a closure of that particular topic in which the student has to prove that he/she can apply the theory individually. After every project, which takes two weeks, students receive individual feedback. Compared to the previous schedule, the sequence of theory processing through ex-cathedra block, tutorials and projects is more continuous and, hence, contributes to a more blended learning environment. The work presented by the PhD students as kinetic modelling examples has been spread out more evenly at the end of the semester. Lastly, some additional topics have been introduced: contribution and reaction pathway analysis (F) and 'reading and analyzing a scientific publication' in regression analysis (G) which are well considered and relevant in the field of kinetic modelling.

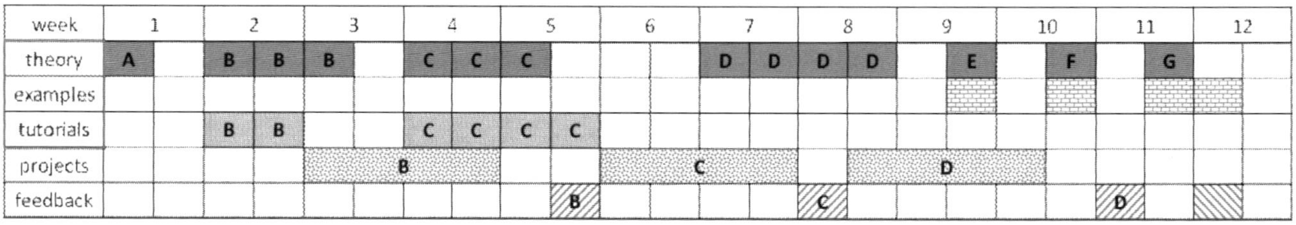

Figure 2: Schedule of the course Kinetic Modelling and Simulation after reorganization

Feedback and grading

Every student receives feedback in various ways throughout the semester. Firstly, all 6 tutorial sessions are led by a number of assistants, i.e., one assistant per 6 to 7 students. Hence, during these tutorials, every student has a close interaction with an experienced assistant. In addition, after every project, each student has personal feedback, not only on their hard skills, i.e., the project report content, but also on their soft skills, i.e., writing, visualizing and summarizing. At the end of the semester, there is a common feedback session in which the students give direct feedback to the teacher and assistants.

The final grade for the KMS course is composed of two parts. One third is earned via the three projects made during the semester. The remaining two thirds are obtained viva voce. Typically, this examination comprises a theoretical question and a case study or a scientific publication analysis. A student acquires the credits for the course when they obtain a weighted score exceeding 50% (10 out of 20) with minimum score of 40% on both components individually.

Benefits and challenges

An overview of the number of contact hours and sessions is given in Table 1. Concerning theory, the contact hours decreased from 24 to 22.5 hours while the number of contact sessions almost doubled. This yielded more time for the students in which they can process the concepts taught. Surprisingly, the time spent on exercises/tutorials stayed identical while the number of sessions increased with 100%. Since these sessions are planned directly after the corresponding theory session, the tutorials are tailored specifically to this theory. Finally, more time was allocated for the PhD students to explain their work within the framework of kinetic modelling to the students. Although it could already be implied from a comparison of Figures 1 and 2, this reorganization did not result in a significant increase in workload for the teacher and assistants. Moreover, the total number of contact hours remained equal while the number of contact sessions doubled. However, having a more strict schedule, requires the teacher and assistants to have a more strict schedule.

Table 1: Number of contact hours and sessions of the course Kinetic Modelling and Simulation before and after reorganization

	before reorganization		after reorganization	
	number of hours	number of sessions	number of hours	number of sessions
theory	24	8	22.5 (↘)	14 (↗)
exercises/tutorials	9	3	9 (=)	6 (↗)
examples	4.5	2	6 (↗)	4 (↗)
total	37.5	13	37.5 (=)	24 (↗)

Conclusions

By critically assessing and reorganizing the schedule for the course Kinetic Modelling and Simulation according to blended learning principles, the student/teacher/assistant interaction has significantly increased, as reflected by the increase in number of contact sessions. This resulted in a significantly augmented student motivation. Although the amount of time spent by the teacher and assistants during the course stays equal, it requires more flexibility. The immediate follow-up of theoretical by the corresponding tutorial sessions allows for a more direct learning process of the students, remediating their previous ill-preparedness for the exercises. In addition, the workload has been spread out more evenly over the semester.

References

[1] M. Oliver, K. Trigwell, Can 'Blended Learning' Be Redeemed, E-Learning, 2 (2005).

[2] N. Hoic-Bozic, V. Mornar, I. Boticki, A Blended Learning Approach to Course Design and Implementation, Ieee T Educ, 52 (2009) 19-30.

[3] M.V. Lopez-Perez, M.C. Perez-Lopez, L. Rodriguez-Ariza, Blended learning in higher education: Students' perceptions and their relation to outcomes, Comput Educ, 56 (2011) 818-826.

[4] V. Woltering, A. Herrler, K. Spitzer, C. Spreckelsen, Blended learning positively affects students' satisfaction and the role of the tutor in the problem-based learning process: results of a mixed-method evaluation, Adv Health Sci Educ, 14 (2009) 725-738.

[5] J.H. Wu, R.D. Tennyson, T.L. Hsia, A study of student satisfaction in a blended e-learning system environment, Comput Educ, 55 (2010) 155-164.

471386 An International Summer School for Undergraduate Chemical Engineers: The Imperial College Experience

Abstract:

The Imperial College Chemical Engineering Summer School programme is provided to international sophomore, junior and senior students studying chemical engineering and offers a schedule combining extensive laboratory experiments, a rig-building team exercise and a Pilot Plant project. The students are trained using a state-of-the-art carbon capture pilot plant which provides students with an experience that is as close to real industrial work as practical. As well as learning how to start up, stop and run efficiently pilot scale chemical engineering plant, the students become familiar with plant safety, shift handover protocols, permits to work, working is small shift teams, alarm management, PID diagrams and many other industrially relevant procedures. Summer school students will also take part in a bespoke range of short and long undergraduate laboratory experiments based on our current MEng. curriculum during their six week programme.

The paper will review the development of the summer school curriculum at Imperial College over the past 5 years, feedback from students and faculty from cohorts from 3 US colleges, and the future opportunities for summer schools as a way of enriching the studentâ€™s academic and cultural learning experiences. Comments on the use of UTA-undergraduate teaching assistants, group laboratory exercises and the overall importance of hands on teaching and learning will be discussed in the context of the current trend towards virtual and computer aided teaching. Â Â

459134 Global Engagement Strategies for Engineering Students at Western Michigan University

Abstract:
Global Engagement Strategies for Engineering Students at Western Michigan University

James Springstead, PhD, Assistant Professor

Said AbuBakr, PhD, Professor, AIChE Fellow

College of Engineering and Applied Sciences

Department of Chemical and Paper Engineering

Kalamazoo, MI

Abstract

Western Michigan University (WMU) has established global engagement as an important part of its vision and strategic objectives. To help implement these global strategic objectives, the faculty senate votes on a memorandum of action to encourage colleges to establish global learning outcomes. Instead of reinventing the wheel, the College of Engineering and Applied Sciences (CEAS) used already established ABET student outcomes that have global elements, with the WMU Department of Chemical and Paper Engineering serving as a leader in establishing clearly linked ABET performance indicators that are used to assess student outcomes.

Introduction

The visionÂ of Western Michigan University is:

Nationally and internationally recognized, the University aspires to distinguish itself as learner centered, discovery driven, and globally engaged.

Learner centered: Western Michigan University is a university where every member of our community is responsive to and responsible for the education of our students. We challenge and engage all members of our community with a university experience that creates skilled, lifeâ€long learners.

Â Discovery driven: Western Michigan University offers experiences that enable discovery, and promote creativity and research. We are committed to pursuing inquiry, disseminating knowledge, and fostering critical thinking that encourages lifeâ€long learning. Our scholarship creates new knowledge, forms a basis for innovative solutions, leads to economic development, and makes substantial contributions to society.

Globally engaged: Western Michigan University impacts the globe positively. We are a community of learners committed to human dignity, sustainability, social responsibility, and justice. Our campus embraces a diverse population of students, faculty and staff who develop learners and leaders who are locally oriented and globally competent, culturally aware and ready to contribute to world knowledge (1).

To implement this vision, WMU developed the following global objectives:

ObjectiveÂ 1.4:Â InternationalÂ learningÂ experiencesÂ prepareÂ allÂ learnersÂ forÂ theÂ globalizedÂ world.

Strategies

a. Increase feasibility of broadâ€based participation in study abroad programs and other globally focused learning opportunities.

b. Implement the facultyâ€adopted internationalized education initiative.

c. Provide enhanced livingâ€learning support for the growing number of international students at WMU.

d. Increase recruitment of University employees with international experience and expertise.

e. Facilitate intercultural understanding through greater interaction between domestic and international students.

ObjectiveÂ 2.1:Â WMUâ€™sÂ academicÂ programsÂ respondÂ toÂ studentÂ andÂ globalÂ needsÂ throughÂ innovative,Â highÂ qualityÂ curricula.

Strategies

a. Increase flexibility within the curriculum development process to better respond to a rapidly changing world.

b. Revise general education curricula to respond to the needs of the 21st century student.

Â c. Develop and utilize Universityâ€wide student learning outcomes for University planning and assessment.

d. Develop stronger connections across colleges and among external constituents to facilitate interdisciplinary opportunities.

e. Identify and support growth in the number of programs that achieve national or international distinction.

Objective 3.3: WMUâ€™s outreach is attentive to local and world needs.

Strategies

a. Develop and implement a renewed Campus Compact Action Plan, dedicating the University to work with other institutions to advance the public purposes of higher education.

b. Review and improve outreach practices and policies to achieve the Carnegie Community Engagement Reclassification by 2020.

c. Operationalize a unified vision to support the Universityâ€™s public engagement mission.

d. Develop rigorous, systematic evaluations of all outreach and community engagement programs to increase impact in conjunction with community agenciesâ€™ planning initiatives.

Â e. Know, measure, tell, and increase WMUâ€™s economic impact and social value.

f. Assist all students with developing meaningful connections in their local and global communities.

The international Education Council of Western Michigan University passed a resolution that was approved by the Faculty Senate asking colleges to create global Learning outcomes (GLOs)

The response of our college was that all our engineering, engineering technology, and computer science programs are ABET accredited. Global learning outcomes are already included in ABET student learning outcomes (3c), (3h), and (3j) and are periodically measured, assessed, and evaluated as required to maintain ABET accreditation (2).

Here are the ABET student outcomes that emphasize global engagements:

(c) an ability to design a system, component, or process to meet desired needs within realistic constraints such as economic, environmental, social, political, ethical, health and safety, manufacturability and sustainability.

The Department of Chemical and Paper Engineering measured metrics (performance indicators):

C3 Understand contemporary environmental and global issues influencing the chemical and biological industries

(h) the broad education necessary to understand the impact of engineering solutions in a global, economic, environmental, and social context

The Department of Chemical and Paper Engineering measured metrics (performance indicators):

H1 Understand contemporary environmental and global issues influencing the chemical and biological industries

H2 Understand safety and regulatory issues related to the hazards associated with processes and the importance of ethical decisions related to these areas, real life examples

(j) a knowledge of contemporary issues

The Department of Chemical and Paper Engineering measured metrics (performance indicators):

J1 Conduct independent learning (outside classroom)

J2 Conduct international literature search using electronic library and utilize resources such as reference books, journal articles, and internet sources to analyze and design processes

By directly integrating performance indicators that assess global issues, the College of Engineering and Applied Sciences directly assesses the success of our globally-focused strategic vision and objectives by integrating this focus into ABET performance indicators. It is necessary and appropriate to build assessment of global engagement and context into ABET performance indicators, especially when student outcomes involve environmental, ethical, economic, safety, and contemporary knowledge, as the world is increasingly integrated with more and more issues being relevant across all countries. Furthermore, global awareness and education of our students will prepare them for this new integrated world and open doors for opportunity for collaboration and involvement.

As demonstrated, the College of Engineering and Applied Sciences has in place a well-established Knowledge, Skills, and Perspectives in the their Global Learning Outcomes as required for ABET accreditation and this MOA.

References

The Gold Standard 2010, Western Michigan University Strategic Plan, www.wmiche.edu

Students learning outcomes, abet.org

453451 Chemical Engineering Education for the Millennial Generation

Abstract:

The millennial generation are grown with the advance in technology and computation . They are the group of people who are faithful fans of Facebook, iPhone and iPad. They have also become the major talent group of chemical industry. Therefore the chemical engineering education for the millennial generation has become a very important topic in most of the countries. This presentation will provide you the challenge and opportunity of chemical engineering education for the millennial generation. Furthermore, the leadership of AIChE Young Professional Committee in this field will also be shared.

449538 More about the *Tendencies of Chemical Engineering in the Last 45 Years. What Is the Future?*

Abstract:

This contribution is a continuation of that of the AICHE 2015, and now the studied span of time was extended since 1970 to the date. The general statements of the 2015-material were that, according to the consulted sources, before the year 2000, chemical engineering received the beneficial of the advances in computation and internet, but in general terms, the traditional topics of chemical engineering were studied, such as: Chemical kinetics and reactors, catalysis and surface phenomena, unit operations, chemical thermodynamics, transport phenomena, simulation and optimization, chemical processes, process control, among others.

More or less after 2000, it is notorious that researchers started to have interest in new topics, where chemical engineering can have influence or can take advantage of the knowledge of other careers. Each time chemical engineers are eager and available to face and deal with topics from other careers and disciplines, and this fact makes chemical engineering stronger, due it extends its influence on other careers. Besides, other disciplines started to have interest in using the principles of chemical engineering and as a result, it is reinforcing and sharing its topics with other careers. In other words, the apparent walls and boundaries of chemical engineering with other careers started to fall or disappear.

The new lines, among others, studied in the last fifteen years or more have been: Environment, bioengineering, nuclear engineering, material science, thin films, renewable energies, food science, pharmacy, medicine, fuel cells, hydrogen power, nanotechnology, biotechnology, etc. Now, using the principles and theories of thermodynamics, transport phenomena, numerical methods, fluid dynamics, unit operations, chemical reactors, balances of mass, energy and entropy, etc, it is possible to have contribution in the purification of the environment, the production of alternative energies, food, artificial kidneys, etc.

In this 2016 - contribution, a deeper research was carried out, inasmuch as more information from internet was collected and analyzed, contents of more indexed reviews were searched, and this allowed distinguishing the tendency and the focus of each review.

The academic programs of chemical engineering of more universities in the five continents were consulted, and in this moment, although roughly, it is possible to confirm that from one geographic region to other, the focus of chemical engineering can change. The curriculum vitae of more researches were consulted; the research lines of more institutes were also investigated.

All of this allows saying that chemical Engineering has been changing and broadening its tendency and influencing in the last fifteen years in a significant form. Of course these new lines started before. Chemical Engineers, researchers, institutions, etc., are focusing their branches on the use of alternative feedstock such as biomass and waste to produce chemicals and energy. Their goal is to use green chemistry with less carbon in the industrial processes and sustainable processing.

The reasons of all these changes are because time is very dynamic, each time the social needs demand more energy, new food, different entertainment, new cloth, etc. The way to obtain and process the natural resources each time is different and requests more technologies and the interaction among the different disciplines must be enhanced because the industrial processes are each time more complex. Now the contamination must be avoided, there are stricter regulations about the environment. Search for new energies must be performed due hydrocarbons are going to be depleted. These external natural and social factors have been and will be the guide of the activities and research in chemical engineering for the next years.

AUTHOR INDEX

AUTHOR INDEX

AUTHOR INDEX

AUTHOR INDEX

AUTHOR INDEX

AUTHOR INDEX

AIChE
120 Wall Street, FL 23
New York, NY 10005-4020

ISBN 978-1-5108-3433-0

CURRAN ASSOCIATES INC.
proceedings.com